DIGITAL SIGNAL PROCESSING:

A System Design Approach

DIGITAL SIGNAL PROCESSING:

A System Design Approach

David J. DeFatta
The Analytic Sciences Corporation (TASC)
formerly with IBM Corporation

Joseph G. Lucas
Federal Systems Division
IBM Corporation

William S. Hodgkiss
Marine Physical Laboratory
Scripps Institution of Oceanography
University of California, San Diego

WILEY

JOHN WILEY & SONS
New York   Chichester   Brisbane
Toronto   Singapore

Copyright © 1988. By John Wiley & Sons, Inc.

All rights reserved. Published simultaneously in Canada.

Reproduction or translation of any part of
this work beyond that permitted by Sections
107 and 108 of the 1976 United States Copyright
Act without the permission of the copyright
owner is unlawful. Requests for permission
or further information should be addressed to
the Permissions Department, John Wiley & Sons.

Library of Congress Cataloging-in-Publication Data

DeFatta, David J.
  Digital signal processing.

  Includes bibliographies and indexes.
    1. Signal processing—Digital techniques.
2. System design.   I. Lucas, Joseph G.
II. Hodgkiss, William S.   III. Title
TK5102.5.D39   1988   621.38′043   87-25425
ISBN 0-471-83788-1

Printed in the United States of America
10 9 8 7 6 5 4 3 2

About the
Authors

DAVID J. DEFATTA

Mr. DeFatta received a B.S. degree in Electrical Engineering from Cooper Union College of Engineering in 1967 and an M.S. degree in Applied Mathematics from the State University of New York in 1972. He worked at IBM Federal Systems Division in Manassas, Virginia. Mr. DeFatta spent two years on a technical sabbatical at the University of Virginia as a lecturer and researcher. He is currently employed at The Analytic Sciences Corporation (TASC), McLean, Virginia. His 20 years of experience include design and implementation of sonar systems including digital signal processing design techniques, finite word length effects, and ocean propagation modeling.

JOSEPH G. LUCAS

Mr. Lucas received a B.S. degree in Electrical Engineering from Pennsylvania State University in 1966 and an M.S. degree in Applied Mathematics from the State University of New York in 1971. He is currently managing a system engineering organization responsible for signal processor design and implementation for Department of Defense applications. His 21 years of experience include system design and implementation of signal processing systems.

WILLIAM S. HODGKISS

Dr. Hodgkiss received a B.S. degree in Electrical Engineering from Bucknell University, Lewisburg, Pennsylvania, in 1972 and the M.S. and Ph.D. degrees from Duke University, Durham, North Carolina, in 1973 and 1975, respectively.

From 1975 to 1977 he worked with the Naval Ocean Systems Center, San Diego, California. From 1977 to 1978 he was a faculty member in the electrical engineering department, Bucknell University. Since 1978 he has been a member of the faculty of the Scripps Institution of Oceanography and on the staff of the Marine Physical Laboratory, University of California—San Diego. His research interests are in the areas of adaptive digital signal processing, adaptive array processing, application of modern signal processing concepts to problems in underwater acoustics, very low frequency propagation of acoustic energy in the oceans and sediments beneath the ocean, and the statistical properties of ambient ocean noise.

Preface

Digital signal processing started to gain popularity in the mid- to late 1960s with the development of integrated circuits (ICs). Since this period, the technology has evolved from medium-scale integrated (MSI) circuits to very large-scale integrated (VLSI) circuits and very high speed integrated circuits (VHSIC). This advancement in digital technology has made the implementation of sophisticated algorithms (e.g., the fast Fourier transform) oriented toward performing real-time digital signal processing tasks feasible.

The development of several more efficient algorithms including multirate processing techniques and fast algorithms for filtering coupled with the latest technologies has made complex real-time digital signal processing systems possible. Programmable signal processors capable of performing the complex tasks for high data-rate applications within limited space and stressing environments can be achieved. Examples of applications that can be implemented efficiently using digital signal processing cover a broad spectrum including sonar, radar, seismic, communications, navigation, telephony, speech, image, and audio processing.

The material in this text evolved over several years from working in the field, from teaching digital signal processing courses to practicing electrical and computer science engineers at IBM, and from teaching senior undergraduate and graduate university students at the University of Virginia. Based on this experience, it was evident that a need existed for a text that goes beyond the presentation of the individual signal processing concepts and develops a systems approach showing the interrelationships between the individual processing elements in solving application problems.

The material in this text has been designed for a broad spectrum of technical users including advanced undergraduate and graduate students in engineering and computer science curriculums and practicing engineers and scientists. The text introduces the digital signal processing concepts, develops design and analysis expressions that can be easily implemented on a personal computer, and illustrates the use of the concepts in designing systems. The material is presented from a system engineering perspective. A major feature of the text is the development of a signal processing system design methodology as a systematic approach to solving complex application design problems.

Chapters 1 through 6 present the basic concepts of linear systems, digital filters, Z-transform analysis, discrete Fourier transforms (DFT), and fast Fourier transform (FFT) algorithms. The advanced concepts of multirate processing, power spectrum estimation, finite arithmetic analysis, system design, and adaptive processing are presented in Chapters 7 through 11. A summary of each of the chapters is provided in the following paragraphs.

Chapter 1 presents an overview of digital signal processing and discusses key concepts and approaches to performing the overall signal processing system design and implementation. Discrete linear systems are defined in Chapter 2—along with the associated theory of convolution, frequency response, and sampling—which are key to the remaining material presented in the text. The general difference equation for a digital filter is defined, and the relationship of two filter types: infinite impulse response (IIR) and finite impulse response (FIR) are developed.

Chapter 3 defines the Z-transform and inverse Z-transform pair and the application of the transform pair to the analysis of discrete linear systems. The definition of a discrete linear system transfer function is developed. Basic digital network concepts and filter realizations are presented. Important Z-transform properties are defined and the relationship between the Z-transform and frequency response is described.

IIR and FIR design techniques are presented in Chapters 4 and 5, respectively. A systematic approach is developed for the design of IIR digital filters using classic analog design functions. This approach is demonstrated by using flow diagrams to determine the poles, zeros, and coefficients of elliptic filters. Examples of lowpass, highpass, bandpass, and bandstop designs are presented. The method has been used to implement the filter designs on a personal computer. IIR design techniques are continuing to evolve. Approaches to direct design of the IIR digital filter are discussed.

FIR filter properties are presented. A systematic approach to designing FIR filters using the Fourier series method is presented with emphasis on the Kaiser window design. The most widely used MINMAX FIR design optimization approach is described. That approach uses the Remez exchange algorithm and was developed by Parks and McClellan. Potential pitfalls in the design are discussed. Again, detailed examples together with computer results illustrate the merit of the systematic approach to the design.

The DFT and the FFT are presented in Chapter 6. The properties of the DFT are presented with emphasis on the spectral leakage and circular shift resulting from the finite length transform. Key characteristics relating to the application of the DFT to spectrum analysis are presented based on the frequency response. The use of weighting functions is described. The need for redundancy processing when averaging successive DFT outputs is defined.

The FFT decimation-in-time and decimation-in-frequency radix-2 algorithms are developed showing the efficiency gained in computational requirements over the DFT. The FFT development stresses the definition of a basic computational unit that is repeatedly executed to perform the FFT. An efficient algorithm for the implementation of an N-point FFT using a radix-4 computational unit is derived and illustrated. BASIC subroutines for the computational unit are presented.

Multirate digital signal processing techniques are presented in Chapter 7. The key concepts of interpolation and decimation multirate processes are defined. Multirate techniques offer significant computational savings in the design of systems; desired bandwidth is small compared to the input bandwidth. Methods of optimizing the design for minimum computational requirements are presented. Filter design approaches for multirate systems are described.

Chapter 8 presents the basic concepts of discrete linear system responses to random signal inputs and power spectrum estimation of the random processes. The autocorrelation and cross-correlation functions are presented and their relationship to the power spectral density and cross-power spectral density of the random process are defined. These functions provide the basis for interpreting linear system performance to random processes. Estimation of the power spectral density by averaging modified periodograms are described. Finally, a discussion of the basic concepts of detection theory is presented with application to a narrowband detection system.

Chapter 9 covers the effects of finite wordlength implementations on the IIR, FIR, and FFT processing algorithms previously presented. The effects of the finite length filter and FFT coefficients are presented. Finite length arithmetic effects for implementation of recursive structures in cascade are presented in detail with respect to system implementation and performance. Fixed-point implementations are emphasized. Floating-point implementations are discussed. The approach presented is easily programmed on a personal computer.

Chapter 10 develops a methodology for performing a signal processing system design. The methodology is used to solve a spectral analysis application. Key signal processor architectural factors are addressed. A thorough analysis of the processor resource requirements is discussed. Results are presented in personal computer spreadsheet form to illustrate the resource analysis process. An acceptable design is achieved, and its performance is estimated. The performance calculations can be programmed and used to vary parameters such as probability of detection, probability of false alarm, bandwidth, and weighting functions to parametrically determine the performance versus system design choices.

In Chapter 11, digital filters that adapt to a changing environment are discussed. The stochastic Wiener filtering and deterministic least-squares problems are set up, and the similarity between them is pointed out. First, the block-processing approach is taken in the solution of these two problems. Second, a fading memory (recursive) approach is taken. Finally, an application of these algorithms to the adaptive beamforming (ABF) problem is considered.

The goal in writing this textbook was to provide a clear description of the application of digital signal processing concepts in solving system engineering signal processing application problems. A methodology was provided as an aid in performing the complex design and analysis processes required to determine an efficient implementation. We hope that we have achieved our goal and that you will find the material useful in your signal processing related work.

For Instructors

The material provides the basis for a two-semester course. Chapters 1 through 6 make up an undergraduate senior level or first-year graduate level digital signal processing course. They may be presented in order or the instructor may prefer to present the DFT portion of Chapter 6 prior to Chapters 4 and 5. Chapters 7 through 11 offer a variety of options for an advanced course and for defining independent course study topics.

A substantial set of problems are provided at the end of each chapter. Some of these involve derivations and exercises to the text material; while others are concerned with computer solution of problems encountered in real applications. In many cases the problems extend the concepts already developed in the text. These problems provide the instructor with an extra degree of freedom to use different approaches in presenting the material.

A comprehensive instructors manual is available that includes solutions to the problems, and suggested additional exercises and projects which enhance the concepts developed in the text. We feel that the usefulness of the book is directly proportional to a problem set which motivates the reader to further investigate the technical concepts. As a result the instructor is provided with a comprehensive treatment of solutions to all problems, where in many cases computer programs or computer generated spreadsheets are used to describe the approach. Finally, subject matter is included in the instructors manual as a guide in presenting the material.

Acknowledgments

Numerous people have contributed to the material and publication of this textbook. The many papers and publications by A. V. Oppenheim, R. W. Schafer, L. R. Rabiner, B. Gold, J. F. Kaiser, T. W. Parks, L. Jackson, A. Antoniou, R. Crochiere, E. O. Brigham, C. S. Burrus, J. W. Cooley, and J. W. Tukey were a source of much of the material presented in this book.

We gratefully acknowledge the helpful comments made by the following reviewers: Wadie Sirgany, Rao Yarlagadda, John Saunders, Ken Moore, Joe Fanto, and Tom Butash, who assisted by reading various portions of the manuscript and offering many helpful comments.

DJD and JGL would like to acknowledge the consultation over many years of colleagues at IBM to whom they are indebted: Wouter Vanderkulk, Gordon Demuth, Dale Bachman, Stu Adams, Wadie Sirgany, Bob Kettelkamp, Bob Kelly, and José Rio. They also would like to thank IBM for allowing them the opportunity to work in this exciting field and their managers Tony Maloney and Ken Baxter for their support. A special thanks to Don Thomson for recommending Dave DeFatta for his faculty loan assignment at the University of Virginia, which was the motivating factor toward starting this endeavor. We would like to thank all UVA students who participated in the digital signal processing courses and provided comments that have made this material significantly better. A special thanks to Professor E. Parrish for his support and for his advice and consultation. We would like to express our appreciation to The Analytic Sciences Corporation (TASC) for supporting Dave DeFatta in the manuscript preparation.

WSH would like to acknowledge the stimulating association with a number of colleagues who have had a substantial influence on his approach to digital signal processing: Prof. Bernard Widrow, John McCool, Dr. John Treichler, Dr. Frank Symons, Dr. Leon Sibul, and Dr. Ayhan Vural. In addition, he would like to thank the faculty, graduate students, and staff of the

Marine Physical Laboratory who have been enthusiastic participants in the application of modern signal processing and data analysis concepts to problems in underwater acoustics. Special mention goes to Prof. Victor Anderson and Dr. Dimitri Alexandrou.

Finally, we sincerely thank Bob Beckley, Paul Kapcio and our families for their support and participation in generating the electronic manuscript and the many revisions through which it has gone.

Contents

chapter 8
RESPONSE OF LINEAR SYSTEMS TO DISCRETE-TIME RANDOM PROCESSES, POWER SPECTRUM ESTIMATION, AND DETECTION OF SIGNALS IN NOISE 378

1

Introduction to Digital Signal Processing

Increasing demands for digital signal processors with large processing capacities utilizing advanced technologies and algorithms have made the design process complex. We have developed a signal processing system design methodology that uses computer-aided design and provides a structured approach for the analysis and definition of complex system applications.

1.0 BACKGROUND

Digital signal processing (DSP) gained popularity in the 1960s with the introduction of solid state technology. Early DSP systems consisted of unique hardware elements, where each element performed a specific signal processing operation (Fig. 1.1a). These implementations were not programmable and required only simple control for turning the hardware on or off. Subsequent advances in digital integrated circuits resulted in microprogrammable signal processing elements. Microprogramming allowed the hardware elements to be time-shared to implement multiple system functions or to be redirected by the system controller to implement a new function (Fig. 1.1b). Medium scale integrated (MSI) circuits and large scale integrated (LSI) circuits led to the development of time-shared programmable signal processors to implement a complete real-time signal processing application. Figure 1.1c is a functional representation of a programmable signal processor. Subsequent very large

1

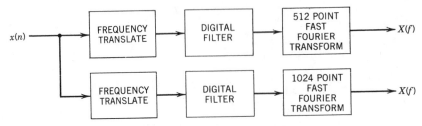

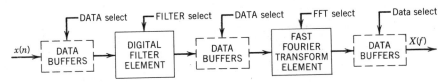

a. Dedicated hardware signal processor.

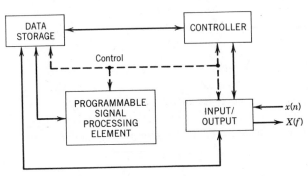

b. Microprogrammable signal processor hardware elements.

c. Programmable signal processor.

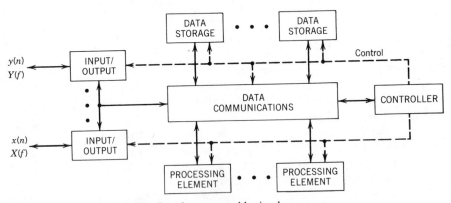

d. Distributed programmable signal processor.

Figure 1.1 Evolution of DSP implementations.

scale integrated (VLSI) circuit and very high speed integrated circuit (VHSIC) technologies have provided the capability to develop distributed signal processor architectures with increased capacity (Fig. 1.1d).

In parallel with the advances in hardware technology, a great deal of research was conducted on developing new signal processing algorithms. The DSP algorithmic research concentrated on improved performance and reduced complexity. Performance studies led to the development of new processing techniques, including adaptive algorithms, and better filter and transform realizations. Reduced complexity studies resulted in reduced complexity algorithms[1-3] and multirate DSP.[4-6] Algorithms to exploit specific hardware capabilities were developed. For example, the Winograd discrete Fourier transform (WDFT)[7] was developed to reduce the multiply requirements in favor of increased additions. Also, optimization techniques for implementing filters that provide significant reductions in multiply requirements and reduced arithmetic requirements were developed.

As the hardware and algorithmic DSP capabilities have advanced, so have the processing demands. Signal processing now encompasses a broad set of real-time applications. **Real-time DSP** is processing of an input sequence to produce an output sequence with minimum delay. A basic requirement of real-time processing is that the outputs occur at the input rate or a reduced rate, dependent on the processing, and can be used to make real-time decisions such as changing the flight path. Non-real-time processing records the input sequence for processing at some later time. Applications such as navigation, communication, radar, and sonar fall into the real-time category. Real-time signal processing places stringent demands on the hardware and software design. Generally, these demands require that several different processing modes operate concurrently.

Analyzing and designing an efficient DSP system to meet the application requirements is a complex task. A structured approach to DSP system design using computer-aided design tools is necessary. Development of a structured design approach to solve complex DSP applications is a key objective of this book. We believe that an understanding of this design methodology will help you develop the skills necessary to work productively in the DSP field.

1.1 SCOPE OF THE BOOK

The material in this book has been designed for a broad spectrum of technical users, including advanced undergraduate and graduate students in engineering and computer science curriculums and practicing engineers, programmers, and scientists. In Chapters 2 through 6, we present the basic concepts of discrete linear systems, Z-transform analysis, digital filters, discrete Fourier transforms, and fast Fourier transforms. Problems that are included in each chapter embellish the chapter material and present meaningful applications. Chapters 7 through 11 cover advanced material, including finite arithmetic effects, power spectrum estimation, multirate processing, system design, and adaptive processing.

Chapter 2 discusses a signal analysis methodology and some important properties of discrete linear systems: linearity, time-invariance, causality, and stability. From the previously defined properties, the convolution sum and its product relationship in the frequency domain are discussed. The general digital filter difference equation is defined, and definitions of infinite impulse response (IIR) and finite impulse response (FIR) filters are presented. Frequency response concepts are shown. The effects of sampling are defined and related to input signal characteristics.

Chapter 3 is devoted to the definition of the Z-transform and the inverse Z-transform and the use of the transform pair in solving DSP problems. The Fourier transform is developed from the Z-transform. Digital filter realizations are developed using the Z-transform properties, and several examples are given to provide insight into the analysis of complex DSP system applications.

IIR and FIR filter theory and design techniques are presented in Chapters 4 and 5, respectively. Emphasis is placed on classic approaches, and we have developed a systematic approach to the filter design process. We have used this approach to program the filter designs on an IBM personal computer. Using the design approach, you will be able to develop your own program from the material presented. Detailed examples are presented, and computer-aided designs are shown. Since filter design techniques and filter realizations are a major area of continuing research, advanced filtering design approaches are discussed and the potential benefits to applications are addressed.

Chapter 6 covers the discrete Fourier transform (DFT) and fast Fourier transform (FFT). Properties of the DFT and key system performance design issues in the use of the DFT (circular convolution, spectral leakage, *picket fence* effect, effective noise bandwidth, and frequency resolution) are discussed. In addition, the concepts of frequency weighting and redundancy processing are introduced. The efficient implementation of the DFT via the FFT is thoroughly covered. FFT decimation-in-time (DIT) and decimation-in-frequency (DIF) algorithms are defined, and in-place versus not-in-place implementations are discussed. A four-point primitive FFT operation, together with a data addressing algorithm, is developed. A personal computer program is provided for the implementation of the FFT. The inverse FFT (IFFT) is shown to be directly implementable using the FFT. Implementation of FIR filters by use of the FFT/IFFT are described (i.e., fast convolution) and computational comparisons are shown as a function of the filter length.

Multirate digital signal processing techniques are presented in Chapter 7. The key concepts of interpolation and decimation are described. Optimization techniques for multirate system design approaches and the use of FIR and IIR filters are discussed. Application examples of multirate techniques are given.

Chapter 8 covers the classical power spectrum estimation techniques. In the chapter we primarily discuss estimating the power spectral density by averaging modified periodograms. The computation of the redundancy processing gain resulting from processing overlapped modified periodograms is discussed. Signal detection concepts are presented, and probability of detection and false alarm performance as a function of threshold settings is discussed. A procedure for computing the detection performance for a sinusoidal signal embedded in Gaussian noise is presented. The procedure can be easily programmed on a personal computer.

Chapter 9 describes the effects of finite wordlength and arithmetic in digital signal processing. A systematic approach to the analysis of the errors resulting from the finite arithmetic effects is developed. Detailed methods of analysis are described and illustrated using personal computer simulation and modeling results.

Chapter 10 provides an introduction to signal processing design and the analysis to configure a signal processor that implements the processing efficiently. A methodology for performing signal processing system design, and a spectral analysis design example using the methodology, is discussed. System design architectural and performance issues are addressed. System performance is estimated, and the results are presented using computer simulation, whereby parameters such as probability of detection, probability of false alarm, and bandwidth can be varied to ascertain the effect performance has on system design.

The material in Chapter 10 bridges the gap between pure signal processing design and the hardware capability to implement the processing. It is essential for the designer to have a structured methodology in performing the system

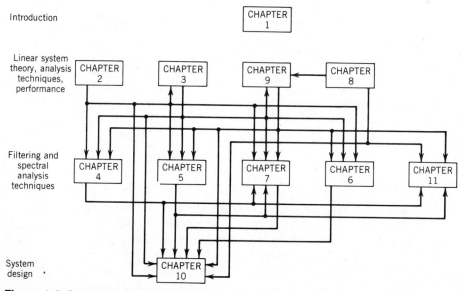

Figure 1.2 Chapter relationships.

design. This ensures that the hardware will meet all requirements. The concepts and signal processing techniques from all preceding chapters are drawn on to demonstrate the design approach.

Chapter 11 covers adaptive DSP. The least squares method to time updating the transfer function parameter estimates is considered. It is shown that the recursive least squares method presented is similar in structure to the Kalman filter. Some applications of sequential parameter estimation are adaptive beamforming, power spectral density estimation, and system identification. The concepts are illustrated using computer-simulated results. Special filter structures that are suitable for adaptive implementations are addressed.

Figure 1.2 illustrates the relationship between the chapters. Each chapter has a set of problems. The problems illustrate the use of the chapter theory and concepts based on real applications. Several useful analysis routines that can be adapted to personal computers are described. A feature of the book is the inclusion of numerous examples to illustrate the theory and concepts.

1.2 SYSTEM DESIGN

Signal processing system design consists of developing an understanding of the application problem and designing a signal processing system that satisfies all the application requirements. The application requirements consist of signal processing requirements and non-signal-processing requirements. This book presents a system design approach to develop the signal processing requirements' definition and design. Non-signal-processing requirements such as volume, power, weight, maintainability, and reliability, which are key to meeting the overall application, must be considered as part of the final design. In order to perform the design effectively, a structured methodology must be followed by a design team including all relevant disciplines. The design methodology described is developed thoroughly in Chapter 10.

Developing specialized computer programs for assisting in the analysis procedures provides convenient tools for the designer. Personal computers when combined with available programs provide a powerful analysis aid. Examples of tools that can be developed and used in the system design are shown throughout the book.

1.2.1 Methodology

Figure 1.3 presents a five-step structured procedure for the design of a signal processing system. We present the theory, concepts, and designs of DSP using a system design approach so that you will obtain a better understanding of the application of DSP to real problems. A brief definition of each step in the system design procedure is provided for introductory purposes. The relationship of the design steps to each of the chapters is addressed. A detailed discussion of each step is given in Chapter 10.

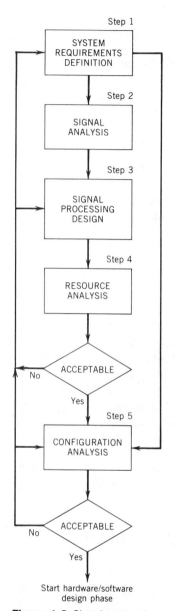

Step 1

Step 1
User/customer driven
Develop system level requirements
Signal processing
Non-signal processing
System level documentation
Requirements specifications
Interface design specifications

Step 2
Define input signal characteristics
Types
Parameters
Noise sources and distributions
Data rates

Step 3
Develop signal processing graphs
for each processing mode
Specify primitive operations
Initial partitioning of processing
element types
Arithmetic analysis
Iterative process
Results in architecture approach
for new signal processor design

Step 4
Final partitioning of process to
element types
Compute all resource utilizations for
each mode
Memory, control, bandwidth, throughput
Iterative process

Step 5
Define configuration
Based on resource analysis
Based on non-signal-processing
requirements
Partition multimode processing to
configuration
Perform resource analysis
Iterative process

Figure 1.3 Signal processing system design.

STEP 1. SYSTEM REQUIREMENTS DEFINITION

This step consists of defining and quantifying the system level requirements. Generally, this step starts with a top-level set of customer-defined requirements. These include human-machine interfaces, signal types, signal-characteristics, required processing options and modes, processing performance, and required performance verification testing. All system level performance requirements, including non-signal-processing requirements, are completely specified. This specification effort is a cooperative effort between the customer and contractor that results in a set of system level specifications. As shown in Figure 1.3, step 1 feeds steps 2—signal analysis—and step 5—configuration analysis.

STEP 2. SIGNAL ANALYSIS

Signal analysis is the complete definition of the input signal types and characteristics. Since most applications deal with stochastic input signals, a signal model is usually developed to approximate the input signal. Frequency bandwidths are defined and characterized. Expected signal-to-noise ratio (SNR) ranges and the maximum and minimum input levels are estimated. Potential interference sources are characterized. The signal analysis results form the basis for the definition of the signal processing algorithms in step 3. The signal analysis procedure is presented in Chapter 2.

STEP 3. SIGNAL PROCESSING DESIGN

In step 3 we develop signal processing data flow diagrams for each signal class defined in step 2 and for each required processing mode defined in step 1. A signal processing data flow diagram graphically shows all functional operations, input and output data, and parameters necessary to execute a processing mode. Signal classes include acoustics and electromagnetics, where each class is subdivided into processing modes. For example, sonar processing of ocean acoustic signals includes several modes; including, spectral analysis detection, broadband energy detection, active pulse detection, frequency tracking, and bearing tracking. As part of this process, the algorithm processing flow is partitioned to system components. We deal only with the signal processor portion of the system, but must be sensitive to the alternative choices between input data sampling rates and front-end signal conditioning hardware complexities. A key portion of this step is to assure that processing errors due to finite arithmetic effects are kept to an acceptable level. A methodology of performing this step will be described in Chapter 10. The results of this step will be iteratively applied to step 4—resource analysis—until an acceptable design is defined.

STEP 4. RESOURCE ANALYSIS

Three high-level categories of resource parameters (throughput, memory, and input/output bandwidth) generally apply across a signal processor's functional elements. Functional elements can be categorized into four primary groups: (1) control processing; (2) signal processing; (3) data storage; and (4) data communication. Figure 1.4 presents a resource parameter diagram for a generic signal processor. Each of the system resource parameters must be fully quantified for the DSP data flow diagrams established in step 3.

In Step 4, the percentage utilizations of each of the system resources are determined and provide the necessary inputs to the configuration analysis, the final step of the procedure. For a project where a new signal processor is being

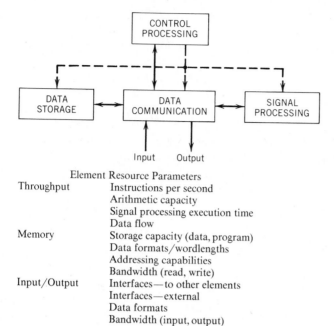

Element Resource Parameters

Throughput	Instructions per second
	Arithmetic capacity
	Signal processing execution time
	Data flow
Memory	Storage capacity (data, program)
	Data formats/wordlengths
	Addressing capabilities
	Bandwidth (read, write)
Input/Output	Interfaces—to other elements
	Interfaces—external
	Data formats
	Bandwidth (input, output)

Figure 1.4 Signal processor resource parameters.

designed, the resource analysis will drive design choices. For existing signal processors, the resource analysis may dictate that the algorithm data flow be reworked to fit into the available processing resources. A methodology for performing the resource analysis is presented in Chapter 10.

STEP 5. CONFIGURATION ANALYSIS

In the configuration analysis step we assign a hardware configuration and partition the data flow graphs to the elements of the configuration. The processing resource requirements are calculated for the defined configuration based on the partitioned processing. Various partitionings are tried, and the partitioning with the lowest resource utilization is selected. If none of the partitioning alternatives are acceptable, the configuration is modified based on the resources required for the case that most closely fits the configuration. Concurrently, the configuration's non-signal-processing performance is estimated. If all requirements are met, the design is finished; otherwise, the analysis must be iterated until an acceptable design is achieved. This can result in a reduction in performance requirements to achieve an acceptable design.

RELATIONSHIP OF CHAPTERS
TO SYSTEM DESIGN METHODOLOGY

Chapter 2 is key to step 2 (signal analysis) and step 3 (signal processing design) of the system design methodology. Chapter 3 provides a signal analysis tool for analyzing signals for step 2 (signal analysis) and step 3 (signal processing design). Chapters 4 and 5 are key to being able to carry out step 3 (signal processing design) of the system design methodology. Chapter 6 provides a good analysis tool (FFT) for use in step 2 (signal analysis) and is key to step 3 (signal processing design) of the system design for many applications. The multirate techniques are key to step 3 (signal processing design). Chapter 8 provides the analysis approaches used on random processes that are required to develop the requirements in step 1, to perform the signal analysis in step 2, and to establish the signal processing design in step 3. Finite arithmetic analysis, Chapter 9, is based on the use of the Z-transform pair and is essential to establishing the signal processing design in step 3. Chapter 11 addresses the signal processing design, step 3, for signals that require adaptive processes in order to achieve performance requirements.

1.2.2 Implementation Methodology

Implementation of the signal processing system design requires a team of professionals from all disciplines, including systems, hardware, software, oper-

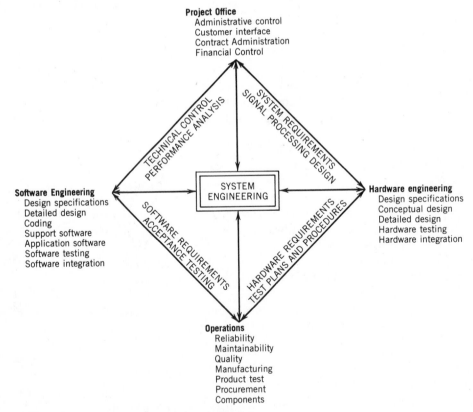

Figure 1.5 Signal processing system design team.

ations, and the project office. This team forms a nucleus for defining, analyzing, implementing, and testing the system. Although implementation is usually not presented as part of DSP theory and concepts, we believe it is important to the understanding of the interrelationships of the different disciplines necessary to develop a DSP system. Design errors because of poor or insufficient requirements specifications or lack of participation from a team member are very costly to correct after the hardware and software have been developed. Utilization of the design methodology requires an understanding of each team member's responsibilities to ensure that the DSP system developed meets all its application requirements.

Figure 1.5 illustrates the DSP system design-team concept. The system engineering organization has technical responsibility for the overall system design. This includes performing the analysis methodology discussed in Section 1.2.1, coordinating and reviewing all system development specifications, performing system integration, and conducting formal customer acceptance tests. The hardware organization is responsible for the electrical and mechanical design, development, and test. The software organization has the responsibility for generating the computer program design specifications, coding of the application programs, and software integration and testing. We include several important groups under the heading of operations, including reliability, maintainability, quality, product test, procurement, components, and manufacturing. These groups are responsible for assuring the non-signal-processing requirements are met and that the system can be manufactured within cost and schedule. The project office is responsible for the customer interface, administration of the contract, data and hardware deliveries, and financial management.

1.3 MOTIVATION

The intent of this section is to interest you in pursuing the details of the summary information presented on DSP. The reader without prior introduction to DSP should not expect to understand the material fully. Instead, we hope the summary gives a general understanding of the relationships between key DSP concepts that will help to tie together the many topics covered throughout the remaining chapters. For those who already have a good

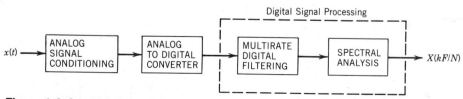

Figure 1.6 Spectral analysis system.

understanding of the basics of DSP, Chapters 1 through 6 provide a reference to the basic DSP material. Chapters 7 through 11 cover advanced DSP topics.

DSP is primarily based on the theory of discrete linear systems. In general, we will discuss systems excited by discrete signals sampled in time. Figure 1.6 represents a narrowband spectral analysis signal processing flow. **Narrowband spectral analysis** is the process of discriminating between input frequencies in very small frequency increments across a specified frequency band of interest. Since this application uses a broad class of DSP algorithms, it is used to give a brief discussion of some key signal processing concepts: sampling, digital filtering, multirate processing, and spectral analysis.

Many DSP applications have analog inputs consisting of signals embedded in random noise. Before any DSP can be performed, the analog input is sampled and digital representations of each sample are developed. **Sampling** is the process of converting a continuous analog signal into discrete amplitudes at specified sampling intervals. These discrete amplitudes are quantized into digital values based on the wordlength used. An analog-to-digital (A/D) converter samples and quantizes a continuous input signal. We will deal with uniform sampled signals over time, as illustrated in Figure 1.7. A continuous signal is shown in Figure 1.7a, and the corresponding discrete signal sampled with time interval T is shown in Figure 1.7b.

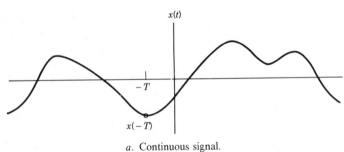

a. Continuous signal.

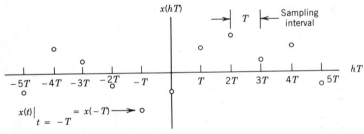

b. Discrete signal.

Figure 1.7 Sampled signal.

The sampling interval, T, is determined from the input signal frequency content based on the Nyquist sampling theorem.[8] The theorem states: **A continuous input signal with frequency bandwidth limited to $f \leq f_h$, can be reconstructed accurately from a sampled signal, provided the sampling rate is greater than twice the input bandwidth.** The sampling rate, F (samples per second), is equal to the reciprocal of the sampling interval T. The choice of input sampling rate and sampling rate choices throughout the signal processing flows are key to an efficient implementation. The sampling process will be covered in Chapter 2. The number of discrete amplitude levels used to represent the signal at each sampling interval introduces quantization noise and saturation noise into the signal. The selection of the number of quantization levels (bits) to minimize the noise level is an important topic that is discussed in Chapter 9.

Digital signal processing is defined as any digital operation performed on a digital input sequence. **Modulation** is the process of multiplying the input

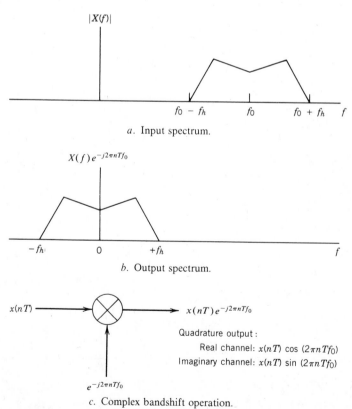

a. Input spectrum.

b. Output spectrum.

c. Complex bandshift operation.

Figure 1.8 Complex bandshift by f_0 Hz.

signal by a sampled function. Modulation is often used in spectral analysis applications to translate a frequency band to simplify subsequent processing. A **quadrature modulation** process translates a frequency band by multiplying the input signal by a complex exponential, as shown in Figure 1.8. A band-limited input signal spectrum of bandwidth $2f_h$ and center frequency f_0 is shown in Figure 1.8a. The resultant signal spectrum after the quadrature modulation by frequency f_0 is shown in Figure 1.8b. The quadrature channel outputs are given in Figure 1.8c; note that a complex data output (real and imaginary) results. Subsequent processing must process both real and imaginary (quadrature) channels.

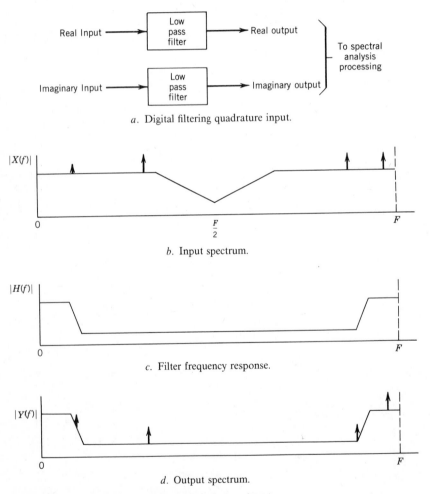

a. Digital filtering quadrature input.

b. Input spectrum.

c. Filter frequency response.

d. Output spectrum.

Figure 1.9 Digital filter process for quadrature input.

A **digital filter** process modifies the input signal samples based on the frequency band specification. Four classes of digital filters can be specified: lowpass (LP), highpass (HP), bandpass (BP), and bandstop (BS). The design of each of these filters will be discussed in Chapters 4 and 5. For the spectral analysis example chosen, the center of the input signal spectrum is translated to baseband (i.e., 0 Hz or DC) by the quadrature modulation process.

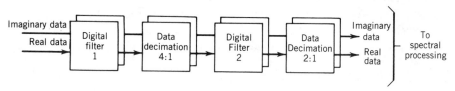

a. Multirate digital filter processing flow.

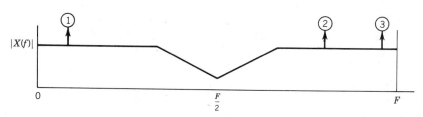

b. Quadrature data input spectrum.

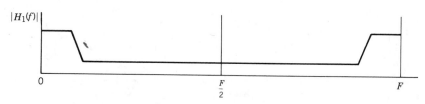

c. Digital filter frequency response (filter 1).

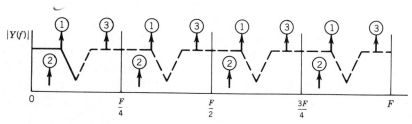

d. Output spectrum after 4 : 1 data decimation.

Figure 1.10 Multirate digital filtering.

Therefore, the filtering process is performed using lowpass digital filters with a frequency cutoff of one-half the frequency band specification. Two lowpass filters are required, as shown in Figure 1.9, one for each quadrature channel. Figure 1.9a shows the two filter channels. The spectrum at the input to the filter process is presented in Figure 1.9b, and the digital filter frequency response is shown in Figure 1.9c. The filter output spectrum is given in Figure 1.9d. The filter outputs are input data to the spectral analysis process.

A **multirate** process means that the sampling rate is changed during the signal processing operations. Figure 1.10 depicts a multirate digital filter implemented as two filters with two decimators to reduce the sampling rate by an amount given by the product of the individual decimators ($4 \times 2 = 8$). The data rate at each filter stage input is equal to the rate of the preceding stage input divided by the decimation factor for the preceding stage. All other samples are discarded. This approach to filtering takes advantage of the fact that as the bandwidth of the input signal is reduced the data rate can be reduced proportionately, provided the Nyquist criteria is preserved. Figure 1.10b illustrates the digital filter input spectrum. The filter frequency response shown in Figure 1.10c provides the necessary rejection of all signals that would alias into the filter passband following the subsequent data decimation operation. **Aliasing errors**, where high frequencies take on the identity of low frequencies, are the result of undersampling the signal. The spectrum following the first decimation stage is shown in Figure 1.10d. Since it is impossible to completely limit all frequencies above some specified value, aliasing will result. The design process must assure that the level of aliasing does not degrade system performance below specified requirements. Digital filters are covered in Chapters 4 and 5, and multirate systems are covered in Chapter 7. Performance impacts due to aliasing are addressed in Chapter 10.

The spectral analysis of the filtered data is performed by the processes shown in Figure 1.11. The fast Fourier transform (FFT) is a computationally efficient algorithm to compute a set of N discrete Fourier transforms (DFTs) efficiently for N uniform frequency intervals from 0 Hz to the FFT input data rate. The frequency interval between FFT outputs equals F/N Hz. F is the FFT input sample rate and N is the FFT data length that is equal to the number of FFT frequency outputs. Various definitions of the FFT resolution exist and are covered in Chapter 6.

Each FFT frequency output, or bin, is a complex value representing the even and odd Fourier components of the input signal. The weighting shown in Figure 1.11 is used to reduce spectral leakage into a bin from signal frequencies outside the bin. The detection process shown in Figure 1.11 computes the

Figure 1.11 Spectral analysis processing.

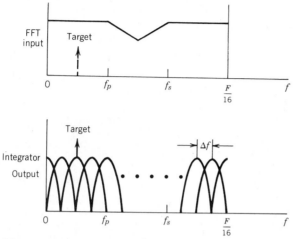

Figure 1.12 Spectral analysis processing input / output illustration.

magnitude of each of the FFT outputs. The integration is a smoothing operation on successive FFT outputs. This reduces the variability of the outputs, which contain noise only, whereas the bins, which contain signal-plus-noise, will have a value related to the signal-plus-noise root mean square (rms) level in the limit. Therefore, bins containing signals are enhanced, as illustrated in Figure 1.12. The output is shown ideally for clarity. The frequency response is derived rigorously in Chapter 6.

Figure 1.13 presents a diagram of a computer model used to compute the spectral outputs outlined in the preceding discussion. The data output in Figure 1.13b is generated by the input signal model of Figure 1.13a and contains two signal components centered on FFT bin outputs 8 and 16. The sinusoidal components are added to noise from a Gaussian random noise generator. Successive groups of 128 samples are generated and processed by the spectral analysis operations. Figure 1.13c is the output for the input given in Figure 1.13b. Figures 1.13d and e show the outputs after integrating 4 and 16 groups of 128 samples, respectively. These figures clearly illustrate the value of this approach to finding spectral components embedded in noise. The system performance for a given input signal-to-noise ratio (SNR) is related to the FFT resolution and the number of samples integrated, and is developed thoroughly in Chapters 8 and 10.

We hope that this introductory discussion on system design has given you some insight into the complete task of designing a DSP system. The concepts discussed from the perspective of a narrowband spectral analysis detection application are intended to interest you in pursuing this material in subsequent chapters. We hope we have been successful to that end and that you will find the subsequent material interesting and useful in the design, analysis, and implementation of digital signal processing systems.

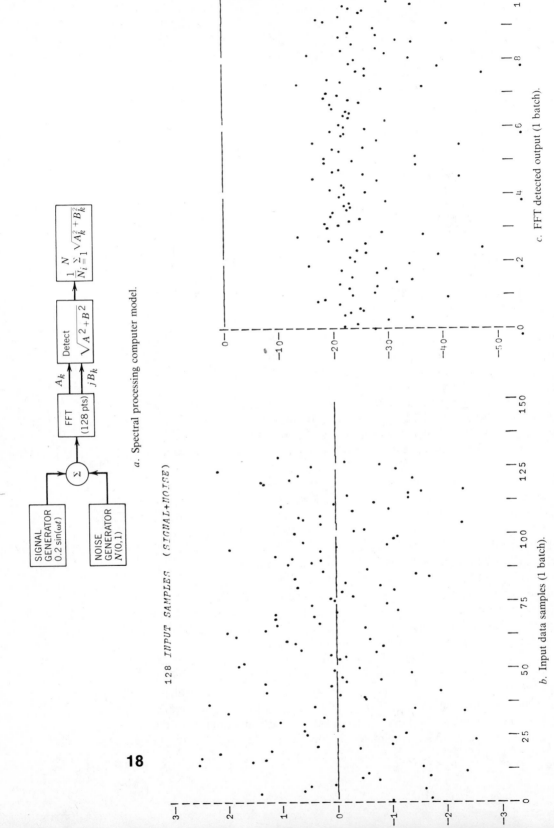

a. Spectral processing computer model.

b. Input data samples (1 batch).

c. FFT detected output (1 batch).

Figure 1.13 Spectral analysis example.

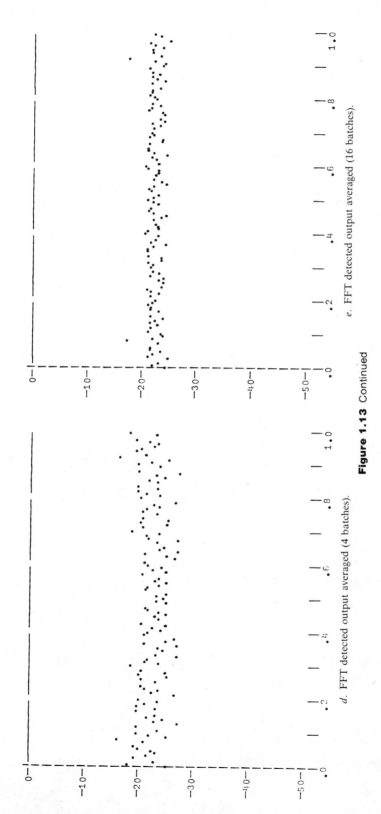

d. FFT detected output averaged (4 batches).

e. FFT detected output averaged (16 batches).

Figure 1.13 Continued

19

References

1. R. E. Blahut, *Fast Algorithms for Digital Signal Processing*, Addison-Wesley, Reading, Mass., 1984.

2. N. J. Nussbaumer, *Fast Fourier Transform and Convolution Algorithms*, 2nd ed., Springer-Verlag, Berlin, 1982.

3. D. F. Elliot and R. Rao, *Fast Transforms: Algorithms, Analysis, and Applications*, Academic Press, New York, 1983.

4. R. E. Crochiere and L. R. Rabiner, *Multirate Digital Signal Processing*, Prentice-Hall, Englewood Cliffs, N.J., 1983.

5. M. G. Bellanger, "Computation Rate and Storage Estimation in Multirate Digital Filtering with Half-Band Filters, *IEEE Trans. ASSP*, ASSP-25(4):344–346, 1977.

6. H. G. Martinez and T. W. Parks, "A Class of Infinite-Duration Impulse Response Digital Filters for Sampling Rate Reduction," *IEEE Trans. ASSP*, ASSP-24:109–114, April 1976.

7. S. Winograd, "On Computing the Discrete Fourier Transform," *Math. Comp.*, 32:175–199, 1978.

8. D. A. Linden and N. M. Abramson, "A Generalization of the Sampling Theorem," *Information and Control*, 3(1):26–31, 1960.

Additional Readings

ANTONIOU, A. *Digital Filters Analysis and Design*. McGraw–Hill, New York, 1979.

BELLANGER, M. *Digital Processing of Signals, Theory and Practice*. Wiley, New York, 1984.

BLANCHARD, B. S., and FABRYCKY, W. J. *Systems Engineering and Analysis*. Prentice-Hall, Englewood Cliffs, N.J. 1981.

BOWEN, B. A., and BROWN, W. R. *VLSI Systems Design for Digital Signal Processing*. Prentice-Hall, Englewood Cliffs, N.J., 1981.

BOWEN, B. A., and BROWN, W. R. . SYSTEMS DESIGN: *Volume II of VLSI Systems Design for Digital Signal Processing*. Prentice-Hall, Englewood Cliffs, N.J., 1985.

BRIGHAM, E. O. *The Fast Fourier Transform*. Prentice-Hall, Englewood Cliffs, N.J. 1974.

Digital Signal Processing Committee. *Selected Papers in Digital Signal Processing, II*. IEEE ASSPS Society, IEEE Press, New York, 1975.

Digital Signal Processing Committee. *Programs for Digital Signal Processing*. IEEE ASSPS Society, IEEE Press, New York, 1979.

GOLD, B., and RADER, C. M. *Digital Processing of Signals*. McGraw–Hill, New York, 1969.

JUNG, M. T. *Methods of Discrete Signal and System Analysis*. McGraw–Hill, New York, 1982.

OPPENHEIM, A. V. (ed). *Papers on Digital Signal Processing*. MIT Press, Cambridge, Mass., 1969.

OPPENHEIM, A. V. (ed). *Applications of Digital Signal Processing.* Prentice-Hall, Englewood Cliffs, N.J., 1978.

OPPENHEIM, A. V., and SCHAFER, R. W. *Digital Signal Processing.* Prentice-Hall, Englewood Cliffs, N.J., 1975.

PARKS, T. W., and BURRUS, C. S. *Digital Filter Design*, Wiley, New York, 1987.

PELED, A., and LIU, B. *Digital Signal Processing: Theory, Design, and Implementation.* Wiley, New York, 1976.

RABINER, L. R., and GOLD, B. *Theory and Application of Digital Signal Processing.* Prentice-Hall, Englewood Cliffs, N.J. 1975.

RABINER, L. R., and RADER, C. M. (eds). *Digital Signal Processing.* IEEE Press, New York, 1972.

STEARNS, S. D. *Digital Signal Analysis.* Hayden, Rochelle Park, N.J., 1975.

TREICHLER, J., JOHNSON JR., R., and LARIMORE, M. *Theory and Design of Adaptive Filters*, Wiley, New York, 1987.

TRETTER, S. A. *Introduction to Discrete-Time Signal Processing.* Wiley, New York, 1976.

2

Discrete-Time Signal Analysis and Linear Systems

Understanding discrete-time signals and linear systems is key to developing the skills required to design digital signal processing systems.

2.0 INTRODUCTION

This chapter develops the theory and concepts of discrete-time linear systems required to address digital signal processing applications. A discrete-time linear system performs an operation on a discrete input signal based on a predefined criteria to produce a modified output. In this chapter we develop a signal analysis methodology, define a general class of discrete-time signals, present the sampling theorem, and define the general classes of systems dealing with digital filtering and discrete Fourier analysis processes. This will provide the necessary background for understanding the material presented in subsequent chapters.

2.1 SIGNAL ANALYSIS

Signals convey specific information that the DSP system is attempting to interpret. One goal of a DSP system is to provide the best approach to analyzing or estimating the information content of the signal. **Signal analysis,** in this context, is the process of defining and quantifying all signal characteristics for the application being addressed. A broader definition generally used for signal analysis is the actual processing performed to detect the signal and subsequently to estimate the parameters of the detected signal. The broader definition is encompassed by the total system design methodology. The signal analysis step of the system design given in Section 1.1 is discussed in the following section. The methodology is presented, signal characteristics are described, and important signals are defined.

2.1.1 Methodology

Signal analysis, the second step of the DSP system design procedure, results in the complete definition of the input signal types and characteristics. Since most applications deal with stochastic input signals, a signal model that adequately approximates the input signal must be developed. All signal transmission controls and formats must be defined. Frequency bandwidths must be defined and characterized. The signal-to-noise ratio ranges and the maximum and minimum input levels must be determined. Any potential interference sources must be characterized. The signal analysis results are used as the basis for definition of signal processing algorithms in step 3. Figure 2.1 presents the signal analysis methodology used to assure that the information required to perform the DSP system design is established.

The signal analysis methodology starts with a definition of all possible signal characteristics/parameters by reviewing a set of 14 items (Fig. 2.1). Many of the items will be given by the requirements specifications for the particular application from the first step of the system design procedure. In some cases, the requirements will be at a high level and need to be decomposed to arrive at the signal characteristics. Computer/hardware analysis can be used at this point when the actual signal data are available. The signals can be stored on tape and analyzed to determine the characteristics.

Using the signal characteristics information, the procedure of Figure 2.1 is followed step-by-step. Note that active signals (i.e., signals that must be generated and transmitted by the system) require definition of the signal characteristics to be transmitted in addition to the receive signal characteristics. For analog signals the signal conditioning requirements are developed and the minimum sampling rate is determined. Next the digital requirements, including the number of bits, data formats, data codes, and data rates, are defined. Finally, the results are used in the subsequent signal processing analysis and design performed in step 3 of the DSP design.

The design methodology presented is flexible; it is basically used to guarantee that all the signal factors are considered during the DSP design. You will develop your own preferred approach as you gain experience in DSP system design.

2.1.2 Signal Characteristics

A list of items was presented and briefly discussed in the preceding section to assist in the characterization of a signal. Generally we deal with signals that

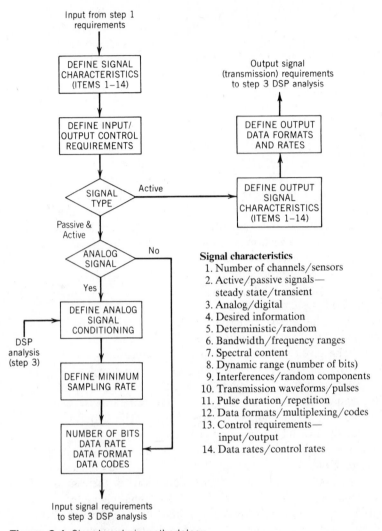

Figure 2.1 Signal analysis methodology.

are a function of one independent variable, x. In DSP applications the variable x is a discrete sequence, written as

$$\{x(nT)\}, \qquad n = \{,\dots, -4, -3, -2, -1, 0, 1, 2, 3, 4\dots\} \qquad (2.1)$$

where T is the uniform interval (typically a time interval) between input samples. $\{x(nT)\}$ can represent a sequence of numbers generated on a computer, data taken from an experiment, or numbers from some natural phenomena that have been sampled. The nth member of the sequence is

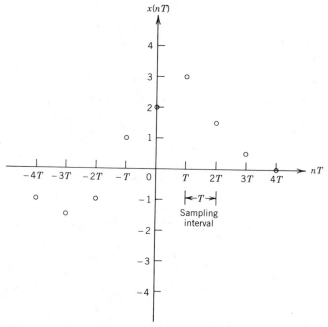

Figure 2.2 Discrete signal x with independent variable nT.

denoted by $x(nT)$. For simplicity we will denote the sequence by $x(nT)$ or $x(n)$, where the uniform sampling period is implied. A discrete sequence is shown in Figure 2.2. The variable x represents a composite of all data sources/phenomena of the signal. For example, an ocean acoustic sensor (i.e., hydrophone) receives information impinging on it from many sources, including wave motion, fish, ships, oil rigs, and submarines. These sources produce a variety of signal components: random, transient, and periodic (see Fig. 2.3).

Random signals are characterized by a nonpredictable sample-to-sample output. Random signals are produced by natural phenomena such as the waves in the ocean. Quantizing a continuous signal to a finite number of bits during the analog-to-digital conversion and the subsequent DSP finite arithmetic operations also introduces random signal components. Random signals are characterized by probability density functions (pdf), expected values (i.e., mean or average), variances, and correlation functions. **White noise** is defined as a random signal with a uniform frequency spectrum. Random signals and probability distributions commonly used to model white noise are discussed in Chapter 8.

Transient signal components are generated by finite duration phenomena. The signal output can be characterized each instant by a similar sample-to-sample output. The periodic signal is a key component produced by many phenomena such as rotating machinery; it is used in active systems such as radar for detection and tracking of aircraft. Periodic signal components, which

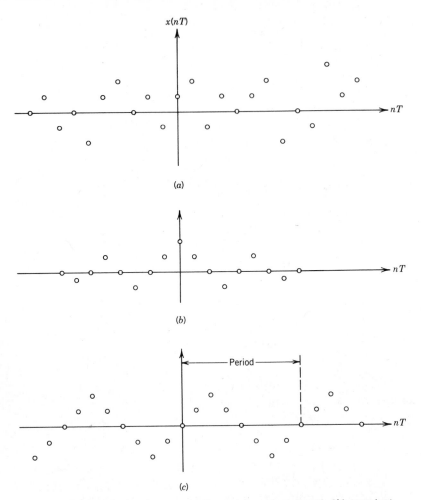

Figure 2.3 Typical signal representations: (*a*) random signal; (*b*) transient signal; (*c*) periodic signal.

will be shown to play a key role in discrete linear systems, are characterized by their spectral content.

Generally we deal with signals where time is the independent variable, although other independent variables such as space are also valid. Each discrete-time signal is represented by a quantized digital value

$$x'(nT) = Q\{x(nT)\} \tag{2.2}$$

Quantization, represented by operator Q in Eq. 2.2, converts a value of a continuous amplitude sampled at time nT to one of a finite number of values by a predefined rule, see Figure 2.4. Rounding and truncation are specific

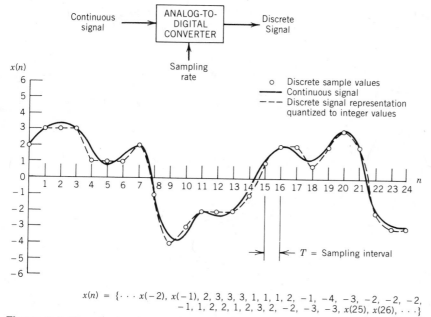

$$x(n) = \{ \cdots x(-2), x(-1), 2, 3, 3, 3, 1, 1, 1, 2, -1, -4, -3, -2, -2, -2,$$
$$-1, 1, 2, 2, 1, 2, 3, 2, -2, -3, -3, x(25), x(26), \cdots \}$$

Figure 2.4 Discrete signal representation of continuous signal.

quantization rules that are presented in Chapter 9.

Since we are dealing with digital equipment, the number of levels is usually a binary representation of the continuous amplitude. The error due to quantization is given by

$$e(nT) = x'(nT) - x(nT) \tag{2.3}$$

Quantization errors will be covered in Chapter 9.

Sample refers to one value from the sequence of numbers representing the discrete signal and neither requires nor implies that the sequence was generated by sampling an analog waveform. Throughout this book the class of signals, represented by a sequence of numbers, that have only a finite set of values are referred to as a digital signal or a signal. The term signal in some instances is used to describe only the desired component of the input, noise being an undesired component. **Waveform** is sometimes used in DSP literature to refer to the total input sequence. Where nomenclature would be ambiguous, we will clarify terms.

Signal characteristics and application goals play a key role in specifying the required signal processing. Several questions about the signal must be answered to assure that the proper representation of the signal is maintained. What type of signal is present? What is the desired information content? Is it deterministic? Is it random? What is the independent variable? What inherent noise sources or interferences must be dealt with? What is the frequency spectrum? What are the maximum and minimum ranges of the input signal levels? What are the expected signal-to-noise ratios? Signal norms, frequency response, and Parseval's theorem all play a key role in characterization of a signal.

SIGNAL MEASURES

An important signal characteristic is a measure of the magnitude of the signal. The L_p signal norm[1,2] is given by

$$L_p = \|x(n)\|_p = \left(\sum_{n=-\infty}^{\infty} |x(n)|^p \right)^{1/p} \tag{2.4}$$

where p is a positive integer. The L_p norm represents a single nonnegative number that provides an overall measure of the size of the signal. The **signal norm** of $x(n)$ satisfies the following axioms.

1. $\|x(n)\| > 0$ for $x(n) \neq 0$ for all n and $\|x(n)\| = 0$ if and only if $x(n) = 0$ for all n.
2. $\|ax(n)\| = |a|\,\|x(n)\|$ for any scalar a.
3. $\|x(n) + y(n)\| \leq \|x(n)\| + \|y(n)\|$ (The triangle inequality).

Three norms have particular significance in defining and analyzing signals in DSP applications: L_1, L_2, and L_∞. The L_1 norm is equal to the sum of the magnitudes of each signal sample

$$L_1 = \|x(n)\|_1 = \sum_{n=-\infty}^{\infty} |x(n)| \tag{2.5a}$$

This measure is used in the determination of system stability of discrete linear systems. The L_2 norm provides a measure of the signal power

$$L_2 = \|x(n)\|_2 = \left(\sum_{n=-\infty}^{\infty} |x(n)|^2 \right)^{1/2} \tag{2.5b}$$

The square of the L_2 norm is used throughout this book in the analysis of DSP signals and systems. The L_∞ norm gives the peak magnitude of the signal

$$L_\infty = \|x(n)\|_\infty = \max |x(n)|, \qquad \text{for all } n \tag{2.5c}$$

This norm provides a useful bound for determining the dynamic range requirements of DSP systems.

SIGNAL FREQUENCY SPECTRUM

For many applications, the class of signals of interest are periodic. Fortunately, for the periodic class of signals, an efficient method of analysis is available using Fourier transforms to obtain the frequency spectrum of the signal. The frequency spectrum of a signal shows its periodic components and is an important signal characteristic in DSP analysis and design. Many analysis techniques were developed for analog signals using Fourier analysis[3] and Laplace transforms.[4] The Fourier transform pair are given by the Fourier transform of $x(t)$

$$X(f) = \int_{-\infty}^{\infty} x(t)e^{-j2\pi ft}\, dt \qquad (2.6a)$$

and the inverse Fourier transform of $X(f)$

$$x(t) = \frac{1}{2\pi}\int_{-\infty}^{\infty} X(f)e^{j2\pi ft}\, df \qquad (2.6b)$$

The Fourier integral maps a signal in one domain (time) to another domain (frequency). Equation 2.6a provides a continuous output as a function of frequency variable f for the continuous input signal as a function time variable t. Although the Fourier transform pair provides a powerful analytic tool for the analysis of continuous signals and systems, the implementation of the transform requires a mathematical representation of the function to be transformed and the calculation of the integral expression. It is desirable to perform a similar operation on discrete signals to facilitate implementation with digital hardware. An approximation of the input signal, to a set of discrete sample values $x'(nT)$, and the truncation of the signal to a finite duration using a sampling period of T are given by

$$x'(nT) \doteq \begin{cases} 0 & \text{for } t < 0 \\ x(nT) & \text{for } nT < t < (n+1)T \\ 0 & \text{for } t \geq (n+1)T \end{cases}$$

Based on the finite set of discrete samples, a discrete Fourier transform (DFT) pair can be developed[5]

$$X\left(\frac{kF}{N}\right) = \frac{1}{N}\sum_{n=0}^{N-1} x(nT)e^{-j2\pi nTkF/N} \qquad (2.7a)$$

$$x(nT) = \sum_{k=0}^{N-1} X\left(\frac{kF}{N}\right)e^{j2\pi nTkF/N} \qquad (2.7b)$$

where

$n = \{0, 1, 2, \ldots, N-1\}$ (time index)

$k = \{0, 1, 2, \ldots, N-1\}$ (frequency index)

$T = 1/F$ (sampling interval = seconds/sample)

$F = 1/T$ (sampling rate = samples/second)

N = number of samples used in computing the DFT

From the DFT (Eq. 2.7a) the output is a complex value that can be written as a magnitude and phase

$$X\left(\frac{kF}{N}\right) = M\left(\frac{kF}{N}\right)e^{j\phi(kF/N)}$$

where

$$M\left(\frac{kF}{N}\right) = \left|X\left(\frac{kF}{N}\right)\right| \quad \text{and} \quad \phi\left(\frac{kF}{N}\right) = \arg X\left(\frac{kF}{N}\right)$$

Calculation of $M(kF/N)$ for all k will map out the frequency amplitude spectrum of $x(nT)$. Likewise, evaluation of $\phi(kF/N)$ generates the frequency phase response. The amplitude and phase responses are periodic with period F. The DFT can be viewed as a truncated sampled Fourier transform where the truncated time function is periodically extended to plus and minus infinity. It will be shown in Chapter 6 that nearly all the continuous Fourier transform pair properties apply with some simple modifications. Although no approximations are implied by the DFT pair, the signal cannot be band-limited in the strict sense owing to its finite duration. A **band-limited** signal has frequency components over a finite set of frequencies.

PARSEVAL'S THEOREM

Parseval's theorem relates the total energy contained in the input signal to the total energy density spectrum calculated from the Fourier transform of the signal. The discrete Parseval theorem is given by

$$\sum_{n=0}^{N-1}|x(n)|^2 = \frac{1}{N}\sum_{k=0}^{N-1}|X(k)|^2 \tag{2.8}$$

Using the definition for the L_2 norm, the theorem can be rewritten

$$\|x(n)\|_2^2 = \frac{1}{N}\|X(k)\|_2^2$$

This relation is very useful in analyzing the effects of finite arithmetic, and provides a method for computing the signal power using the frequency spectral output.

2.1.3 Typical Discrete-Time Signals

A discrete-time signal is represented by a sequence of numbers

$$x = \{x(nT)\}, \qquad \begin{array}{l} n = \text{set of all integers} \\ T = \text{time interval between samples} \end{array}$$

where $x(nT)$ is the nth number from the set of numbers making up the sequence, x. For convenience, we let $x(nT)$ represent the sequence x, even though from a mathematical viewpoint $x(nT)$ represents only the value of the sequence at time nT. We adopt the tabular format for explicitly listing the elements of a sequence.[1,6]

$$x = \left\{ \ldots, x(-3), x(-2), x(-1), x(0), x(1), x(2), x(3), \ldots \right\} \quad (2.9a)$$

or

$$x = \left\{ x(0), x(1), x(2), x(3), \ldots \right\} \quad (2.9b)$$

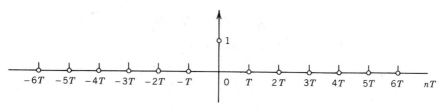

a. Unit impulse $\delta(nT)$

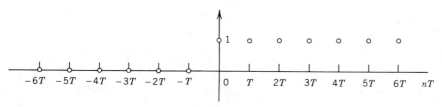

b. Unit step $(u(nT)$

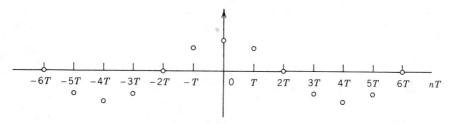

c. Sinusoidal $\cos(2\pi n/8)$

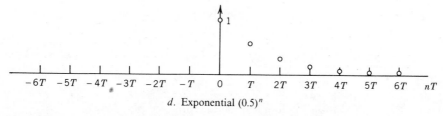

d. Exponential $(0.5)^n$

Figure 2.5 Discrete signals commonly encountered in DSP analysis and design.

where T was set equal to unity and the arrow represents the location of the sequence value at time zero. Positive increments in time are given to the right of the arrow and negative increments to the left, as shown in Eq. 2.9a. If the negative argument sequence values are all equal to zero, then the arrow may be omitted and the sequence formatted as shown in Eq. 2.9b. The sequences shown in Fig. 2.5 are encountered frequently in DSP analysis and design.

UNIT-IMPULSE SEQUENCE
The unit-impulse sequence is unity when its argument is zero and is zero elsewhere.

$$\delta(nT) = \begin{cases} 1 & n = 0 \\ 0 & n \text{ not equal to zero} \end{cases} \tag{2.10}$$

or

$$\delta(nT) = \left\{ \ldots, 0, 0, \underset{\uparrow}{1}, 0, 0, \ldots \right\} = \{1, 0, 0, \ldots\}$$

The application of the unit-impulse to linear discrete-time systems provides the system impulse response, a very useful characterization of the system. Also, any arbitrary sequence, x, can be defined as a summation of the sequence values multiplied by appropriately shifted unit-impulse sequences

$$x(nT) = \sum_{k=-\infty}^{\infty} x(kT)\delta(nT - kT) \tag{2.11}$$

where $\delta(nT - kT)$, is unity for $n = k$ and zero for all other terms. This process is illustrated in the following example.

Example 2.1
Any arbitrary sequence can be represented as a summation of the sequence values with shifted unit impulses. The arbitrary sequence values are given by

$$x = \left\{ 2, 1, 0.5, 0, \underset{\uparrow}{1.5}, 0, 2.5, -3, 0, 0, 4 \right\}$$

Using Eq. 2.11, this sequence is represented by the following unit-impulse summation

$$x(n) = 2\delta(n + 4) + 1\delta(n + 3) + 0.5\delta(n + 2) + 1.5\delta(n) + 2.5\delta(n - 2)$$
$$- 3\delta(n - 3) + 4\delta(n - 6)$$

The unit impulse is also referred to in the literature as the unit sample, delta, and impulse sequence. ∎

UNIT-STEP SEQUENCE

The unit-step sequence is unity for all arguments greater than or equal to zero and zero elsewhere, as shown in Figure 2.5b.

$$u(nT) = \begin{cases} 1 & \text{for } n \geq 0 \\ 0 & \text{for } n < 0 \end{cases} \tag{2.12}$$

or

$$u(nT) = \left\{ \ldots 0, 0, \underset{\uparrow}{1}, 1, 1, 1, \ldots \right\} = \{1, 1, 1, 1, 1, \ldots\}$$

An arbitrary sequence can also be represented as a summation of appropriately shifted and scaled unit-step sequences.

$$x(nT) = \sum_{k=-\infty}^{n} x(kT)[u(nT) - u(nT - 1)]$$

The unit-step sequence is used to make an arbitrary sequence zero for all arguments less than zero by forming the product of the unit-step with the sequence

$$y(nT) = x(nT)u(nT)$$

$$= \left\{ \ldots, 0, 0, \underset{\uparrow}{x}(0), x(1T), x(2T), x(3T), \ldots \right\}$$

The unit step is also used to form a rectangular sequence that is nonzero over a desired range of m samples

$$y(nT) = x(nT)u(nT) - x(nT)u((n - m)T)$$

$$= \left\{ \ldots, 0, 0, \underset{\uparrow}{x}(0), x(1T), x(2T), \ldots, x((m - 1)T), 0, 0, \ldots \right\}$$

SINUSOIDAL SEQUENCE

The sinusoidal function is defined by either the sine or cosine functions, which are shown in Figure 2.5c.

$$x(nT) = \sin(2\pi fnT + \phi)$$
$$x(nT) = \cos(2\pi fnT + \phi) \tag{2.13}$$

Sinusoidal sequences play a key roll in many DSP applications. The parameter f determines the frequency of oscillations within the periodic signal, and ϕ is the phase parameter that determines the value of the sequences at the origin. For radar and sonar applications, the transmission and detection of sinusoidal signals is a primary method of determining whether targets exist.

COMPLEX EXPONENTIAL SEQUENCE

The complex exponential sequence is made up of a real (cosine) component and an imaginary (sine) component times a magnitude parameter

$$x(nT) = (ae^{i2\pi fT})^n = a^n(\cos 2\pi fnT + j\sin 2\pi fnT) \qquad (2.14a)$$

$$x(nT) = a^n \qquad \text{for } f = 0 \qquad (2.14b)$$

$$x(nT) = (\cos 2\pi fnT + j\sin 2\pi fnT) \qquad (2.14c)$$

Equation 2.14b is a real exponential sequence. If a is equal to a value less than unity, then this sequence represents the system response to a unit impulse for a simple class of discrete-time linear systems frequently encountered in DSP. The complex exponential sequences defined by Eq. 2.14a are the basis for developing a Fourier representation of a sequence as a linear combination of complex exponential sequences. In addition, the complex sinusoids given in Eq. 2.14c are used to perform frequency translation by multiplication of the input signal with a complex exponential. Also, the coefficients used in the fast Fourier transform will be shown to be complex exponentials.

RANDOM SEQUENCE

Most real applications are characterized by input signals that contain random components. These random components are the result of natural phenomena such as the acoustic noise of the ocean environment, a major component of acoustic signals processed by sonar systems. A typical approach used to model the random input is to generate a Gaussian random sequence, a selectable routine in most computer systems. For simplicity, the theory and concepts of DSP will be presented using deterministic signals. Keep in mind the need to address signals degraded due to noise sources inherent in real applications. Random sequences, and their impact on system design and performance, will be covered in Chapters 8, 9, and 10 of this book.

2.1.4 Operations of Signals

Digital signal processing is a group of basic operations applied to an input signal resulting in another signal as the output. Formally we describe these operations as a mathematical transformation from one signal to another by some predefined rule, R

$$y(nT) = R[x(nT)]$$

or

$$y(n) = R[x(n)], \qquad \text{where } T = 1$$

For simplicity, T is set equal to unity. It will be important to remember that a sampling interval is implied and must be accounted for in multirate systems where the sampling interval varies.

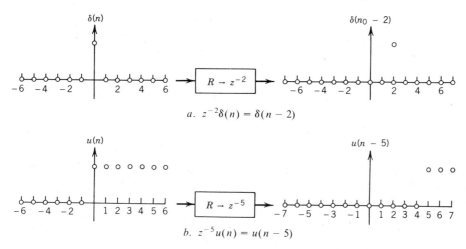

Figure 2.6 Shift operation.

Mathematically, these rules are defined and meaningful for all discrete sample times, nT. In applications involving real-time processing, the output at time nT cannot be a function of samples for times greater than nT. This is intuitively obvious since future samples have not been received yet and therefore cannot be used to determine the output of the system. This limitation on real-time systems will be formally defined in Section 2.2.

The basic set of operations used in DSP are described in the following paragraphs.

SHIFT OPERATION

The shift operation takes the input sequence values and shifts the values by an integer increment of the independent variable (i.e., delays or advances the sequence in time)

$$y(n) = z^{-m}x(n) = x(n - m) \qquad (2.15)$$

The unit impulse of a shift operation with m equal to 2 is shown in Figure 2.6a. The output is simply a delayed version of the input by two sample intervals. This operation can be implemented using a set of m registers. At each sampling interval, the input is advanced one register. The input-output relationship of the unit-step signal with a shift operation equal to 5 is shown in Figure 2.6b. The shift operation is used in all digital filter implementations and is discussed in Chapter 3. The case when $m = 1$ is referred to as the **unit-delay operator.**

TRANSPOSE OPERATION

The transpose of a signal is performed by changing the sign of the argument.

$$y(n) = x(-n) \qquad \text{for all } n$$

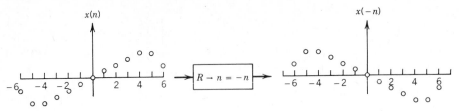

Figure 2.7 Transpose operation.

This operation results in a mirror image of the input signal, as shown in Figure 2.7. This operation is used in the development of the DSP unit-impulse system response.

SCALAR MULTIPLICATION OPERATION

Multiplication of a signal by a scalar results in an output signal where each output sample point is given by the product of the input value times the scalar multiplier.

$$y(n) = ax(n)$$

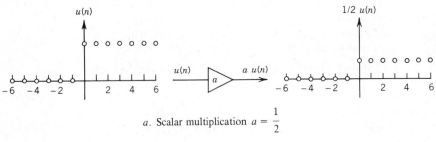

a. Scalar multiplication $a = \dfrac{1}{2}$

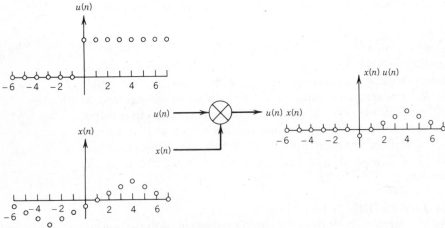

b. Signal vector multiplication

Figure 2.8 Multiplication operation: (*a*) scalar; (*b*) vector.

This operation is shown in Figure 2.8a. In digital circuitry, this operation is implemented by multiplier logic or can be accomplished by shifts and adds of the binary data.

SIGNAL (VECTOR) MULTIPLICATION OPERATION

The signal multiplication results in the product of two signals on a sample-by-sample basis.

$$y(n) = w(n)x(n)$$

This operation is defined as **modulation** of one signal by another and is depicted as shown in Figure 2.8b with $w(n) = u(n)$ a unit step.

SCALAR ADDITION OPERATION

Scalar addition adds a scalar value to each input sample sequence to produce the output sequence. Scalar addition changes the mean value of the signal by the amount of the scalar value. This operation can be used to produce a zero

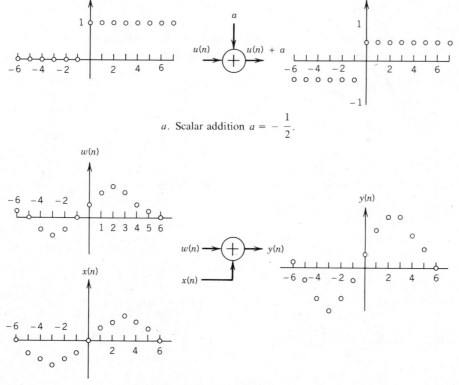

a. Scalar addition $a = -\dfrac{1}{2}$.

b. Signal (vector) addition.

Figure 2.9 Addition operation: (a) scalar; (b) vector.

mean output by adding a scalar equal to the negative of the mean of the input. The operation is given by

$$y(n) = a + x(n)$$

where a is a scalar as shown in Figure 2.9a and $x(n) = u(n)$.

SIGNAL (VECTOR) ADDITION OPERATION

Signal addition is the sample-by-sample summation of input signals as the output

$$y(n) = w(n) + x(n)$$

Figure 2.9b illustrates the addition of two signals using standard diagram notation.

SAMPLING RATE DECREASE OPERATION

Sampling rate decrease is the process of taking every Dth input sample as the output.

$$y(mT_2) = x(mDT_1), \qquad \text{where } T_2 = DT_1 \quad \text{and} \quad F_2 = \frac{F_1}{D}$$

This results in an increase in the sampling interval; therefore the independent variable T must be considered. The operation is represented as shown in Figure 2.10a.

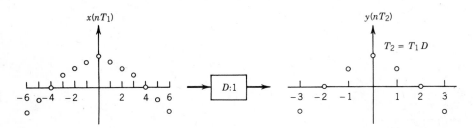

a. Sampling rate decrease (decimation) $D = 2$.

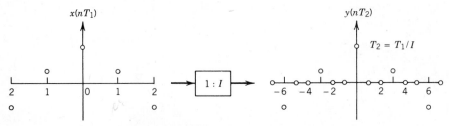

b. Sampling rate increase (interpolation) $I = 3$.

Figure 2.10 Sampling rate modification: (a) decrease; (b) increase.

This is a basic operation for many multirate systems, called decimators, and is covered in Chapter 7. Since the process results in a reduction in the sampling rate, the frequency content of the input signal must be modified to assure that the Nyquist criteria is preserved. A procedure for determining the filter requirements to meet specified aliasing levels will be presented in Chapter 10.

SAMPLING RATE INCREASE OPERATION

The sampling rate increase process is shown in Figure 2.10b. As illustrated in the figure, each input sample is appended with zeros based on the sampling rate increase factor, I.

$$y(mT_2) = \begin{cases} x\left(\dfrac{mT_1}{I}\right) & \text{for } m = \{\,\ldots -3I,\,-2I,\,-I,\,0,\,I,\,2I,\,3I,\ldots\} \\ 0 & \text{for } m \text{ otherwise} \\ \text{where} \quad T_2 = \dfrac{T_1}{I} \text{ and } F_2 = IF_1 \end{cases}$$

This is also a basic operation for many multirate systems, called interpolators, and is covered in Chapter 7. Interpolators require filters to eliminate images created by the expansion of the sampling rate.

2.2 LINEAR TIME-INVARIANT DIGITAL SYSTEMS

The theory of discrete-time, linear, time-invariant systems forms the basis for DSP. A discrete-time system performs an operation on the input signal based on a predefined criteria to produce a modified output signal. The input signal, $x(n)$, is the system excitation, and $y(n)$ is the response of the system to the excitation. The system output is a function of the algorithm used to transform the input signal. The transformation operation is represented by the operator R, as shown in Figure 2.11.

Systems that meet the linearity and time-invariance requirements satisfy a broad class of DSP operations. Such systems are completely characterized by their unit-impulse response, $h(n)$.

$$h(n) = R[\delta(n)]$$

Once the unit-impulse response is determined, the output of the system for any input is given by

$$y(n) = R[x(n)] = \sum_{k=-\infty}^{\infty} x(k)h(n-k) \qquad (2.16a)$$

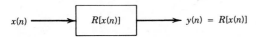

Figure 2.11 System operator R.

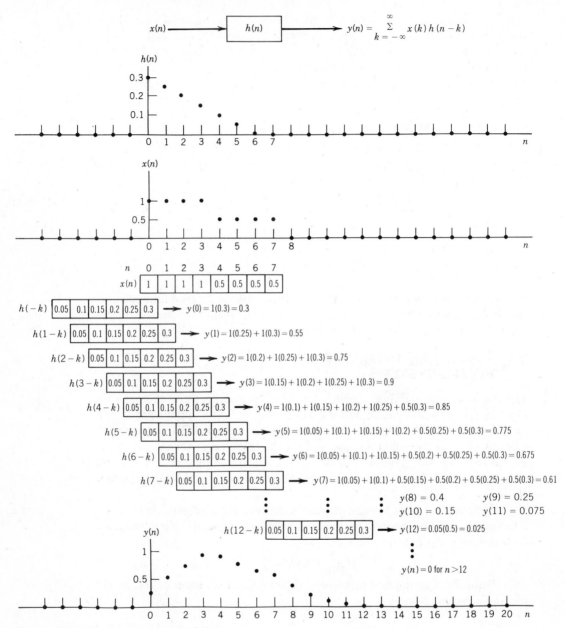

Figure 2.12 Convolution-sum calculation,

or

$$y(n) = \sum_{k=-\infty}^{\infty} h(k)x(n-k) \qquad (2.16b)$$

This equation is the convolution between the input signal and the system's unit-impulse response (Figure 2.12). From the figure, for each time sample, $x(n)$, an output, $y(n)$, is produced by transposing and shifting $h(n)$ such that the transposed $h(n)$ shifts one sample to the right for each successive output. The summation is performed over all nonzero values $x(n)$ from minus to plus infinity. The length of the convolution output for sequences x and h, $N(x, h)$, is equal to the sum of the finite lengths of the sequences, $L(x)$ and $M(h)$, minus one

$$N(x, h) = L(x) + M(h) - 1$$

From Figure 2.12 $M(h) = 6$ and $L(x) = 8$, therefore $N(x, h) = 13$.

The development of Eq. 2.16 relies on the properties of linearity and time-invariance presented in the following paragraphs. We will also present the properties of causality and stability required for the system to be physically realizable.

2.2.1 Linearity

A system is **linear** if and only if the system's response to the sum of two signals, each multiplied by arbitrary scalar values, is equal to the sum of the system's responses to the two signals, each multiplied by the same arbitrary scalar values. Let the system's response be represented by R, then the response to the sum of the two inputs x_1 and x_2 multiplied by arbitrary constants a_1 and a_2 is given by

$$y(n) = R[a_1x_1(n) + a_2x_2(n)]$$

The sum of the system's responses to the individual inputs, and then multiplied by the constants, is

$$y(n) = a_1R[x_1(n)] + a_2R[x_2(n)]$$

The system, operator R, is linear if and only if the two outputs are equal

$$a_1R[x_1(n)] + a_2R[x_2(n)] = R[a_1x_1(n) + a_2x_2(n)] \qquad (2.17)$$

A procedure for determining if a system is linear is shown in Figure 2.13. The process computes the right-hand side (RHS) and the left-hand side (LHS) of Eq. 2.17 and checks to see if they are equal. If the RHS equals the LHS, then the system is linear; otherwise the system is nonlinear. This process applies to any number of signals.

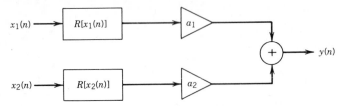

a. Linear system LHS calculation: $a_1 R[x_1(n)] + a_2 R[x_2(n)]$.

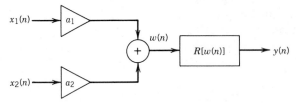

b. Linear system RHS calculation: $R[a_1 x_1(n) + a_2 x_2(n)]$.

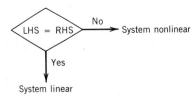

c. Linearity test.

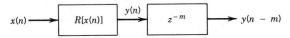

d. Time-invariant LHS calculation.

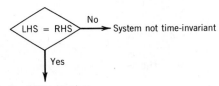

e. Time-invariant RHS calculation.

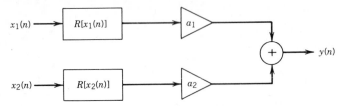

f. Time-invariant test.

Figure 2.13 Linear system and time-invariance system tests.

Example 2.2

The test for linearity is easily applied to a system algorithm rule. Given the following system rule

$$y(n) = x(n) - bx(n-1)$$

determine if the system is linear.

Solution:

First the LHS of Eq. 2.17 is calculated as shown in Figure 2.13a

$$a_1 R[x_1(n)] = a_1 x_1(n) - a_1 bx_1(n-1)$$
$$a_2 R[x_2(n)] = a_2 x_2(n) - a_2 bx_2(n-1)$$

Next the RHS of Eq. 2.17 is calculated as shown in Figure 2.13b

$$R[a_1 x_1(n) + a_2 x_2(n)] = a_1 x_1(n) + a_2 x_2(n) - a_1 bx_1(n-1)$$
$$- a_2 b_2 x(n-1)$$

Since the LHS and RHS of Eq. 2.17 are equal and the system is linear. ∎

Several system expressions are given as problems at the end of the chapter in order to give you insight into linear and nonlinear systems.

2.2.2 Time-Invariance

A time-invariant system always produces the same output-to-input signal relationship. A system is **time-invariant** if the response to a shifted version of the input is identical to a shifted version of the response based on the unshifted input. A system operation R is time-invariant if

$$R[x(n-m)] = z^{-m} R[x(n)] \qquad (2.18)$$

for all values of m. This is true for all possible input signal excitations. The operator z^{-m} represents a signal delay of m samples.

Example 2.3

Time-invariance is easily checked by applying Eq. 2.18 to a system rule. Using the same system rule from Example 2.2,

$$y(n) = R[x(n)] = x(n) - bx(n-1)$$

determine if the system is time-invariant.

Solution:
First we substitute $x(n - m)$ in the system and compute the response to a delayed excitation, the LHS of Eq. 2.18

$$y(n - m) = R[x(n - m)] = x(n - m) - bx(n - m - 1)$$

Next we compute the RHS of Eq. 2.18 as shown in Figure 2.13e

$$z^{-m}R[x(n)] = z^{-m}x(n) - bz^{-m}x(n - 1) = x(n - m) - bx(n - m - 1)$$

Applying the test shown in Figure 2.13d, e, and f, this system is time-invariant. Several exercises are given in the problem set on the process of determining if a system is time-invariant. ∎

2.2.3 Unit-Impulse System Response

The convolution sum of Eq. 2.16 was defined as the summation of the product of a transposed and shifted system unit-impulse response with the system's input signal. Starting with the general representation of a signal given by Eq. 2.11, we develop the system output response

$$y(n) = R\left[\sum_{k=-\infty}^{\infty} x(k)\,\delta(n - k)\right]$$

Because the system is linear, the response of the system to a sum of inputs is the same as the sum of the system's responses to each of the individual inputs. Thus, the foregoing response can be computed as the sum of the product of the scalars with the responses to the shifted unit impulse

$$y(n) = \sum_{k=-\infty}^{\infty} x(k)R[\delta(n - k)]$$

By definition, $R[\delta(k)] = h(k)$. If the system is also time-invariant, then the output response to a shifted version of the input, $R[\delta(n - k)]$, is equivalent to a shifted output response, $h(n - k)$

$$y(n) = \sum_{k=-\infty}^{\infty} x(k)h(n - k)$$

or

$$y(n) = \sum_{k=-\infty}^{\infty} h(k)x(n - k)$$

Therefore, provided the system is linear and time-invariant, the convolution sum applies. This is an extremely powerful result that allows us to compute the system output for any input signal excitation by the convolution of the input signal with the system unit-impulse response. The usefulness of the convolution sum will become more evident with the introduction of the system frequency response in Section 2.4.

2.2.4 Causality

Causality refers to a system that is realizable in real time. A causal system is a system that at time m produces a system output that is dependent only on current and past inputs, $n \leq m$, and past outputs, $n < m$. This will always be true for a unit-impulse response, which is zero for $n < 0$. Therefore, a discrete-time, linear, time-invariant system is causal if and only if $h(n) = 0$ for $n < 0$.

2.2.5 Stability

A system is **stable** if it produces a bounded output signal for every bounded input signal. Using the alternate form of Eq. 2.16b, the output is bounded by taking the magnitude of the output

$$|y(n)| = \left| \sum_{k=-\infty}^{\infty} h(k)x(n-k) \right|$$

$$\leq \sum_{k=-\infty}^{\infty} |h(k)||x(n-k)|$$

Setting the largest value of the input equal to B, the expression can be rewritten

$$|y(n)| \leq B \sum_{k=-\infty}^{\infty} |h(k)|, \quad \text{where } B = \|x(n-k)\|_{\infty} \quad (2.19)$$

If B is bounded, then a necessary and sufficient condition for stability is that the sum of the magnitudes of the system unit-impulse response is finite

$$|h(n)| < \infty$$

This is the L_1 norm defined by Eq. 2.5a of the unit-impulse response.

Example 2.4
A typical system unit-impulse response is used to illustrate the system stability analysis. The system unit-impulse response for a first order IIR filter is given by

$$y(n) = x(n) + by(n-1)$$
$$h(n) = b^n u(n) = \{b^0, b^1, b^2, \dots\}$$

where $|b| < 1$.

Solution:

Calculating the L_1 norm using the geometric series

$$\|h(n)\|_1 = \sum_{n=0}^{\infty} |b^n| = \frac{1}{1 - |b|}$$

and therefore, provided the magnitude of b is less than 1, the system is stable. ∎

2.3 DIGITAL FILTERS

A broad class of digital filters are described by linear, constant coefficient, difference equations

$$\sum_{k=0}^{K} D_k y(n - k) = \sum_{l=0}^{L} C_l x(n - l) \tag{2.20}$$

where C_l and D_k are coefficients that define the system, and $y(n)$ and $x(n)$ are used to represent the output and input signals, respectively. Although this

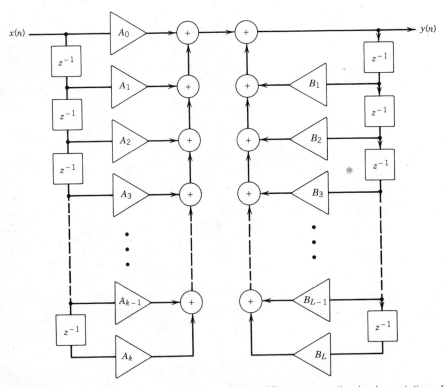

Figure 2.14 Direct form linear constant coefficient difference equation implementation of Eq. 2.21.

expression is general, only causal systems are discussed with both $y(n)$ and $x(n)$ equal to zero for $n < 0$. Rewriting Eq. 2.20 to express the present output in terms of present and past inputs and past outputs yields

$$y(n) = \sum_{l=0}^{L} A_l x(n-l) - \sum_{k=1}^{K} B_k y(n-k) \qquad (2.21)$$

where

$$A_l = C_l/D_0 \quad \text{and} \quad B_k = D_k/D_0$$

This expression can be implemented as a set of multiplications, summations, and delays. Figure 2.14 illustrates a particular direct form implementation of Eq. 2.21. The coefficients for Eq. 2.21 have been normalized by D_0.

The task of designing digital filters requires determination of the coefficients of Eq. 2.21 to meet the criteria specified for the input-output relationship. Two main classes of digital filters exist: infinite impulse response (IIR); and finite impulse response (FIR). Each class is defined in the following sections. Digital filter design techniques are covered in Chapters 4 and 5; realizations are covered in Chapter 3, and finite arithmetic effects are covered in Chapter 9.

2.3.1 Infinite Impulse Response (IIR) Filter

The response of an IIR filter is a function of current and past input signal samples and past output signal samples. The general difference equation given by Eq. 2.21 represents an IIR filter since it is a function of both elements of the excitation and the response. The dependency on past outputs (i.e., recursive) gives rise to the infinite duration of the filters output response even when the input values have stopped.

Example 2.5
A simple first-order difference equation illustrates the IIR filter. Determine the impulse response for the filter defined by the first-order difference equation given by

$$y(n) = x(n) + by(n-1)$$

Solution:
Exciting the system with an impulse signal gives the system unit-impulse response for the first-order expression

$$h(n) = R[\delta(n)] = b^n u(n)$$

The computation of the impulse response is shown in Figure 2.15a, along with

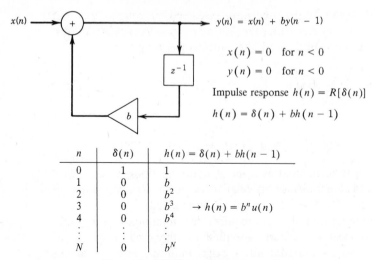

$$y(n) = x(n) + by(n - 1)$$

$$x(n) = 0 \quad \text{for } n < 0$$
$$y(n) = 0 \quad \text{for } n < 0$$

Impulse response $h(n) = R[\delta(n)]$

$$h(n) = \delta(n) + bh(n - 1)$$

n	$\delta(n)$	$h(n) = \delta(n) + bh(n - 1)$
0	1	1
1	0	b
2	0	b^2
3	0	b^3 $\quad \to h(n) = b^n u(n)$
4	0	b^4
$\vdots$	$\vdots$	$\vdots$
N	0	b^N

a. First-order IIR filter unit-impulse response.

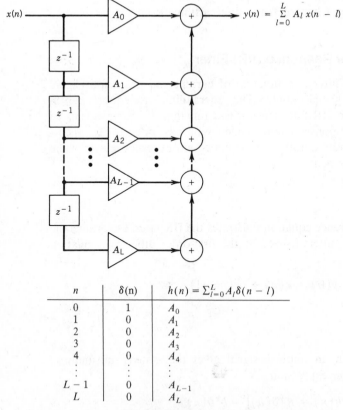

$$y(n) = \sum_{l=0}^{L} A_l\, x(n - l)$$

n	$\delta(n)$	$h(n) = \sum_{l=0}^{L} A_l \delta(n - l)$
0	1	A_0
1	0	A_1
2	0	A_2
3	0	A_3
4	0	A_4
$\vdots$	$\vdots$	$\vdots$
$L - 1$	0	A_{L-1}
L	0	A_L

b. $L + 1$ order FIR filter unit-impulse response.

Figure 2.15 IIR and FIR filter unit-impulse responses.

an implementation diagram. The feedback term [i.e., past output $y(n - 1)$ is fed back, multiplied by b, and summed with the input] at each successive sample time, results in a system output of infinite duration. ■

IIR COMPUTATIONAL COMPLEXITY

Digital filters play a key role in DSP. Therefore, the number of computations required to implement the filter is directly proportional to the amount of hardware/software required. We briefly address the computational complexity of the IIR filter in this section. Much more is presented in Chapter 7 (Multirate DSP) and Chapter 10 (System Design).

From Eq. 2.16 the computational requirements of the IIR filter are computed.

$$\text{Multiplies/second (MPS)} = F(L + K + 1)$$
$$\text{Adds/second (APS)} = F(L + K)$$
$$\text{Storage registers} = L + K.$$
$$\text{Storage coefficients} = L + K$$

F is the filter input sampling rate, and L and K are the number of filter coefficients. IIR filters must be computed at the input sample rate regardless of subsequent data decimation operations that ignore some of the outputs. This is a result of the fact that each output is required in order to compute subsequent outputs. The MPS and APS requirements are based on the rate at which the operations are performed, F, and the number of multiplies or additions that are performed to produce each output.

Alternative implementations that allow feedback and feed-forward storage registers to be shared will be shown in Chapter 3. Also, decimating IIR filters that take advantage of subsequent decimation, thereby reducing the computational requirements of the filter, will be discussed in Chapter 7.

IIR filters exhibit noise buildup because the noise terms created by arithmetic implementation errors are fed back into the system. Errors due to the approximation of the filter coefficients using finite registers must be analyzed for alternative filter implementations in order to select the best design approach. This will be covered in detail in Chapter 9.

2.3.2 Finite Impulse Response (FIR) Filter

If the output samples of the system depend only on the present input, and a finite number of past input samples, then the filter has a finite impulse response. Equation 2.21, with the B_k coefficients equal to zero, represents a FIR filter of duration L.

$$y(n) = \sum_{l=0}^{L} A_l x(n - l) \tag{2.22}$$

The coefficients of the FIR filter are equivalent to the filter's impulse response.

Example 2.6
Given the general FIR filter expression, Eq. 2.22, determine the impulse response.

Solution:
The general FIR filter difference equation given by Eq. 2.22 is excited by a unit impulse

$$h(n) = R[\delta(n)] = A_n \quad \text{for } n = 0, 1, 2, \ldots, L$$

Computation of the impulse response is shown in Figure 2.15*b* along with an implementation diagram. As shown, a FIR filter has an impulse response equal to the coefficients of the difference equation. ■

FIR COMPUTATIONAL COMPLEXITY
The multiplication, addition, and storage requirements for the FIR filter are given by

$$\text{Multiplies/second (MPS)} = F(L + 1)$$
$$\text{Adds/second (APS)} = F(L)$$
$$\text{Storage registers} = L$$
$$\text{Storage coefficients} = L$$

The MPS and APS requirements were computed assuming that the filter output rate is equal to the input rate. Generally, the FIR filter coefficients are symmetric, and therefore the symmetric components can be added prior to the multiplication. For many applications, the output rate can be reduced (i.e., decimated). A useful FIR feature is that FIR filters followed by decimation to a lower data rate need only be computed at the decimated rate. This is a common approach used in DSP multirate designs to reduce the computational requirements. Therefore, the computation rate (multiply and add requirements per second) can be decreased by the decimation factor if a sampling rate decrease is applied at the FIR filter output. This is covered in detail in Chapter 7.

Nonrecursive FIR filter implementations have a finite impulse response where the output noise does not get fed back. The noise term is therefore limited by the number of coefficients in Eq. 2.22. The finite arithmetic effects and the filter coefficient approximation errors will be covered in detail in Chapter 9.

2.4 SYSTEM FOURIER TRANSFORM RELATIONSHIPS

The frequency spectrum of a signal was discussed in Section 2.1. Sections 2.2 and 2.3 dealt with the time domain representations of discrete-time linear systems. Fourier representations of input signals provide a useful system interpretation, since the output of a discrete-time linear system to a complex exponential is equal to the complex exponential modified by a complex value. This leads to the frequency-response interpretation of a system.

The time domain system representation used the convolution sum as a method for computing the system output for any input once the impulse response was determined. Likewise, in the frequency domain representation, the system output is the product of the system frequency response (i.e., the Fourier transform of the impulse response) with the input signal frequency response. This leads to the equality between the convolution of two signals in the time domain and the inverse Fourier transform of the product of their Fourier transforms. Also, the product of two signals in the time domain is equal to the inverse Fourier transform of the convolution of their Fourier transforms. These transform relationships provide a powerful system analysis method, which is developed in Chapter 3 using Z-transforms. In addition, the product relationship can be used to implement filters in the frequency domain which is covered in Chapter 6.

2.4.1 System Frequency Response

The **frequency response** of a system is defined by the input-output relationship for the set of all possible complex sinusoidal inputs.

$$y(n) = R\left[A_i e^{j(2\pi f_i nT)}\right] \tag{2.23}$$

A signal at arbitrary frequency f_i with amplitude A_i has been used to represent the set of all possible complex sinusoidal inputs. Using Eq. 2.16b, we compute the convolution sum for the complex sinusoidal system input.

$$y(n) = \sum_{k=-\infty}^{\infty} A_i h(k) e^{j(2\pi f_i (n-k)T)}$$

Since the summation is over k, the term involving n may be brought outside the summation, resulting in

$$y(n) = A_i e^{j(2\pi f_i nT)} \sum_{k=-\infty}^{\infty} h(k) e^{-j(2\pi f_i kT)}$$

The first part is seen to be identical to the input signal. The second part is a summation that results in a complex multiplier times the input signal. This summation is called the **system frequency response.** Therefore, given a complex sinusoidal input signal to a linear time-invariant system, the output is a sinusoid of the same frequency modified by

$$H(f_i) = \sum_{k=-\infty}^{\infty} h(k) e^{-j(2\pi f_i kT)} \tag{2.24}$$

This causes a change in the amplitude and phase of the input signal. Each frequency, f_i, applied to the system has a response given by Eq. 2.24. If Eq.

2.24 is evaluated for all input frequencies, then the output response is the sum of the individual responses for a linear system. Since the complex sinusoidal signal is a basic signal encountered in DSP (i.e., general signals can be made up of a linear combination of sinusoids), this result is very useful in the analysis and design of DSP systems. The filter output can be represented as the superposition of responses of the system to each of the K frequency components at the input to the system.

$$y(n) = \sum_{i=0}^{K} A_i e^{j(2\pi f_i nT)} H(f_i) \tag{2.25}$$

Example 2.7
The frequency response of the system is computed from Eq. 2.24 based on the system impulse response. The system impulse response for a simple first-order IIR filter was given by (see Ex. 2.5 and Fig. 2.15)

$$h(n) = b^n u(n), \qquad \text{for} \qquad |b| < 1$$

Solution:
Evaluating Eq. 2.24, the system frequency response is

$$H(f_i) = \sum_{k=-\infty}^{\infty} b^k e^{-j(2\pi f_i kT)} u(k)$$

$$= \sum_{k=0}^{\infty} \left[b e^{-j(2\pi f_i T)} \right]^k$$

$$= \frac{1}{1 - b e^{-j(2\pi f_i T)}} \qquad \text{(from geometric series)}$$

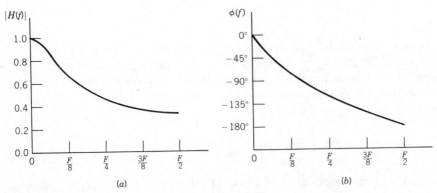

Figure 2.16 Frequency response of first-order IIR filter.

The magnitude response for $H(f_i)$ is plotted in Figure 2.16, along with the impulse response.

$$|H(f_i)|^2 = \frac{1}{1 + b^2 - 2b \cos(2\pi f_i T)} \qquad \text{Magnitude}$$

$$\phi(f_i) = 2\pi f_i T - \tan^{-1}\left[\frac{\sin(2\pi f_i T)}{\cos(2\pi f_i T) - b}\right] \qquad \text{Phase}$$

2.4.2 Fourier Transform System Input / Output Relationship

The Fourier transform pair of a continuous signal was given in Eqs. 2.6a, b. The discrete Fourier transform pair was defined by Eqs. 2.7a, b. The Fourier transform for the unit-impulse system response given by Eq. 2.24 is general and can be applied to represent the Fourier transform of any discrete signal.

$$X(f_i) = \sum_{k=-\infty}^{\infty} x(k)e^{-j(2\pi f_i kT)} \tag{2.26}$$

Equation 2.26 is a Fourier series representation where $x(k)$ are viewed as the Fourier series coefficients. If $X(f_i)$ is given, then the coefficients $x(k)$ can be determined using an inverse transform relationship given by

$$x(k) = \frac{1}{F}\int_{-F/2}^{F/2} X(f_i)e^{-j(2\pi f_i kT)}\, df \tag{2.27}$$

Provided the series in Eq. 2.26 converges, Eqs. 2.26 and 2.27 form a Fourier transform pair. The discrete transform pair given in Eq. 2.7 are derived from Eqs. 2.26 and 2.27 assuming an ideal finite duration input signal and a band-limited frequency spectrum. The DFT pair given in Eq. 2.7 are the transform pair realized in DSP applications.

Using the Fourier transform, a relationship between the convolution sum in the time domain and its frequency domain counterpart shown in Figure 2.17 is developed. The Fourier transform of the system output is given by

$$Y(f_i) = \sum_{k=-\infty}^{\infty} y(k)e^{-j(2\pi f_i kT)}$$

Substituting the convolution sum for $y(k)$ results in

$$Y(f_i) = \sum_{k=-\infty}^{\infty} \sum_{m=-\infty}^{\infty} h(m)x(k-m)e^{-j(2\pi f_i kT)}$$

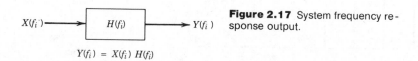

Figure 2.17 System frequency response output.

Changing the order of summation and multiplying by unity (i.e., complex exponentials with positive and negative arguments of $2\pi f_i mT$),

$$Y(f_i) = \sum_{m=-\infty}^{\infty} h(m) \sum_{k=-\infty}^{\infty} x(k-m) e^{-j2\pi f_i(k-m)T} e^{-j2\pi f_i mT}$$

and rewriting with $l = k - m$, the product relationship results:

$$Y(f_i) = \sum_{m=-\infty}^{\infty} h(m) e^{-j2\pi f_i mT} \sum_{l=-\infty}^{\infty} x(l) e^{-j2\pi f_i lT}$$

Thus from the definition of the Fourier transform, the output transform is equal to the product of the transforms of the input signal $x(n)$ and the system impulse response $h(n)$.

$$Y(f_i) = H(f_i) X(f_i) \leftrightarrow x(n) * h(n) = y(n) \qquad (2.28)$$

where $*$ denotes a convolution-sum operation.

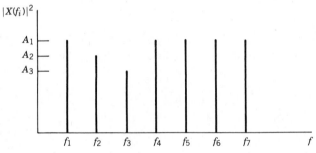

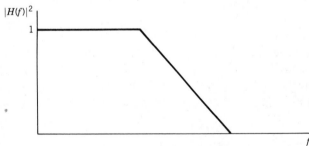

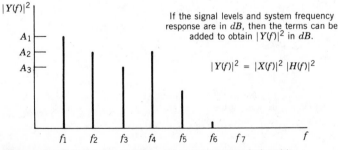

Figure 2.18 Input / output frequency response relationship.

The reverse is also true (i.e., if two signals are multiplied in the time domain modulation process, the frequency domain is represented as a convolution).

$$y(n) = x(n)h(n) \leftrightarrow H(f_i) * X(f_i) = Y(f_i) \qquad (2.29)$$

The input signal frequency response, the system frequency response, and the corresponding system output frequency response using Eq. 2.28 is shown in Figure 2.18. This is a powerful result often used when dealing with finite length signals and transforms and to perform high-speed convolution. The time domain output signal is obtained via an inverse Fourier transform.

The Fourier transform pair relationship for realizations with finite length signals results in circular effects that must be compensated for. Also, special algorithms have been developed to process contiguous sets of finite length inputs. These techniques are covered in Chapter 6.

2.5 SAMPLING ANALOG SIGNALS AND SAMPLING RATE CONVERSION

Most applications have continuous signals that require processing in order to obtain specific information for subsequent decision/presentation processing. Generally, prior to system processing the signals are converted from the transmission medium into an analog signal via a transducer. Before the advances in digital technology made DSP practical, the majority of processing was performed using analog circuitry. This was especially true for wide bandwidth applications that required high sampling rates to preserve the information content of the signal.

Digital hardware technology has advanced to the point where digital processing of signals from applications requiring high sampling rates is practical. In this section we develop the theory associated with the sampling process and present the concepts of aliasing. The theory will show that the maximum sampling interval is a function of the spectral content of the input signal. The higher the frequency, the smaller the sampling interval required (i.e., smaller sampling intervals are required to capture the variations in the signal).

2.5.1 Analog Signal Conditioning

Analog signal conditioning prepares the signal for sampling and subsequent DSP without corruption of the signal's information content. This process consists of three primary functions: analog presampling filtering (APF); gain control; and analog-to-digital (A/D) conversion.

The APF process band-limits the input signal spectral content to application-required frequency ranges. Higher bandwidths require less APF processing, but higher sampling rates result, which translate into higher processing requirements on the digital system components. Therefore, the design must consider the alternatives between digital and analog implementa-

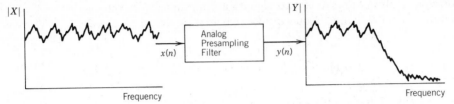

a. APF band-limits the input spectrum to the desired frequency range to minimize the subsequent sampling rate of the A/D process.

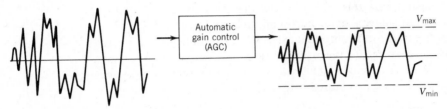

b. AGC estimates variation of input and adjusts inputs within specified output range based upon desired probability that output will not exceed range.

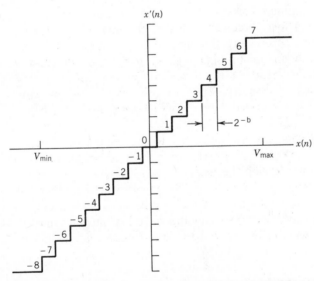

c. Analog-to-digital converter (A/D). The A/D samples and quantizes the input analog voltage into 2^b equal digital levels.

Figure 2.19 Analog signal conditioning: (a) APF; (b) AGC; (c) A / D.

tions in order to select the partitioning that provides an efficient implementation. Modulation/demodulation processes (based on signal transmission techniques) are sometimes required and must be allocated to the analog or digital processing based on hardware alternatives. Figure 2.19*a* illustrates the APF process. The criteria used to establish design choices for the APF is addressed in Chapter 10.

The gain control operation adjusts the input signal level within a specified voltage range of the A/D. The gain control is necessary because of the variability of the input signal random components. Generally, an automatic gain control (AGC) process is implemented that makes an estimate of the input signal energy and adjusts the gain based on optimum criteria. The optimization is based on the minimization of the noise due to the A/D conversion process to a finite number of bits. Figure 2.19*b* illustrates the operation of the gain control.

The A/D converter samples the analog signal and performs the signal conversion from an analog signal to a digital signal. We consider only A/D converters that sample the input at equal intervals.[7] At each sampling instant, the voltage level of the waveform is converted to a digital approximation based on the number of bits used to represent the signal.[8] Figure 2.19*c* illustrates the A/D process for a four-bit converter. Fewer bits require less hardware, but result is larger quantization errors. The number of required bits will be based on the input signal characteristics and is addressed in Chapters 9 and 10.

2.5.2 Sampling

An analog signal $a(t)$ is a continuous function of the independent continuous time variable t. We discuss sampling the analog signal with a uniform time interval T to produce the corresponding digital signal

$$x(nT) = a(t)|_{t=nT} \qquad -\infty < n < \infty$$

T is the reciprocal of the sampling rate, F. Sampling is often represented as a modulation of the analog signal with a summation of delayed unit-impulse functions $m(t)$,[9] as shown in Figure 2.20.

$$x(nT) = a(t) \times m(t)$$

$$m(t) = \sum_{n=-\infty}^{\infty} \delta(t - nT)$$

The spectrum of the analog signal, the modulation signal, and the sampled output signal are illustrated in Figure 2.21. Note that the input signal is shown as a band-limited function.

$$A(f) = 0, \qquad \text{for } -f_h < f < f_h$$

The spectrum of the modulation signal is also a set of impulses with repetition F. F is the reciprocal of the sampling interval, T. The output spectrum is equal to the convolution of the analog signal spectrum with the modulation signal spectrum

$$X(f) = A(f) * M(f)$$

This is a direct result of the dual relationship between multiplication and convolution of signals in the time domain and frequency domain.

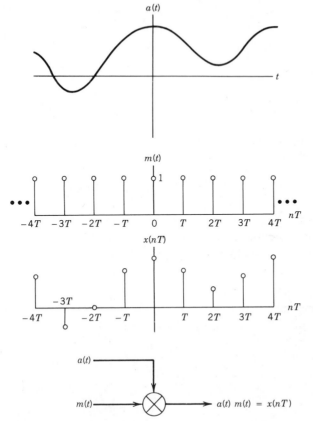

Figure 2.20 Sampling process: modulating a continuous signal with a summation of unit impulses.

The modulation signal spectrum is a set of impulse functions separated by F Hz,

$$M(f) = \sum_{k=-\infty}^{\infty} \delta(f - kF)$$

The output signal spectrum is visualized as a repetition of the analog signal spectrum with intervals of F Hz. This is the result of each impulse sliding by the input signal. Figure 2.21 illustrates the output spectra for the analog signal with two different sampling intervals. Note that only one of the outputs preserves the spectrum of the input after sampling. In order to assure that the input is preserved when sampled, the sampling function must sample at a rate greater than or equal to twice the highest input frequency

$$F \geq 2f_h$$

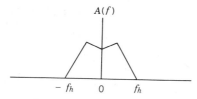

a. Input signal frequency spectrum.

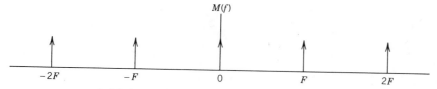

b. Modulation sampling frequency spectrum $f > 2f_h$.

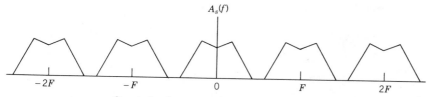

c. Ouput signal spectrum for $F > 2f_h$—no aliasing.

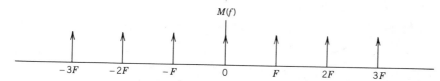

d. Modulation sampling frequency spectrum $F < 2f_h$.

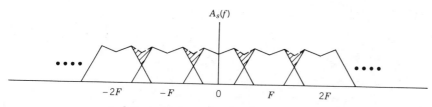

e. Output signal spectrum for $F < 2f_h$—aliasing.

Figure 2.21 Output signal spectrum for sampling signals at greater than and less than the Nyquist rate.

This is the **Nyquist criteria**. If this rate is not maintained, the repetitions of the input signal overlap the main signal, distorting the output signal. The overlap is referred to as the aliased output spectra.

Practically, the preceding discussion breaks down for two reasons: the modulation signal is realized with a pulse train of finite widths instead of the ideal Dirac delta function; and the input signal will be time-limited and therefore can not be band-limited. The spectrum of the finite width pulse train is given by[10]

$$M(f) = \sum_{k=-\infty}^{\infty} K \operatorname{sinc} \{ku\pi F\} \, \delta(f - kF) \qquad (2.31)$$

where

$$\operatorname{sinc} \{x\} = \sin(x)/x$$

$$u = \text{pulse width}$$

$$K = A(u)F, \text{ main pulse amplitude}$$

$$A = \text{pulse amplitude}$$

$$F = \text{pulse rate, width between pulses}$$

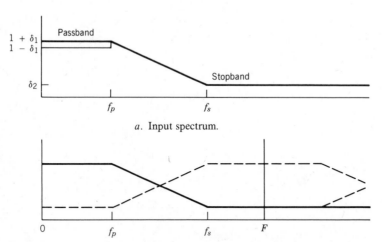

a. Input spectrum.

b. Select sampling rate F such that the lowest frequency that will fold (alias) into passband equals f_s.

$$f_s = F - f_p$$

Therefore,

$$F = f_s + f_p$$

Aliasing is maintained below δ_2 stopband level provided sampling rate is selected as shown.

Figure 2.22 Aliasing for an input signal that is not band-limited in the strict sense.

This results in a frequency pulse train with decreasing amplitudes that must be convolved with the input. Since the input signal is not band-limited, the spectrum images resulting from the convolution with the impulse train will overlap, creating aliasing errors. The error due to this overlap is a function of T and the spectrum of the input. Since signals from real applications have spectrums that decrease as frequencies increase above F, the aliasing error is reduced as T is decreased. In the remainder of the book, we will refer to a band-limited signal $A(f)$ as one whose amplitude above f_h is less than or equal to a specified amplitude level. The level will be based on a specified performance criteria that will be covered in Chapter 10. Figure 2.22 illustrates the resultant aliasing for an input signal with a lowpass spectrum.

2.5.3 Aliasing

Aliasing is higher frequencies impersonating lower frequencies due to the sampling rate not satisfying the Nyquist sampling criteria. For real signals folding of frequencies above one half the sampling rate, $F/2$, into frequencies below $F/2$ occurs. Aliasing is illustrated in Figure 2.23.

A sinusoidal analog signal of frequency, ΔF, is shown in Figure 2.23a. The analog signal is sampled with a sampling rate of $F = 4\Delta F/3$ in Figure 2.23b. The resultant samples are illustrated by connecting the discrete samples. The resampled signal appears to represent a lower sinusoidal signal. This is a direct result of violating the Nyquist sampling criteria. The aliased frequency equals $\Delta F/3$. Obviously, a higher sampling rate is required to preserve the frequency. Figure 2.23c shows a signal sampled at exactly twice the input frequency rate. This results in a constant output if the samples fall exactly at the crossover point or alternating positive and negative equal values with amplitudes based on the relation of the input signal phase to the sampling instant. Sampling at a rate of $8\Delta F/3$ satisfies the Nyquist sampling criteria, providing a signal representation that is unambiguous, as shown in Figure 2.23d. Sampling at higher rates (i.e., smaller intervals) more closely approximates the continuous signal. Provided the Nyquist criteria is satisfied, the continuous signal can be reconstructed

$$x(t) = \sum_{n=-\infty}^{\infty} x(nT)\frac{\sin\{2\pi f_h(t - nT)\}}{\{2\pi f_h(t - nT)\}} \tag{2.32}$$

where $T \leq 1/2f_n$. Equation 2.32 is referred to as Whittaker's reconstruction formula.[10] The accuracy of the reconstruction is dependent on the number of terms used to perform the summation in Eq. 2.32. The formula can be rewritten using the definition given for sinc in Eq. 2.31.

$$x(t) = \sum_{n=-\infty}^{\infty} x(nT)\operatorname{sinc}\{2f_h(t - nT)\}$$

The derivation of Eq. 2.32 is given as an exercise in Problem 2.5.5.

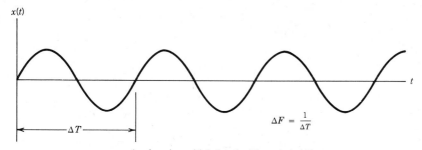

a. Analog sinusoidal signal with period ΔT.

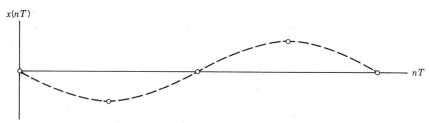

b. Sampling at $T = 0.75\Delta T$. Therefore $F = \dfrac{4\Delta F}{3} < 2\Delta F$.

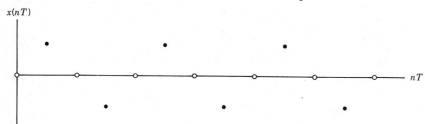

c. Sampling at $T = \dfrac{\Delta T}{2}$. Therefore $F = \dfrac{1}{T} = \dfrac{2}{\Delta T} = 2\Delta F$.

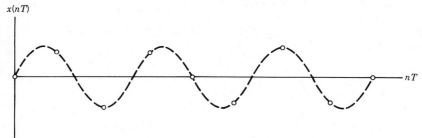

d. Sampling at $T = \dfrac{3\Delta T}{8}$. Therefore $F = \dfrac{1}{T} = \dfrac{8}{3\Delta T} = \dfrac{8\Delta F}{3} > 2\Delta F$.

Figure 2.23 Aliasing.

The aliased frequency components are given by

$$f_{\text{aliased}} = \begin{cases} kF - f_{\text{signal}}, & \text{for } \left(k - \frac{1}{2}\right)F < f_{\text{signal}} < kF \\ f_{\text{signal}} - kF, & \text{for } kF < f_{\text{signal}} < \left(k + \frac{1}{2}\right)F \end{cases} \quad (2.33a)$$

For real signals the impersonated frequency, shown in Figure 2.24b, is obtained by folding the figure over at one-half the sampling frequency and noting where each frequency appears in the region from zero to one-half of the sampling rate. Note that for complex signals, integer multiples of the sampling rate can be viewed as falling on top of the spectrum from 0 to F given by

$$f_{\text{aliased}} = f_{\text{signal}} - kF, \quad \text{for } kF < f_{\text{signal}} < (k + 1)F \quad (2.33b)$$

2.5.4 Sampling Rate Conversion

Once the analog signal has been sampled, the resultant digital signal is ready for subsequent DSP operations. Several DSP operations require a change in the sampling rate. The operations of decimation (sampling rate decrease) and interpolation (sampling rate increase) were defined in Chapter 1. These operations can be used in cascade to implement a conversion of the input sampling rate by any rational factor. In order to preserve the information content of the signal, aliasing must be kept to acceptable levels during the conversion process. One way to interpret the process of sampling rate conversion is to convert the signal back to the analog domain and to resample the signal at the new rate. Since it is desirable to perform this operation completely in the digital domain, this approach is briefly described in the following paragraphs. Sampling rate conversion is described in detail in Chapter 7.

The process of decimation is shown in Figure 2.25, along with the spectral interpretation of the process. For the purpose of this discussion, only integer decimation factors D will be considered.

$$D = \frac{\text{Input sampling rate}}{\text{Output sampling rate}}$$
$$= F_{\text{in}}/F_{\text{out}}$$

Assuming that the input is properly band-limited for the input sampling rate, the signal must be further band-limited to avoid aliasing prior to the data decimation operation. The antialiasing filter shown in Figure 2.25 restricts the input band.

Passband: $20 \log \left(1 - \delta_p\right) < H(f) < 20 \log \left(1 + \delta_p\right), \qquad f_p \leq F/2D$

Stopband: $H(f) < 20 \log \left(\delta_s\right), \qquad\qquad\qquad\qquad f_s = F/D - f_p$

Following the antialiasing filter operation, every Dth sample is taken as the output $y(m)$.

$$y(m) = \sum_{n=0}^{\infty} h(Dm - n)x(n)$$

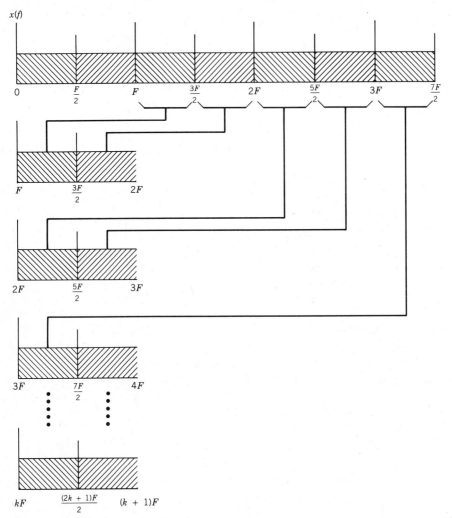

a. Multiples of sampling rate overlay baseband (output is summation of baseband plus aliased terms).

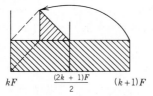

b. Folding for real signals (upper half folds onto lower half).

Figure 2.24 Aliasing — frequency spectrum illustration.

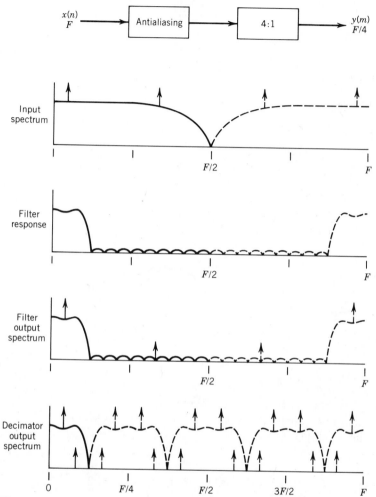

Figure 2.25 Multirate processing decimation ($D = 4$).

This process is time-varying since only versions of the output shifted by multiple factors of the decimation factor, D, will match the response to shifted versions of the input.

The process of interpolation is shown in Figure 2.26, along with the spectral interpretation of the process. For the purpose of this discussion, only integer interpolation factors I will be considered.

$$I = \frac{\text{Output sampling rate}}{\text{Input sampling rate}}$$

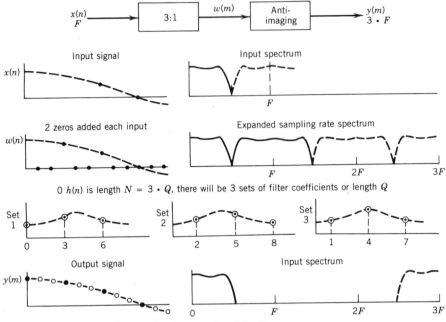

Figure 2.26 Multirate processing interpolation ($I = 3$).

$$= F_{out}/F_{in}$$

Assuming that the input is properly band-limited, for the input sampling rate, the signal must be filtered to eliminate the images created by the insertion of the $I - 1$ zeros into the data. The filter passband and stopband frequencies are given by

$$\text{Passband: } 20 \log \left(1 - \delta_p\right) < H(f) < 20 \log \left(1 + \delta_p\right), \qquad f_p < F/2$$
$$\text{Stopband: } H(f) < 20 \log \left(\delta_s\right), \qquad f_s = F - f_p$$

Following the anti-imaging filter operation, the output is given by

$$y(m) = \sum_{n=0}^{N} h(nI + m \, \text{MOD} \, I) \times ([m/I] - n)$$

where MOD represents the modulo operation and $[e]$ represents the integer less than or equal to e. Also, this filter is time-varying.

Both the decimators and interpolators can be implemented with significant reductions in processing requirements. Decimators using FIR filters can be implemented at the rate following the decimation. For interpolators the implementation control can be designed to avoid the multiplications by the $I - 1$ zeros inserted to increase the sampling rate. In addition, implementing a filter using decimators and interpolators can result in significant computational savings over a straightforward implementation of the filter. The computational benefits of these filters are addressed in Chapters 7 and 10.

2.6 SUMMARY

Several important DSP concepts have been introduced in this chapter. Understanding the input signal characteristics is necessary in order to develop the proper DSP system. A procedure was presented for performing a signal analysis. Definitions of linearity, time-invariance, causality, and stability were given for discrete systems. It was shown that for a discrete linear, time-invariant system the system is completely characterized by the impulse response. The frequency response of the system was developed, and its product relationship to the convolution sum was shown. This relationship will be extended in the following chapter on Z-transforms. The general difference equation for implementation of a digital filter was given, and IIR and FIR filters were introduced. The Nyquist sampling theorem was defined, and aliasing resulting from sampling was illustrated.

References

1. J. A. Cadzow and H. F. Van Landingham, *Signals, Systems, and Transforms*, Prentice–Hall, Englewood Cliffs, N.J., 1985.
2. B. Noble, *Applied Linear Algebra*, Prentice–Hall, Englewood Cliffs, N.J., 1969.
3. E. O. Brigham, *The Fast Fourier Transform*, Prentice–Hall, Englewood Cliffs, N.J., 1974.
4. G. Doetsch, *Guide to Applications of Laplace Transform*, (translated), Van Nostrand, Princeton, N.J., 1963.
5. F. J. Harris, "On the Use of Windows for Harmonic Analysis with the Discrete Fourier Transform," *Proc. IEEE*, 66(1): 1978.
6. C. D. McGillen and G. R. Cooper, *Continuous and Discrete Signal and System Analysis*, Holt, Rinehart & Winston, New York, 1984.
7. D. A. Linden, "A Discussion of Sampling Theorems," *Proc. IRE*, 47(7): 1219–1226, 1959.
8. W. R. Bennett, "Spectra of Quantized Signals," *Bell System Tech. J.*, 27:446–472, 1948.
9. S. D. Stearns, *Digital Signal Analysis*, Hayden, Rochelle Park, N.J., 1975.
10. R. J. Mayham, *Discrete-Time and Continuous-Time Linear Systems*, Addison-Wesley, Reading, Mass., 1984.

Additional Readings

JENKINS, G. M., and WATTS, D. G. *Spectral Analysis and Its Applications*, Holden-Day, San Francisco, 1969.

PAPOULIS, A. *Signal Analysis*, McGraw–Hill, New York, 1977.

PAPOULIS, A. *The Fourier Integral and its Applications*, McGraw–Hill, New York, 1962.

STEIGLITZ, K. *An Introduction to Discrete Systems*, John Wiley and Sons, New York 1974.

PROBLEMS

2.1.1   Generate the plot sequences for the following signals over the range of the independent variable indicated.

a. b^n, $n = 0, 1, 2, 3, 4, 5, 6, 7, 8$; $b = 0.9$.

b. part a with $b = 1.1$.

c. $\sin(2\pi f n/F)$, $f = 10$; $F = 100$;
$n = 0, 1, 2, 3, 4, 5, 6, 7, 8, 9, 10, \ldots 19, 20$.

d. $\cos(2\pi f n/F)$, $f = 10$; $F = 100$;
$n = 0, 1, 2, 3, 4, 5, 6, 7, 8, 9, 10, \ldots 19, 20$.

e. $\sin(2\pi f n/F)$, $f = 13$; $F = 100$;
$n = 0, 1, 2, 3, 4, 5, 6, 7, 8, 9, 10, \ldots 19, 20$.

f. $\sin(2\pi f_1 n/F) + \sin(2\pi f_2 n/F)$, $f_1 = 10$; $f_2 = 5$; $F = 100$;
$n = 0, 1, 2, \ldots, 19, 20$.

If the sequences are extended to infinity, are they periodic?

2.1.2   Compute the L_1, L_2, and L_∞ norms for the following sequences.

a. $b^n u(n)$, $n = 0, 1, 2, \ldots$; $b = 0.5$.

b. $\sin(2\pi f n/F)$, $f = 10$; $F = 100$; $n = 0, 1, \ldots$.

c. $x = \{\ldots, 0, 0, 2, -1, 1, 3, 2, -2, -1, 0, 2, 1, 0, 0, \ldots\}$.

2.1.3   Compute the frequency response for the sequences given in Problem 2.1.2.

2.1.4   Develop a signal representation for the following sequence values using the unit-impulse arbitrary signal representation.

a. $w = \{0, 2, 3, 1, 0, 3.5, 1, 0, -1, -2, 0, 3\}$
$\uparrow$

b. $x = \{-4, 3, 2, -3, -4, 2, 0, 3, -1, 3\}$
$\uparrow$

c. $y = \{3, 3, 1, 0, 0, -1, -2, 2, 3, 0\}$
$\uparrow$

d. $x = \{2, 1, 1, 0, -1, -1, -2, -1, 0, 1, 3\}$

e. $y = \{0, 4, 2, -2, 0, 1, 3, 0, -3, -1, 0\}$

2.1.5   Express the following signals using the unit-step sequence.

a. $x(n) = 0$ for $n < -3$
$x(n) = 2$ for $-3 \le n \le 0$

$$x(n) = 1 \quad \text{for} \quad 1 \le n \le 5$$
$$x(n) = 0.5 \quad \text{for} \quad 6 \le n \le 12$$
$$x(n) = 1 \quad \text{for} \quad 13 \le n \le 20$$
$$x(n) = 0 \quad \text{for} \quad n \ge 21$$
b. $x(n) = 0 \quad \text{for} \quad n < 2$
$$x(n) = 4 \quad \text{for} \quad 3 \le n \le 7$$
$$x(n) = 1 \quad \text{for} \quad 8 \le n \le 15$$
$$x(n) = 5 \quad \text{for} \quad 16 \le n \le 20$$
$$x(n) = -2 \quad \text{for} \quad 21 \le n \le 30$$
$$x(n) = -4 \quad \text{for} \quad 31 \le n \le 35$$
$$x(n) = 0 \quad \text{for} \quad n \ge 36$$

2.1.6 Show that the sampled function given by

$$x(n) = a^n [u(n) - u(n - 21)]$$

equals a^n for $0 \le n \le 20$ and zero otherwise.

2.1.7 Apply the shift operator to the sequence developed for Problem 2.1.4 part b resulting in a delay of three samples.

2.1.8 Apply the shift operator to the sequence developed for Problem 2.1.4 part d resulting in an advance of five samples.

2.1.9 Draw a diagram to implement the following expressions:

a. $y(n) = x(n) + 1.5y(n - 1) + 0.75y(n - 2)$

b. $y(n) = 0.2x(n - 4) + 0.4x(n - 3) + 0.5x(n - 2) + 0.4x(n - 1) + 0.2x(n)$

c. $y(n) = x(n) + 2x(n - 1) + x(n - 2) + 1.6y(n - 1) - 0.8y(n - 2)$

d. $y(n) = x(n)x(n)$

e. $y(n) = 0.2x(n - 2) + 0.4x(n - 1) + 0.5x(n) + 0.4x(n + 1) + 0.2x(n + 2)$

Comment on the realizability of each of the expressions in a real-time sense.

2.2.1 Use the appropriate test to determine if the systems given in Problem 2.1.9 are linear.

2.2.2 For each of the following system expressions, use the appropriate tests to determine if the system operations are linear.

a. $y(n) = x(n) + bx(n - 1)$ f. $y(n) = x(n) + c$

b. $y(n) = x(n) + bx(n + 1)$ g. $y(n) = x(n)e^n$

c. $y(n) = nx(n)$ h. $y(n) = a^{x(n)}$

d. $y(n) = nx(n)x(n)$ i. $y(n) = z^{-2}x(n)$

e. $y(n) = \sin(2\pi fn/F)x(n)$ j. $y(n) = |x(n)|$

2.2.3 Show that the systems described below are linear and time invariant.

a. $y(n) = \sum_{k=0}^{N-1} h(k)x(n-k)$

b. $y(n) = \sum_{k=0}^{m} a(k)x(n-k) - \sum_{k=1}^{m} b(k)y(n-k)$

2.2.4 Check each of the expressions given in Problem 2.2.2 using the appropriate test for time-invariance.

2.2.5 Check each of the expressions given in Problem 2.2.2 using the appropriate test for causality.

2.2.6 Check each of the expressions given in Problem 2.2.2 using the appropriate test for stability, and determine range of coefficient values for stability if required.

2.2.7 One definition for a system to be stable is that a bounded input produces a bounded output. Show that for the system described by

$$y(n) = \sum_{m=-\infty}^{\infty} x(m)h(n-m)$$

a necessary and sufficient condition for stability is that

$$\lim_{n \to \infty} h(n) = 0$$

Show that in the case of a system with rational transfer function, stability is equivalent to the absence of poles outside the unit circle.

2.2.8 Develop the impulse response to the system given in Problem 2.1.9 part b and perform a convolution sum with the input sequence $x(n)$.

$$x = \left\{ \dots 0, 1, 0.75, \underset{\uparrow}{0.5}, 0.25, 0 \dots \right\}$$

What is the nonzero length of the convolution-sum output? What is the general relationship for determining the nonzero convolution-sum output length from two nonzero finite length input sequences of length N and M.

2.2.9 Given two linear time-invariant systems that are implemented in parallel:

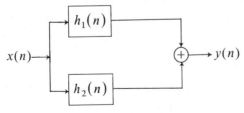

Determine the equivalent serial system unit-impulse response $h_3(n)$.

$$x(n) \longrightarrow \boxed{h_3(n)} \longrightarrow y(n)$$

2.2.10 A cascade of two linear time-invariant systems is shown here.

$$x(n) \longrightarrow \boxed{h_1(n)} \longmapsto \boxed{h_2(n)} \longmapsto y(n)$$

What is the equivalent single unit-impulse response?

$$x(n) \longrightarrow \boxed{h_3(n)} \longmapsto y(n)$$

Does the order of implementing the cascade stages affect the result?

$$x(n) \longrightarrow \boxed{h_2(n)} \longmapsto \boxed{h_1(n)} \longmapsto y(n)$$

2.2.11 Perform the convolution between the input sinusoidal sequence and the unit-impulse response of the following system.

$$y(n) = \sum_{k=0}^{6} A_k x(n-k)$$

where

$$A_0 = 0.1 = A_6 \qquad x(n) = \sin(2\pi fn/F)$$
$$A_1 = 0.2 = A_5$$
$$A_2 = 0.3 = A_4 \qquad f = 5; \ F = 25$$
$$A_3 = 0.4$$

Plot the result. Is the input signal periodic? Is the output signal periodic? Repeat the response calculation for $f = 0, 5, 6.25, 7.5,$ and 10. What observation do you have about the magnitude and phase response of the output? Is there a more efficient way to calculate the magnitude and phase frequency response of a linear time-invariant system to a sinusoidal excitation?

2.2.12 Calculate the convolution sum for the input excitations and the system unit-impulse responses given in Figure P2.2.12.

2.3.1 Draw the direct form implementation for a second-order IIR difference equation given by

$$y(n) = x(n) + A_1 x(n-1) + A_2 x(n-2) + B_1 y(n-1) + B_2 y(n-2)$$

2.3.2 Given the system diagrams shown in Figure P2.3.2, develop the difference equations for each of the filter networks.
Comment on the computational requirements in terms of multiplies per second (MPS), adds per second (APS), and storage for each of the networks. Assume the input sampling rate is F.

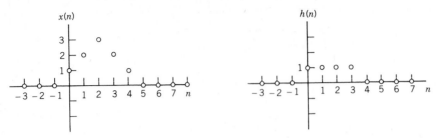

a. Input and impulse response—set 1.

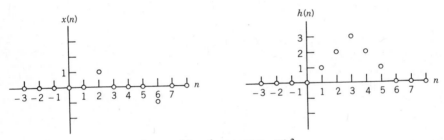

b. Input and impulse response—set 2.

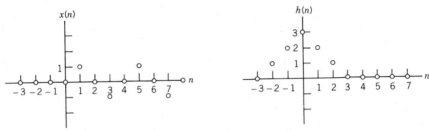

c. Input and impulse response—set 3.

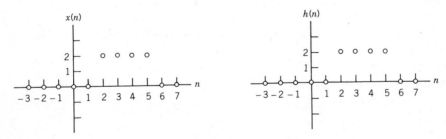

d. Input and impulse response—set 4.

Figure P2.2.12 Input and system responses for convolution-sum calculations.

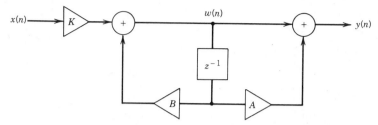

a. Filter diagram 1: First-order IIR.

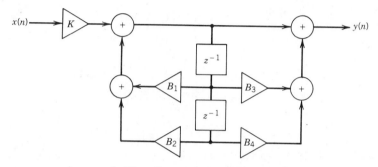

b. Filter diagram 2: Second-order IIR.

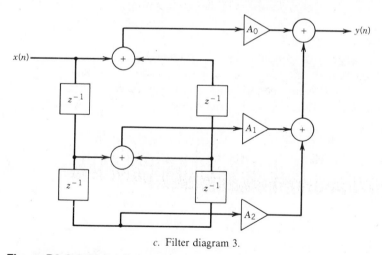

c. Filter diagram 3.

Figure P2.3.2 Digital filter diagrams.

2.3.3 The impulse response for a FIR filter on length 7 is given by

$$h(n) = \{A_0, A_1, A_2, A_3, A_4, A_5, A_6\}$$

where the coefficients are symmetric: $A_0 = A_6$, $A_1 = A_5$, A2 = A4. This is the same filter used in Problem 2.2.11. Draw a diagram implementing the filter such that only four multiplies are required for each output computation.

Comment on the computational requirements (MPS, APS, and storage) as a function of the input sampling rate F.

2.3.4 A digital filter is defined by the following difference equation

$$y(n) = x(n) + Ax(n - 1) + By(n - 1), \qquad y(n) = 0 \text{ for } n < 0$$

  a. Determine the unit-impulse response of the system.

  b. For what values of A and B is the system stable?

  c. Determine the unit-step response of the system.

  d. Determine the response of the system to a complex exponential $x(n) = \exp(j2\pi fn/F)$ using the convolution sum and the impulse response determined in part a. Plot the magnitude and phase response for f from 0 to $F/2$ in increments of $0.05(F)$.

2.3.5 A filter defined by cascading two filters, each defined by the system expression in Problem 2.3.4, is implemented.

  a. Determine the unit-impulse response of the system.

  b. For what values of A and B is the system stable?

  c. Determine the unit-step response of the system.

  d. Determine the response of the system to a complex exponential $x(n) = \exp(j2\pi fn/F)$ using the convolution sum and the impulse response determined in part a. Plot the magnitude and phase response for f from 0 to $F/2$ in increments of $0.05(F)$.

2.4.1 The relationship between the convolution sum of two sequences and the Fourier transform of the convolution sum was derived in Section 2.4. Show that the reverse relationship exists between the convolution sum of the Fourier transforms of two sequences and the inverse Fourier transform of the convolution sum as given in Eq. 2.29.

2.4.2 Several properties apply to the Fourier transform pair defined by Eqs. 2.26 and 2.27. Show that the relationships given in Table P2.4.2 apply.

2.4.3 Determine the frequency response of the system defined in Problem 2.3.4.

TABLE P2.4.2
FOURIER TRANSFORM RELATIONSHIPS

Given $x(n)$ the Fourier Transform equals $X(e^{jw})$	
Sequence	**Fourier Transform**
$x^*(n)$	$X^*(e^{jw})$
Real $[x(n)]$	$X_{even}(e^{jw})$
j Imaginary $[x(n)]$	$X_{odd}(e^{jw})$
$x_{even}(n)$	Real$[X(e^{jw})]$
$x_{odd}(n)$	j Imaginary $[X(e^{jw})]$
$x^*(-n)$	$X^*(e^{jw})$

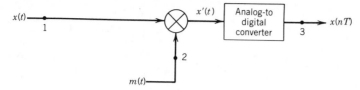

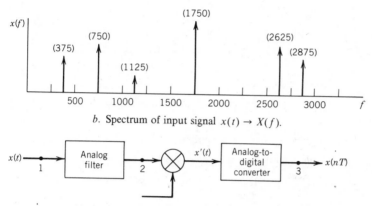

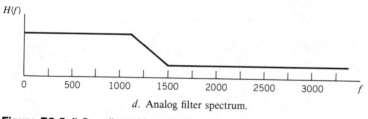

a. Sampling process without analog filter.

b. Spectrum of input signal $x(t) \rightarrow X(f)$.

c. Sampling process with analog filter.

d. Analog filter spectrum.

Figure P2.5.1 Sampling with and without an analog presampling filter.

2.4.4 Determine the frequency response of the system defined in Problem 2.3.5.

2.5.1 The signal processing configuration shown in Fig. P2.5.1*a* is used to sample the analog input signal $x(t)$ with frequency spectrum illustrated in Figure P2.5.1*b*.

a. Determine the spectrum at node 3 given a sampling rate $F = 3000$. Comment on the results with respect to aliasing.

b. What sampling rate is required to assure that no aliasing results.

If an analog filter is added prior to the sampling process as shown in Fig. P2.5.1*c*, the input spectrum is modified and the resultant output spectrum is changed.

$x(nT_1)$———

$F = 24$

4:1

———▶ $y(nT_2)$

$\dfrac{F}{4} = 6$

a. Sampling rate decrease—4 : 1.

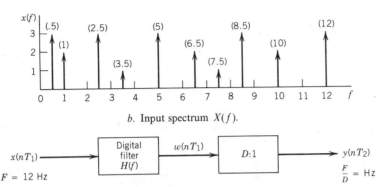

b. Input spectrum $X(f)$.

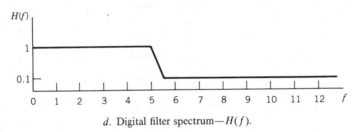

$x(nT_1)$ ———▶

$F = 12$ Hz

Digital
filter
$H(f)$

$w(nT_1)$ ———▶

D:1

———▶ $y(nT_2)$

$\dfrac{F}{D} =$ Hz

c. Digital filter followed by sampling rate decrease.

$H(f)$

1

0.1

0   1   2   3   4   5   6   7   8   9   10   11   12 f

d. Digital filter spectrum—$H(f)$.

Figure P2.5.3 Sampling rate decrease operation.

c. Determine the spectrum at node 2 and node 3 with the analog filter inserted into the processing.

2.5.2 Use the reconstruction formula given by Eq. 2.32 to produce the analog signal from the following sequence.

$$x(n) = \{ \cdots 0, 1, 2, 3, 2, 1, 0 \cdots \}$$

Discuss the realization of the reconstruction using a digital-to-analog (D/A) converter.

2.5.3 The sampling rate decrease operation shown in Figure P2.5.3*a* is performed on the signal spectrum shown in Figure P2.5.3*b*.

a. If no filtering is applied prior to a decimation of 4 : 1, calculate and plot the output signal spectrum.

A filter with spectrum $H(f)$ shown in Figure 2.5.3*c* is used to filter the spectrum prior to the decimation operation.

b. Determine the maximum integer decimation factor that can be used without aliasing of the input spectrum.

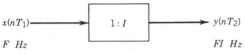

a. Sampling rate increase—$1:I$.

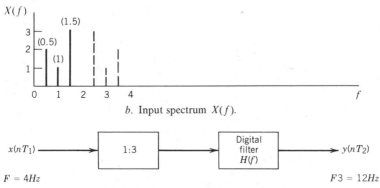

b. Input spectrum $X(f)$.

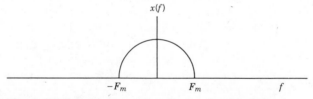

c. Digital filtering performed to reduce images.

Figure P2.5.4 Sampling rate increase operation.

2.5.4 A sampling rate increase operation (Fig. P2.5.4*a*) is performed on the signal shown in Figure P2.5.4*b*.

a. Plot the spectrum of the output from the sampling rate increase with $I = 3$ and $F = 4$.

b. Comment on the filter spectrum required at the output of the sampling rate increase process in order to delete the image spectrums shown in part a.

2.5.5 Derive the reconstruction formula given in Eq. 2.32. Assume that the input signal spectrum is band-limited as shown in Figure P2.5.5. Use the continuous time Fourier transform given by Eq. 2.27.

HINT: Use the fact that

$$X'(f) = \sum_{n=-\infty}^{\infty} x'(nT) e^{-j2\pi fn/F}$$

where X' is the sampled spectrum from x', the sampled input.

Figure P2.5.5 Spectrum of continuous-time input signal.

3

The Z-Transform

As a result of the advancements made in digital technology since the early 1960s, considerable interest has been aroused in discrete-time system analysis and design. In this chapter, the Z-transform method, which provides the mathematical basis for the analysis of discrete-time systems, is developed.

3.0 INTRODUCTION

The Z-transform provides the engineer with a powerful method for the analysis of linear time-invariant discrete systems. As discussed in Chapter 2, a discrete-time system can be represented by a difference equation relating the input and output signals in the time domain. The Z-transform provides a method for the analysis of discrete-time systems in the frequency domain, which is generally more efficient than is time domain analysis. The efficiency in performing system analysis using the Z-transform method will become evident when the concept of the system *transfer function* $H(z)$ is described. We will see that the frequency response of a discrete-time system can readily be determined by evaluating the transfer function on the unit circle in the z-plane.

In this chapter we define the Z-transform and the inverse Z-transform and describe the more important properties of the Z-transform. Finally, the Z-transform method is applied to the analysis of discrete-time systems where we investigate the system transfer function, system stability, and digital

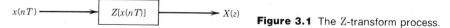

Figure 3.1 The Z-transform process.

network realizations of the system. Z-transforms are used extensively in steps 2 and 3 of the system design methodology, which is presented in Chapter 10.

3.1 THE Z-TRANSFORM

Consider the discrete-time sequence $x(nT)$, for $n = 0, \pm 1, \pm 2, \ldots$. This sequence is considered two-sided since the time index n is defined for both positive and negative values. The **two-sided Z-transform** of this sequence is defined as

$$X(z) = Z[x(nT)] = \sum_{n=-\infty}^{\infty} x(nT)z^{-n} \qquad (3.1)$$

We can interpret z^{-n} as a delay operator, that is, a delay of nT seconds for each element in the sequence $x(nT)$ is equivalent to multiplication of the Z-transform by z^{-n}. Therefore, the Z-transform of sequences of finite duration can be found by direct application of Eq. 3.1, where the sample $x(nT)$ is the coefficient of z^{-n} in the power series expansion of $X(z)$.

More formally, z is a complex variable that can assume any value in the **complex z-plane** at which the infinite series converges. In polar form z can be expressed as $z = re^{j\omega T}$, where it should be noted that on the unit circle in the z-plane $r = 1$, that is, for $|z| = 1$, the Z-transform is equivalent to the *discrete Fourier transform* (*DFT*) introduced in Chapter 2. The DFT will be developed in detail in Chapter 6.

Since causal sequences form the basis of most physical systems, **right-sided Z-transforms** will be emphasized, that is, sequences for which the time index n is defined only for positive values. In this form we have

$$X(z) = Z[x(nT)] = \sum_{n=0}^{\infty} x(nT)z^{-n} \qquad (3.2)$$

It is instructive to view the Z-transform as a transformation that maps an input sequence $x(nT)$ into a complex function $X(z)$. Therefore, the Z-transform process is an operation on the input signal, as shown in Figure 3.1. We will see in the following sections that this transformation to the z-domain allows the analysis of linear time-invariant discrete systems to be performed with relative ease.

3.2 PROPERTIES OF THE Z-TRANSFORM

In the following sections we develop the basic properties of the Z-transform. Causal sequences will be emphasized, since they form the basis of **real-time signal processing systems**; that is, the system response is determined as the samples arrive in a continuing stream at the sampling period T (seconds/sample).

For the case where $X(z)$ is a rational function of z, that is, a ratio of polynomials in z, we can define the transform in terms of the roots of the system. The roots of the numerator polynomial are referred to as the **zeros** of $X(z)$, and the roots of the denominator polynomial are referred to as the **poles** of $X(z)$. It will be shown in the following section that there is a direct relationship between the location of the poles of the discrete function (or system) and the **region of convergence** (ROC) in the z-plane for which the magnitude of $X(z)$ is finite.

3.2.1 Region of Convergence (ROC)

If $x(nT)$ is to have a Z-transform, then the magnitude of $X(z)$ must be finite. We now define the **region of convergence (ROC)** to be the set of all z in the complex z-plane for which the magnitude of $X(z)$ is finite. Representing z in polar form, we require that

$$|X(z)| = \left| \sum_{n=-\infty}^{\infty} x(nT)z^{-n} \right| \leq \sum_{n=-\infty}^{\infty} |x(nT)| r^{-n}$$

$$= \sum_{n=-\infty}^{-1} |x(nT)| r^{-n} + \sum_{n=0}^{\infty} |x(nT)| r^{-n}$$

$$= \sum_{n=1}^{\infty} |x(-nT)| r^{n} + \sum_{n=0}^{\infty} |x(nT)| r^{-n} < \infty \qquad (3.3)$$

For the infinite sum of Eq. 3.1 to be absolutely summable, each of the two sums of Eq. 3.3 must be finite. For the sums of Eq. 3.3 to be finite, we find three positive constants R_1, R_2, and M. For causal sequences we must satisfy the condition $|x(nT)| \leq MR_1^n$ for $n \geq 0$, and for noncausal sequences we must satisfy the condition $|x(nT)| \leq MR_2^n$ for $n < 0$. The series of negative powers of the complex variable z converges outside a circle of radius R_1, and the series of positive powers of z converges inside a circle of radius R_2. On substituting these bounds into Eq. 3.3, we obtain

$$\sum_{n=-\infty}^{\infty} |x(nT)z^{-n}| \leq M \left[\sum_{n=1}^{\infty} R_2^{-n} r^n + \sum_{n=0}^{\infty} R_1^n r^{-n} \right] \qquad (3.4)$$

We now observe that the sums in Eq. 3.4 are finite if and only if $r/R_2 < 1$ for the first sum and $R_1/r < 1$ in the second sum, that is, Eq. 3.1 converges absolutely for all z in the ring of convergence $R_1 < |z| < R_2$.

We now proceed to describe the ROC for *causal* and *noncausal* sequences. On re-examination of Eq. 3.1 the Z-transform is a power series where both negative and positive powers of z are involved. This can be seen by

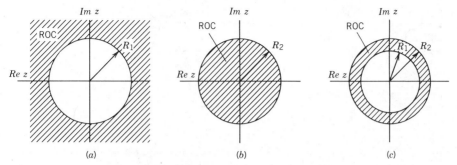

Figure 3.2 Region of convergence (ROC): (*a*) right-sided sequence: $|z| > R_1$; (*b*) left-sided sequence: $|z| < R_2$; (*c*) two-sided sequence: $R_1 < |z| < R_2$.

expressing $X(z)$ in the form

$$X(z) = \sum_{n=0}^{\infty} x(nT)z^{-n} + \sum_{n=-\infty}^{-1} x(nT)z^{-n} = X_1(z) + X_2(z) \quad (3.5)$$

For causal sequences, the powers of z are negative and $x(nT)$ is nonzero only in the region $0 \le N_1 \le n < \infty$, where N_1 is an integer. For this case, the ROC for $X(z)$ exists everywhere outside a circle of radius R_1 where the value of R_1 depends on the location of the poles of $X(z)$. If the poles of $X(z)$ are inside the unit circle in the z-plane, the system is also stable (see Section 3.4.2). Thus, for this case the ROC for $X(z)$ in the complex z-plane is shown in Figure 3.2a. We refer to these sequences as **right-sided sequences**. It should further be noted that the Z-transform of causal sequences are unique, that is, given $X(z)$, we can recover $x(nT)$ by the inverse Z-transform.

For noncausal sequences, the powers of z are positive, and $x(nT)$ is nonzero in the region $-\infty < n < N_2 < 0$. For this case, the ROC exists everywhere inside a circle of radius R_2, where R_2 depends on the poles of $X(z)$. Since noncausal sequences exist for positive n, future samples are assumed to be available. Generally, this condition does not occur in real-time signal processing systems; however, it can be achieved by the use of a time delay. Noncausal sequences are of interest in system analysis problems. For this case the ROC for $X(z)$ is shown in Figure 3.2b. We refer to these sequences as **left-sided sequences**.

When the sequence has values for both positive and negative n, its ROC is given by Eq. 3.4, that is, $R_1 < |z| < R_2$, where $R_2 \ge R_1$. That is, the ROC results in a **ring of convergence**, as illustrated in Figure 3.2c. These sequences are referred to as **two-sided sequences**. The inner circle bounds in terms in negative powers of z away from the origin. The outer circle bounds the terms in positive powers of z away from large $|z|$ values.

In summary, for causal sequences the ROC is exterior to a circle passing through the pole farthest from the origin in the z-plane. Also, no poles of $X(z)$

TABLE 3.1
Z-TRANSFORM REGION OF CONVERGENCE

Sequence	Transform	ROC	
Two-sided	$X_1(z) + X_2(z)$	$R_1 < \|z\| < R_2,$	(for $R_1 < R_2$)
Right-sided	$X_1(z)$	$\|z\| > R_1$	
Left-sided	$X_2(z)$	$\|z\| < R_2$	

can occur within the ROC since the Z-transform does not converge at a pole. If the sequence is also stable, then all poles lie inside the unit circle and the ROC includes the unit circle. For noncausal two-sided sequences, the ROC must be bounded by poles, that is, we have a ring of convergence. Also, if the ROC does not extend to zero or infinity, the sequence is two-sided. The results of the preceding paragraphs are summarized in Table 3.1.

The ROC is of interest to the system analyst when there is a need to know the region in which the Z-transform is defined for the general two-sided sequence. Since most signal processing systems are causal, for subsequent discussions, we will consider primarily the right-sided Z-transform. This restriction will allow us to focus on realizable real-time systems, that is, systems that depend on past input and output values as well as present input values. Finally, it should be emphasized that if the discrete sequence is causal, the ROC is implied; that is, the ROC is outside a circle centered at the origin in the z-plane.

Example 3.1
Find the Z-transform of the following finite duration sequences.

    a. $x(nT) = x(-2T), \; x(-T), \; x(0), \; x(T), \; x(2T)$
                                       ↑

    b. $x(nT) = 1, 2, 3, \; 4 \,, 3, 2, 1$
                       ↑

    c. $x(nT) = \; 1 \,, 2, 3, 4, 3, 2, 1$
              ↑

    where the arrow corresponds to the time index $n = 0$.

Solution:

    a. $X(z) = x(-2T)z^2 + x(-T)z + x(0) + x(T)z^{-1} + x(2T)z^{-2}$
    b. $X(z) = z^3 + 2z^2 + 3z^1 + 4 + 3z^{-1} + 2z^{-2} + z^{-3}$
    c. $X(z) = 1 + 2z^{-1} + 3z^{-2} + 4z^{-3} + 3z^{-4} + 2z^{-5} + z^{-6}$

Thus, sequences a and b converge absolutely for all z except $z = 0$ and $z = \infty$. Sequence c converges absolutely for all z except $z = 0$. It should be noted that sequences a and b are noncausal sequences, whereas sequence c is a causal sequence. Also observe sequences b and c, where sequence c is shifted in time by $(N - 1)/2 = 3$ samples with respect to sequence b. This time shift results in converting sequence b into a causal sequence. It will be shown in Chapter 5 that this technique is used in the design of linear phase FIR digital filters. Another observation you should make is that the Z-transform of sequences b and c results in **mirror-image polynomials**. More will be said about this property in Chapter 5 where **finite impulse filters** (FIR) are discussed. ∎

Example 3.2

Find the Z-transform and the ROC of the following sequences.

   a. $x(nT) = u(nT)$

   b. $x(nT) = \delta(nT)$

Solution:

   a. From the basic definition of the unit-step function, we can write

$$Z[u(nT)] = \sum_{n=0}^{\infty} z^{-n}$$

This infinite series converges for $|z| > 1$.

   A closed-form expression of this infinite series can be obtained by noting that $Z[u(nT)]$ is a geometric series. Since the geometric series is used frequently in digital signal analysis, you should review the properties of this series. Finally, we obtain

$$Z[u(nT)] = \frac{z}{z - 1} \qquad \text{ROC: } |z| > 1 \qquad (3.6)$$

The unit step function is used frequently in system analysis. For example, the step response of a digital system can be obtained by multiplying $Z[u(nT)]$ by the transfer function of the system and then taking the inverse Z-transform. It should further be noted that multiplying a discrete-time sequence $x(nT)$ by $u(nT)$ defines a causal sequence since it forces all values of $x(nT)$ to zero for $n < 0$.

   b. From the definition of the unit-impulse function, we can write that

$$Z[\delta(nT)] = 1 \qquad \text{ROC: } |z| \geq 0 \qquad (3.7)$$

Since $Z[\delta(nT)]$ is independent of z, the ROC is the entire z-plane. It will be shown in Section 3.5 that since $Z[\delta(nT)] = 1$, that is, $X(z) = 1$, then the unit-impulse response of a discrete system can be obtained by finding the inverse Z-transform of the system transfer function.   ∎

Example 3.3

Find the Z-transforms and the ROC of the following two sequences.

a. $x(nT) = b^n u(nT)$

b. $x(nT) = -b^n u(-nT - T)$

Solution:

a. For a causal sequence, the right-sided transform yields

$$X(z) = \sum_{n=0}^{\infty} (bz^{-1})^n = \frac{z}{z - b} \qquad \text{ROC: } |z| > b$$

since the infinite series converges for $|b/z| < 1$.

b. For a noncausal sequence, the left-sided transform yields

$$X(z) = -\sum_{n=-\infty}^{-1} (bz^{-1})^n = -\sum_{n=1}^{\infty} (b^{-1}z)^n$$

$$= 1 - \sum_{n=0}^{\infty} (b^{-1}z)^n = 1 + \frac{b}{z - b}$$

$$= \frac{z}{z - b} \qquad \text{ROC: } |z| < b$$

since the infinite series converges for $|z/b| < 1$.

We observe that the Z-transforms of sequences a and b are equal. Therefore, this example demonstrates that if we are given the Z-transform of a sequence, we must know the ROC in order to determine the time domain sequence.   ∎

Example 3.4

The unit-impulse response of an all-pole second-order digital filter is given by

$$h(nT) = \frac{r^n \sin[(n + 1)\omega T]}{\sin \omega T} u(nT) \qquad 0 < r < 1$$

Find the Z-transform and the ROC. Also, determine system stability from the location of the poles in the z-plane.

Solution:
Since the sequence is causal, the right-sided Z-transform is given by

$$H(z) = \sum_{n=0}^{\infty} \frac{r^n \sin\left[(n+1)\omega T\right] z^{-n}}{\sin \omega T}$$

$$= \sum_{n=0}^{\infty} \frac{r^n z^{-n}}{2j \sin \omega T} \left(e^{j(n+1)\omega T} - e^{-j(n+1)\omega T}\right)$$

$$= \sum_{n=0}^{\infty} \left(rz^{-1}e^{j\omega T}\right)^n \frac{e^{j\omega T}}{2j \sin \omega T} - \sum_{n=0}^{\infty} \left(rz^{-1}e^{-j\omega T}\right)^n \frac{e^{-j\omega T}}{2j \sin \omega T}$$

Both series converge if the ROC is given by

$$\left|rz^{-1}e^{j\omega T}\right| = \left|rz^{-1}e^{-j\omega T}\right| < 1, \quad \text{or } |z| > |r|$$

Also, if $|z| = 1$, then for $r < 1$ the system is stable. Finally, the closed-form solution of the geometric series is given by

$$H(z) = \frac{1}{2j \sin \omega T} \left(\frac{e^{j\omega T}}{1 - rz^{-1}e^{j\omega T}} - \frac{e^{-j\omega T}}{1 - rz^{-1}e^{-j\omega T}}\right)$$

$$= \frac{1}{\left(1 - rz^{-1}e^{j\omega T}\right)\left(1 - rz^{-1}e^{-j\omega T}\right)} \tag{3.8}$$

$$= \frac{z^2}{z^2 - 2r(\cos \omega T)z + r^2} \tag{3.9}$$

where the filter coefficients are $B_1 = 2r\cos \omega T$ and $B_2 = r^2$. The poles are at $z = re^{\pm j\omega T}$ and a double zero at $z = 0$. ∎

Equation 3.9 describes the **transfer function** of an all-pole infinite impulse response (IIR) second-order digital filter, which forms a transform pair with the unit-impulse response $h(nT)$.

The stability condition requires that the poles of $H(z)$ lie inside the unit circle, and the ROC includes the unit circle in the z-plane, that is, $|r| < 1$. Since the two poles of Eq. 3.9 occur as a complex conjugate pair, the coefficients of the transfer function are real.

It should be emphasized that two different sequences can have the same analytical form in the z-domain and the same pole-zero configuration. Problem 3.2.3 demonstrates this concept by showing that the Z-transform of the following system impulse response is equivalent to Eq. 3.9.

$$h(nT) = \frac{-r^n \sin (n+1)\omega T}{\sin \omega T} u(-nT - T) \qquad \text{ROC: } |z| < r \tag{3.10}$$

TABLE 3.2
Z-TRANSFORMS OF SAMPLED TIME FUNCTIONS

$x(nT)$ $(n \geq 0)$	$X(z)$	Region of Convergence
$\delta(nT)$	1	$\|z\| < \infty$ (Entire z-plane)
1. $u(nT)$	$\dfrac{z}{z-1}$	$\|z\| > 1$
2. nT	$\dfrac{zT}{(z-1)^2}$	$\|z\| > 1$
3. a^{nT}	$\dfrac{z}{z-a^T}$	$\|z\| > \|a^T\|$
4. e^{-anT}	$\dfrac{z}{z-e^{-aT}}$	$\|z\| > \|e^{-aT}\|$
5. ne^{-anT}	$\dfrac{ze^{-aT}}{(z-e^{-aT})^2}$	$\|z\| > \|e^{-aT}\|$
6. $\sin \omega nT$	$\dfrac{z \sin \omega T}{z^2 - 2z \cos \omega T + 1}$	$\|z\| > 1$
7. $\cos \omega nT$	$\dfrac{z(z - \cos \omega T)}{z^2 - 2z \cos \omega T + 1}$	$\|z\| > 1$
8. $r^{nT} \sin \omega nT$	$\dfrac{r^T z \sin \omega T}{z^2 - 2r^T z \cos \omega T + r^{2T}}$	$\|z\| > \|r\|^T$
9. $r^{nT} \cos \omega nT$	$\dfrac{z(z - r^T \cos \omega T)}{z^2 - 2r^T z \cos \omega T + r^{2T}}$	$\|z\| > \|r\|^T$
10. $e^{-anT} \sin \omega nT$	$\dfrac{ze^{-aT} \sin \omega T}{z^2 - 2e^{-aT} z \cos \omega T + e^{-2aT}}$	$\|z\| > \|e^{-aT}\|$
11. $e^{-anT} \cos \omega nT$	$\dfrac{z(z - e^{-aT} \cos \omega T)}{z^2 - 2e^{-aT} z \cos \omega T + e^{-2aT}}$	$\|z\| > \|e^{-aT}\|$

Conversion of the sample time functions to sample functions can be accomplished by making the following substitutions:

$$a^{nT} = (a^T)^n = b^n$$
$$e^{-anT} = (e^{-aT})^n = c^n$$

Therefore, to obtain a unique sequence in the time domain for a given Z-transform of a sequence, the ROC must be specified.

A list of frequently encountered causal Z-transform pairs with the associated ROC is presented in Table 3.2.

3.2.2 Linearity

Given that the Z-transforms of $x(n)$ and $y(n)$ are denoted by $X(z)$ and $Y(z)$, then

$$Z[ax(n) + by(n)] = aX(z) + bY(z) \qquad (3.11)$$

Equation 3.11 states that given the transform of the sum of two sequences multiplied by arbitrary constants, its transform may also be found by summing the corresponding transform multiplied by the arbitrary constants. The ROC for the transform of a sum of sequences is the intersection of the ROC of the transforms of the two sequences.

3.2.3 Delay Property

The delay theorem for causal sequences can be expressed as

$$Z[x(nT - iT)] = z^{-i}X(z) + z^{-i} \sum_{m=-i}^{-1} x(mT)z^{-m}, \qquad i \geq 0$$

$$= z^{-i}X(z) + x(-iT) + x(-iT + T)z^{-1}$$

$$+ \cdots + x(-T)z^{-(i-1)} \tag{3.12}$$

where i is an arbitrary delay integer.

Proof: From the definition of the one-sided Z-transform of causal sequences, the Z-transform of a delayed sequence can be expressed by

$$Z[x(nT - iT)] = \sum_{n=0}^{\infty} x(nT - iT)]z^{-n}$$

letting $m = n - i$, we obtain

$$Z[x(nT - iT)] = \sum_{m=-i}^{\infty} x(mT)z^{-m-i}$$

$$= z^{-i} \sum_{m=0}^{\infty} x(mT)z^{-m} + z^{-i} \sum_{m=-i}^{-1} x(mT)z^{-m}$$

$$= z^{-i}X(z) + z^{-i} \sum_{m=-i}^{-1} x(mT)z^{-m}$$

This property is useful when considering a system with nonzero initial conditions. For the case when $x(nT) = 0$, $n < 0$, that is, for zero initial conditions we obtain

$$Z[x(nT - iT)] = z^{-i}X(z) \tag{3.13}$$

This property is used extensively to transform difference equations to the z-domain when the initial conditions are zero.

Example 3.5

Consider the following third-order difference equation relating the input and delayed output discrete signals of the system.

$$y(nT) = A_0x(nT) - B_1y(nT - T) - B_2y(nT - 2T) - B_3y(nT - 3T)$$

Determine the system response $Y(z)$ assuming nonzero initial conditions.

Solution:

Taking the Z-transform of both sides of $y(nT)$ gives

$$Y(z) = A_0 X(z) - B_1 \sum_{n=0}^{\infty} y(nT - T)z^{-n} - B_2 \sum_{n=0}^{\infty} y(nT - 2T)z^{-n}$$

$$- B_3 \sum_{n=0}^{\infty} y(nT - 3T)z^{-n}$$

We now proceed to determine the Z-transform of the delayed response by the variable substitution $m = n - i$, $i = 1, 2, 3$. Setting $m = n - 1$, the first summation is evaluated as

$$\sum_{n=0}^{\infty} y(nT - T)z^{-n} = \sum_{m=-1}^{\infty} y(mT)z^{-(m+1)} = y(-T) + \sum_{m=0}^{\infty} y(mT)z^{-(m+1)}$$

$$= y(-T) + Y(z)z^{-1}$$

By similar procedure, we can show that for $m = n - 2$ and $m = n - 3$ we obtain

$$\sum_{n=0}^{\infty} y(nT - 2T)z^{-n} = \sum_{m=-2}^{\infty} y(mT)z^{-(m+2)}$$

$$= y(-2T) + y(-T)z^{-1} + Y(z)z^{-2}$$

$$\sum_{n=0}^{\infty} y(nT - 3T)z^{-n} = \sum_{m=-3}^{\infty} y(mT)z^{-(m+3)}$$

$$= y(-3T) + y(-2T)z^{-1} + y(-T)z^{-2} + Y(z)z^{-3}$$

Finally,

$$Y(z) = \frac{A_0 X(z)}{1 + B_1 Y(z)z^{-1} + B_2 Y(z)z^{-2} + B_3 Y(z)z^{-3}}$$

$$- \frac{\left[B_1 + B_2 z^{-1} + B_3 z^{-2} \right] y(-T) + \left[B_2 + B_3 z^{-1} \right] y(-2T) + B_3 y(-3T)}{1 + B_1 Y(z)z^{-1} + B_2 Y(z)z^{-2} + B_3 Y(z)z^{-3}}$$

$Y(z)$ is seen to contain a steady-state response due to the input signal $X(z)$ and a transient response resulting from the initial conditions. If the poles of the system lie inside the unit circle in the z-plane, the transient response eventually decays to zero. This can be seen in the time domain by determining the inverse Z-transform. We will demonstrate this result in Section 3.4.4. It should also be noted that for zero initial conditions, that is, $y(-T) = y(-2T) = y(-3T) = 0$, $Y(z)$ is only a function of the steady-state response, which can be obtained by direct application of Eq. 3.13. ■

3.2.4 Time Scaling by a Complex Exponential Sequence

It can be shown that multiplying a sequence $x(nT)$ by a complex exponential sequence α^n, where $\alpha = re^{j\omega T}$, corresponds to a cyclic rotation of the poles and zeros in the z-plane. This rescaling of the z-plane is expressed by

$$Z[\alpha^n x(nT)] = X(\alpha^{-1}z) \qquad (3.14)$$

The proof of this property is left as an exercise in Problem 3.2.2. This property can also be demonstrated by replacing $\alpha^{-1}z$ for z in pairs 6 and 7 to obtain pairs 8 and 9 in Table 3.2. We will see in Chapter 4 that digital filter design by pole-zero placement is a direct application of this property.

3.2.5 Differentiation of $X(z)$ or Multiplication of $x(nT)$ by (nT)

This property states that if $x(nT)$ is multiplied by time index (nT) the result is equivalent to the derivative of the Z-transform of $x(nT)$ multiplied by $-(Tz)$, that is,

$$Z[(nT)x(nT)] = -Tz\frac{d}{dz}X(z) \qquad (3.15)$$

The proof of this property is left as an exercise in Problem 3.3.2.

Example 3.6
Find the Z-transform of the sequence

$$x(nT) = nTa^n$$

Solution:
The Z-transform of a function linearly weighted by nT is given by

$$Z[nTa^n] = -Tz\frac{d}{dz}\left(\frac{z}{z-a}\right) = \frac{Tza}{(z-a)^2} \qquad \text{ROC: } |z| > a$$

Note that differentiation results in a double pole at $z = a$. ∎

3.2.6 Convolution Property

The convolution sum was developed in Chapter 2. We now consider the Z-transform of the convolution sum, that is,

$$Z[y(nT)] = Z[h(nT) * x(nT)] = Z\left[\sum_{m=-\infty}^{\infty} x(nT - mT)h(mT)\right]$$

$$= Z\left[\sum_{m=-\infty}^{\infty} x(mT)h(nT - mT)\right] = X(z)H(z) \qquad (3.16)$$

Therefore, the Z-transform of the convolution sum is equal to the product of the Z-transform of the input sequence $x(n)$ and the impulse sequence $h(n)$.

Proof: The Z-transform of the convolution sum is given by

$$Y(z) = \sum_{n=-\infty}^{\infty} \left[\sum_{m=-\infty}^{\infty} x(mT)h(nT - mT) \right] z^{-n}$$

Let $i = n - m$, we then obtain

$$Y(z) = \sum_{i=-\infty}^{\infty} \sum_{m=-\infty}^{\infty} x(mT)h(iT)z^{-(i+m)}$$

$$= \sum_{i=-\infty}^{\infty} \left[\sum_{m=-\infty}^{\infty} x(mT)z^{-m} \right] h(iT)z^{-i}$$

$$= X(z) \sum_{i=-\infty}^{\infty} h(iT)z^{-i} = X(z)H(z)$$

It was shown in Chapter 2 that the response of an LTI discrete system was given by $y(nT) = x(nT) * h(nT)$. This operation is called discrete linear convolution. Equation 3.16 states that linear convolution of two discrete signals is equal to the product of the Z-transforms. In Section 3.4 we will show the application of this theorem to discrete system analysis.

In Chapter 6 we will investigate another type of convolution called circular (periodic) convolution and demonstrate its important application in the field of DSP. The properties and application of circular convolution will be presented in detail in Chapter 6.

3.2.7 Initial Value Theorem

If $x(nT)$ is a causal signal, then the initial value of the signal is given by

$$x(0) = \lim_{z \to \infty} X(z) \tag{3.17}$$

Proof:

$$\lim_{z \to \infty} X(z) = \lim_{z \to \infty} \sum_{n=0}^{\infty} x(nT)z^{-n}$$

$$= x(0)z^0 + x(1)z^{-1} + x(2)z^{-2} + \cdots = x(0)$$

that is, all terms in the summation except $x(0T)$ approach zero as z approaches infinity.

3.2.8 Final Value Theorem

If $x(nT)$ is causal and stable with Z-transform $X(z)$, then the final value theorem is given by

$$\lim_{n \to \infty} x(nT) = \lim_{z \to 1} (1 - z^{-1})X(z) \tag{3.18}$$

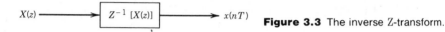

Figure 3.3 The inverse Z-transform.

In order for the limit of $(1 - z^{-1})X(z)$ to exist as z approaches unity, we must have $z = 1$ included in the ROC, the system must be stable, and therefore all poles of the system must lie inside the unit circle.

3.3 THE INVERSE Z-TRANSFORM

In the preceding sections we were given a sequence in the time domain and obtained the Z-transform of the sequence. We now investigate methods of obtaining the inverse Z-transform, that is, obtain the discrete-time sequence $x(nT)$ given $X(z)$. Operationally, the inverse Z-transform can be represented by the system block diagram shown in Figure 3.3.

The three basic methods that can be employed to recover the original sequence from its Z-transform are

1. Complex inversion integral.

2. Partial-fraction expansion.

3. Inversion by division.

Each of these methods is described. However, since the complex integration method is the most general method and is essentially the same as the partial-fraction method, our discussion will focus on the application of the method of residues for the evaluation of the complex inversion integral.

3.3.1 Complex Inversion Integral

The Z-transform inversion integral can be derived from Eq. 3.1 and using the Cauchy integral theorem from the theory of complex variables.[3,5] Multiplying both sides of Eq. 3.1 by z^{m-1} and then integrating around a closed contour in the z-plane, it follows that

$$\oint_C X(z)z^{m-1}\,dz = \oint_C \sum_{n=0}^{\infty} x(nT)z^{m-n-1}\,dz$$

According to the *Cauchy integral theorem*, if the integration path is within the ROC, and if the ROC includes the unit circle, the series $x(nT)$ is absolutely convergent. It is then valid to interchange the summation and integration, yielding

$$\oint_C X(z)z^{m-1}\,dz = \sum_{n=0}^{\infty} x(nT)\oint_C z^{m-n-1}\,dz$$

If the path of integration encloses the origin, then, according to the *Cauchy integral theorem*,[1] the integral on the right side is zero except for $m = n$, where for this case the integral reduces to $2\pi j$. We finally obtain the Z-transform *inversion integral*

$$x(nT) = Z^{-1}[X(z)] = \frac{1}{2\pi j} \oint_C X(z) z^{n-1} \, dz \qquad (3.19)$$

Equation 3.19 represents a contour C, within which $X(z)$ is analytic; that is, let the poles of the system, p_i, be inside the closed contour C. For rational Z-transforms, the contour integral given by Eq. 3.19 can be evaluated using Cauchy's *residue theorem*, which is a fundamental result obtained from complex variable theory. First define $X_0(z)$ as a rational function with the denominator expanded in a product of pole factors

$$X_0(z) = X(z) z^{n-1} = \frac{N(z)}{\displaystyle\prod_{i=1}^{N} (z - p_i)^{m_i}} \qquad (3.20)$$

where N is a positive integer representing the total number of poles and m_i is the pole order. Then by the residue theorem we obtain for poles inside the contour of integration

$$x(nT) = \sum_{i=1}^{N} \operatorname*{Res}_{z = p_i} [X_0(z)] \qquad n \geq 0 \qquad (3.21)$$

where for simple poles, that is, $m = 1$, the residue of $X_0(z)$ at p_i is given by

$$\operatorname*{Res}_{z = p_i} [X_0(z)] = \lim_{z \to p_i} [(z - p_i) X_0(z)]$$

$$= (z - p_i) X_0(z)|_{z = p_i} \qquad (3.22)$$

For an mth-order pole, $(m = 2, 3, \ldots)$ the residue is given by

$$\operatorname*{Res}_{z = p_i} [X_0(z)] = \frac{1}{(m-1)!} \lim_{z \to p_i} \frac{d^{m-1}}{dz^{m-1}} [(z - p_i)^m X_0(z)]$$

$$= \frac{1}{(m-1)!} \frac{d^{m-1}}{dz^{m-1}} [(z - p_i)^m X_0(z)] \Bigg|_{z = p_i} \qquad (3.23)$$

For the poles of $X(z)$ outside the contour of integration, the sum of the residues of $X_0(z)$ is given by

$$x(nT) = - \sum_{i=1}^{N} \operatorname*{Res}_{z = p_i} [X_0(z)] \qquad n < 0 \qquad (3.24)$$

It should be noted that for the case where $X_0(z)$ has a simple pole at the origin when $n = 0$, $x(0)$ is determined independently. We also note that the terms in the inverse Z-transform are determined by the poles of the transform function, with the zeros affecting only the magnitude of the terms.

Example 3.7
Determine the inverse Z-transform of the single-pole function

$$X(z) = \frac{1}{1 - bz^{-1}} \qquad \text{ROC: } |z| > |b|, \quad \text{and} \quad 0 < b < 1$$

Solution:
Since the ROC is outside the pole b and includes the unit circle in the z-plane, $X(z)$ is both stable and causal (see Section 3.2.1). Therefore, from Eqs. 3.21 and 3.22 we obtain, using the residue theorem,

$$X_0(z) = X(z)z^{n-1} = \frac{z^n}{z - b}$$

$$\lim_{z \to b} (z - b)X_0(z) = b^n u(n)$$

Finally, the discrete-time sequence is given by

$$x(nT) = b^n u(nT) \quad \blacksquare$$

Example 3.8
Find the inverse Z-transform of the function given by

$$H(z) = \frac{z^2}{z^2 - B_1 z + B_2} = \frac{z^2}{(z - p_1)(z - p_1^*)}, \qquad \text{ROC: } |z| > r$$

where for real coefficients the poles occur in complex conjugate pairs at $re^{\pm j\omega T}$ in the z-plane and $0 < r < 1$.

Solution:
Let $p_1 = a + jb$, and $p_1^* = a - jb$. For $n \geq 0$ the sum of the residues is given by Eqs. 3.21 and 3.22, that is,

$$h(nT) = \operatorname*{Res}_{z=a+jb} [H_0(z)] + \operatorname*{Res}_{z=a-jb} [H_0(z)]$$

$$= \frac{z^{n+1}}{z - a + jb}\bigg|_{z=a+jb} + \frac{z^{n+1}}{z - a - jb}\bigg|_{z=a-jb}$$

$$= \frac{1}{j2b}\left[(a + jb)^{n+1} - (a - jb)^{n+1}\right]$$

Since

$$a \pm jb = re^{\pm j\omega T}$$

where

$$r^2 = a^2 + b^2, \qquad \phi = \omega T = \arctan\left(\frac{b}{a}\right)$$

we have

$$h(nT) = \frac{r^n}{j2b}\left[re^{j(n+1)\omega T} - re^{-j(n+1)\omega T}\right]$$

Finally, since $b = r \sin \omega T$, we obtain

$$h(nT) = \frac{r^n}{\sin \omega T} \sin\left[(n+1)\omega T\right]u(nT)$$

which is the impulse response of the second-order system given in Example 3.4. Since the poles of the system are defined to be inside the unit circle $|r| < 1$, then $h(nT)$ represents a damped sinusoid decaying exponentially to zero. ∎

Example 3.9
Find the inverse Z-transform of the following function.

$$X(z) = \frac{A}{(z - 0.25)(z - 0.5)} \qquad \text{ROC: } |z| > 0.5$$

Solution:
From Eq. 3.20 we obtain

$$X_0(z) = X(z)z^{n-1} = \frac{Az^n}{z(z - 0.25)(z - 0.5)}$$

Since $X_0(z)$ has a pole at the origin when $n = 0$, $x(0)$ must be obtained independently as follows: For $n = 0$ the sum of the residues, given by Eqs. 3.21 and 3.22, is

$$x(0) = \operatorname*{Res}_{z=0}\left[X_0(z)\right] + \operatorname*{Res}_{z=0.25}\left[X_0(z)\right] + \operatorname*{Res}_{z=0.5}\left[X_0(z)\right]$$

$$= \frac{A}{(z - 0.25)(z - 0.5)}\bigg|_{z=0} + \frac{A}{z(z - 0.5)}\bigg|_{z=0.25} + \frac{A}{z(z - 0.25)}\bigg|_{0.5} = 0$$

For $n > 0$

$$x(nT) = \frac{Az^{n-1}}{(z - 0.25)}\bigg|_{z=0.5} + \frac{Az^{n-1}}{(z - 0.5)}\bigg|_{z=0.25}$$
$$= A\left[8(0.5)^n - 16(0.25)^n\right]$$

Alternatively, for $n \geq 0$

$$x(nT) = A\left[8(0.5)^n - 16(0.25)^n\right]u(nT - T) \quad \blacksquare$$

Example 3.10

Find the inverse Z-transform of the following system transfer function. It should be noted that this function is actually a cascade-form realization of two first-order digital filters:

$$H(z) = \frac{A^2z^2}{(z - b)^2}, \quad \text{ROC: } |z| > b, \quad \text{and} \quad 0 < b < 1$$

Solution:

From Eqs. 3.20 and 3.23 we obtain

$$H_0(z) = H(z)z^{n-1} = \frac{A^2z^{n+1}}{(z - b)^2}$$

$$h(n) = \operatorname*{Res}_{z=b}\left[H_0(z)\right] = \lim_{z \to b}\frac{d}{dz}\left[(z - b)^2 H_0(z)\right]$$

Finally,

$$h(n) = A^2(n + 1)b^n \quad \blacksquare$$

3.3.2 Inversion by Partial Fractions

In this section we briefly examine the Z-transform inversion by the method of partial fraction expansion. Consider a rational function $H(z)$ given by

$$H(z) = \frac{A_0z^m + A_1z^{m-1} + A_2z^{m-2} + \cdots + A_M}{(z - p_1)(z - p_2)\ldots(z - p_N)} \tag{3.25}$$

Since the coefficients of the polynomial of $H(z)$ are assumed real, all the complex poles and zeros will occur in complex conjugate pairs, where z_i and p_i are the zeros and poles of the rational function (ratio of polynomials).

For distinct poles the function $H(z)$ is expanded in the form

$$H(z) = \frac{C_1}{z - p_1} + \frac{C_2}{z - p_2} + \cdots + \frac{C_{N-1}}{z - p_n} \tag{3.26}$$

Multiplying both sides of $H(z)$ by $(z - p_i)$ results in the residue equation

$$C_i = (z - p_i)H(z)|_{z=p_i} \qquad i = 1, 2, \ldots, N \qquad (3.27)$$

which gives the value of the coefficient for any distinct real or complex pole. For the inverse Z-transform of $H(z)$ with repeated poles see Reference 10.

The following example demonstrates that the computation involved in obtaining the inverse Z-transform using the partial fraction method (Eq. 3.27) is essentially the same as the contour integral method (Eq. 3.21).

Example 3.11

Let us consider Example 3.8 to illustrate the method of obtaining the inverse Z-transform using partial fractions. The transfer function of the system is given by

$$H(z) = \frac{z^2}{(z - 0.5 - j0.5)(z - 0.5 + j0.5)}$$

where for this example the real and imaginary parts of the complex pole are given by $a = b = 0.5$.

Solution:
For convenience we expand the function $H(z)/z$. Then from Eq. 3.26 we obtain

$$\frac{H(z)}{z} = \frac{C_1}{(z - 0.5 - j0.5)} + \frac{C_2}{(z - 0.5 + j0.5)}$$

Equation 3.27 gives

$$C_1 = \frac{z}{(z - 0.5 + j0.5)}\bigg|_{z=0.5+j0.5} = \frac{0.5 + j0.5}{j}$$

$$C_2 = \frac{z}{(z - 0.5 - j0.5)}\bigg|_{z=0.5-j0.5} = \frac{0.5 - j0.5}{-j}$$

The transform $H(z)$ can now be written as

$$H(z) = \frac{(0.5 + j0.5)z}{j(z - 0.5 - j0.5)} - \frac{(0.5 - j0.5)z}{j(z - 0.5 + j0.5)}$$

Finally, the inverse Z-transform can be obtained from pair 3 of Table 3.2, or we can proceed as described in Example 3.8.

$$h(n) = \frac{1}{j}\left[(0.5 + j0.5)^{n+1} - (0.5 - j0.5)^{n+1}\right]$$

which is equivalent to the result obtained in Example 3.8. ∎

3.3.3 Inversion by Division

From Eq. 3.2 the right-sided Z-transform can be expanded into an infinite series in powers of z^{-1} as follows.

$$H(z) = h(0T) + h(T)z^{-1} + \cdots + h(kT)z^{-k} + \cdots + h(nT)z^{-n} + \cdots$$

Thus, the values of $h(nT)$ at any instant of time are the coefficients of z^{-k}. If $H(z)$ is given as a ratio of two polynomials, the coefficients $h(0T), h(T), \ldots, h(nT)$ can be obtained by synthetic division of the numerator by the denominator as follows.

$$H(z) = \frac{A_0 + A_1 z^{-1} + A_2 z^{-2} + \cdots + A_M z^{-M}}{1 + B_1 z^{-1} + B_2 z^{-2} + \cdots + B_N z^{-N}}$$

$$= h(0T) + h(T)z^{-1} + h(2T)z^{-2} + \cdots \tag{3.28}$$

Therefore, the values of the sequence that represent the inverse Z-transform are $h(nT)$ for $n \geq 0$. From Eq. 3.28 it can be seen that the expansion does not result in a closed-form solution. Instead, the inverse Z-transform obtained by this method results in a sequence of numbers. Unless the sequence is simple enough to deduce a closed-form solution, this method does not seem to have much value. However, it does provide insight into the meaning of the Z-transform, and in some cases an open form may be sufficient.[4]

3.3.4 Complex Convolution Theorem

In Section 3.2.6 we demonstrated that the product of two Z-transforms corresponds to the convolution of the sequences. In this section we investigate the Z-transform of the product of two sequences. Let the Z-transform of the product of two sequences be given by

$$U(z) = \sum_{n=0}^{\infty} x(nT)y(nT)z^{-n}$$

Since $X(z)$ is the Z-transform of $x(nT)$, and $Y(z)$ is the Z-transform of $y(nT)$, we have from Eq. 3.19

$$x(nT) = \frac{1}{2\pi j} \oint_C X(v)v^{n-1}\,dv$$

$$y(nT) = \frac{1}{2\pi j} \oint_C Y(z)z^{n-1}\,dz$$

where z and y are complex variables. Then the product of two sequences can be expressed in the form

$$x(nT)y(nT) = \frac{1}{2\pi j} \oint_C \left[\frac{1}{2\pi j} \oint_C X(v)Y(z)v^{n-1} \, dv \right] z^{n-1} \, dz$$

Finally, the Z-transform of $x(nT)y(nT)$ yields (see Problem 3.3.5)

$$U(z) = \frac{1}{2\pi j} \oint_C X(v)Y\left(\frac{z}{v}\right)v^{-1} \, dv \qquad (3.30)$$

Equation 3.30 is the **complex convolution theorem**, where the contour of integration is a closed contour inside the overlap ROC for $X(v)$ and $Y(z/v)$. To see that Eq. 3.30 has the form of a convolution, make the following substitutions:

$$v = \rho e^{j\theta} \quad \text{and,} \quad z = re^{j\phi}$$

where the path of integration is a circle, and noting that $dv = j\rho e^{j\theta} \, d\theta$, Eq. 3.30 then becomes

$$U(re^{j\varnothing}) = \frac{1}{2\pi} \int_0^{2\pi} X(\rho e^{j\theta})Y\left(\frac{r}{\rho}e^{j(\varnothing-\theta)}\right) d\theta \qquad (3.31)$$

which is recognized as the convolution integration of two Fourier transforms. We will return to this equation when we discuss FIR filter design techniques in Chapter 5.

An important special case of Eq. 3.30 is obtained when $x(nT) = y(nT)$ and the path of integration is the unit circle. Making the substitution $z = e^{j\theta}$, where $dz = je^{j\theta} \, d\theta$ and $\theta = \omega T$, we obtain

$$\sum_{n=0}^{\infty} y^2(nT) = \frac{1}{2\pi j} \oint_C Y(z)Y(z^{-1})z^{-1} \, dz = \frac{1}{2\pi} \int_0^{2\pi} |Y(e^{j\theta})|^2 \, d\theta \quad (3.32)$$

Equation 3.32 is a form of Parseval's relation.[1] This relation equates the energy in the signal to the energy in the spectrum. It should be emphasized that $Y(z)$ is assumed to have poles inside the unit circle, then $Y(1/z)$ has poles outside the unit circle. For stable systems the integral of Eq. 3.32 is only evaluated for poles inside the unit circle in the z-plane. For a closed form solution of Eq. 3.32, see Reference 5, pages 168–172.

Since the system response $Y(z)$ to a unit-impulse sequence is the system transfer function $H(z)$, then Eq. 3.32 can be expressed in terms of $H(z)$. The result of expressing Eq. 3.32 as an infinite sum of squares of the unit-impulse response can be applied to obtaining the computational noise variance due to fixed-point finite wordlength arithmetic. A detailed analysis of the effects of finite wordlength arithmetic will be given in Chapter 9.

3.4 ANALYSIS OF DISCRETE-TIME SYSTEMS

Analysis of discrete-time systems centers around the concept of the *system transfer function*. Section 3.2.6 has already shown how the transfer function $H(z)$ relates the response of the system to an excitation $X(z)$, that is,

$$Y(z) = X(z)H(z) \tag{3.33}$$

When the excitation is a unit impulse $\delta(nT)$, then $X(z) = 1$ and $Y(z) = H(z)$. Thus, the inverse Z-transform of the transfer function $H(z)$ yields the unit impulse response of the system, that is,

$$h(nT) = Z^{-1}[H(z)] \tag{3.34}$$

We now proceed with the analysis of linear discrete systems by describing the following system concepts.

1. Digital filter transfer function representation for infinite impulse response (IIR) and finite impulse response (FIR) digital filters.
2. System stability as a function of the z-plane pole locations.
3. Time domain and frequency domain system analysis.

3.4.1 Discrete-Time Transfer Function

It has been shown in Section 2.2 that the general difference equation for a discrete-time system can be expressed by

$$y(nT) = \sum_{i=0}^{M} A_i x(nT - iT) - \sum_{i=1}^{N} B_i y(nT - iT) \tag{3.35}$$

Taking the right-sided Z-transform of both sides of Eq. 3.35 using Eq. 3.13 yields

$$Y(z) = X(z) \sum_{i=0}^{M} A_i z^{-i} - Y(z) \sum_{i=1}^{N} B_i z^{-i}$$

Finally, the system transfer function becomes

$$H(z) = \frac{Y(z)}{X(z)} = \frac{\displaystyle\sum_{i=0}^{M} A_i z^{-i}}{1 + \displaystyle\sum_{i=1}^{N} B_i z^{-i}} \tag{3.36}$$

The excitation of $y(nT)$ by a unit-impulse sequence yields the unit-impulse response $h(nT)$ of the system. Since $H(z)$ is in the form of a rational function with real coefficients (ratio of polynomials) $h(nT)$ is of infinite duration. This system is commonly referred to as an *infinite impulse response* (IIR) digital filter. In the literature, some authors prefer to call IIR filters *autoregressive-moving average* (ARMA) filters. It should be emphasized that Eq. 3.36 represents a causal filter only if $M \leq N$.

A closed-form solution for the IIR unit-impulse response can be obtained by evaluating the contour integral, that is,

$$h(nT) = \frac{1}{2\pi j}\oint_C H(z)z^{n-1}\,dz \qquad (3.37)$$

Since for a stable system $h(nT)$ is a sum of damped sinusoids, $h(nT)$ approaches zero as n approaches infinity. Equation 3.16 showed that the system response $y(nT)$ is the convolution of the input $x(nT)$ and the unit-impulse response $h(nT)$. Thus, Eqs. 3.36 and 3.37 define the linear difference equation that characterizes the IIR digital filter.

By finding the roots of the numerator and denominator polynomials, Eq. 3.36 can be expressed as the ratio of a product of factors representing the poles and zeros of $H(z)$. Thus, $H(z)$ can be put in the form

$$H(z) = \frac{Y(z)}{X(z)} = C_0 \frac{\prod\limits_{i=1}^{M}(z - z_i)}{\prod\limits_{i=1}^{N}(z - p_i)} \qquad (3.38)$$

where z_i are the zeros and p_i are the poles of the system. The factor C_0 is the system *normalization constant*, computed to provide unity gain at the frequency where the magnitude response is maximum. Computation of the normalization (or scaling) constants will be covered in Chapters 4 and 9. Finally, since the coefficients of Eq. 3.36 are real, the poles and zeros can be real or occur in complex conjugate pairs.

Let us now express Eq. 3.38 in the form

$$Y(z) = H(z)X(z) = \frac{N_{hi}(z)}{\prod\limits_{i=1}^{N}(z - p_{hi})} * \frac{N_{xi}(z)}{\prod\limits_{i=1}^{M}(z - p_{xi})} \qquad (3.39)$$

where p_{hi} are the poles of the transfer function and p_{xi} are the poles of the excitation function.[2] A partial fraction expansion of Eq. 3.39 results in the general summations

$$Y(z) = \sum\limits_{i=1}^{N}\frac{a_{hi}z}{z - p_{hi}} + \sum\limits_{i=1}^{M}\frac{a_{xi}z}{z - p_{xi}} \qquad (3.40)$$

Equation 3.40 demonstrates that the system response function depends on both the poles of the transfer function and the input function. Taking the inverse Z-transform of Eq. 3.40 produces terms of the form $a_{hi}(p_{hi})^n$. Using this form, we can infer that if the poles of the transfer function lie inside the unit circle in the z-plane the system is stable. The inverse Z-transform of the first term of Eq. 3.40 yields the *transient response* and the second term will yield the *steady-state response* of the system. System stability will be covered in detail in the following sections.

Setting the coefficients B_i equal to zero, Eq. 3.36 becomes

$$H(z) = \sum_{i=0}^{M} A_i z^{-i} = \sum_{i=0}^{M} h(iT) z^{-i} \tag{3.41}$$

The difference equation for this case is obtained by evaluating the inverse Z-transform of Eq. 3.41

$$y(nT) = \sum_{i=0}^{M} h(iT) x(nT - iT) \tag{3.42}$$

Equation 3.42 defines a *finite impulse response* (FIR) digital filter. These filters are also known as *moving-average* or *transversal digital filters*. The right side of Eq. 3.42 is the convolution sum of the impulse response, which for this case is identical to the filter coefficients A_i. This filter type is widely used in *multirate digital signal processing systems*, discussed in Chapters 5, 7, and 10.

Example 3.12
Find the system transfer function and unit-impulse response of the second-order difference equation given by

$$y(nT) = x(nT) - 0.25 y(nT - 2T)$$

Assume zero initial conditions.

Solution:
From the delay property (Eq. 3.12) we obtain

$$Y(z) = X(z) - 0.25 Y(z) z^{-2}$$

The transfer function, using Eq. 3.33, is

$$H(z) = \frac{1}{1 + 0.25 z^{-2}}$$

The unit-impulse response of the system is obtained from the residue theorem, that is,

$$H_0(z) = \frac{z^{n+1}}{(z + j0.5)(z - j0.5)}$$

$$h(nT) = \operatorname*{Res}_{z=-j0.5}\left[H_0(z)\right] + \operatorname*{Res}_{z=j0.5}\left[H_0(z)\right]$$

$$= \frac{z^{n+1}}{(z - j0.5)}\bigg|_{z=-j0.5} + \frac{z^{n+1}}{(z + j0.5)}\bigg|_{z=j0.5}$$

$$= \frac{1}{j}\left[(j0.5)^{n+1} - (-j0.5)^{n+1}\right]$$

$$= \frac{(0.5)^n}{j2}\left(e^{j(n+1)\pi/2} - e^{-j(n+1)\pi/2}\right) = 0.5^n \sin\left(n + 1\right)\frac{\pi}{2}$$

For $n < 0$, one can show that $h(nT) = 0$.
Hint: Substitute $n = -1, -2, \ldots$ into $H_0(z)$ and solve for the residues. ■

It is left as an exercise for you to show that the closed-form solution of the impulse response is equivalent to the results obtained by exciting the difference equation with a unit impulse sequence. The location of the poles of the system, and the convergence of the impulse response are key characteristics to investigate along with the frequency response of the system.

3.4.2 System Stability

From Eq. 3.21 the unit-impulse response of a filter can be expressed as

$$h(nT) = \sum_{i=1}^{N} \operatorname*{Res}_{z=p_i}\left[H(z)\right] p_i^{n-1}$$

where $p_i = r_i e^{j\phi}$, that is, r_i and ϕ_i specify the magnitude and angular location of the poles inside the unit circle (see Section 3.4.6) and N is the number of poles inside the unit circle. As a result, a necessary and sufficient condition that the foregoing summation converges is $|p_i| < 1$, for $i = 1, 2, \ldots, N$. Therefore, if the system is causal, the system response will remain bounded if the poles of the transfer function are all inside the unit circle. Formally, we can state that a linear time-invariant discrete system with unit-impulse response $h(nT)$ is stable if and only if

$$\sum_{n=0}^{\infty} |h(nT)| < \infty \qquad (3.43)$$

Finally, the *stability criterion* can be evaluated by finding the poles of the transfer function. For example, the transfer function given in Example 3.12 is stable since the poles are inside the unit circle in the z-plane. Notice that since the poles of FIR digital filters are all located at the origin in the z-plane these filters are always stable.

3.4.3 Response of Linear Discrete Systems to Sinusoidal Inputs

In this section we investigate the response of linear systems to discrete signals that can be expressed as linear combinations of sinusoids. According to the convolution theorem, the Z-transform of the response of a linear system is given by the product of the Z-transforms of the transfer function and the excitation function. The time-domain response of a system can then be evaluated using the residue theorem given by Eq. 3.21,

$$y(nT) = Z^{-1}[H(z)X(z)] = \sum_{i=1}^{N} \operatorname*{Res}_{z=p_i} [H(z)X(z)z^{n-1}] \qquad (3.44)$$

It will be shown that for a sinusoidal input with phase θ and amplitude A, Eq. 3.44 can be expressed as the sum of the *transient response* due to the poles of $H(z)$ and the *steady-state response* due to the poles of $X(z)$. We present these terms as follows (see Eq. 3.39):

$$y(nT) = \sum_{i=1}^{N} \operatorname*{Res}_{z=p_i} [H(z)] X(p_i) p_i^{n-1}$$

$$+ \frac{A}{j2} [H(e^{j\omega T}) e^{j(\omega nT + \theta)} - H(e^{-j\omega T}) e^{-j(\omega nT + \theta)}] \qquad (3.45)$$

where

$$X(z) = Z[x(nT)] = Z[A \sin(\omega nT + \theta)] \qquad (3.46)$$

From Eq. 3.45, it is evident that for system stability the poles of $H(z)$ must lie inside the unit circle, that is, for $|p_i| < 1$, $i = 1, 2, \ldots, N$, the first term represents the transient response, which approaches zero as n approaches infinity. The second term is the steady-state response of the system, which can be expressed as

$$y(nT) = AM(\omega) \sin[\omega nT + \phi(\omega) + \theta] \qquad (3.47)$$

where $\phi(\omega)$ is the phase response of the transfer function, and θ is the phase angle of the signal. The derivation of this result is left as an exercise problem (see Problem 3.4.5). The magnitude and phase-response functions of Eq. 3.47 can be expressed by

$$M(\omega) = |H(e^{j\omega T})|, \qquad \phi(\omega) = \tan^{-1} \frac{\operatorname{Im}[H(e^{j\omega T})]}{\operatorname{Re}[H(e^{j\omega T})]} \qquad (3.48)$$

where $H(e^{j\omega T})$ is an even function of ω, whereas the phase-response function is an odd function of ω.

From Eq. 3.47 we can see that the steady-state response of a system to a sinusoidal signal is another sinusoid of the same frequency shifted in phase and scaled in amplitude. Finally, it can be seen that given the system transfer function and a sinusoid of amplitude A and phase ϕ, the response of the system can be obtained by evaluating the steady-state frequency response of the system transfer function, that is,

$$H(e^{j\omega T}) = H(z)\big|_{z=e^{j\omega T}} = M(\omega)e^{j\phi(\omega)} \qquad (3.49)$$

which is the **frequency-response function**. When the coefficients of $H(z)$ are real (i.e., poles and zeros occur in complex conjugate pairs), then the magnitude-squared function is given by

$$M^2(\omega) = \left|H(e^{j\omega T})\right|^2 = H(e^{j\omega T})H(e^{-j\omega T}) = H(z)H(z^{-1})\big|_{z=e^{j\omega T}} \qquad (3.50)$$

It should be noted that the poles and zeros of $H(z^{-1})$ are the reciprocals of those of $H(z)$ with respect to the unit circle in the z-plane. The *magnitude-squared function* can be expressed as a product of M *second-order IIR digital filter sections*, that is,

$$M^2(f_k) = \prod_{i=1}^{M} C_i^2 \frac{A_{0i}^2 + A_{1i}^2 + A_{2i}^2 + 2(A_{0i}A_{1i} + A_{1i}A_{2i})\cos 2\pi f_k T + 2A_{0i}A_{2i}\cos 4\pi f_k T}{B_{0i}^2 + B_{1i}^2 + B_{2i}^2 + 2(B_{0i}B_{1i} + B_{1i}B_{2i})\cos 2\pi f_k T + 2B_{0i}B_{2i}\cos 4\pi f_k T} \qquad (3.51)$$

where the second-order section transfer function is given by

$$H(z) = C_i \frac{A_{0i} + A_{1i}z^{-1} + A_{2i}z^{-2}}{1 + B_{1i}z^{-1} + B_{2i}z^{-2}} \qquad (3.52)$$

In Eq. 3.51 the filter order is $2M$ where M equals the number of second-order sections and C_i is the section normalization factor. The derivation of Eq. 3.51 is left as an exercise problem (see Problem 3.4.6). The product of IIR second-order sections is known as the cascade-form realization of the system function. The frequency-response function for FIR digital filters is covered in Chapter 5.

It is instructive to consider the discrete frequency parameter f_k in Eq. 3.51, that is

$$f_k = \frac{kF}{N} \qquad k = 0, 1, \ldots, \frac{N}{2} \qquad (3.53)$$

where k is the frequency index, N is the number of points between 0 and F, and F/N is the frequency resolution given by

$$\Delta F = \frac{F}{N} \qquad (3.54)$$

Since the poles of $H(e^{j\omega T})$ are either real or occur in complex conjugate pairs, the frequency-response function is uniquely determined in the range $0 \le f_k \le F/2$. The frequency $F/2$ is referred to as the Nyquist or folding frequency.

The following example illustrates the analysis of linear systems by Z-transform methods.

Example 3.13

Find the response of the digital filter with transfer function

$$H(z) = \frac{1}{z^2 + 0.81}$$

to a sinusoidal input given by

$$x(nT) = \sin(\omega T)u(nT)$$

Show that the resultant solution can be put into the form of Eq. 3.45; that is, determine the transient and steady-state responses.

Solution:

From Table 3.2 the Z-transform of $x(nT)$ is given by

$$X(z) = \frac{z \sin \omega T}{z^2 - 2z \cos \omega T + 1}$$

then

$$Y(z) = H(z)X(z) = \frac{z \sin \omega T}{(z - j0.9)(z + j0.9)(z - e^{j\omega T})(z - e^{-j\omega T})}$$

The time-domain response is given by the sum of the residues, that is,

$$y(nT) = \sum_{i=1}^{} \operatorname*{Res}_{i=p_i}[Y_0(z)]$$

$$= \frac{z^n \sin \omega T}{(z + j0.9)(z - e^{j\omega T})(z - e^{-j\omega T})}\bigg|_{z=j0.9}$$

$$+ \frac{z^n \sin \omega T}{(z - j0.9)(z - e^{j\omega T})(z - e^{-j\omega T})}\bigg|_{z=-j0.9}$$

$$+ \frac{z^n \sin \omega T}{(z - j0.9)(z + j0.9)(z - e^{-j\omega T})}\bigg|_{z=e^{j\omega T}}$$

$$+ \frac{z^n \sin \omega T}{(z - j0.9)(z + j0.9)(z - e^{j\omega T})}\bigg|_{z=e^{-j\omega T}}$$

Solving for $y(nT)$ yields

$$y(nT) = \frac{(j0.9)^n \sin \omega T}{j1.8(0.19 - j1.8 \cos \omega T)}$$
$$+ \frac{(-j0.9)^n \sin \omega T}{-j1.8(0.19 + j1.8 \cos \omega T)}$$
$$+ \frac{e^{j\omega nT}}{j2(0.81 + \cos 2\omega T + j \sin 2\omega T)}$$
$$+ \frac{e^{-j\omega nT}}{-j2(0.81 + \cos 2\omega T - j \sin 2\omega T)}$$

Notice that the first two terms of $y(nT)$ represent the *transient response* of the system and correspond to the first term of Eq. 3.45, whereas the last two terms of $y(nT)$ are equivalent to the *steady-state response* of the system and are given by the second term of Eq. 3.45. The response in polar form can be expressed as

$$y(nT) = \frac{(0.9)^n e^{j\pi n/2} \sin \omega T}{j1.8(0.0361 + 3.24 \cos^2 \omega T)^{0.5} e^{j\theta_1}}$$
$$- \frac{(0.9)^n e^{-j\pi n/2} \sin \omega T}{j1.8(0.0361 + 3.24 \cos^2 \omega T)^{0.5} e^{-j\theta_1}}$$
$$+ \frac{e^{j\omega nT}}{j2\left[(0.81 + \cos 2\omega T)^2 + \sin^2 2\omega T\right]^{0.5} e^{j\theta_2}}$$
$$- \frac{e^{-j\omega nT}}{j2\left[(0.81 + \cos 2\omega T)^2 + \sin^2 2\omega T\right]^{0.5} e^{-j\theta_2}}$$

where the phase terms are given by

$$\theta_1 = -\tan^{-1}\left(\frac{1.8}{0.19}\right) \cos \omega T$$

$$\theta_2 = \tan^{-1} \frac{\sin 2\omega T}{0.81 + \cos 2\omega T}$$

Finally, the time-domain system response is given by

$$y(nT) = \left[\frac{(0.9)^n \sin \omega T \sin\left(\frac{\pi n}{2} - \theta_1\right)}{0.9\left[(0.0361 + 3.24 \cos^2 \omega T)\right]^{0.5}} \right.$$
$$\left. + \frac{\sin(\omega nT - \theta_2)}{\left[(0.81 + \cos 2\omega T)^2 + \sin^2 2\omega T\right]^{0.5}} \right] u(nT)$$

The steady-state solution is obtained by observing that in the limit as n approaches infinity the first term approaches zero. The steady-state response can also be obtained directly from Eq. 3.49, that is,

$$H(e^{j\omega T}) = \frac{1}{e^{j2\omega T} + 0.81} = \frac{1}{(\cos 2\omega T + 0.81) + j \sin 2\omega T}$$

$$= \frac{e^{-j\theta_2}}{\left[(\cos 2\omega T + 0.81)^2 + \sin^2 2\omega T\right]^{0.5}}$$

where the phase angle θ_2 was given previously. Therefore, for an input signal $x(nT) = \sin \omega nT$, the steady-state sinusoidal response is given by

$$y(nT) = M(\omega) \sin(\omega nT - \theta_2)$$

Thus, the input signal is multiplied by the magnitude of the transfer function at frequency ω and shifted in phase by the phase angle of the transfer function. ∎

3.4.4 Discrete Systems with Nonzero Initial Conditions

The development of the system transfer function assumed the system with zero initial conditions, that is, all registers set to zero. However, for transient effects nonzero initial conditions must be considered. To determine the effects of nonzero initial conditions, consider Example 3.13 with initial conditions $x(-T) = x(-2T) = y(-2T) = 0$, and $y(-T) = 1$. It is left as an exercise problem to show that the response due to the foregoing initial conditions is given by (see Problem 3.4.8, and Example 3.5).

$$y(nT) = 0.9(0.9)^n \sin\left(\frac{n\pi}{2}\right), \qquad n > 0$$

Therefore, the total response of a linear system to a sinusoidal excitation can be separated into three components: the *transient response under zero initial conditions*, the *transient response due to nonzero initial conditions*, and the *steady-state (forced) response*.

3.4.5 Frequency Axis Scaling for Discrete-Time Systems

For discrete-time systems the frequency variable f_k is normally expressed in the range, $0 \leq f_k \leq F$, where the sampling frequency is indicated as F samples per second (s/s). Figure 3.4, demonstrates several ways of conveniently scaling the frequency axis for discrete-time systems.

In addition to the frequency axis scaling conventions shown in Figure 3.4, the axis can be equivalently scaled in the interval $-F/2 \leq f \leq F/2$,

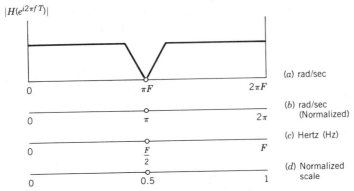

Figure 3.4 Equivalent frequency axis scaling units.

where the frequencies ranging from $-F/2 \leq f < 0$ represent the frequency interval $F/2 < f < F$.

3.4.6 System Frequency Response Using z-plane Pole-Zero Pattern

Section 3.4.3 showed that a discrete-time system can be represented in the frequency domain by an *amplitude response* and a *phase response*. Evaluating Eq. 3.38 on the unit circle, that is, $z = e^{j\omega T}$ yields

$$H(e^{j\omega T}) = M(\omega)e^{j\theta(\omega)} = \frac{C_0 \prod_{i=1}^{N}\left(e^{j\omega T} - z_i\right)}{\prod_{i=1}^{N}\left(e^{j\omega T} - p_i\right)} = \frac{C_0 \prod_{i=1}^{N} M_{zi}e^{j\phi_{zi}}}{\prod_{i=1}^{N} M_{pi}e^{j\phi_{pi}}} \quad (3.55)$$

which can be expressed in the form,

$$M(\omega) = \frac{C_0 \prod_{i=1}^{N} M_{zi}}{\prod_{i=1}^{N} M_{pi}} \quad (3.56)$$

$$\theta(\omega) = \sum_{i=1}^{N} \phi_{zi} - \sum_{i=1}^{N} \phi_{pi} \quad (3.57)$$

Thus, from Eqs. 3.56 and 3.57 the magnitude and phase response of the system can be obtained by determining the magnitude and the phase angles of the vectors at a given frequency on the unit circle. This geometric evaluation of $M(\omega)$ and $\theta(\omega)$ is illustrated in Fig. 3.5.

Points A and C in Fig. 3.5 correspond to frequencies 0 and $F/2$, and one complete revolution of the phasor $e^{j\omega T}$ about the origin corresponds to a frequency revolution of F Hz. As illustrated in the figure, since the transfer function $H(z)$ has real coefficients, the poles occur in complex conjugate pairs,

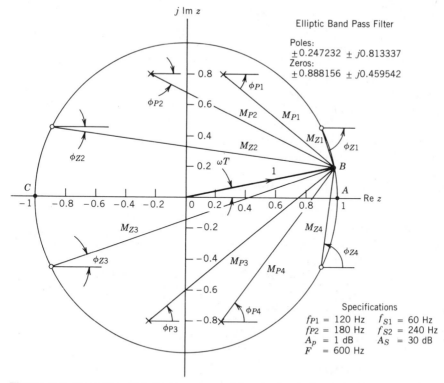

Figure 3.5 Frequency and phase response determination in the z-plane.

which result in the behavior of $H(e^{j\omega T})$ being uniquely determined in the range $0 \le f \le F/2$. Equivalently, the frequency response has the property $H(e^{j\omega T}) = H(e^{-j\omega T})$. Therefore, the magnitude function $M(\omega)$ is an even function of ω, and the phase function $\theta(\omega)$ is an odd function of ω. Note that F is the sampling frequency in units of samples per second (s/s); $T = 1/F$ is the sampling period in units of seconds per sample. The frequency $F/2$ is called the Nyquist or folding frequency since the frequency response in the range $0 \le f \le F/2$ is folded about $F/2$ into the range $F/2 < f \le F$. This result can be illustrated by referring to Fig. 3.5, that is, the frequency response of the filter can be plotted using Eq. 3.56 by rotating the phasor $e^{j\omega T}$ in the interval $0 \le f \le F/2$ since the response is unique in this interval. It can be seen that continuing to rotate the phasor in the interval $F/2 < f \le F$ will reproduce the *mirror image* of the lower-half frequency response.

3.5 REALIZATION OF DIGITAL LINEAR SYSTEMS

The realization of discrete systems is accomplished by converting the system transfer function into an algorithm for implementation via a digital network. The implementation of the algorithm represents the selected form of the difference equation considering computational requirements and finite word-length effects on system performance. The process of determining the optimum realization for the digital network is iterative and is initiated in step 3, *Signal Processing Design*, of the system design methodology. The resources of the signal processor hardware are analyzed for the proposed implementation in step 4, *Resource Analysis*. When an acceptable design is achieved, the iteration is stopped. In this chapter we introduce basic digital linear system realizations and show the use of the Z-transform in the realization analysis. The iterative analysis considering finite wordlength effects and computational requirements will be left to Chapters 9 and 10.

3.5.1 Network Definitions and Operations

In Chapter 2 we introduce three basic operations—addition, multiplication, and unit delay—that are used to implement the systems described here. In addition, sampling rate decrease, sampling rate increase, and modulation operations were defined. We now consider the general solution of the response of the system, $y(nT)$, in terms of the input and the network. In Section 3.4.1 we developed the general Nth-order system transfer function given by Eq. 3.36, that is, the system response function can be written as

$$Y(z) = X(z)H(z) \tag{3.58}$$

Thus, the Z-transform of the response $Y(z)$ is determined as the product of the Z-transform of the excitation function $X(z)$ and the transfer function $H(z)$. We have shown that the response of the system can be efficiently determined using z-domain analysis.

Example 3.14
A second-order difference equation is given by

$$y(n) = x(n) + A_1 x(n-1) + A_2 x(n-2) - B_1 y(n-1) - B_2 y(n-2)$$

Draw a digital network to implement the difference equation and develop the filter transfer function.

Solution:
First we draw the digital network shown in Fig. 3.6a for input $x(n)$ and output $y(n)$ based on the direct implementation of the difference equation. Another form that implements the same difference equation is shown in Fig. 3.6b. These forms are generalized in the following section. Next we take the Z-transform of $y(n)$

$$Y(z) = X(z) + A_1 X(z)z^{-1} + A_2 X(z)z^{-2} - B_1 Y(z)z^{-1} - B_2 Y(z)z^{-2}$$

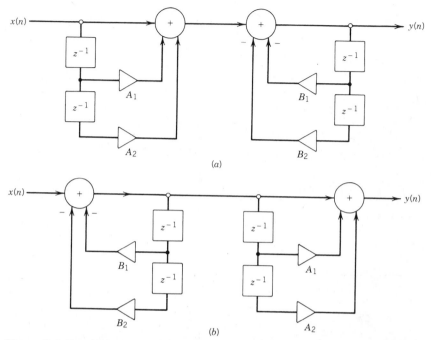

Figure 3.6 Example of IIR digital network forms: (*a*) direct-form realization of second-order IIR filter; (*b*) second-order IIR filter with feed-forward and feedback portions switched (note that adjacent delay elements have common samples).

and the transfer function on the system is given by

$$H(z) = \frac{1 + A_1 z^{-1} + A_2 z^{-2}}{1 + B_1 z^{-1} + B_2 z^{-2}}$$

The frequency response is obtained by substituting $z = e^{j2\pi fT}$ and evaluating $H(z)$ as a function of f. Now we can take the inverse Z-transform of $H(z)$ to obtain the system impulse response. This will be left as a problem. ■

The **transpose** of a network is defined by the following operations:

Reverse the direction of all branches in the signal flow graph.

Interchange the inputs and outputs.

Reverse the roles of all nodes in the flow graph.

• Summing points become branching points.

• Branching points become summing points.

Linear time-invariant branch operations remain unchanged in the process of transposition, where only the direction of the branches are reversed. Finally, the system transfer function is unchanged by transposition.

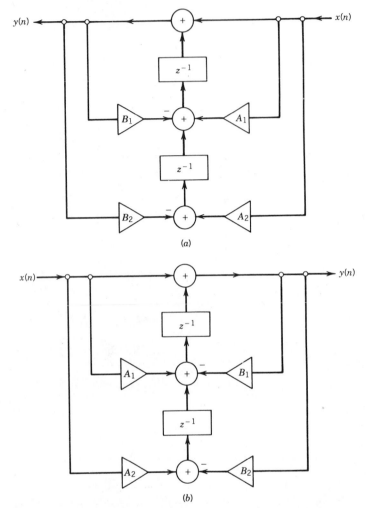

Figure 3.7 Example of IIR network transpose: (a) transpose of Figure 3.6b — second-order IIR filter with common delay elements; (b) Figure 3.7a redrawn to show input on left in standard network form.

Figure 3.7a is the transpose of the network shown in Fig. 3.6b showing the input on the right and the output on the left. The figure is redrawn in Fig. 3.7b in the normal form with the input on the left. It will be shown in Chapter 7 that the transpose operation provides a method for obtaining filter networks that allow more efficient multirate implementations. Also, in Chapter 9 we show that the network implementation effects the errors that result from finite arithmetic effects.

The **dual** of a network performs a complementary function to the original network. For example, modulators and demodulators are dual oper-

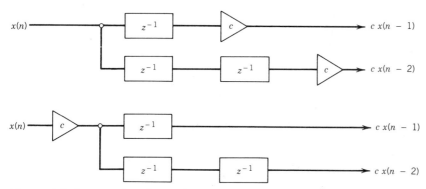

Figure 3.8 Commutation of network operations.

ations. Given a network, its dual is found by transposing the network. Any linear time-invariant or time-varying network has a dual. Interpolators and decimators are dual systems. In Chapter 7 we will see that the dual of a decimator (interpolator) will often result in an efficient form for an interpolator (decimator).

Commutation is the changing of the order of network operations, as shown in Fig. 3.8 without changing the network results. This can provide for computational savings, as shown in the figure by performing the multiplication prior to the branch operation.

3.5.2 IIR Direct-Form Filter Realizations

The structure of Figure 3.6a is referred to as a **direct form 1** realization of a second-order IIR filter. Note that the feed-forward portion is implemented first followed by the feedback portion. Figure 3.6b defines a **direct form 2** realization, which is **canonic** (i.e., the number of storage delays equals the order of the filter when implemented with common delay elements). This structure implements the feedback portion prior to the feed-forward section.

The digital network for implementing the general difference equation of Eq. 3.35 is shown in Figure 3.9a. This network is a **direct form 1** realization of the system characterized by Eq. 3.36. We have used an alternative form of network representation where the multiplications are represented by arrows and summing nodes are given by any node with two or more branch inputs. A **direct form** implementation is obtained by converting the system transfer function into the corresponding difference equation and using our basic operations to implement the difference equation directly. Note that this implementation requires N unit delays for both the feedback portion and the feed-forward portion of the network. Each unit delay represents a requirement for a register to store the data. The multiplications and additions must be performed at the input sample rate. Figure 3.9b is the corresponding canonic **direct form 2** IIR filter implementation.

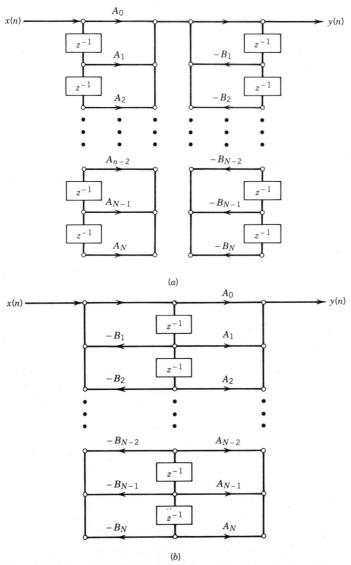

Figure 3.9 General IIR digital network direct forms: (*a*) direct realiza -
tion of an IIR filter; (*b*) direct form 2 canonic filter.

It has been shown,[6] that when N is large the coefficient accuracy
required to implement the network and the arithmetic implementation requirements are high. Leland Jackson[6] also analyzed several alternative form digital
filter implementations and showed that parallel and cascade realizations result
in lower arithmetic requirements than does the direct form.

3.5.3 IIR Cascade and Parallel Filter Realizations

It can be seen from Figure 3.9a that the realization of $H(z)$ was factored into the realization of two transfer functions, that is

$$H(z) = H_D(z)H_N(z) = \left(\frac{1}{\sum\limits_{i=0}^{N} B_i z^{-i}} \right) \left(\sum\limits_{i=0}^{N} A_i z^{-i} \right) \tag{3.59}$$

Equation 3.59 gives $H(z)$ as the product of the denominator portion of the transfer function, $H_D(z)$, with the numerator portion given by $H_N(z)$. This is referred to as a **cascade** factorization of the transfer function into the feed-forward portion resulting from the zeros (the numerator) and the feedback portion resulting from the poles (the denominator). Each of these portions is shown in Figure 3.9.

We can reverse the order (i.e., **commutation rule**) of the cascade operations without changing the transfer function. The direct form is shown in Figure 3.9a with the numerator and denominator terms commuted. Since the unit delays (storage elements) for both portions contain identical data, the delays can be combined, thereby reducing the storage required for implementing the filter. Thus Figure 3.9b requires only one half the number of unit delay operations as required by Figure 3.9a. The resulting digital network is called **direct form 2** or canonic form, that is, the number of delays equal the filter order.

It can be seen from Figure 3.9 that if instant nT is taken to be the present, the present response of an IIR filter is a function of the present and past N values of the input as well as the past values of the response. If we let the coefficients $B_i = 0$, then Eq. 3.60 simplifies to a FIR digital filter. For this case the output samples of the system $y(nT)$ depend only on the present and past input samples.

Figures 3.10a and b illustrate a sixth-order filter realization by three second-order sections in cascade and parallel, respectively. The coefficients for the cascade or parallel realization are obtained by factoring or partial fraction expansion of the direct-form transfer function, $H(z)$.

$$H(z) = \frac{1 + A_1 z^{-1} + A_2 z^{-2} + A_3 z^{-3} + A_4 z^{-4} + A_5 z^{-5} + A_6 z^{-6}}{1 + B_1 z^{-1} + B_2 z^{-2} + B_3 z^{-3} + B_4 z^{-4} + B_5 z^{-5} + B_6 z^{-6}}$$

The cascade expression is obtained by factoring the transfer function, resulting in a product relationship given by

$$H(z) = \prod_{i=1}^{N} H_i(z) \tag{3.60}$$

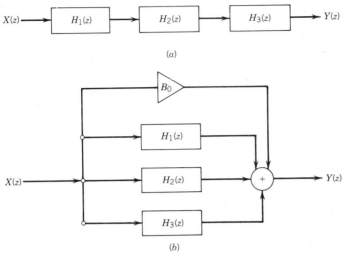

(a)

(b)

Figure 3.10 Cascade and parallel sixth-order IIR realization: (a) cascade sixth-order realization using second-order sections [$H(z)$ is given by Eq. 3.61]; (b) parallel sixth-order realization using second-order sections [$H(z)$ is given by Eq. 3.63].

where the transfer function for the ith stage in the cascade for second-order sections is given by

$$H_i(z) = \frac{1 + A_{1i}z^{-1} + A_{2i}z^{-2}}{1 + B_{1i}z^{-1} + B_{2i}z^{-2}} \tag{3.61}$$

For an odd-order filter, one of the stages will be a first-order section with a real pole and zero. The second-order sections will have complex conjugate or real pole and zero pairs. The cascade sixth-order transfer function is therefore the product of the three second-order sections.

$$H(z) = H_1(z)H_2(z)H_3(z)$$

The parallel form is given by a summation of transfer functions.

$$H(z) = B_0 + \sum_{i=1}^{N} H_i(z) \tag{3.62}$$

where B_0 is a constant and the direct form is divided into N transfer functions. Again, the individual sections are given by $H_i(z)$, which are obtained by partial fraction expansion of the overall transfer function. A second-order section is given by

$$H_i(z) = \frac{A_{0i} + A_{1i}z^{-1}}{1 + B_{1i}z^{-1} + B_{2i}z^{-2}} \tag{3.63}$$

For a first-order section, A_{1i} and B_{2i} are set equal to zero in Eq. 3.63. For the sixth-order transfer function the parallel form transfer function is given by

$$H(z) = H_1(z) + H_2(z) + H_3(z) + B_0$$

We will show in Chapter 9, when finite wordlength effects are considered, that the amount of error introduced will depend on the factorization performed, the order in which the sections are implemented, and the pairing of the numerator and denominator terms.

3.5.4 FIR Filter Realizations

A direct form FIR filter implementation is shown in Figure 3.11. Note that this network is equivalent to the feed-forward portion of Figure 3.9a for the direct form 1 IIR filters. It will be shown that the FIR filters have symmetrical coefficients, and therefore they can be implemented as shown in Figures 5.10 and 5.11. This provides a factor of two savings in the computational requirements. Again as we stated for the IIR filter, the direct form FIR can result in significant accuracy requirements for high-order filters, and therefore cascaded implementations may be desirable, as shown in Figure 3.12.

3.5.5 Other Filter Realizations

Since this work, other forms have been analyzed such as Lattice networks,[7] which require reduced arithmetic implementations. This approach is covered in Chapter 11.

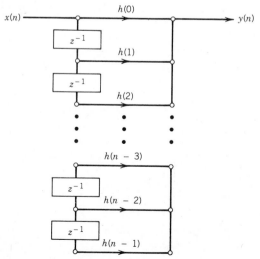

Figure 3.11 Direct realization of a FIR digital filter.

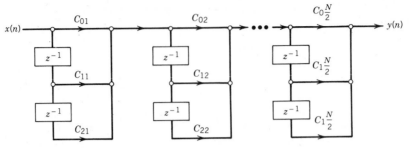

Figure 3.12 Cascade realization of a FIR digital filter.

3.5.6 Network Analysis

As an example of the use of the Z-transform in digital network analysis, consider Figure 3.13. In the digital network signal flow diagram the unit delay of T seconds per sample represented by z^{-1} is equal to the reciprocal of the sampling rate, F.

From Figure 3.13 we obtain the response functions

$$U(z) = X(z) + A_1 U(z)z^{-1} - A_2 U(z)z^{-2}$$

$$Y(z) = A_2 U(z) - A_1 U(z)z^{-1} + U(z)z^{-2}$$

Hence solving for $U(z)$,

$$U(z) = \frac{X(z)}{1 - A_1 z^{-1} + A_2 z^{-2}}$$

Finally, the transfer function of the system is expressed as

$$H(z) = \frac{Y(z)}{X(z)} = \frac{A_2 - A_1 z^{-1} + z^{-2}}{1 - A_1 z^{-1} + A_2 z^{-2}} = \frac{z^{-2}(A_2 z^2 - A_1 z + 1)}{1 - A_1 z^{-1} + A_2 z^{-2}}$$

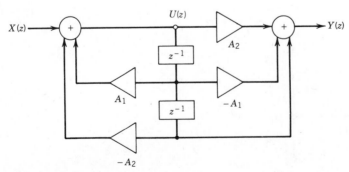

Figure 3.13 Example of digital network.

We can see that $H(z)$ is of the form

$$H(z) = \frac{z^{-2}A(z^{-1})}{A(z)} \tag{3.64}$$

The filter of the form of Eq. 3.64 represents an **all-pass** filter, that is, the magnitude of the frequency response is constant at all frequencies, and only the phase response changes as the pole and zero positions vary. Since the coefficients of $A(z)$ are real, the zeros of $A(z^{-1})$ are the reciprocals of the poles of $A(z)$. Therefore, the necessary condition for $H(z)$ to be an all-pass filter is that the poles and zeros occur in reciprocal complex conjugate pairs,[8,9] that is,

$$H(z) = \frac{\left[z - (1/r)e^{j2\pi fT}\right]\left[z - (1/r)e^{-j2\pi fT}\right]}{(z - re^{j2\pi fT})(z - re^{-j2\pi fT})}$$

Finally, the all-pass second-order section transfer function is given by

$$H(z) = \frac{1}{r^2}\frac{r^2 - (2r\cos 2\pi fT)z^{-1} + z^{-2}}{1 - (2r\cos 2\pi fT)z^{-1} + r^2z^{-2}}$$

The actual implementation can be in the form of a cascade of second-order sections.

3.6 APPLICATION OF THE Z-TRANSFORM TO THE ANALYSIS OF DISCRETE-TIME SYSTEMS

The basic engineering application of the Z-transform is in the field of discrete-time system analysis. In this section, we perform a complete systems analysis of a second-order digital filter in the time domain and frequency domain. It will be shown in Chapter 10 that the analysis techniques can be applied to multichannel systems where we will further extend the analysis to multirate DSP systems. Consider the second-order difference given by

$$y(nT) = a\{x(nT) - x[(n-2)T]\} - ry[(n-2)T]$$

The transfer function of the system is obtained by taking the Z-transform of the difference equation, that is

$$H(z) = \frac{a(z^2 - 1)}{z^2 + r}$$

The magnitude and phase response functions determined from Eq. 3.49 are

$$M^2(\omega) = \frac{2a^2(1 - \cos 2\omega T)}{1 + r^2 + 2r\cos 2\omega T} \tag{3.65}$$

$$\phi(\omega) = \phi_1(\omega) - \phi_2(\omega)$$

where the individual phase terms are given by

$$\phi_1(\omega) = \tan^{-1}\frac{\sin 2\omega T}{\cos 2\omega T - 1}, \qquad \phi_2(\omega) = \tan^{-1}\frac{\sin 2\omega T}{\cos 2\omega T + r} \quad (3.66)$$

You are asked to continue the solution of this example as an exercise in Problem 3.4.3.

References

1. A. V. Oppenheim and W. S. Schafer, *Digital Signal Processing*, Prentice–Hall, Englewood Cliffs, N.J., 1975.

2. A. D. Poularikas and S. Seely, *Signals and Systems*, PWS Publishers, PWS Engineering, Boston, 1985.

3. N. Ahmed and T. Natarajan, *Discrete-Time Signals and Systems*, Reston Publishing, Reston Va., 1983.

4. M. T. Jong, *Methods of Discrete Signal and System Analysis*, McGraw–Hill, New York, 1982.

5. E. I. Jury, *Theory and Application of the Z-transform*, Krieger, Huntington, N.Y., 1963.

6. L. B. Jackson, "An Analysis of Roundoff Noise in Digital Filters," Sc.D. Thesis, Stevens Institute of Technology, Hoboken, N.J., 1969.

7. A. H. Gray Jr., and J. D. Markel, "Digital Lattice and Ladder Filter Synthesis," *IEEE Trans. Audio Electroacoustics*, Vol. AU-21, December 1973.

8. S. A. Tretter, *Introduction to Discrete-Time Signal Processing*, Wiley, New York, 1976.

9. L. R. Rabiner and B. Gold, *Theory and Application of Digital Signal Processing*, Prentice–Hall, Englewood Cliffs, N.J., 1975.

10. R. J. Mayhan, *Discrete-Time and Continuous-Time Linear Systems*, Addison-Wesley, Reading Mass., 1984.

PROBLEMS

3.2.1   Find the Z-transform and the ROC of the following discrete signals.

  a. $u(nT - T)$

  b. $u(nT) - u(nT - T)$

  c. $\delta(nT - T)$

  d. $(0.25)^{nT}u(nT) + (0.5)^{nT}u(nT)$

  e. $\cos(n\omega T)u(nT)$

  f. $r^{nT}\cos(n\omega T)u(nT)$

  g. $(nT)^2 a^{nT}u(nT)$

  h. $Ar^{nT}\cos(n\omega T + \phi)u(nT), 0 < r < 1$

  i. $a^{|n|}, \qquad 0 < |a| < 1$

3.2.2 Prove Eqs. 3.14, and 3.15.

3.2.3 Determine the Z-transform and the ROC of Eq. 3.10.

3.2.4 Find the initial value of the Z-transform pairs 8 and 9 given in Table 3.2 using the initial value theorem. Show that your result agrees with the associated time-domain function.

3.2.5 Determine the Z-transform of the autocorrelation function given by

$$c(nT) = \sum_{k=0}^{\infty} x(kT)x(nT + kT)$$

3.2.6 Find the Z-transform of the sequence given by

$$x(nT) = \begin{cases} 1 & \text{for } 0 \le n \le k \\ 0 & \text{otherwise} \end{cases}$$

Evaluate $X(z)$ around the unit circle, that is, let $z = e^{j\omega T}$

3.2.7 Show that the Z-transforms of the two unit step functions $u(n)$ and $-u(-n - 1)$ have the same Z-transform. However, the corresponding regions of convergence are different. Describe the ROC for both functions.

3.2.8 Show that the Z-transform of the following time sequences are the same; however, the ROC are different.

a. $x(nT) = \begin{cases} e^{anT} & \text{for } nT \ge 0 \\ 0 & \text{for } nT < 0 \end{cases}$

b. $x(nT) = \begin{cases} -e^{anT} & \text{for } nT < 0 \\ 0 & \text{for } nT \ge 0 \end{cases}$

3.2.9 In Example 3.8 we found that the inverse Z-transform of $H(z)$ was given by

$$h(nT) = \frac{r^n}{j2b}\left[(re^{j\omega(n+1)T}) - (re^{-j\omega(n+1)T})\right]$$

a. Determine $Z[e^{j\omega_0 nT}h(nT)]$. Show that the same result could have been obtained by application of Eq. 3.14.

b. As stated in Section 3.2.4, multiplication of a function by a complex exponential results in a rotation of the poles and zeroes in the z-plane, and for $r < 1$ a change of scale along radial lines. From the results of part a, determine whether the resultant filter is real or complex. If the filter has complex coefficients, what modifications can be made to obtain a real filter. Let it be noted that this technique of **pole-zero placement** is the basis of an iterative approach to filter design. In Chapter 4 we will return to this technique of filter design and explore the possibilities of an iterative approach to obtain an arbitrary specified frequency response.

3.3.1 Let $x(n)$ be the input to the cascade of two linear discrete LTI systems as shown in Figure P3.3.1, where $h_1(n) = a^n u(n)$ and $h_2(n) = b^n u(n)$, and

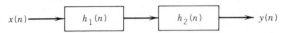

Figure P3.3.1

$0 < a < 1$, $0 < b < 1$. Determine the overall unit impulse response of the system by finding

    a. $h(n) = h_1(n) * h_2(n)$

    b. $h(n) = \sum_{i=1}^{2} \operatorname{Res}_{z=p_i}[H_0(z)]$ where $H(z) = H_1(z)H_2(z)$.

3.3.2 Find the inverse Z-transform of the digital filter represented by the right-sided transfer function

$$H(z) = \frac{1 + A_1 z^{-1} + A_2 z^{-2}}{1 + B_1 z^{-1} + B_2 z^{-2}}$$

for the following conditions.

    a. $A_1 = 2,$ $A_2 = 1$

    b. $A_1 = -2,$ $A_2 = 1$

    c. $A_1 = 0,$ $A_2 = -1$

Deduce the filter type type, that is, state whether the transfer function represents a lowpass, highpass, or bandpass filter for each of these conditions.

3.3.3 The following Z-transforms correspond to causal signals. Determine their inverse Z-transforms using the complex inversion integral.

    a. $X(z) = \dfrac{z^2}{z^2 - z + 0.5}$

    b. $X(z) = \dfrac{1}{(z - 0.25)^4}$

    c. $X(z) = \dfrac{-(z - a)}{a(z - 1/a)}$

3.3.4 Find the inverse Z-transform of the function $X(z)$ obtained in Problem 3.2.1i.

3.3.5 Complete the derivation of Eq. 3.30.

3.3.6 Parseval's relation for discrete systems is given by Eq. 3.32. From the overall transfer function of Problem 3.3.1 determine the infinite sum of the squares of the unit impulse response. It should be emphasized that only the poles inside the unit circle need to be evaluated.

    Given $a = 0.9$ and $b = 0.5$, using a computer program compute the sum

$$\sum_{n=0}^{N} h^2(n)$$

where N is a large number determined by the convergence of the series.

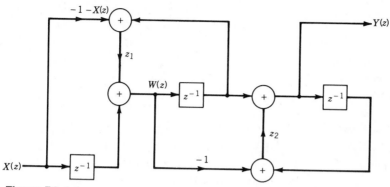

Figure P3.4.1

Compare this result with the closed form solution obtained from Problem 3.3.1.

3.3.7 Given the transfer function of Example 3.8, evaluate Parseval's relation given by Eq. 3.32.

3.4.1 For the realization shown in Figure P3.4.1, determine the following responses.

   a. $W(z)/X(z)$

   b. $Y(z)/X(z)$

3.4.2 Figure P3.4.2 shows a digital network for a second order IIR digital filter determine the transfer functions at the indicated nodal positions, i.e., find the following transfer functions: $U_1(z)/X(z)$, $U_2(z)/X(z)$, $U_3(z)/X(z)$, and $Y(z)/X(z)$.

3.4.3 The analysis of a discrete-time system was considered in Section 3.6. In this problem we continue the analysis as follows:

   a. Show that the magnitude and phase response are given by Eqs. 3.65 and 3.66.

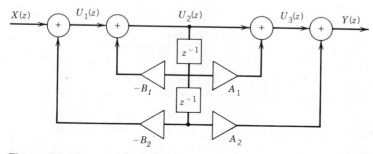

Figure P3.4.2

b. Calculate the frequency response for the following frequencies: $f_k = 0$, $F/8$, $F/4$, $F/2$, $3F/4$, $7F/8$, F, where F is the sampling rate in Hz. Let $r = 0.9$, and sketch the response. What type of filter does the system correspond to, that is, determine if the filter response is lowpass, highpass, bandstop, or bandpass.

c. Determine the value of the normalizing factor as a function of r. Note that the response should be normalized to unity at the peak of the frequency response.

d. Plot the poles and zeros of the transfer function on the unit circle in the z-plane. Show the ROC and determine from the plot if the system is stable and causal; explain. Also show how the magnitude response can be obtained from the pole-zero plot.

e. Determine a closed form expression of the impulse response, and compare this result with the sequence generated by exciting the difference with a unit impulse sequence.

f. Realize the transfer function in a canonic digital network. Check your results by showing the network yields the difference of the system.

3.4.4 An IIR digital filter is defined by the transfer function

$$H(z) = \frac{z}{z^2 + 0.25}$$

a. Determine the difference equation of the filter.

b. Compute the unit impulse response for $n = 0, 1, \ldots, 5$.

c. Determine the unit impulse response of the system using the complex inversion integral. Compute the first five values of $h(nT)$ and compare with part b. Also comment on the stability of the system from the expression of $h(nT)$.

d. Determine the steady state magnitude and phase response functions to a sinusoidal input.

e. Realize the transfer function in a canonic digital network.

3.4.5 Show that the steady response given by Eq. 3.47 can be obtained from the steady state term of Eq. 3.45.

3.4.6 Derive Eq. 3.51.

3.4.7 Equation 3.51 determines the magnitude-squared response of the system transfer function for a cascade of second-order sections. From the definition of the transfer function given in Eq. 3.61, derive the general expression for the phase response function of a cascade realization of second-order sections.

3.4.8 Derive the transient response due to the nonzero initial condition obtained in Section 3.4.4.

4

Infinite Impulse Response Digital Filter Design

Discrete-time signal processing is concerned with the transformation of an input signal to an output signal that has been modified according to some prescribed specification. In this chapter we are concerned with the design of infinite impulse response digital filters, which are a special class of discrete-time linear systems. The digital filter design process concerns procedures for determining filter transfer function coefficients that satisfy specifications in the frequency domain.

4.0 INTRODUCTION

This chapter emphasizes the most widely used technique for the design of linear time-invariant infinite impulse response (IIR) digital filters. The design method considered is based on the bilinear transformation of classic analog filters. The bilinear transformation uniquely maps the entire left half of the s-plane into the interior of the unit circle in the z-plane. The design procedure for converting classic analog filters to digital filters includes the following:

1. Design formulas for the generation of analog Butterworth, Chebyshev, and elliptic lowpass poles and zeros.

2. Frequency band transformation formulas for the conversion of analog lowpass filters to analog highpass, bandpass, and bandstop filters.

3. The bilinear transformation that maps poles in the s-plane to poles in the z-plane.

125

The systematic approach developed in this chapter, for the design of IIR digital filters, readily lends itself to the generation of computer simulation programs. Closed-form equations will be derived for the design of digital lowpass, highpass, bandpass, and bandstop filters. This approach will be demonstrated by using logical flow diagrams to determine the poles, zeros, and coefficients of elliptic digital filters.

It is instructive to consider some of the advantages and disadvantages of digital versus analog filters. A digital filter performs the same filtering function as its analog counterpart, except that it operates with digital samples of the signal together with digital samples of the filter characteristics. Whereas analog filter theory is based on linear differential equations, digital filter theory is based on linear difference equations. Analog filters operate on analog signals and consists of resistors, capacitors, and inductor components. Digital filters operate on discrete sample sequences and consist of adders, multipliers, and delays (shift registers) implemented in digital logic.

Digital filter design requires the determination of the difference equation coefficients such that the specified frequency or time response is obtained. Since this requirement is similar to the approximation problem for analog filters analog-to-digital mapping techniques were developed to satisfy the design specifications.[2] This method is the most popular for the design of standard lowpass, highpass, bandpass, and bandstop filters. The analog approximations used for digital filter design are the well-known Butterworth, Chebyshev (types 1 and 2), and elliptic filters.

As an example of a digital and corresponding analog filter consider the single-pole lowpass filter shown in Figure 4.1 where it can be seen that

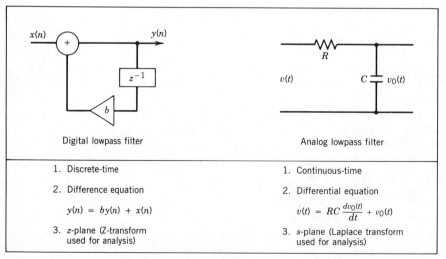

Digital lowpass filter	Analog lowpass filter
1. Discrete-time	1. Continuous-time
2. Difference equation	2. Differential equation
$y(n) = by(n) + x(n)$	$v(t) = RC\dfrac{dv_0(t)}{dt} + v_0(t)$
3. z-plane (Z-transform used for analysis)	3. s-plane (Laplace transform used for analysis)

Figure 4.1 Comparison between a digital lowpass filter and an analog lowpass filter.

the digital filter consists of an adder, multiplier, and a unit delay, whereas the analog filter consists of a resister and capacitor.

Some advantages and disadvantages of digital filters are listed here.

ADVANTAGES

1. Thermal stability: Temperature changes affecting resistors, capacitors, and inductors are eliminated since digital filters are implemented with adders, multipliers, and shift registers.

2. Precision: Filter performance enhancement can be obtained, with respect to accuracy, dynamic range, stability, and frequency response tolerance, by increasing the processor register bit-length. An extensive discussion of these effects will be treated in Chapter 9.

3. Adaptability: Digital filter frequency response can efficiently be changed by reading a new set of filter coefficients from memory into the coefficient registers. The design can be programmable and provide for implementation of any filter order.

4. Multiplexing: If the input bit rate (sampling rate times bits per sample) is significantly below the capability of the digital networks, the digital filter can be multiplexed to use the networks more efficiently. For example, it may be required to process multiple channels of different discrete signals simultaneously. For this case the input samples from the different channels are multiplexed and fed (serially) into the arithmetic unit, which is operating at the appropriate higher rate.

DISADVANTAGES

1. Limited bandwidth: As a result of the sampling process of the analog-to-digital (A/D) converter, the bandwidth for real discrete signals is limited to half the sampling frequency.

2. Finite register length effects: The implementation of a discrete time system in special-purpose hardware can result in performance degradation. This effect results from using data and coefficient registers with a finite number of bits (see Chapter 9).

As discussed in Chapter 3, the system transfer function of a digital filter can be expressed as a rational polynomial in z^{-i}, that is,

$$H(z) = \frac{\displaystyle\sum_{i=0}^{M} A_i z^{-i}}{1 + \displaystyle\sum_{i=1}^{N} B_i z^{-i}} \tag{4.1}$$

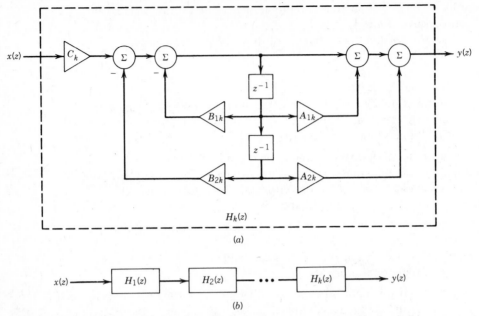

(a)

(b)

Figure 4.2 IIR digital filter second-order section realization: (*a*) canonic form of a second-order section; (*b*) cascade realization of second-order sections.

It will be shown in Chapter 9 that filter coefficient quantization errors are less severe for a cascade realization of Eq. 4.1. It is, therefore, advantageous to factor the Nth-order transfer function into a product of second-order sections given by

$$H(z) = \prod_{k=1}^{N/2} C_k H_k(z) \tag{4.2}$$

where

$$H_k(z) = \frac{1 + A_{1,k} z^{-1} + A_{2,k} z^{-2}}{1 + B_{1,k} z^{-1} + B_{2,k} z^{-2}} \tag{4.3}$$

The transfer function given by Eq. 4.2 represents a cascade of second-order sections, as shown in Figure 4.2. This realization requires the implementation of the difference equation obtained by taking the inverse Z-transform of Eq. 4.3, that is,

$$y_k(nT) = C_k[x(nT) + A_{1,k} x(nT - T) + A_{2,k} x(nT - 2T)]$$
$$- B_{1,k} y(nT - T) - B_{2,k} y(nT - 2T) \tag{4.4}$$

where C_k is the second-order section scaling constant and the section coefficients are given by $A_{i,k}$ and $B_{i,k}$. This realization represents the canonical form digital network since it has the minimum number of multiplier, adder, and delay elements. Note that the number of delays equals the section order.

Since the poles and zeros of the second-order sections are either real or occur in complex conjugate pairs, the coefficients of the transfer function are real. It should be emphasized that the ordering of the numerator and denominator second-order factors are important when considering finite wordlength effects on filter performance. A complete discussion of this problem is presented in Chapter 9.

4.1 DIGITAL FILTER DESIGN

A digital filter is a discrete-time system that is designed to pass the spectral content of the input signal in a specified band of frequencies; that is, the filter transfer function forms a spectral window through which only the desired portion of the input spectrum is allowed to pass. Based on the response of the transfer function, filters are classified into four types: lowpass (LP), highpass (HP), bandpass (BP), and bandstop (BS). The idealized frequency response characteristics, which are indicated by the solid line curves, are shown in Figure 4.3 for $f \geq 0$. The shaded areas in the figures show the tolerance regions of the actual frequency response to be approximated, and the frequencies f_p and f_s represent the **passband** and **stopband edge frequencies**. Finally, the frequency range $f_p \leq f \leq f_s$ is referred to as the **transition band**. Because of the discontinuity at the passband frequencies, these ideal frequency characteristics are not physically realizable. Therefore, a filter approximation function that approaches the ideal responses within the specified tolerance passband

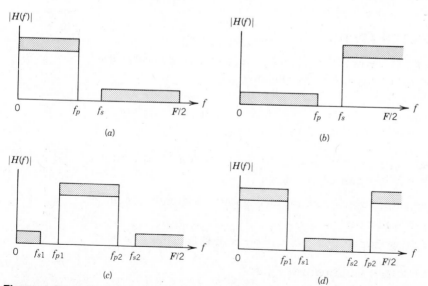

Figure 4.3 Idealized and actual filter magnitude responses demonstrating critical digital frequencies: (*a*) lowpass (LP) filter, (*b*) highpass (HP) filter, (*c*) bandpass (BP) filter, (*d*) bandstop (BS) filter.

and stopband regions must be found. The approximation problem for digital filter design is conceptually no different than that for analog filter design. The approach to analog filter design involves the analytic approximation of the filter specifications by a transfer function, and then developing an analog network that realizes this desired transfer function. The solution to the approximation problem for analog filter design is well developed[2] and includes procedures for the design of Butterworth, Chebyshev, and elliptic filters.

A realizable transfer function is one that characterizes a stable and causal linear network. Stability and causality can be accomplished by requiring that the transfer function be a rational function of s with real coefficients, the poles of the analog filter lie in the left-half s-plane, and the degree of the numerator be equal to or less than that of the denominator polynomial.

The digital filter design problem requires the determination of the difference equation coefficients to meet desired filter characteristics, such as time or frequency-domain response. Since classic analog filter approaches to filter design already existed, transformations were developed that map the analog s-plane poles and zeros into the z-plane to achieve the desired digital filter characteristics. This approach is developed in detail in this chapter; therefore the following sections address the analog filter approximations.

Alternative procedures have been developed for the direct design of digital filters. One method of direct design is the placement of the poles and zeros in the z-plane to meet an arbitrary frequency response specification. This approach is discussed in Section 4.9.1. A practical method for designing IIR digital filters with arbitrary, prescribed magnitude characteristics is discussed in Section 4.9.3.

4.2 ANALOG FILTER APPROXIMATIONS

The approximation problem in the design of IIR digital filters is usually solved by using the following analog filter approximations.

1. Butterworth
2. Chebyshev
3. Elliptic

The derivation and properties of these approximations form the subject of this section. The discussion is organized as follows.

1. Define the amplitude response of the analog lowpass approximation.
2. Derive the filter order equations.
3. Derive the analog lowpass pole and zero-generation equations.

4.2.1 Analog Lowpass Butterworth Filters

Butterworth filters are defined by the property that the magnitude response is maximally flat in the passband. The basic design formulas for the lowpass filter are presented in this section.

The magnitude squared function for an Nth-order analog Butterworth filter is of the form

$$|H(j\omega)|^2 = \frac{1}{1 + \varepsilon^2\left(\dfrac{\omega}{\omega_p}\right)^{2N}} \qquad (4.5)$$

As shown in Figure 4.4, the function is monotonically decreasing, where the maximum response is unity at $\omega = 0$. It should further be noted that the magnitude response approaches the ideal lowpass characteristic of Figure 4.3 for increasing N. Since the first $(2N - 1)$ derivatives of a Maclaurin series expansion of the magnitude function $(|H(j\omega)|)$ evaluated at $\omega = 0$ are zero, the Butterworth response is said to be maximally flat in the passband region. From Figure 4.4 we define the following filter design parameters.

Passband Tolerance: $|\omega| \leq \omega_p$,

$$|H(j\omega)|^2 > \frac{1}{1 + \varepsilon^2} \qquad (4.6)$$

Stopband Tolerance: $|\omega| \geq \omega_s$

$$|H(j\omega)|^2 < \frac{1}{1 + \lambda^2} \qquad (4.7)$$

On examination of Figure 4.4, it can be seen that the ideal lowpass response of Figure 4.3 is obtained when ε approaches zero, ω_s approaches ω_p, and λ approaches infinity.

We can now proceed to determine the order equation given the filter specifications just defined. Letting $\omega = \omega_s$, equating Eq. 4.5 to Eq. 4.7, and

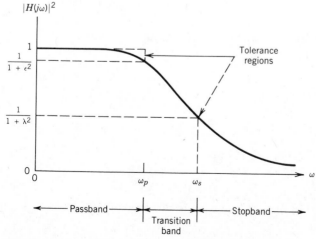

Figure 4.4 Lowpass Butterworth filter magnitude-squared response.

solving for N we obtain

$$N \geq \frac{\log\left(\dfrac{\lambda}{\varepsilon}\right)}{\log\left(\dfrac{\omega_s}{\omega_p}\right)} \qquad (4.8)$$

where from Eqs. 4.6 and 4.7 we obtain

$$\varepsilon = \left(10^{0.1A_p} - 1\right)^{0.5} \qquad (4.9a)$$

$$\lambda = \left(10^{0.1A_s} - 1\right)^{0.5} \qquad (4.9b)$$

where A_p is the maximum passband attenuation in positive dB, and A_s is the minimum stopband attenuation in positive dB. For simplicity of notation, we now define the parameters A and K_0 as follows.

$$A = \frac{\lambda}{\varepsilon} = \left(\frac{10^{0.1A_s} - 1}{10^{0.1A_p} - 1}\right)^{0.5} \qquad (4.10)$$

and

$$K_0 = \frac{\omega_p}{\omega_s} \qquad (4.11)$$

Finally, the order equation for the lowpass Butterworth analog filter is given by

$$N > \frac{\log A}{\log\left(\dfrac{1}{K_0}\right)} \qquad (4.12)$$

The order equations for the highpass, bandpass, and bandstop digital filters will be given in Table 4.4.

The s-plane normalized poles are found by setting the denominator of Eq. 4.5 equal to zero. To normalize the result, let $\omega_p = 1$, and $\varepsilon = 1$, then

$$1 + \omega^{2N} = 0$$

substituting $s = j\omega$, we obtain

$$\left(-s^2\right)^N + 1 = 0$$

Then expressing -1 in polar notation for multiple roots

$$\left(-1\right)^N s^{2N} = e^{j(2k-1)\pi} = -1 \qquad k = 1, 2, \ldots, N$$

Now the kth root for the poles in the left-half s-plane can be expressed by

$$s_k = \sigma_k + j\omega_k = e^{j(2k+N-1)\pi/2N} = je^{j(2k-1)\pi/2N}$$

Finally, the normalized analog poles for the Butterworth lowpass filter can be obtained from

$$s_k = -\sin\left(\frac{2k-1}{2N}\right)\pi + j\cos\left(\frac{2k-1}{2N}\right)\pi \qquad (4.13)$$

$$k = \begin{cases} 1, 2, \ldots, (N+1)/2, & \text{for } N\text{-odd} \\ 1, 2, \ldots, N/2, & \text{for } N\text{-even} \end{cases}$$

In general, the unnormalized poles are given by

$$s_k' = s_k \omega_p^{-1/N} \qquad (4.14)$$

In Eq. 4.14, the poles are uniformly distributed π/N radians apart in the s-plane on a circle of radius $\omega_p \varepsilon^{-1/N}$. The normalized analog poles are given by Eq. 4.13. For this case, $\omega_p = 1$, $A_p \approx -3.01$ dB, and the poles lie on a unit circle spaced π/N apart.

Exercise 4.1

Determine the normalized lowpass Butterworth analog poles for $N = 10$.

Solution:
The solution to this problem is given in Example 4.A.1 of Appendix 4.A. You are requested to check the results presented in the computer listing. ∎

4.2.2 Analog Lowpass Chebyshev Filters

The Chebyshev filter has the property that the magnitude of the frequency response is either equiripple in the passband and monotonic in the stopband (type 1), or monotonic in the passband and equiripple in the stopband (type 2). A detailed discussion of the Chebyshev type 2 filter can be found in either Stearns[6] or Rabiner and Gold.[8]

The type 1 filter magnitude response is shown in Figure 4.5. Notice that the passband ripple at zero frequency starts at the minimum of the peak-to-peak tolerance for N-even, whereas the ripple curve for N-odd stars at the maximum of the ripple tolerance. The analytic form of the squared magnitude function is defined by

$$|H(j\omega)|^2 = \frac{1}{1 + \varepsilon^2 V_N^2\left(\dfrac{\omega}{\omega_p}\right)} \qquad N = 1, 2, \ldots \qquad (4.15)$$

where ε is a design parameter given by Eq. 4.9a and $V_N(x)$ is the Chebyshev

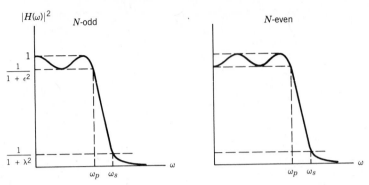

Figure 4.5 Lowpass Chebyshev filter magnitude-squared response.

polynomial of degree N. Closed-form expressions for the Chebyshev polynomial are given by

$$V_N(x) = \cosh(N\cosh^{-1}x), \qquad |x| > 1 \qquad \text{(Stopband)} \qquad (4.16a)$$

and

$$V_N(x) = \cos(N\cos^{-1}x), \qquad |x| < 1 \qquad \text{(Passband)} \qquad (4.16b)$$

It is instructive to consider the Chebyshev polynomial defined by the recursive formula

$$V_N(x) = 2xV_{N-1}(x) - V_{N-2}(x), \qquad N > 1 \qquad (4.17)$$

where $V_0(x) = 1$, and $V_1(x) = x$. The Chebyshev polynomials described by Eq. 4.17 or 4.16b have the following properties.

1. $V_N(x) = -V_N(-x)$ for N-odd
   $V_N(x) = V_N(-x)$ for N-even
   $V_N(0) = (-1)^{N/2}$ for N-even
   $V_N(0) = 0$ for N-odd
   $V_N(1) = 1$ for all N.
   $V_N(-1) = 1$ for N even
   $V_N(-1) = -1$ for N odd
2. $V_N(x)$ oscillates with equal ripple between ± 1 for $|x| \le 1$.
3. For all N $0 \le |V_N(x)| \le 1$ for $0 \le |x| \le 1$
               $|V_N(x)| > 1$ for $|x| \ge 1$
4. $V_N(x)$ is monotonically increasing for $|x| > 1$ for all N (approaches infinity as $2^{N-1}x^N$).

These properties can be demonstrated by plotting Eq. 4.16b for several values of N.

In the passband, $0 \le \omega \le \omega_p$, the magnitude deviations have a peak-to-peak ripple, in decibels, of

$$A_p = 10 \log_{10} (1 + \varepsilon^2) \qquad (4.18)$$

This magnitude deviation is illustrated in Figure 4.5.

To find the required order N of the Chebyshev filter, let $\omega = \omega_s$ and equate Eq. 4.15 to the stopband tolerance as shown in Figure 4.5, that is,

$$\frac{1}{1 + \lambda^2} = \frac{1}{1 + \varepsilon^2 V_N^2 \left(\dfrac{\omega_s}{\omega_p} \right)} \qquad (4.19)$$

Solving Eq. 4.19 for N, using Eq. 4.16a for the Chebyshev polynomial, yields

$$N > \frac{\cosh^{-1} \left(\dfrac{\lambda}{\varepsilon} \right)}{\cosh^{-1} \left(\dfrac{\omega_s}{\omega_p} \right)} = \frac{\cosh^{-1} A}{\cosh^{-1} \left(\dfrac{1}{K_0} \right)} \qquad (4.20)$$

The design parameters ε and λ are given by Eqs. 4.9a and 4.9b. Finally, the parameter K_0 is given by Eq. 4.11. In Eq. 4.20 $\cosh^{-1}(x)$ can be evaluated using the identity

$$\cosh^{-1} x = \ln \left[x + (x^2 - 1)^{0.5} \right] \qquad (4.21)$$

It should be emphasized that Eq. 4.20 is applicable to both type 1 and type 2 Chebyshev lowpass filters.

We now proceed to derive the analog lowpass pole-generating equations for the Chebyshev filter. The poles for the type 1 filter are obtained by setting the denominator of Eq. 4.15 equal to zero and normalizing the passband frequency equal to unity. For the normalized case we obtain

$$1 + \varepsilon^2 V_N^2 (-js) = 0 \qquad (4.22)$$

then the roots s_k are the solutions to

$$V_N(-js) = \pm \frac{j}{\varepsilon} = \cos \left[N \cos^{-1} (-js) \right] \qquad (4.23)$$

Noting that the argument of the cosine term of Eq. 4.23 is complex, we define

$$\cos^{-1} (-js) = x - jy \qquad (4.24)$$

Then using trigonometric identities for the imaginary arguments of the cosine and sine functions, yields

$$\pm \frac{j}{\varepsilon} = \cos(Nx - jNy)$$

$$= \cos(Nx)\cos(jNy) + \sin(Nx)\sin(jNy)$$

$$= \cos(Nx)\cosh(Ny) + j\sin(Nx)\sinh(Ny) \qquad (4.25)$$

Equating the real and imaginary parts of both sides of Eq. 4.25 results in

$$\cos(Nx)\cosh(Ny) = 0 \qquad (4.26a)$$

$$\sin(Nx)\sinh(Ny) = \pm\frac{1}{\varepsilon} \qquad (4.26b)$$

Since $\cosh(Ny) > 0$ for y real, then in order to satisfy Eq. 4.26a we have

$$x = (2k - 1)\frac{\pi}{2N} \qquad k = 1, 2, \ldots, N \qquad (4.27)$$

Using this result and Eq. 4.26b we can now solve for y, where $\sin Nx = \pm 1$, we then obtain

$$y = \pm\frac{1}{N}\sinh^{-1}\left(\frac{1}{\varepsilon}\right) \qquad (4.28)$$

Combining Eqs. 4.24, 4.27, and 4.28, we obtain the normalized left-half plane pole locations given by

$$s_k = j\cos(x - jy) = -\sin x \sinh y + j\cos x \cosh y \qquad (4.29)$$

Eq. 4.29 can be simplified using the identity

$$\sinh^{-1}(\varepsilon^{-1}) = \ln\left(\varepsilon^{-1} + \sqrt{\varepsilon^{-2} + 1}\right)$$

or

$$\alpha = e^{\sinh^{-1}(\varepsilon^{-1})} = \varepsilon^{-1} + \sqrt{\varepsilon^{-2} + 1} \qquad (4.30)$$

We then obtain, using the hyperbolic trigonometric identities for hyperbolic sine and cosine,

$$\sinh y = \sinh\left(\frac{1}{N}\sinh^{-1}\frac{1}{\varepsilon}\right) = \frac{1}{2}\left(\alpha^{1/N} - \alpha^{-1/N}\right) \qquad (4.31a)$$

$$\cosh y = \cosh\left(\frac{1}{N}\sinh^{-1}\frac{1}{\varepsilon}\right) = \frac{1}{2}\left(\alpha^{1/N} + \alpha^{-1/N}\right) \qquad (4.31b)$$

Finally, the desired normalized ($\omega_p = 1$) lowpass analog pole-generating equation is given by

$$s_k = -\sin x \sinh y + j \cos x \cosh y = \sigma_k + j\omega_k \qquad (4.32a)$$

$$k = \begin{cases} 1, 2, \ldots, \dfrac{N+1}{2} & \text{for } N\text{-odd} \\[2mm] 1, 2, \ldots, \dfrac{N}{2} & \text{for } N\text{-even} \end{cases}$$

where the argument x is given by Eq. 4.27. The functions $\sinh y$ and $\cosh y$ are given by Eqs. 4.31a and 4.31b, respectively. The parameter α is obtained from Eq. 4.30. The Chebyshev normalized analog lowpass poles may now be found, for a given ε and N, by using Eqs. 4.27, 4.31, and 4.32. It should be noted that the unnormalized pole equation is given by

$$s_k' = s_k \omega_p \qquad (4.32b)$$

It can be shown that the poles of the Chebyshev transfer function are located on an ellipse in the s-plane. The ellipse is tangent to the circles of radii $\omega_s \cosh y$ and $\omega_s \sinh y$, respectively. The equation of the ellipse is given by

$$\frac{\sigma_k^2}{\sinh^2 y} + \frac{\omega_k^2}{\cosh^2 y} = 1 \qquad (4.33)$$

Exercise 4.2

Determine the normalized Chebyshev analog lowpass poles for $N = 6$.

Solution:

The solution to this example is given in Example 4.A.2 of Appendix 4.A. You are requested to check the results presented in the computer listing. ■

4.2.3 Analog Lowpass Elliptic Filters

Elliptic filters' are characterized by a magnitude response that is equiripple in both the passband and stopband. Elliptic filters are optimum in the sense that for a given order and for a given ripple specification no other filter achieves a faster transition between the passband and stopband, that is, has a narrower transition bandwidth. The magnitude response is shown in Figure 4.6. This approximation is given by the magnitude response function

$$|H(j\omega)|^2 = \frac{1}{1 + \varepsilon^2 F_N^2(\omega)} \qquad (4.34)$$

where $F_N(\omega)$ is a Chebyshev rational function, that is, a ratio of polynomials. It can be shown that the poles and zeros are related to the Jacobi elliptic sine functions. The derivation of the pole and zero-generating equations are quite involved and require a knowledge of elliptic functions. In this text we will not undertake the derivation of these equations; however, the interested reader is referred to an excellent treatment of this subject by Antoniou.[1]

As already stated, the normalized elliptic lowpass s-plane poles and zeros are derived with the aid of elliptic sine functions. Then, given the filter design specifications, that is, passband and stopband frequencies, the passband ripple in decibels, and the minimum attenuation in the stopband in decibels, the poles and zeros can be calculated from the following set of design equations:[1]

$$s_{p,k} = a_{p,k} \pm j b_{p,k}, \qquad \text{and } \sigma_0 \qquad (4.35a)$$

$$= \frac{\sigma_0 V_k \pm j\Omega_k W}{1 + \sigma_0^2 \Omega_k^2} \qquad k = \begin{matrix} 1, 2, \ldots, N/2, & N\text{-even} \\ 1, 2, \ldots, (N-1)/2, & N\text{-odd} \end{matrix} \qquad (4.35b)$$

The real pole, σ_0 given by Eq. 4.35a, applies only for odd-order filters (N-odd). We then define the pole-zero design equations as

$$\sigma_0 = \frac{-2q^{0.25} \sum\limits_{m=0}^{\infty} (-1)^m q^{m(m+1)} \sinh\left[(2m+1)\Lambda\right]}{1 + 2 \sum\limits_{m=1}^{\infty} (-1)^m q^{m^2} \cosh(2m\Lambda)} \qquad (4.36)$$

$$\Omega_k = \frac{2q^{0.25} \sum\limits_{m=0}^{\infty} (-1)^m q^{m(m+1)} \sin\left[(2m+1)\pi u/N\right]}{1 + 2 \sum\limits_{m=1}^{\infty} (-1)^m q^{m^2} \cos(2m\pi u/N)} \qquad (4.37)$$

$$u = \begin{cases} k - 0.5 & \text{for } N\text{-even} \\ k & \text{for } N\text{-odd} \end{cases}$$

$$\Lambda = \frac{1}{2N} ln\left(\frac{10^{0.05 A_p} + 1}{10^{0.05 A_p} - 1}\right) \qquad (4.38)$$

$$W = \left[(1 + K\sigma_0^2)(1 + \sigma_0^2/K)\right]^{0.5} \qquad (4.39)$$

$$V_k = \left[(1 - K\Omega_k^2)(1 - \Omega_k^2/K)\right]^{0.5} \qquad (4.40)$$

$$q = q_0 + 2q_0^5 + 15q_0^9 + 150q_0^{13} \qquad (4.41)$$

$$q_0 = \frac{1 - (1 - K^2)^{0.25}}{2\left[1 + (1 - K^2)^{0.25}\right]} \qquad (4.42)$$

The analog lowpass zeros are given by

$$s_{z,k} = \frac{\pm j}{\Omega_k} \qquad k = \begin{matrix} 1, 2, \ldots, N/2 & \text{for } N\text{-even} \\ 1, 2, \ldots, (N-1)/2 & \text{for } N\text{-odd} \end{matrix} \qquad (4.43)$$

It should be noted that the series in Eqs. 4.36 and 4.37 converge rapidly, and three or four terms are sufficient for most applications.

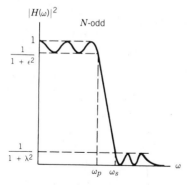
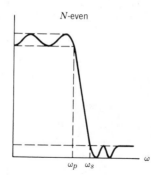

Figure 4.6 Lowpass elliptic filter magnitude-squared response.

The elliptic order equation for the lowpass approximation[1] is given by

$$N \geq \frac{\log(16A)}{\log\left(\dfrac{1}{q}\right)} \tag{4.44}$$

where for the analog lowpass filters, the parameters A, K_0, q, and q_0 are given by Eqs. 4.10, 4.11, 4.41, and 4.42, respectively. The order equations and parameter K for the lowpass, highpass, bandpass, and bandstop digital filters are given in Table 4.4.

Exercise 4.3

Determine the normalized elliptic analog poles and zeros given by Eqs. 4.35 and 4.43. The approximate filter parameters are given as follows.

$$N = 4 \qquad A_p = 0.1737 \text{ dB} \qquad q = 0.0204$$
$$V_1 = 0.8907 \qquad V_2 = 0.3098 \qquad W = 1.2855$$

Solution:
The solution to this problem is given in Example 4.A.3 of Appendix 4.A. You are requested to verify the computer results by solving Eqs. 4.35, 4.36, 4.37, and 4.43. The solution to Eqs. 4.36 and 4.43 can be approximated with the first term of the infinite series. Since the zeros of the elliptic lowpass filter are equal to the reciprocal of Ω_k, the results of the first-term approximation of Eq. 4.43 can be checked against the entries in the computer listing. ∎

It should be noted that from Eqs. 4.39, 4.40, and 4.41 the parameters W, V, and q require the parameter K given in Table 4.4. Section 4.4 will show that the bilinear transformation of the tangent of the critical frequencies is required in order to obtain a specified digital filter. We therefore defer this calculation until the bilinear transformation is covered.

4.3 FREQUENCY BAND TRANSFORMATIONS

The preceding sections have discussed techniques for designing analog normalized (1-Hz passband frequency) lowpass filters. Conversion of normalized analog lowpass filters to unnormalized lowpass, highpass, bandpass, and bandstop filter types is well documented.[2,3] There are two techniques for performing the transformation to obtain the desired digital filter. The first technique, shown in Fig. 4.7a, transforms a normalized analog lowpass filter to an unnormalized analog filter that can be digitized to give the desired digital filter type. The second technique, shown in Fig. 4.7b, first maps the normalized analog lowpass filter from the s-plane to the z-plane and then applies a digital frequency band transformation to give the desired digital filter type. It has been shown[4] that digitizing analog filters using the bilinear transformation produces similar results using either of the frequency transformations. As a result, the first approach will be used in this book for designing digital filters. The frequency transformation equations to be used in developing the closed form expressions for designing IIR digital filters are given in Table 4.1.

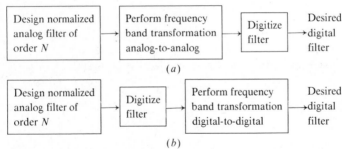

Figure 4.7 Frequency band transformations for lowpass filters: (a) analog-to-analog transformation; (b) digital-to-digital transformation.

TABLE 4.1
ANALOG FREQUENCY BAND TRANSFORMATION

Filter Type	Transformation of Normalized Analog Lowpass Filter to Specified Unnormalized Analog Filter Type
lowpass → lowpass	$s \rightarrow \dfrac{s}{\omega_u}$
lowpass → highpass	$s \rightarrow \dfrac{\omega_u}{s}$
lowpass → bandpass	$s \rightarrow \dfrac{s^2 + \omega_l \omega_u}{s(\omega_u - \omega_l)}$
lowpass → bandstop	$s \rightarrow \dfrac{s(\omega_u - \omega_l)}{s^2 + \omega_l \omega_u}$

ω_l = lower passband frequency.
ω_u = upper passband frequency.

The digital-to-digital transformation equations can be obtained by referring to reference 3. The frequency transformations given in Table 4.1 will be used in Section 4.5 when we consider digital filter design techniques using the bilinear transformation.

4.4 BILINEAR TRANSFORMATION

The bilinear transformation is a mapping, or transformation, from the analog s-plane to the digital z-plane. The conversion maps analog poles and zeros into digital poles and zeros, where each point in the s-plane is mapped to a unique point in the z-plane. The bilinear transformation has the property of preserving the magnitude response of the piecewise constant analog filter with a resultant phase distortion. If both linear phase as well as a specified magnitude response are desired, the designer can resort to an FIR digital filter design (see Chapter 5). It should be noted, however, that linear phase may not be required in all applications; for example, detection processing does not require linear phase, whereas correlation processing, to implement time of arrival, requires a linear phase characteristic.

The bilinear transformation is the simplest unique mapping of the extended s-plane onto the extended z-plane. This transformation, based on a numerical integration technique used to simulate an integrator of an analog filter,[7] is defined by

$$s = \frac{z - 1}{z + 1} \tag{4.45}$$

For $s = j\omega'$ Eq. 4.45 yields

$$j\omega' = \frac{e^{j\omega T} - 1}{e^{j\omega T} + 1} = \frac{e^{j\omega T/2} - e^{-j\omega T/2}}{e^{j\omega T/2} + e^{-j\omega T/2}} \tag{4.46}$$

where ω' is the analog frequency variable and ω is the digital frequency variable. Letting $\omega = \omega_i$, and the sampling frequency $F = 1/T$, Eq. 4.46 is used to obtain a tangent relationship between the analog frequency variable and the digital frequency variable given by the expression

$$\omega_i' = \tan\left(\omega_i T/2\right) = \tan\left(\pi f_i/F\right) \tag{4.47}$$

The frequencies f_i are the critical digital passband and stopband frequencies relating to the filter design specifications. From Eq. 4.47, an analog frequency ω_i' on the ω-axis maps to a frequency f_i on the unit circle. As a result, an analog filter transfer function is related to a corresponding digital filter transfer function by the relation $H(j\omega') = H(e^{j2\pi f_i T})$. Observe that the frequency relationship of Eq. 4.47 is nonlinear. This nonlinearity is a consequence of the bilinear transformation and is referred to as frequency warping. Frequency warping compensation is accomplished by prewarping the critical digital frequencies according to Eq. 4.47, where the critical frequencies will be shifted back to the desired values after application of the bilinear transformation.

TABLE 4.2
BILINEAR TRANSFORMATION MAPPING

s-plane	z-plane
$0 + j0$	$1 + j0$
$\infty + j0$	$-1 + j0$
$0 + j1$	$0 + j1$
$0 - j1$	$0 - j1$
$-1 + j0$	$0 + j0$
$0 + j\infty$	$-1 + j0$
point on real axis	point on real axis
points on $j\omega$-axis from $0 \rightarrow \infty$	points on unit circle from $0 \rightarrow \pi$
points on $j\omega$-axis from $0 \rightarrow -\infty$	points on unit circle from $0 \rightarrow -\pi$
point on any line	point on circle passing through $-1 + j0$
$\sigma < 0$ (stability condition)	$\|z\| < 1$
$\sigma = 0$	$\|z\| = 1$
$\sigma > 0$	$\|z\| > 1$

The properties of the bilinear transformation can easily be shown by solving Eq. 4.45 for z, in terms of s, and letting $s = \sigma + j\omega'$. The results of this inverse mapping are shown in Table 4.2.

$$z = \frac{1 + s}{1 - s} = \frac{1 + \sigma + j\omega'}{1 - \sigma - j\omega'} \tag{4.48}$$

Table 4.2 gives a mapping of some critical points in the s and z-planes.

From Table 4.2, it is apparent that poles in the left-half s-plane ($\delta < 0$) map into the interior of the unit circle in the z-plane ($|z| < 1$). Poles in the right-half s-plane ($\delta > 0$) map into the exterior of the unit circle in the z-plane ($|z| > 1$). Finally, the $j\omega$ ($\sigma = 0$) maps to the boundary of the unit circle ($|z| = 1$). The bilinear transformation "compresses" infinite $j\omega$-axis onto the unit circle. This compression is known as the frequency warping discussed earlier. The nonlinear effects can be demonstrated by plotting Eq. 4.47, where $f = 0$ is mapped to $f' = 0$ and $f_i = F/2$ is mapped to $f' = \infty$. This frequency mapping agrees with Table 4.2; for example, $\infty + j0$ maps to -1, which corresponds to $F/2$. Eq. 4.47 is plotted in Figure 4.8. The frequency warping effect is evident from the plot; that is, the amplitude response of the digital filter is expanded at the lower frequencies and compressed at the higher frequencies in comparison to the analog filter. Consequently, if the critical design frequencies are to be achieved in the digital filter, the analog filter frequencies must be prewarped using Eq. 4.47 before application of the bilinear transformation.

System stability requires that all the poles of $H(z)$ lie within the z-plane unit circle. This requirement is assured since the bilinear transformation maps

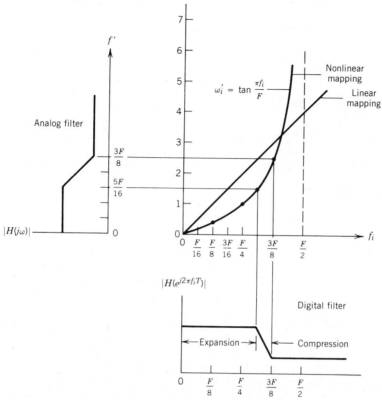

Figure 4.8 Frequency warping of the bilinear transformation.

the open left-half s-plane onto the interior of the unit circle. Therefore, a stable analog filter will yield a stable digital filter. The s-plane to z-plane mapping is shown in Figure 4.9. Notice that the points C and D in the s-plane are mapped on the curve line in the z-plane passing through the point $-1 + j0$.

The design of digital IIR filters from analog prototype filters using the bilinear transformation involves the following design procedure.

1. Specify the critical parameters of the digital filter: passband and stop-band frequencies in Hz, passband and stopband attenuation in dB, and sampling frequency.
2. Prewarp the critical analog frequencies using Eq. 4.47.
3. Apply the appropriate frequency band transformation from Table 4.1.
4. Apply the bilinear transformation of Eq. 4.45.

This design procedure is developed in the following sections, where simple closed-form equations applicable to computer simulations are derived.

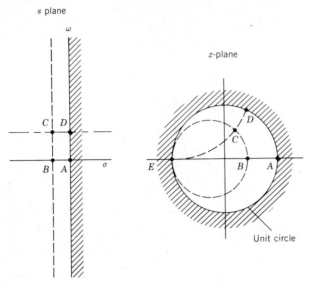

Figure 4.9 *s*-plane to *z*-plane bilinear transformation mapping.

4.5 DIGITAL FILTER DESIGN EQUATIONS

The foregoing material for the design of IIR digital filters using classic analog filter design techniques represents a standard approach to IIR filter design. The balance of this chapter discusses a design methodology for the design of Butterworth, Chebyshev, and elliptic digital filters. This design methodology can readily be used to generate efficient computer programs.

The design equations developed in this section will be based on the bilinear transformation. Design equations for the digital poles and zeros, second-order section filter coefficients, and scaling factors will be derived for all the filter approximations and types.

4.5.1 Lowpass Digital Filter Design Equations

The normalized to unnormalized frequency band transformation for the lowpass-to-lowpass design is given in Table 4.1 as

$$s = \frac{s}{\alpha} \tag{4.49}$$

where we have replaced the frequency scaling variable ω_p by the frequency scaling parameter α. The parameter α for the different filter approximations is given in Table 4.3.

The frequency scaling parameter α for the Butterworth filter was obtained from Eq. 4.14, where the analog frequency ω_p was replaced by Eq. 4.47. The specified attenuation of the response at the critical frequency ω_p is

TABLE 4.3
FREQUENCY SCALING PARAMETER

Filter Approximation	α
Butterworth (lowpass)	$(10^{0.1A_p} - 1)^{-1/2N} \tan(\pi f_p / F)$
Butterworth (highpass)	$(10^{0.1A_p} - 1)^{1/2N} \tan(\pi f_p / F)$
Chebyshev	$\tan(\pi f_p / F)$
Elliptic	$[\tan(\pi f_p / F) \tan(\pi f_s / F)]^{0.5}$

reflected by the factor $\varepsilon^{\pm 1/N}$. Likewise, for the Chebyshev filter the parameter α is obtained from Eq. 4.32. For the elliptic filter the parameter α can be deduced by referring to reference 1.

Consider the transfer function of a single-pole analog filter, where the normalized lowpass poles, $s_{p,k}$ for the Butterworth, Chebyshev, and elliptic filters, can be obtained from Eqs. 4.13, 4.32a, and 4.35, respectively,

$$H(s) = \frac{-s_{p,k}}{s - s_{p,k}} \qquad k = 1, 2, \dots, N \qquad (4.50)$$

Substituting the lowpass-to-lowpass band transformation yields the frequency scaled transfer function

$$H(s) = \frac{-\alpha s_{p,k}}{s - \alpha s_{p,k}} \qquad (4.51)$$

where the parameter α for the lowpass approximation is obtained from Table 4.3. We now find $H(z)$ corresponding to $H(s)$ using the bilinear transform, that is, from Eq. 4.45 we obtain

$$H(z) = H(s)\big|_{s = z - 1/z + 1} = \frac{\dfrac{-\alpha s_{p,k}}{1 - \alpha s_{p,k}}(z + 1)}{z - \dfrac{1 + \alpha s_{p,k}}{1 - \alpha s_{p,k}}} \qquad (4.52)$$

From Eq. 4.52 we can see that the pole locations in the z-plane are a function of the normalized lowpass analog poles scaled by the parameter α, that is,

$$z_{p,k} = \frac{1 + \alpha s_{p,k}}{1 - \alpha s_{p,k}} \qquad (4.53)$$

The general form of the IIR transfer function for the Butterworth and Chebyshev digital lowpass filter is given by

$$H(z) = \frac{N(z)}{D(z)} = \frac{C_0(z + 1)^N}{\displaystyle\prod_{k=1}^{N} (z - z_{p,k})(z - z_{p,k}^*)} \qquad (4.54)$$

where $z_{p,k}^*$ is the complex conjugate of $z_{p,k}$. It is noted that the zeros are all located at $z = -1$ on the unit circle in the z-plane. It should be emphasized that throughout this chapter the notation $z_{p,k}$ and $z_{z,k}$ refer to the z-plane digital poles and zeros for the respective second-order section k.

In general, the zeros of the elliptic digital filter are complex and the numerator polynomial, $N(z)$, is of the form

$$N(z) = C_0 \prod_{k=1}^{N} (z - z_{z,k})(z - z_{z,k}^*) \tag{4.55}$$

and the denominator is as given in Eq. 4.54. For the kth second-order section, let $s_{p,k} = a_{p,k} \pm jb_{p,k}$ represent the lowpass analog poles, and $s_{z,k} = a_{z,k} \pm jb_{z,k}$ represent the lowpass analog zeros. Then the second-order section real coefficients are given by

$$A_{1,k} = -2\operatorname{Re}[z_{z,k}]$$
$$A_{2,k} = \operatorname{Re}[z_{z,k}]^2 + \operatorname{Im}[z_{z,k}]^2 \tag{4.56a}$$
$$B_{1,k} = -2\operatorname{Re}[z_{p,k}]$$
$$B_{2,k} = \operatorname{Re}[z_{p,k}]^2 + \operatorname{Im}[z_{p,k}]^2 \tag{4.56b}$$

where $\operatorname{Re}[z_{i,k}]$ refers to the real part of $z_{i,k}$, and $\operatorname{Im}[z_{i,k}]$ refers to the imaginary part of $z_{i,k}$. The z-plane pole or zero locations are found from

$$z_{i,k} = \frac{1 - \left(a_{i,k}^2 + b_{i,k}^2\right)\alpha^2}{D_{i,k}} + j\frac{2b_{i,k}\alpha}{D_{i,k}} \tag{4.57}$$

where

$$D_{i,k} = 1 + 2|a_{i,k}|\alpha + \left(a_{i,k}^2 + b_{i,k}^2\right)\alpha^2$$

and

$$k = \begin{cases} 1,2,\ldots,N/2 & \text{for } N\text{-even} \\ 1,2,\ldots,(N+1)/2 & \text{for } N\text{-odd} \end{cases}$$

For Eq. 4.57 the index i equals p or z, depending on whether poles or zeros are being calculated. It should be noted that the quantity $a_{i,k}^2 + b_{i,k}^2 = 1$ for the Butterworth approximation.

LOWPASS TRANSFER FUNCTION
NUMERATOR COEFFICIENTS

From Eq. 4.54, the Butterworth and Chebyshev lowpass approximation has double zeros at $z = -1$. Then the second-order section numerator digital coefficients are given by

$$(z + 1)(z + 1) = z^2 + 2z + 1 \tag{4.58a}$$

or equivalently

$$1 + 2z^{-1} + z^{-2} \qquad (4.58b)$$

Thus, for the Butterworth and Chebyshev filters, $A_{1,k} = 2$ and $A_{2,k} = 1$.

For the elliptic lowpass filter we have from Eq. 4.43, the analog lowpass zero given by

$$s_{z,k} = \frac{j}{\Omega_k} = jb_{z,k} \qquad (4.59)$$

Since the analog zero is purely imaginary, Eq. 4.57 reduces to

$$z_{z,k} = \frac{1 - b_{z,k}^2 \alpha^2}{1 + b_{z,k}^2 \alpha^2} + j\frac{2b_{z,k}\alpha}{1 + b_{z,k}^2 \alpha^2} \qquad (4.60)$$

where the coefficients $A_{1,k}$ and $A_{2,k}$ are obtained from Eq. 4.56a. It should be noted, as shown in Table 4.5, that $A_{2,k} = 1$ for all second-order sections.

LOWPASS DIGITAL FILTER: FIRST-ORDER SECTION

For N-odd, the generalized transfer function of Eq. 4.2 can be expressed in terms of normalized first- and second-order sections by

$$H(z) = H_1'(z) \prod_{k=2}^{(N+1)/2} H_k(z) \qquad (4.61)$$

where the unnormalized first-order section transfer function (see Sections 4.6 and 4.7) is given by

$$G_1'(z) = \frac{1 + z^{-1}}{1 + B_{1,k}z^{-1}} \qquad (4.62)$$

Since the analog pole for the first-order section is purely real, Eq. 4.57 reduces to

$$z_{p,k} = \frac{1 - (a_{p,k}\alpha)^2}{1 + 2|a_{p,k}|\alpha + (a_{p,k}\alpha)^2} = \frac{1 - |a_{p,k}|\alpha}{1 + |a_{p,k}|\alpha} \qquad (4.63)$$

The coefficient $B_{1,k}$ can now be determined from

$$B_{1,k} = -z_{p,k} \qquad (4.64)$$

For the Butterworth, Chebyshev, and elliptic approximations $a_{p,k}$ is given by Eqs. 4.14, 4.32, and 4.36, respectively. For the Butterworth first-order section, $a_{p,k} = -1$. The section order index k, in Eq. 4.61, was arbitrarily set equal to unity. In Chapter 9, when considering finite arithmetic effects, the order in which individual sections are cascaded influences the output noise variance due to product roundoff.

TABLE 4.4
DIGITAL FILTER ORDER EQUATIONS

Approximation	Filter Order Equation
Butterworth	$N > \dfrac{\log A}{\log 1/K}$
Chebyshev	$N > \dfrac{\cosh^{-1} A}{\cosh^{-1} 1/K}$
Elliptic	$N > \dfrac{\log 16A^2}{\log 1/q}$

Lowpass	Highpass
$K = K_0 = \dfrac{\tan(\pi f_p/F)}{\tan(\pi f_s/F)}$	$K = 1/K_0$

Bandpass	Bandstop
$K = \begin{cases} K_1, & \text{if } K_C \geq K_B \\ K_2, & \text{if } K_C < K_B \end{cases}$	$K = \begin{cases} 1/K_2, & \text{if } K_C \geq K_B \\ 1/K_1, & \text{if } K_C < K_B \end{cases}$

$$K_A = \tan(\pi f_{p2}/F) - \tan(\pi f_{p1}/F)$$
$$K_B = \tan(\pi f_{p1}/F)\tan(\pi f_{p2}/F)$$
$$K_C = \tan(\pi f_{s1}/F)\tan(\pi f_{s2}/F)$$

$K_1 = \dfrac{K_A\tan(\pi f_{s1}/F)}{K_B - \tan^2(\pi f_{s1}/F)}$	$K_2 = \dfrac{K_A\tan(\pi f_{s2}/F)}{\tan^2(\pi f_{s2}/F) - K_B}$

Elliptic design parameter

$q = q_0 + 2q_0^5 + 15q_0^9 + 150q_0^{13}$	$q_0 = \dfrac{1 - (1 - K^2)^{0.25}}{2\left[1 + (1 - K^2)^{0.25}\right]}$

The parameter A is given by Eq. 4.10.

TABLE 4.5
DIGITAL FILTER NUMERATOR COEFFICIENTS

coef.	Butterworth / Chebyshev Filter Type LP	HP	BP	BS	Elliptic Filter Type LP	HP	BP	BS
$A_{0,k}$	1 (1)	1 (1)	1	1	1 (1)	1 (1)	1 (1)	1
$A_{1,k}$	2 (1)	−2 (−1)	0	$A_{1,k}$	$A_{1,k}$ (1)	$A_{1,k}$ (−1)	$A_{1,k}$ (0)	$A_{1,k}$
$A_{2,k}$	1 (0)	1 (0)	−1	1	1 (0)	1 (0)	1 (−1)	1
Equation				4.99	4.60	4.74	4.91	4.100

The coefficients in parentheses correspond to the first and second-order odd section transfer functions (refer to Eqs. 4.110a-4.110d).

The design equations that determine the order N of the Butterworth, Chebyshev, and elliptic digital filters are summarized in Table 4.4. The critical frequencies of Table 4.4 are the prewarped frequencies as defined by Eq. 4.47 and shown in Figure 4.3.

The second-order section numerator coefficients for the Butterworth, Chebyshev, and elliptic digital filters are in general given by the polynomial

$$1 + A_{1,k}z^{-1} + A_{2,k}z^{-2} \tag{4.65}$$

where k is the second-order section index. The coefficients $A_{1,k}$, which are a function of filter approximation and type, are given in Table 4.5.

Exercise 4.4

From Eqs. 4.56 and 4.57 determine the digital (z-plane) poles, zeros, and second-order section coefficients given in the computer listings of Examples 4.A.1, 4.A.2, and 4.A.3 of Appendix 4.A.

4.5.2 Highpass Digital Filter Design Equations

The analog frequency band transformation for the lowpass-to-highpass design is given by

$$s = \frac{\alpha}{s} \tag{4.66}$$

where the frequency-scaling parameter α is obtained from Table 4.3 for all filter types. The frequency-scaled transfer function of a single pole analog highpass filter is obtained as follows.

$$H(s) = \frac{-s_{i,k}}{s - s_{i,k}}\bigg|_{s=\alpha/s} = \frac{-s_{i,k}s}{\alpha - s_{i,k}s} \tag{4.67}$$

Substituting the bilinear transformation for s, we obtain the digital transfer function in terms of the analog lowpass poles, that is,

$$H(z) = H(s)\big|_{s=z-1/z+1} = \frac{\dfrac{-s_{i,k}}{(\alpha - s_{i,k})}(z - 1)}{z + \dfrac{\alpha + s_{i,k}}{\alpha - s_{i,k}}} \tag{4.68}$$

Therefore, the equation for the location of the z-plane digital poles or zeros is

$$z_{i,k} = \frac{-(\alpha + s_{i,k})}{\alpha - s_{i,k}} \qquad i = p \text{ or } z \tag{4.69}$$

where $s_{i,k}$ are the Butterworth, Chebyshev, or elliptic analog lowpass poles (or zeros).

The transfer function for the Butterworth and Chebyshev digital filter can be expressed as

$$H(z) = \frac{N(z)}{D(z)} = \frac{C_0(z-1)^N}{\displaystyle\prod_{k=1}^{N}(z - z_{p,k})(z - z_{p,k}^*)} \tag{4.70}$$

For the elliptic digital filter, the zeros are complex (see Eq. 4.43), and the numerator polynomial of Eq. 4.70 becomes

$$N(z) = C_0 \prod_{k=1}^{N}(z - z_{z,k})(z - z_{z,k}^*) \tag{4.71}$$

Now let $s_{i,k} = a_{i,k} + jb_{i,k}$ represent the analog lowpass poles or zeros of kth second-order section. Then the z-plane poles and zeros can be expressed by

$$z_{i,k} = \frac{-\left[\alpha^2 - \left(a_{i,k}^2 + b_{i,k}^2\right)\right]}{D_{i,k}} \pm j\frac{2b_{i,k}\alpha}{D_{i,k}} \tag{4.72}$$

$$D_{i,k} = \alpha^2 + 2|a_{i,k}|\alpha + a_{i,k}^2 + b_{i,k}^2$$

$$k = \begin{cases} 1, 2, \ldots, N/2 & \text{for } N\text{-even} \\ 1, 2, \ldots, (N+1)/2 & \text{for } N\text{-odd} \end{cases}$$

$$i = p \text{ or } z$$

The second-order section coefficients can now be determined from Eq. 4.56.

SECOND-ORDER SECTION TRANSFER FUNCTION NUMERATOR COEFFICIENTS

For the Butterworth and Chebyshev highpass approximations, we have a double zero at $z = 1$, that is, the second-order section equation is

$$(z-1)(z-1) = z^2 - 2z + 1 \tag{4.73}$$

Therefore, $A_{1,k} = -2$ and $A_{2,k} = 1$ (see Table 4.5). For the elliptic filter, since the analog lowpass zero is purely imaginary (see Eq. 4.43), Eq. 4.72 reduces to

$$z_{z,k} = \frac{-\left(\alpha^2 - b_{z,k}^2\right)}{\alpha^2 + b_{z,k}^2} + j\frac{2b_{z,k}\alpha}{\alpha^2 + b_{z,k}^2} \tag{4.74}$$

The numerator coefficients $A_{1,k}$ and $A_{2,k}$ can now be computed using Eq. 4.56. Referring to Table 4.5, we can see that $A_{2,k}$ is always equal to unity, and therefore, need not be calculated.

HIGHPASS DIGITAL FILTER: FIRST-ORDER SECTION TRANSFER FUNCTION

For N-odd, a first-order section is required in combination with second-order sections to realize the digital filter with cascade sections. The transfer function for the unnormalized first-order section is given by

$$G_1'(z) = \frac{1 - z^{-1}}{1 + B_{1,k}z^{-1}} \tag{4.75}$$

where

$$B_{1,k} = \frac{\alpha - |a_{p,k}|}{\alpha + |a_{p,k}|} \tag{4.76}$$

For the elliptic highpass filter, $a_{p,k} = \sigma_0$ is given by Eq. 4.36, and for the Butterworth and Chebyshev filters, $a_{p,k}$ are given by Eqs. 4.13 and 4.32, respectively.

A discussion of the transfer functions for all filter approximations and types is given in Section 4.7.

Exercise 4.5

From Eqs. 4.72 and 4.57 determine the digital poles, zeros, and coefficients given in the computer listings of Examples 4.A.4, 4.A.5, and 4.A.6 of Appendix 4.A.

4.5.3 Bandpass Digital Filter Design Equations

The lowpass-to-bandpass transformation, obtained from Table 4.1, is given by

$$s_l = \frac{s_b^2 + \omega_0'}{\beta \Delta\omega' s_b} \tag{4.77}$$

where s_l is the analog lowpass pole and s_b is the analog bandpass pole. Solving Eq. 4.77 for s_b we obtain

$$s_b^2 - \beta \Delta\omega' s_l s_b + \omega_0' = 0 \tag{4.78}$$

where

$$\omega_u' = \tan\left(\pi f_{pu}/F\right)$$
$$\omega_l' = \tan\left(\pi f_{pl}/F\right)$$
$$\Delta\omega' = \left(\omega_u' - \omega_l'\right)$$
$$\omega_0' = \omega_u'\omega_l'$$

where f_{pu} and f_{pl} are the lower and upper passband frequencies. The parameter β, chosen to meet frequency specifications, is obtained from Table 4.6 as a function of filter approximation and type. The geometric center frequency of the passband of the bandpass filter is given by the geometric mean $\sqrt{\omega_0}$. Solving Eq. 4.78 for the analog bandpass poles we obtain

$$s_b = \frac{\beta \Delta \omega'}{2} \left[s_l \pm \left(s_l^2 - \frac{4\omega_0'}{(\beta \Delta \omega')^2} \right)^{0.5} \right] \tag{4.79}$$

Thus we can see that the bandpass transformation maps each analog lowpass pole into two sets of analog bandpass poles. Now let the analog lowpass pole be expressed as

$$s_l = -a_p + jb_p \tag{4.80}$$

where a_p is a positive number. We then obtain

$$s_{bp} = \frac{\beta \Delta \omega'}{2} \left[-|a_p| + jb_p \pm \left(-c_1 - j2|a_p|b_p \right)^{0.5} \right] \tag{4.81}$$

where we define

$$c_1 = b_p^2 + \left[\frac{4\omega_0'}{(\Delta \omega' \beta)^2} \right] - a_p^2 \tag{4.82}$$

It can be shown[2] that the square root of a complex number is given by

$$(-c_1 + jy)^{0.5} = -\left[\frac{\left(c_1^2 + y^2 \right)^{0.5} - c_1}{2} \right]^{0.5}$$
$$+ j \left[\frac{\left(c_1^2 + y^2 \right)^{0.5} + c_1}{2} \right]^{0.5} \tag{4.83}$$

where we define $y = -2a_p b_p$. Substituting Eq. 4.83 into Eq. 4.81 we obtain

$$s_{bp} = \frac{\beta \Delta \omega'}{2} \left\{ -|a_p| + jb_p \pm \left[-\left(\frac{\left(c_1^2 + 4a_p^2 b_p^2 \right)^{0.5} - c_1}{2} \right)^{0.5} \right. \right.$$
$$\left. \left. + j \left(\frac{\left(c_1^2 + 4a_p^2 b_p^2 \right)^{0.5} + c_1}{2} \right)^{0.5} \right] \right\} \tag{4.84}$$

Grouping the real and imaginary parts, and similarly solving for $-a_p - jb_p$, the prewarped analog bandpass poles are expressed as follows.

$$s_{bp1} = \frac{\beta \Delta\omega'}{2}\left[(-|a_p| - c_3) \pm j(b_p + c_4)\right] = a_{bp1} + jb_{bp1}$$

$$s_{bp2} = \frac{\beta \Delta\omega'}{2}\left[(-|a_p| + c_3) \pm j(b_p - c_4)\right] = a_{bp2} + jb_{bp2} \quad (4.85a)$$

where

$$c_2 = \left[c_1^2 + (2a_p b_p)^2\right]^{0.5} \quad (4.85b)$$

$$c_3 = \left(\frac{c_2 - c_1}{2}\right)^{0.5} \quad (4.85c)$$

$$c_4 = \left(\frac{c_2 + c_1}{2}\right)^{0.5} \quad (4.85d)$$

where c_1 is given by Eq. 4.82.

For the elliptic filter it is necessary to consider the analog lowpass zero given by Eq. 4.43. Since the elliptic zero is purely imaginary, we obtain from Eq. 4.85a the analog bandpass zero

$$s_{bz(1,2)} = \pm j\frac{\beta \Delta\omega'}{2}\left[b_{z,k} \pm c_5^{0.5}\right] = \pm jb_{bz(1,2)} \quad (4.86a)$$

where for this case c_5, obtained from Eq. 4.82, is given by

$$c_5 = b_{z,k}^2 + \frac{4\omega_0'}{(\beta \Delta\omega')^2} \quad (4.86b)$$

and $b_{z,k} = 1/\Omega_k$ is the analog lowpass zero given by Eq. 4.43.

For an Nth-order Butterworth or Chebyshev filter the bandpass frequency transformation results in mapping N zeros at the origin in the s-plane. Digitizing the analog bandpass filter using the bilinear transformation places zeros at ± 1 in the z-plane for second-order section realizations. Therefore, as shown in Table 4.5, the second-order section numerator coefficients for Butterworth and Chebyshev digital filters need not be calculated.

As a result of Eq. 4.85, it can be seen that the lowpass-to-bandpass transformation generates N bandpass pole-pairs corresponding to N lowpass analog poles. As a result, the filter order, N, for the bandpass (and bandstop) digital filters indicates the number of second-order sections. For the lowpass and highpass designs, $N/2$ second-order sections are generated. Therefore, the analog bandpass poles and zeros derived from each analog lowpass pole or zero of the elliptic filter are given by

$$s_{bp1} = a_{bp1} \pm jb_{bp1} \quad (4.87a)$$

$$s_{bp2} = a_{bp2} \pm jb_{bp2}$$

$$s_{bz1} = \pm jb_{bz1} \quad (4.87b)$$

$$s_{bz2} = \pm jb_{bz2}$$

The digital bandpass poles can now be obtained from the bilinear transformation.

$$z_{p,k} = \frac{1 + s_{bp,k}}{1 - s_{bp,k}} \qquad k = 1, 2, \ldots, N \qquad (4.88)$$

Substituting Eq. 4.87a into Eq. 4.88 yields a general expression for calculating digital bandpass poles

$$z_{p,k} = \frac{1 - \left(a_{bp,k}^2 + b_{bp,k}^2\right)}{D_{bp,k}} + j\frac{2b_{bp,k}}{D_{bp,k}} \qquad (4.89)$$

where

$$D_{bp,k} = 1 + 2|a_{bp,k}| + a_{bp,k}^2 + b_{bp,k}^2 \qquad (4.90)$$

the second-order section coefficients can now be calculated from Eq. 4.56b.

Since the analog bandpass zeros obtained from Eq. 4.86 are purely imaginary, the elliptic digital bandpass zeros are expressed in the form

$$z_{bz,k} = \frac{1 - b_{bz,k}^2}{1 + b_{bz,k}^2} + j\frac{2b_{bz,k}}{1 + b_{bz,k}^2} \qquad (4.91)$$

The numerator coefficient $A_{1,k}$ is obtained from Eq. 4.56. As shown in Table 4.5, the coefficient $A_{2,k} = 1$; therefore, this coefficient need not be calculated since the elliptic digital zeros lie on the unit circle.

BANDPASS DIGITAL FILTER: SECOND-ORDER SECTION FOR N-ODD

For N-odd, the transfer function results in a second-order section given by

$$H_1'(z) = \frac{1 - z^2}{1 + B_{1,k}z^{-1} + B_{2,k}z^{-2}} \qquad (4.92)$$

where the coefficients $B_{1,k}$ and $B_{2,k}$ are determined by noting that for N-odd the lowpass analog pole is purely real. For the Butterworth, Chebyshev, and elliptic approximations the real analog lowpass poles are obtained from Eqs. 4.13, 4.32, 4.36, respectively, where it is noted that for the Butterworth filter the normalized pole is equal to -1. As an example, consider the elliptic filter approximation where $-a_p = \sigma_0$; we then obtain from Eq. 4.85a the analog bandpass pole, that is,

$$s_{bp,k} = \frac{\beta \Delta\omega'}{2}\left[\sigma_0 \pm jc_1^{0.5}\right] = a_{bp,k} \pm jb_{bp,k} \qquad (4.93)$$

TABLE 4.6
BANDPASS AND BANDSTOP FREQUENCY SPECIFICATION PARAMETER

Filter Approximation / Type	β
Butterworth BPF	$(10^{0.1 A_p} - 1)^{-1/2N}$
Butterworth BSF	$(10^{0.1 A_p} - 1)^{1/2N}$
Chebyshev (BPF/BSF)	1
Elliptic BPF	$K^{-0.5}$
Elliptic BSF	$K^{0.5}$

For the elliptic filter the parameter K is obtained from Table 4.4

where c_1 is obtained from Eq. 4.82, noting that $b_p = 0$. The digital poles and coefficients can now be determined from Eq. 4.89 and 4.56, respectively.

It can be shown that the analog bandpass poles for the elliptic filter Eq. 4.93 can also be expressed by

$$s_{bp,k} = 0.5\beta\sigma_0 \, \Delta\omega' \pm j\Big[\omega_0' - (0.5\beta\sigma_0 \, \Delta\omega')^2 \Big]^{0.5} \qquad (4.94)$$

The digital poles, determined from Eq. 4.89, are given by

$$z_{bp,k} = \frac{1 - \omega_0'}{D_{bp,k}} + j\frac{2\Big[\omega_0' - (0.5\beta\sigma_0 \, \Delta\omega')^2 \Big]^{0.5}}{D_{bp,k}} \qquad (4.95)$$

where $D_{bp,k} = 1 - \beta\Delta\omega'\sigma_0 + \omega_0'$.

Exercise 4.6

  a. Determine the order N of the elliptic filter specified in the computer output listing of Example 4.A.9 of Appendix 4.A.

  b. Determine the s-plane bandpass poles and zeros of Example 4.A.9 by first solving Eqs. 4.82, 4.85b, 4.85c, and 4.85d. The s-plane poles can then be determined from Eq. 4.85a. It should be noted that the calculation for the analog bandpass pole for the second-order odd-section is obtained from Eq. 4.93.

  c. Determine the digital poles, zeros, and second-order section coefficients by solving Eqs. 4.89, 4.91, and 4.56.

4.5.4 Bandstop Digital Filter Design Equations

From Table 4.1 the lowpass-to-bandstop transformation is expressed as

$$s_1 = \frac{\beta \Delta\omega' s_b}{s_b^2 + \omega_0'} \qquad (4.96)$$

Solving for s_b gives

$$s_b = \frac{\beta \, \Delta\omega'}{2}\left[s_1^{-1} \pm \left(s_1^{-2} - \frac{4\omega_0'}{(\beta \, \Delta\omega')^2}\right)^{0.5}\right]$$ (4.97)

It should be observed that Eq. 4.97 is similar to Eq. 4.79 except for the reciprocal analog lowpass pole, s_1. Let the analog lowpass pole be defined by $s_1 = -a_p + jb_p$ then,

$$-a_p' \pm jb_p' = \frac{1}{-a_p + jb_p} = \frac{-|a_p|}{a_p^2 + b_p^2} \pm j\frac{b_p}{a_p^2 + b_p^2}$$ (4.98)

where a_p is a positive number. We can now design digital bandstop filters using Eq. 4.85. Therefore, the design of digital bandstop filters is equivalent to the design of digital bandpass filters by simply using the reciprocal of the analog lowpass poles given by Eq. 4.98 in Eq. 4.85.

BANDSTOP FILTER NUMERATOR COEFFICIENT CALCULATION

The bandstop digital filter requires the computation of the numerator coefficient, $A_{1,k}$ as a function of filter type.

For Butterworth and Chebyshev filters, the analog bandstop zeros are obtained from Eq. 4.97. Since the normalized analog lowpass filter has zeros at infinity in the s-plane, the bandstop frequency transformation maps a pair of zeros at

$$s_{bz} = \pm j(\omega_0')^{0.5}$$ (4.99)

The digital zeros and numerator coefficients are determined from Eqs. 4.91 and 4.56, respectively.

For the elliptic filter, the analog lowpass zero is given by Eq. 4.43; that is, for this case a pair of bandstop analog zeros are located at

$$s_{bz,k} = \pm j\frac{\beta \, \Delta\omega'}{2}\left[\Omega_k \pm \left(\Omega_k^2 + \frac{4\omega_0'}{(\Delta\omega'\beta)^2}\right)^{0.5}\right]$$ (4.100)

Similarly, the digital zeros and numerator coefficients for the elliptic filter are determined from Eqs. 4.91 and 4.56, respectively.

BANDSTOP DIGITAL FILTER: SECOND-ORDER SECTION FOR *N*-ODD

As discussed in Section 4.5.3, the odd lowpass analog poles for all the filter types are purely real.

For the Butterworth and Chebyshev filters we obtain

$$s_{bp} = \frac{\beta \Delta \omega'}{2} \left[-\frac{1}{a_p} + j(c_1)^{0.5} \right] \tag{4.101}$$

where c_1 is obtained from Eq. 4.82 with $b_p = 0$. It is noted that for the Butterworth case $a_p = 1$ for all values of N.

For the elliptic filter the odd section pole is given by

$$s_{bp(1,2)} = \frac{\beta \Delta \omega'}{2} \left[\frac{1}{\sigma_0} \pm \left(\frac{1}{\sigma_0^2} - \frac{4\omega_0'}{(\beta \Delta \omega')^2} \right)^{0.5} \right] \tag{4.102}$$

where the parameters σ_0 and the critical frequencies are given by Eqs. 4.36 and 4.78, respectively. The parameter β is obtained from Table 4.6. Since Eq. 4.102 is purely real, it can be shown that the digital bandstop poles for the odd second-order section are given by

$$z_{p,1} = \frac{1 - |s_{bp,1}|}{1 + |s_{bp,1}|}$$

$$z_{p,2} = \frac{1 - |s_{bp,2}|}{1 + |s_{bp,2}|} \tag{4.103}$$

Finally, the bandstop denominator coefficients are

$$B_{1,k} = -(z_{p,1} + z_{p,2})$$

$$B_{2,k} = z_{p,1} z_{p,2} \tag{4.104}$$

The analog bandstop zeros for all filter types are given by Eq. 4.99. We can now obtain the digital zeros from Eq. 4.89 in the form

$$z_{bz} = \frac{1 - \omega_0'}{1 + \omega_0'} + j \frac{2\omega_0'^{0.5}}{1 + \omega_0'} \tag{4.105}$$

Finally, the numerator coefficients are obtained from Eq. 4.56a.

Exercise 4.7

a. Determine the s-plane bandstop poles and zeros of Example 4.A.12 given in the computer listing of Appendix 4.A by solving Eqs. 4.98, 4.85, and 4.100.

b. Complete the design of the digital filter as described in Section 4.5.4.

4.6 SECOND-ORDER SECTION EXTERNAL NORMALIZATION

A fixed-point implementation often requires normalization (or scaling) of the section output peak frequency response to unity gain (0 dB). This external normalization does not, however, prevent overflows at the internal sum nodes of the filter network. The problem of internal scaling will be addressed in Chapter 9 where dynamic range constraints will be considered.

Consider the cascade realization of second-order sections shown in Figure 4.10, where the unnormalized second-order section has the transfer function

$$G_k(z) = \frac{1 + A_{1,k}z^{-1} + A_{2,k}z^{-2}}{1 + B_{1,k}z^{-1} + B_{2,k}z^{-2}} \tag{4.106}$$

As shown in Figure 4.10, we must determine C_1 such that $\max |G_1(z)| = 1$, C_2 such that $\max |G_1(z)||G_2(z)| = 1$ with C_1 computed, and so on. Formally, the normalization requires the evaluation of the L_∞ norm of $G(e^{j2\pi fT})$, that is, the peak value of $G(e^{j2\pi fT})$ over all f is expressed in the form

$$\|G\|_\infty = \max_{0 \le f \le F} \left| G_k(e^{j2\pi fT}) \right| \tag{4.107}$$

where the normalization coefficients, C_k, are determined as follows:[1]

$$C_1 = \frac{1}{\max\limits_{0 \le f \le F} \left[\left| G_1(e^{j2\pi fT}) \right| \right]}$$

$$C_2 = \frac{1}{C_1 \max\limits_{0 \le f \le F} \left[\left| G_1(e^{j2\pi fT}) \right| \left| G_2(e^{j2\pi fT}) \right| \right]}$$

$$\vdots$$

$$C_M = \frac{1}{\prod\limits_{k=1}^{M-1} C_k \max\limits_{0 \le f \le F} \left[\prod\limits_{k=1}^{M} \left| G_k(e^{j2\pi fT}) \right| \right]} \tag{4.108}$$

For a cascade realization, the normalization factors are a function of section ordering. In Chapter 9 we will show that optimal section ordering minimizes

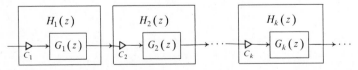

Figure 4.10 Cascade realization of normalized second-order sections.

the output computational noise resulting from fixed point finite wordlength arithmetic.

4.7 GENERALIZED IIR DIGITAL FILTER TRANSFER FUNCTION FORMS

The generalized transfer functions for the Butterworth, Chebyshev, and elliptic digital filters are as follows.

$$H(z) = H_1'(z) \prod_{k=2}^{(N+1)/2} H_k(z) \qquad \text{for LPF/HPF, } N\text{-odd} \qquad (4.109a)$$

$$H(z) = \prod_{k=1}^{(N/2)} H_k(z) \qquad \text{for LPF/HPF, } N\text{-even} \qquad (4.109b)$$

$$H(z) = H_1'(z) \prod_{k=2}^{N} H_k(z) \qquad \text{for BPF/BSF, } N\text{-odd} \qquad (4.109c)$$

$$H(z) = \prod_{k=1}^{N} H_k(z) \qquad \text{for BPF/BSF, } N\text{-even} \qquad (4.109d)$$

where the overall normalization factor is given by

$$C_0 = \prod_{k=1}^{L} C_k$$

$$L = \begin{cases} (N+1)/2 & \text{for LPF/HPF, } N\text{-odd} \\ N/2 & \text{for LPF/HPF, } N\text{-even} \\ N & \text{for BPF/BSF, } N\text{-even or odd} \end{cases} \qquad (4.109e)$$

The unnormalized odd-section transfer functions can be expressed as a function of filter type as follows.

$$G_k'(z) = \frac{1 + z^{-1}}{1 + B_{1,k} z^{-1}} \qquad \text{for LPF} \qquad (4.110a)$$

$$G_k'(z) = \frac{1 - z^{-1}}{1 + B_{1,k} z^{-1}} \qquad \text{for HPF} \qquad (4.110b)$$

$$G_k'(z) = \frac{1 - z^{-2}}{1 + B_{1,k} z^{-1} + B_{2,k} z^{-2}} \qquad \text{for BPF} \qquad (4.110c)$$

$$G_k'(z) = \frac{1 + A_{1,k} z^{-1} + A_{2,k} z^{-2}}{1 + B_{1,k} z^{-1} + B_{2,k} z^{-2}} \qquad \text{for BSF} \qquad (4.110d)$$

For convenience, the filter section index, k, was set equal to unity in Eq. 4.109. From Eqs. 4.110 $G_k'(z)$ is given by a first-order section for lowpass and highpass filters and by second-order sections for bandpass and bandstop filters (see Table 4.5).

4.8 IIR DIGITAL FILTER DESIGN PROCEDURE

This section systematically summarizes the IIR digital filter design procedure developed in this chapter. An elliptic filter design flow diagram readily lends itself as a guide for generating efficient computer programs is presented. The utility of the flow diagram is demonstrated with design examples of the four elliptic filter types. Solutions to the examples were generated with a computer program written for the IBM personal computer. It can be seen on examination of the filter design output listings given in Appendix 4.A that the results are consistent with the flow diagram given in Fig. 4.11. We have concentrated most of the effort on elliptic filter design since elliptic filters provide the

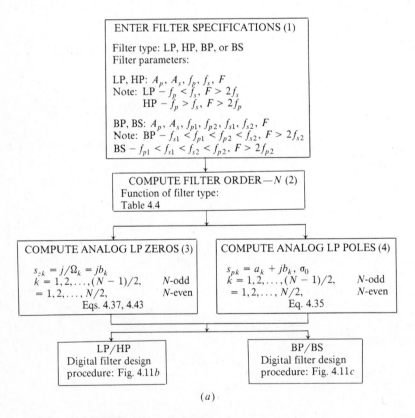

Note: Numbers in parentheses refer to computation unit.

Figure 4.11 IIR elliptic digital filter design flow diagram; (*a*) computations common to all elliptic digital filter types; (*b*) lowpass and highpass computation units; (*c*) bandpass and bandstop computation units.

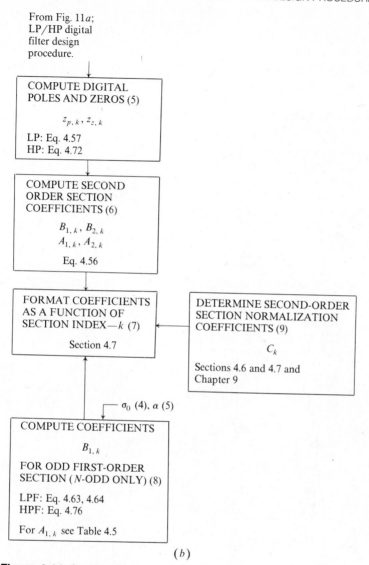

(b)

Figure 4.11 Continued

steepest transition rate from passband to the stopband and therefore are considered the most efficient of the analog filter approximations. However, you can see that the flow diagrams for the Butterworth and Chebyshev can similarly be constructed. Detailed examples for all the filter approximations and types have been included in Appendix 4.A to aid you in understanding the design process and in developing efficient computer programs.

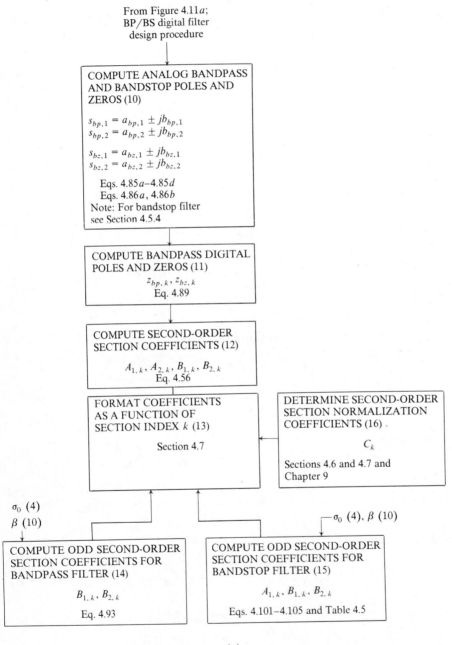

From Figure 4.11a;
BP/BS digital filter
design procedure

COMPUTE ANALOG BANDPASS
AND BANDSTOP POLES AND
ZEROS (10)

$s_{bp,1} = a_{bp,1} \pm jb_{bp,1}$
$s_{bp,2} = a_{bp,2} \pm jb_{bp,2}$

$s_{bz,1} = a_{bz,1} \pm jb_{bz,1}$
$s_{bz,2} = a_{bz,2} \pm jb_{bz,2}$

Eqs. 4.85a–4.85d
Eqs. 4.86a, 4.86b
Note: For bandstop filter
see Section 4.5.4

COMPUTE BANDPASS DIGITAL
POLES AND ZEROS (11)
$z_{bp,k}, z_{bz,k}$
Eq. 4.89

COMPUTE SECOND-ORDER
SECTION COEFFICIENTS (12)

$A_{1,k}, A_{2,k}, B_{1,k}, B_{2,k}$
Eq. 4.56

FORMAT COEFFICIENTS
AS A FUNCTION OF
SECTION INDEX k (13)

Section 4.7

DETERMINE SECOND-ORDER
SECTION NORMALIZATION
COEFFICIENTS (16)

C_k

Sections 4.6 and 4.7 and
Chapter 9

σ_0 (4)
β (10)

σ_0 (4), β (10)

COMPUTE ODD SECOND-ORDER
SECTION COEFFICIENTS FOR
BANDPASS FILTER (14)

$B_{1,k}, B_{2,k}$

Eq. 4.93

COMPUTE ODD SECOND-ORDER
SECTION COEFFICIENTS FOR
BANDSTOP FILTER (15)

$A_{1,k}, B_{1,k}, B_{2,k}$

Eqs. 4.101–4.105 and Table 4.5

(c)

4.9 OTHER TECHNIQUES FOR DESIGNING IIR DIGITAL FILTERS

The previous sections discussed a discrete-time approximation for the design of IIR digital filters based on continuous-time filters. This section investigates other methods for designing transfer functions with desired frequency responses.

4.9.1 IIR Digital Filter Design by Pole-Zero Placement

Section 3.4.6 described a method of evaluating the frequency response of a digital filter using the z-plane pole-zero pattern. We now show that digital filters can be designed by a trial-and-error placement of poles and zeros in the z-plane. The transfer function of any discrete-time system can be expressed in the form

$$H(e^{j\omega T}) = \frac{C_0(z - z_1)(z - z_1^*) \cdots (z - z_M)(z - z_M^*)}{(z - p_1)(z - p_1^*) \cdots (z - p_N)(z - p_N^*)} \quad (4.111)$$

where $z_1, z_2, \ldots$ are the zeros of the filter and $p_1, p_2, \ldots$ are the poles. As discussed in Section 4.1, a realizable transfer function must be both stable and causal. This requires that the following conditions be met.

1. System stability: All poles must lie inside the unit circle in the z-plane.
2. Real filter design: The poles and zeros must occur in complex conjugate pairs or lie on the real axis. Also, the placement angle $\theta_0 = 2\pi f_0/F$ is restricted by the frequency range $0 \le f_0 \le F/2$, that is, $0 \le \theta_0 \le \pi$.
3. Arbitrary zero placement: Zero placement is arbitrary; however, zeros placed on the unit circle will completely attenuate the frequency response at the placement angle.
4. Causality: Requires the number of zeros be less than or equal to the number of poles.

One method of obtaining a desired frequency response is by properly rotating the poles and zeros of a digital filter designed by conventional means, for example, bilinear transformation. Since the poles and zeros of real filters occur in complex conjugate pairs, that is,

$$(z - p_i)(z - p_i^*) = 0$$
$$(z - z_i)(z - z_i^*) = 0 \quad (4.112)$$

frequency response shaping can be obtained by modifying Eq. 4.111 to reflect both magnitude scaling and critical frequency shifting. This pole-zero placement technique is obtained by multiplying the current set of poles and zeros by $re^{\pm j2\pi fT}$. This mapping factor results in an angular rotation and scaling of the complex conjugate poles and zeros. Eq. 4.111 is now expressed as a ratio of modified complex conjugate pole-zero pairs as follows.

$$\frac{\left(z - rp_i e^{j2\pi f_0 T}\right)\left(z - rp_i^* e^{-j2\pi f_0 T}\right)}{\left(z - rz_i e^{j2\pi f_0 T}\right)\left(z - rz_i^* e^{-j2\pi f_0 T}\right)} \tag{4.113}$$

where $0 < |r| < 1$ is the magnitude of the pole or zero and $2\pi f_0 T$ is the rotation angle. The trial-and-error method of pole-zero placement can efficiently be accomplished through iterations performed on a digital computer and with a graphics display of the frequency response.

4.9.2 Digital Filters with Complex Coefficients

Digital filters with complex coefficients find applications in digital processing of analytic signals.[7] Analytic signals are complex signals whose spectral density is one sided and whose real part is the original real signal and imaginary part is in-phase quadrature (90°) with that of the real signal. This section describes a design approach for complex digital filters, which can be used to filter complex signals. For example, complex filters are indicated when the digital signal processing system includes a discrete Fourier transformation.

Complex filters can be developed from real filters by rotating the poles and zeros inside the unit circle of the z-plane $\omega_p T$ degrees, without producing a conjugate response at $-\omega_p$. That is, the complex digital filter is designed to reduce the negative frequencies to zero along the bottom half of the unit circle.

An application of complex digital filters is to produce a bandpass spectrum that is symmetric about a specified center frequency (see Fig. 4.12). This can be accomplished by designing bandpass filters that are linear translations of lowpass filters, thus avoiding the distortions inherent in the lowpass-to-bandpass transformation. The transfer function of a complex digital filter can be expressed by

$$H(z) = \frac{C_0\left(z - z_1 e^{j\omega_0 T}\right)\left(z - z_2 e^{j\omega_0 T}\right) \cdots \left(z - z_M e^{j\omega_0 T}\right)}{\left(z - p_1 e^{j\omega_0 T}\right)\left(z - p_2 e^{j\omega_0 T}\right) \cdots \left(z - p_N e^{j\omega_0 T}\right)} \tag{4.114}$$

where only the poles and zeros associated with the upper half of the unit circle are included. The complex coefficient $e^{j\omega T}$ accounts for the rotation of the poles and zeros along circles centered at the origin. As shown in Figure 4.12, the design of complex filters starts with real filter characteristics. Shifting the center frequency results in the magnitude response centered at f_0, without a conjugate filter at $-f_0$ as would be the result of a conventional lowpass to bandpass transformation.

Consider a single pole complex filter, with a zero at the origin given by[9]

$$H(z) = \frac{Y(z)}{X(z)} = \frac{1}{1 - p_0 z^{-1}} \tag{4.115}$$

where $Y(z)$, $X(z)$, and p_0 are complex.

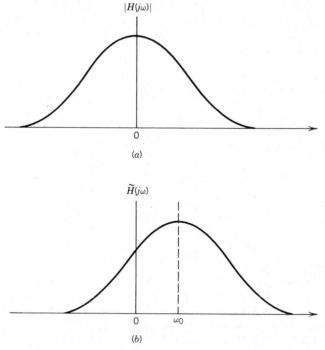

Figure 4.12 Complex digital filters from real filters: (*a*) real digital filter response; (*b*) shifted response of complex digital filter.

Expanding Eq. 4.115, we obtain for the real and imaginary responses the system equations given by[9]

$$\mathrm{Re}\,[Y(z)] = \mathrm{Re}\,[X(z)] + \mathrm{Re}\,[p_0]\,\mathrm{Re}\,[Y(z)]\,z^{-1} - \mathrm{Im}\,[p_0]\,\mathrm{Im}\,[Y(z)]\,z^{-1}$$

$$\mathrm{Im}\,[Y(z)] = \mathrm{Im}\,[X(z)] + \mathrm{Re}\,[p_0]\,\mathrm{Im}\,[Y(z)]\,z^{-1} + \mathrm{Im}\,[p_0]\,\mathrm{Re}\,[Y(z)]\,z^{-1}$$

$$(4.116)$$

The difference equations can be determined by taking the inverse Z-transforms of Eqs. 4.116, that is,

$$\mathrm{Re}\,[y(n)] = \mathrm{Re}\,[x(nT)] + \mathrm{Re}\,[p_0]\,\mathrm{Re}\,[y(n-1)T]$$

$$- \mathrm{Im}\,[p_0]\,\mathrm{Im}\,[y(n-1)T]$$

$$\mathrm{Im}\,[y(n)] = \mathrm{Im}\,[x(nT)] + \mathrm{Re}\,[p_0]\,\mathrm{Im}\,[y(n-1)T]$$

$$+ \mathrm{Im}\,[p_0]\,\mathrm{Re}\,[y(n-1)T] \qquad (4.117)$$

To find the system transfer functions, we take the real and imaginary parts of

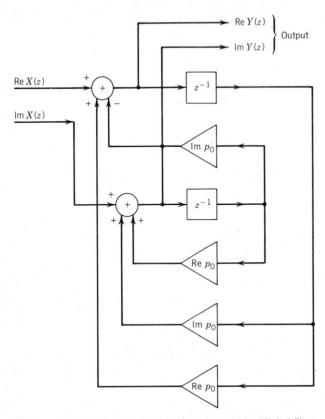

Figure 4.13 Realization of a single-pole complex digital filter.

Eq. 4.115. This gives

$$\text{Re}\,[H(z)] = \frac{1 - \text{Re}\,[p_0]z^{-1}}{1 - 2\,\text{Re}\,[p_0]z^{-1} + |p_0|^2 z^{-2}}$$

$$\text{Im}\,[H(z)] = \frac{\text{Im}\,[p_0]z^{-1}}{1 - 2\,\text{Re}\,[p_0]z^{-1} + |p_0|^2 z^{-2}} \qquad (4.118)$$

From Eqs. 4.116 or 4.117 the realization can now be obtained as shown in Figure 4.13.

4.9.3 Computer-aided Design of IIR Digital Filters

A practical method for designing IIR digital filters with arbitrary, prescribed magnitude characteristics is described by Steiglitz.[11] The method uses the Fletcher-Powell optimization algorithm[12] to minimize a square error criterion in the frequency domain. A strategy is described whereby stability and minimum-phase constraints[7] are observed, while still using the unconstrained optimization algorithm. Cascade second-order sections are used, so that the resultant filters can be realized accurately and simply. The minimum squared error criterion was extended to a higher-order error criteria by Deczky.[13] Computer programs for these design techniques can be found in references 11 and 14.

The foregoing techniques are used to design IIR filters with arbitrary frequency response. It is also possible to design an IIR filter whose impulse response $h(n)$ approximates a desired impulse response $g(n)$. That is, the problem is to find the set of filter coefficients A_i, B_i such that the error

$$\sum_{n=0}^{P-1} [g(n) - h(n)]^2 \omega(n) \qquad (4.119)$$

is minimized over all possible choices of A_i, B_i. The reader who is interested in these techniques is referred to references 8, 15, and 16.

APPENDIX 4.A
COMPUTER OUTPUT LISTINGS
FOR IIR DIGITAL FILTER
DESIGN EXAMPLES

This Appendix presents the computer listings of the design examples presented in this chapter and additional examples that can be used for further study. The design examples not only provide a guide in understanding the digital filter design process, but also will aid you in developing efficient computer programs. Included in this section are the following design examples.

Example Number	Approximation / Type
4.A.1	Butterworth LPF
4.A.2	Chebyshev LPF
4.A.3	Elliptic LPF
4.A.4	Butterworth HPF
4.A.5	Chebyshev HPF
4.A.6	Elliptic HPF
4.A.7	Butterworth BPF
4.A.8	Chebyshev BPF
4.A.9	Elliptic BPF
4.A.10	Butterworth BSF
4.A.11	Chebyshev BSF
4.A.12	Elliptic BSF

EXAMPLE 4.A.1
BUTTERWORTH LOWPASS (ORDER = 10)
LOWPASS FILTER FORMAT

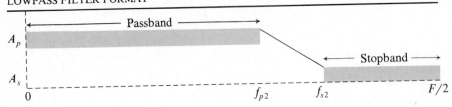

Transition band from f_{p2} to f_{s2}
Sampling rate $F = 2500$ Samples/Second

Passband

f_{p1} =   0.00 HZ

f_{p2} = 500.00 HZ

Ripple =   0.1737 dB

Stopband

f_{s1} =   0.00 HZ

f_{s2} = 750.00 HZ

Ripple =   40.00 dB

s-PLANE POLES AND ZEROS → LOWPASS

No.	Zeros Real	Zeros Imaginary	Poles Real	Poles Imaginary
1	0.0000000	0.0000000	−0.1564345	0.9876885
2	0.0000000	0.0000000	−0.4539906	0.8910065
3	0.0000000	0.0000000	−0.7071068	0.7071067
4	0.0000000	0.0000000	−0.8910066	0.4539905
5	0.0000000	0.0000000	−0.9876884	0.1564344

z-PLANE POLES and ZEROS → LOWPASS

No.	Zeros Real	Zeros Imaginary	Poles Real	Poles Imaginary
1	−1.0000000	0.0000000	+0.1370099	±0.8447676
2	−1.0000000	0.0000000	+0.1092149	±0.6074742
3	−1.0000000	0.0000000	+0.0931414	±0.4111430
4	−1.0000000	0.0000000	+0.0841441	±0.2384710
5	−1.0000000	0.0000000	+0.0800774	±0.0782001

Double zeros for all but odd-order stage.

SECOND-ORDER SECTION COEFFICIENTS

Stage	Numerator Coefficients A_1	A_2	Denominator Coefficients B_1	B_2
1	2.0000000	1.0000000	-0.2740197	0.7324039
2	2.0000000	1.0000000	-0.2184297	0.3809528
3	2.0000000	1.0000000	-0.1862828	0.1777139
4	2.0000000	1.0000000	-0.1682881	0.0639486
5	2.0000000	1.0000000	-0.1601547	0.0125276

IIR NORMALIZING FACTOR: $C_0 = 0.00125$

STAGE 1 NORMALIZING FACTOR: $C_1 = 0.11360$

STAGE 2 NORMALIZING FACTOR: $C_2 = 0.25799$

STAGE 3 NORMALIZING FACTOR: $C_3 = 0.32522$

STAGE 4 NORMALIZING FACTOR: $C_4 = 0.36908$

STAGE 5 NORMALIZING FACTOR: $C_5 = 0.35624$

FREQUENCY RESPONSE OUTPUT

Frequency	Magnitude	Frequency	Magnitude
0.000	-0.000	20.000	-0.000
40.000	-0.000	60.000	-0.000
80.000	-0.000	100.000	-0.000
120.000	-0.000	140.000	-0.000
160.000	-0.000	180.000	-0.000
200.000	-0.000	220.000	0.000
240.000	-0.000	260.000	-0.000
280.000	-0.000	300.000	-0.000
320.000	-0.000	340.000	-0.000
360.000	-0.000	380.000	-0.000
400.000	-0.001	420.000	-0.002
440.000	-0.007	460.000	-0.021
480.000	-0.061	500.000	-0.174
530.000	-0.776	560.000	-2.816
590.000	-7.138	620.000	-12.988
650.000	-19.367	680.000	-25.911
710.000	-32.556	740.000	-39.317
770.000	-46.233	800.000	-53.352
830.000	-60.724	860.000	-68.398
890.000	-76.336	920.000	-83.899
950.000	-88.637	980.000	-89.842
1010.000	-89.987	1040.000	-89.999
1070.000	-90.000	1100.000	-90.000
1130.000	-90.000	1160.000	-90.000
1190.000	-90.000	1220.000	-90.000

EXAMPLE 4.A.2
CHEBYSHEV LOWPASS (ORDER = 6)
LOWPASS FILTER FORMAT

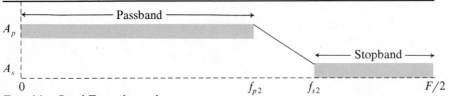

Transition Band From f_{p2} to f_{s2}
Sampling rate $F = 2500$ Samples/Second

Passband		Stopband	
f_{p1}	= 0.00 HZ	f_{s1}	= 0.00 HZ
f_{p2}	= 500.00 HZ	f_{s2}	= 750.00 HZ
Ripple =	0.1737 dB	Ripple =	40.00 dB

s-PLANE POLES AND ZEROS → LOWPASS

	Zeros		Poles	
No.	Real	Imaginary	Real	Imaginary
1	0.0000000	0.0000000	−0.1017847	1.0379357
2	0.0000000	0.0000000	−0.2780809	0.7598216
3	0.0000000	0.0000000	−0.3798656	0.2781140

z-PLANE POLES AND ZEROS → LOWPASS

	Zeros		Poles	
No.	Real	Imaginary	Real	Imaginary
1	−1.0000000	0.0000000	+0.2472978	±0.8758249
2	−1.0000000	0.0000000	+0.3740355	±0.6310338
3	−1.0000000	0.0000000	+0.5290679	±0.2421385

Double zeros for all but odd-order stage.

SECOND-ORDER SECTION COEFFICIENTS

	Numerator Coefficients		Denominator Coefficients	
Stage	A_1	A_2	B_1	B_2
1	2.0000000	1.0000000	−0.4945955	0.8282254
2	2.0000000	1.0000000	−0.7480710	0.5381062
3	2.0000000	1.0000000	−1.0581359	0.3385440

THE IIR NORMALIZING FACTOR: $C_0 = 0.00453$

STAGE 1 NORMALIZING FACTOR: $C_1 = 0.06794$

STAGE 2 NORMALIZING FACTOR: $C_2 = 0.19751$

STAGE 3 NORMALIZING FACTOR: $C_3 = 0.33721$

FREQUENCY RESPONSE OUTPUT

Frequency	Magnitude	Frequency	Magnitude
0.000	− 0.173	20.000	− 0.166
40.000	− 0.145	60.000	− 0.115
80.000	− 0.079	100.000	− 0.044
120.000	− 0.016	140.000	− 0.001
160.000	− 0.003	180.000	− 0.022
200.000	− 0.055	220.000	− 0.096
240.000	− 0.136	260.000	− 0.164
280.000	− 0.173	300.000	− 0.157
320.000	− 0.117	340.000	− 0.064
360.000	− 0.017	380.000	0.000
400.000	− 0.031	420.000	− 0.103
440.000	− 0.167	460.000	− 0.147
480.000	− 0.023	500.000	− 0.173
530.000	− 3.574	560.000	− 10.327
590.000	− 17.009	620.000	− 23.055
650.000	− 28.616	680.000	− 33.856
710.000	− 38.889	740.000	− 43.802
770.000	− 48.664	800.000	− 53.535
830.000	− 58.468	860.000	− 63.515
890.000	− 68.723	920.000	− 74.118
950.000	− 79.615	980.000	− 84.726
1010.000	− 88.247	1040.000	− 89.633
1070.000	− 89.946	1100.000	− 89.994
1130.000	− 90.000	1160.000	− 90.000
1190.000	− 90.000	1220.000	− 90.000

EXAMPLE 4.A.3
ELLIPTIC LOWPASS (ORDER = 4)
LOWPASS FILTER FORMAT

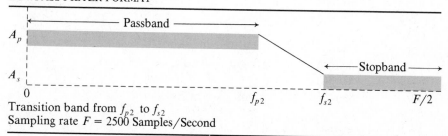

Transition band from f_{p2} to f_{s2}
Sampling rate $F = 2500$ Samples/Second

Passband	Stopband
f_{p1} = 0.00 HZ	f_{s1} = 0.00 HZ
f_{p2} = 500.00 HZ	f_{s2} = 750.00 HZ
Ripple = 0.1737 dB	Ripple = 40.00 dB

S-PLANE POLES AND ZEROS → LOWPASS

No.	Zeros		Poles	
	Real	Imaginary	Real	Imaginary
1	0.0000000	3.3607185	− 0.4304955	0.3741359
2	0.0000000	1.4730365	− 0.1375649	0.7839348

z-PLANE POLES AND ZEROS → LOWPASS

No.	Zeros		Poles	
	Real	Imaginary	Real	Imaginary
1	− 0.8373247	± 0.5467058	+ 0.3086023	± 0.3422556
2	− 0.3690528	± 0.9294085	+ 0.1920363	± 0.8214731

Double zeros for all but odd-order stage.

SECOND-ORDER SECTION COEFFICIENTS

Stage	Numerator Coefficients		Denominator Coefficients	
	A_1	A_2	B_1	B_2
1	1.6746495	1.0000000	− 0.6172046	0.2123743
2	0.7381057	1.0000000	− 0.3840726	0.7116959

IIR NORMALIZING FACTOR: $C_0 = 0.07699$

STAGE 1 NORMALIZING FACTOR: $C_1 = 0.16197$

STAGE 2 NORMALIZING FACTOR: $C_2 = 0.47533$

FREQUENCY RESPONSE OUTPUT

Frequency	Magnitude	Frequency	Magnitude
0.000	− 0.173	20.000	− 0.170
40.000	− 0.162	60.000	− 0.149
80.000	− 0.131	100.000	− 0.110
120.000	− 0.088	140.000	− 0.064
160.000	− 0.042	180.000	− 0.023
200.000	− 0.008	220.000	− 0.000
240.000	− 0.000	260.000	− 0.009
280.000	− 0.027	300.000	− 0.053
320.000	− 0.086	340.000	− 0.120
360.000	− 0.150	380.000	− 0.170
400.000	− 0.170	420.000	− 0.144
440.000	− 0.090	460.000	− 0.024
480.000	− 0.004	500.000	− 0.173
530.000	− 1.359	560.000	− 4.283
590.000	− 8.507	620.000	− 13.250
650.000	− 18.247	680.000	− 23.611
710.000	− 29.757	740.000	− 37.875
770.000	− 56.963	800.000	− 46.562

Frequency	Magnitude	Frequency	Magnitude
830.000	− 42.486	860.000	− 41.678
890.000	− 42.291	920.000	− 43.952
950.000	− 46.780	980.000	− 51.614
1010.000	− 63.875	1040.000	− 57.937
1070.000	− 50.401	1100.000	− 46.815
1130.000	− 44.638	1160.000	− 43.231
1190.000	− 42.338	1220.000	− 41.839

EXAMPLE 4.A.4
BUTTERWORTH HIGHPASS (ORDER = 10)
HIGHPASS FILTER FORMAT

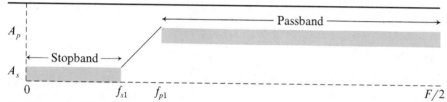

Transition band from f_{s1} to f_{fp1}
Sampling rate F = 2500 Samples/Second

Passband		Stopband	
f_{p1}	= 750.00 HZ	f_{s2}	= 500.00 HZ
f_{p2}	= 0.00 HZ	f_{s2}	= 0.00 HZ
Ripple	= 0.1737 dB	Ripple	= 40.00 dB

s-PLANE POLES AND ZEROS → LOWPASS

No.	Zeros		Poles	
	Real	Imaginary	Real	Imaginary
1	0.0000000	0.0000000	− 0.1564345	0.9876885
2	0.0000000	0.0000000	− 0.4539906	0.8910065
3	0.0000000	0.0000000	− 0.7071068	0.7071067
4	0.0000000	0.0000000	− 0.8910066	0.4539905
5	0.0000000	0.0000000	− 0.9876884	0.1564344

z-PLANE POLES AND ZEROS → HIGHPASS

No.	Zeros		Poles	
	Real	Imaginary	Real	Imaginary
1	+ 1.0000000	± 0.0000000	+ 0.1370101	± 0.8447676
2	+ 1.0000000	± 0.0000000	+ 0.1092151	± 0.6074742
3	+ 1.0000000	± 0.0000000	+ 0.0931416	± 0.4111430
4	+ 1.0000000	± 0.0000000	− 0.0841442	± 0.2384709
5	+ 1.0000000	± 0.0000000	− 0.0800775	± 0.0782001

Double zeros for all but odd-order stage.

SECOND-ORDER SECTION COEFFICIENTS

	Numerator Coefficients		Denominator Coefficients	
Stage	A_1	A_2	B_1	B_2
1	-2.0000000	1.0000000	0.2740203	0.7324040
2	-2.0000000	1.0000000	0.2184302	0.3809528
3	-2.0000000	1.0000000	0.1862832	0.1777139
4	-2.0000000	1.0000000	0.1682884	0.0639486
5	-2.0000000	1.0000000	0.1601550	0.0125277

IIR NORMALIZING FACTOR: $C_0 = 0.00125$

STAGE 1 NORMALIZING FACTOR: $C_1 = 0.11360$

STAGE 2 NORMALIZING FACTOR: $C_2 = 0.25799$

STAGE 3 NORMALIZING FACTOR: $C_3 = 0.32522$

STAGE 4 NORMALIZING FACTOR: $C_4 = 0.36908$

STAGE 5 NORMALIZING FACTOR: $C_5 = 0.35624$

FREQUENCY RESPONSE OUTPUT

Frequency	Magnitude	Frequency	Magnitude
0.000	-90.000	30.000	-90.000
60.000	-90.000	90.000	-90.000
120.000	-90.000	150.000	-90.000
180.000	-90.000	210.000	-89.999
240.000	-89.987	270.000	-89.842
300.000	-88.637	330.000	-83.899
360.000	-76.336	390.000	-68.398
420.000	-60.724	450.000	-53.352
480.000	-46.233	510.000	-39.317
540.000	-32.556	570.000	-25.911
600.000	-19.367	630.000	-12.988
660.000	-7.138	690.000	-2.816
720.000	-0.776	750.000	-0.174
770.000	-0.061	790.000	-0.021
810.000	-0.007	830.000	-0.002
850.000	-0.001	870.000	-0.000
890.000	-0.000	910.000	-0.000
930.000	-0.000	950.000	-0.000
970.000	-0.000	990.000	-0.000
1010.000	-0.000	1030.000	-0.000
1050.000	-0.000	1070.000	-0.000
1090.000	-0.000	1110.000	-0.000
1130.000	-0.000	1150.000	-0.000
1170.000	-0.000	1190.000	-0.000
1210.000	-0.000	1230.000	0.000

EXAMPLE 4.A.5
CHEBYSHEV HIGHPASS (ORDER = 6)
HIGHPASS FILTER FORMAT

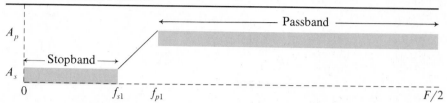

Transition band from f_{s1} to f_{p1}
Sampling rate F = 2500 Samples/Second

Passband		Stopband	
f_{p1} = 750.00 HZ		f_{s1} = 500.00 HZ	
f_{p2} = 0.00 HZ		f_{s2} = 0.00 HZ	
Ripple = 0.1737		Ripple = 40.00 dB	

s-PLANE POLES AND ZEROS → LOWPASS

	Zeros		Poles	
No.	Real	Imaginary	Real	Imaginary
1	0.0000000	0.0000000	− 0.1017847	1.0379357
2	0.0000000	0.0000000	− 0.2780809	0.7598216
3	0.0000000	0.0000000	− 0.3798656	0.2781140

z-PLANE POLES AND ZEROS → HIGHPASS

	Zeros		Poles	
No.	Real	Imaginary	Real	Imaginary
1	+ 1.0000000	± 0.0000000	− 0.2472980	± 0.8758247
2	+ 1.0000000	± 0.0000000	− 0.3740357	± 0.6310337
3	+ 1.0000000	± 0.0000000	− 0.5290681	± 0.2421385

Double zeros for all but odd-order stage.

SECOND-ORDER SECTION COEFFICIENTS

Stage	Numerator Coefficients		Denominator Coefficients	
1	− 2.0000000	1.0000000	0.4945960	0.8282253
2	− 2.0000000	1.0000000	0.7480714	0.5381063
3	− 2.0000000	1.0000000	1.0581361	0.3385441

IIR NORMALIZING FACTOR: C_0 = 0.00453

STAGE 1 NORMALIZING FACTOR: C_1 = 0.06794

STAGE 2 NORMALIZING FACTOR: C_2 = 0.19751

STAGE 3 NORMALIZING FACTOR: C_3 = 0.33721

FREQUENCY RESPONSE OUTPUT

Frequency	Magnitude	Frequency	Magnitude
0.000	−90.000	30.000	−90.000
60.000	−90.000	90.000	−90.000
120.000	−90.000	150.000	−89.994
180.000	−89.946	210.000	−89.633
240.000	−88.247	270.000	−84.726
300.000	−79.615	330.000	−74.118
360.000	−68.723	390.000	−63.515
420.000	−58.468	450.000	−53.535
480.000	−48.664	510.000	−43.802
540.000	−38.889	570.000	−33.856
600.000	−28.616	630.000	−23.055
660.000	−17.009	690.000	−10.327
720.000	−3.574	750.000	−0.173
770.000	−0.023	790.000	−0.147
810.000	−0.167	830.000	−0.103
850.000	−0.031	870.000	0.000
890.000	−0.017	910.000	−0.064
930.000	−0.117	950.000	−0.157
970.000	−0.173	990.000	−0.164
1010.000	−0.136	1030.000	−0.096
1050.000	−0.055	1070.000	−0.022
1090.000	−0.003	1110.000	−0.001
1130.000	−0.016	1150.000	−0.044
1170.000	−0.079	1190.000	−0.115
1210.000	−0.145	1230.000	−0.166

EXAMPLE 4.A.6
ELLIPTIC HIGHPASS (ORDER = 4)
HIGHPASS FILTER FORMAT

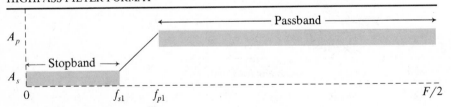

Transition band from f_{s1} to f_{p1}
Sampling rate F = 2500 Samples/Second

Passband		**Stopband**	
f_{p1}	= 750.000 HZ	f_{s1}	= 500.00 HZ
f_{p2}	= 0.00 HZ	f_{s2}	= 0.00 HZ
Ripple =	0.1737 dB	Ripple =	40.00 dB

s-PLANE POLES AND ZEROS → LOWPASS

No.	Zeros		Poles	
	Real	Imaginary	Real	Imaginary
1	0.0000000	3.3607185	−0.4304955	0.3741359
2	0.0000000	1.4730365	−0.1375649	0.7839348

z-PLANE POLES AND ZEROS → HIGHPASS

No.	Zeros		Poles	
	Real	Imaginary	Real	Imaginary
1	+0.8373246	±0.5467060	−0.3086025	±0.3422556
2	+0.3690524	±0.9294086	−0.1920366	±0.8214730

Double zeros for all but odd-order stage.

SECOND-ORDER SECTION COEFFICIENTS

Stage	Numerator Coefficients		Denominator Coefficients	
	A_1	A_2	B_1	B_2
1	−1.6746491	1.0000000	0.6172051	0.2123744
2	−0.7381049	1.0000000	0.3840733	0.7116958

IIR NORMALIZING FACTOR: $C_0 = 0.07699$

STAGE 1 NORMALIZING FACTOR: $C_1 = 0.16202$

STAGE 2 NORMALIZING FACTOR: $C_2 = 0.47517$

FREQUENCY RESPONSE OUTPUT

Frequency	Magnitude	Frequency	Magnitude
0.000	−41.678	30.000	−41.839
60.000	−42.338	90.000	−43.231
120.000	−44.638	150.000	−46.815
180.000	−50.401	210.000	−57.937
240.000	−63.875	270.000	−51.614
300.000	−46.780	330.000	−43.952
360.000	−42.291	390.000	−41.678
420.000	−42.486	450.000	−46.562
480.000	−56.964	510.000	−37.875
540.000	−29.757	570.000	−23.611
600.000	−18.247	630.000	−13.250
660.000	−8.507	690.000	−4.283
720.000	−1.359	750.000	−0.173
770.000	−0.004	790.000	−0.024
810.000	−0.090	830.000	−0.144
850.000	−0.170	870.000	−0.170

Frequency	Magnitude	Frequency	Magnitude
890.000	−0.150	910.000	−0.120
930.000	−0.086	950.000	−0.053
970.000	−0.027	990.000	−0.009
1010.000	0.000	1030.000	−0.000
1050.000	−0.008	1070.000	−0.023
1090.000	−0.042	1110.000	−0.064
1130.000	−0.088	1150.000	−0.110
1170.000	−0.131	1190.000	−0.149
1210.000	−0.162	1230.000	−0.170

EXAMPLE 4.A.7
BUTTERWORTH BANDPASS (ORDER = 8)
BANDPASS FILTER FORMAT

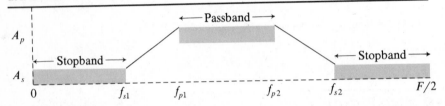

Transition bands from f_{s1} to f_{p1}, and f_{p2} to f_{s2}
Sampling Rate $F = 600$ Samples/second

Passband		Stopband	
f_{p1}	= 120.00 HZ	f_{s1}	= 60.00 HZ
f_{p2}	= 180.00 HZ	f_{s2}	= 240.00 HZ
Ripple =	0.50 dB	Ripple =	35.00 dB

s-PLANE POLES AND ZEROS → LOWPASS

No.	Zeros Real	Imaginary	Poles Real	Imaginary
1	0.0000000	0.0000000	−0.3826835	0.9238796
2	0.0000000	0.0000000	−0.9238796	0.3826833

s-PLANE POLES AND ZEROS → BANDPASS

No.	Zeros Real	Imaginary	Poles Real	Imaginary
1	0.0000000	0.0000000	−0.2211526	±1.4534092
2	0.0000000	0.0000000	−0.4578606	±1.0988795
3	0.0000000	0.0000000	−0.1023239	±0.6724680
4	0.0000000	0.0000000	−0.3230807	±0.7754031

z-PLANE POLES AND ZEROS → BANDPASS

No.	Zeros		Poles	
	Real	**Imaginary**	**Real**	**Imaginary**
1	± 1.0000000	0.0000000	-0.3222618	± 0.8066403
2	± 1.0000000	0.0000000	-0.1251682	± 0.6594147
3	± 1.0000000	0.0000000	$+0.3222615$	± 0.8066400
4	± 1.0000000	0.0000000	$+0.1251678$	± 0.6594146

SECOND-ORDER SECTION COEFFICIENTS

Stage	Numerator Coefficients		Denominator Coefficients	
	A_1	A_2	B_1	B_2
1	0.0000000	-1.0000000	0.6445237	0.7545212
2	0.0000000	-1.0000000	0.2503365	0.4504948
3	0.0000000	-1.0000000	-0.6445229	0.7545205
4	0.0000000	-1.0000000	-0.2503357	0.4504946

IIR NORMALIZING FACTOR: $C_0 = 0.01084$

STAGE 1 NORMALIZING FACTOR: $C_1 = 0.12650$

STAGE 2 NORMALIZING FACTOR: $C_2 = 0.31123$

STAGE 3 NORMALIZING FACTOR: $C_3 = 0.63584$

STAGE 4 NORMALIZING FACTOR: $C_4 = 0.43300$

FREQUENCY RESPONSE OUTPUT

Frequency	Magnitude	Frequency	Magnitude
0.000	-90.000	10.000	-89.935
20.000	-82.805	30.000	-68.946
40.000	-58.033	50.000	-49.007
60.000	-41.022	70.000	-33.569
80.000	-26.287	90.000	-18.880
100.000	-11.181	110.000	-4.008
120.000	-0.500	122.400	-0.253
124.800	-0.120	127.200	-0.052
129.600	-0.021	132.000	-0.007
134.400	-0.002	136.800	-0.001
139.200	-0.000	141.600	-0.000
144.000	-0.000	146.400	-0.000
148.800	-0.000	151.200	-0.000
153.600	-0.000	156.000	0.000
158.400	-0.000	160.800	0.000
163.200	-0.001	165.600	-0.002
168.000	-0.007	170.400	-0.021
172.800	-0.052	175.200	-0.120
177.600	-0.253	180.000	-0.500
189.231	-3.567	198.462	-9.991

Frequency	Magnitude	Frequency	Magnitude
207.692	−17.128	216.923	−24.034
226.154	· −30.766	235.385	−37.540
244.615	−44.615	253.846	−52.319
263.077	−61.136	272.308	−71.870
281.538	−84.953	290.769	−89.966

EXAMPLE 4.A.8
CHEBYSHEV BANDPASS (ORDER = 6)
BANDPASS FILTER FORMAT

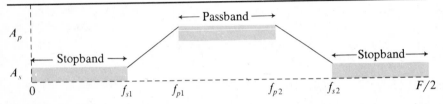

Transition bands from f_{s1} to f_{p1}, and f_{p2} to f_{s2}
Sampling rate $F = 600$ Samples/Second

Passband	Stopband
f_{p1} = 120.00 HZ	f_{s1} = 60.00 HZ
f_{p2} = 180.00 HZ	f_{s2} = 240.00 HZ
Ripple = 0.50 dB	Ripple = 35.00 dB

s-PLANE POLES AND ZEROS → LOWPASS

No.	Zeros Real	Imaginary	Poles Real	Imaginary
1	0.0000000	0.0000000	−0.3132283	1.0219275
2	0.0000000	0.0000000	−0.6264565	0.0000000

s-PLANE POLES AND ZEROS → BANDPASS

No.	Zeros Real	Imaginary	Poles Real	Imaginary
1	0.0000000	0.0000000	−0.1339817	1.3812982
2	0.0000000	0.0000000	−0.2035481	0.9790651
3	0.0000000	0.0000000	−0.0695665	−0.7172092

z-PLANE POLES AND ZEROS → BANDPASS

No.	Zeros		Poles	
	Real	Imaginary	Real	Imaginary
1	± 1.0000000	0.0000000	− 0.2899077	± 0.8649604
2	± 1.0000000	0.0000000	− 0.0000001	± 0.8134822
3	± 1.0000000	0.0000000	+ 0.2899076	± 0.8649612

SECOND-ORDER SECTION COEFFICIENTS

Stage	Numerator Coefficients		Denominator Coefficients	
	A_1	A_2	B_1	B_2
1	0.0000000	− 1.0000000	0.5798153	0.8322030
2	0.0000000	− 1.0000000	0.0000002	0.6617533
3	0.0000000	− 1.0000000	− 0.5798151	0.8322043

IIR NORMALIZING FACTOR: $C_0 = 0.01542$

STAGE 1 NORMALIZING FACTOR: $C_1 = 0.08420$

STAGE 2 NORMALIZING FACTOR: $C_2 = 0.31687$

STAGE 3 NORMALIZING FACTOR: $C_3 = 0.57782$

FREQUENCY RESPONSE OUTPUT

Frequency	Magnitude	Frequency	Magnitude
0.000	− 90.000	10.000	− 87.411
20.000	− 72.435	30.000	− 61.406
40.000	− 53.139	50.000	− 46.273
60.000	− 40.146	70.000	− 34.351
80.000	− 28.571	90.000	− 22.480
100.000	− 15.642	110.000	− 7.522
120.000	− 0.493	122.400	− 0.047
124.800	− 0.013	127.200	− 0.159
129.600	− 0.329	132.000	− 0.448
134.400	− 0.493	136.800	− 0.465
139.200	− 0.383	141.600	− 0.270
144.000	− 0.151	146.400	− 0.054
148.800	− 0.000	151.200	− 0.000
153.600	− 0.054	156.000	− 0.151
158.400	− 0.270	160.800	− 0.383
163.200	− 0.465	165.600	− 0.493
168.000	− 0.448	170.400	− 0.329
172.800	− 0.159	175.200	− 0.013
177.600	− 0.047	180.000	− 0.493
189.231	− 6.850	198.462	− 14.487
207.692	− 20.988	216.923	− 26.748
226.154	− 32.143	235.385	− 37.450
244.615	− 42.911	253.846	− 48.798
263.077	− 55.490	272.308	− 63.632
281.538	− 74.535	290.769	− 88.232

EXAMPLE 4.A.9
ELLIPTIC BANDPASS (ORDER = 6)
BANDPASS FILTER FORMAT

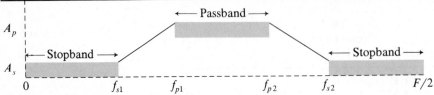

Transition bands from f_{s1} to f_{p1}, and f_{p2} to f_{s2}
Sampling Rate $F = 600$ Samples/Second

	Passband	Stopband	
f_{p1}	= 120.00 HZ	f_{s1}	= 60.00 HZ
f_{p2}	= 180.00 HZ	f_{s2}	= 240.00 HZ
Ripple =	0.50 dB	Ripple =	35.00 dB

s-PLANE POLES AND ZEROS → LOWPASS

	Zeros		Poles	
No.	Real	Imaginary	Real	Imaginary
1	0.0000000	2.3681457	−0.1477768	0.4982435
2	0.0000000	0.0000000	−0.3105073	0.0000000

s-PLANE POLES AND ZEROS → BANDPASS

	Zeros		Poles	
No.	Real	Imaginary	Real	Imaginary
1	0.0000000	±0.2892977	−0.1301877	1.3830704
2	0.0000000	0.0000000	−0.2076488	0.9782036
3	0.0000000	±3.4566479	−0.0674610	−0.7166792

z-PLANE POLES AND ZEROS → BANDPASS

	Zeros		Poles	
No.	Real	Imaginary	Real	Imaginary
1	+0.8455409	±0.5339107	−0.2914646	±0.8670722
2	+1.0000000	±0.0000000	−0.0000001	±0.8100065
3	−0.8455409	±0.5339106	+0.2914641	±0.8670719

SECOND-ORDER SECTION COEFFICIENTS

	Numerator Coefficients		Denominator Coefficients	
Stage	A_1	A_2	B_1	B_2
1	−1.6910818	1.0000000	0.5829291	0.8367658
2	0.0000000	−1.0000000	0.0000002	0.6561106
3	1.6910819	1.0000000	−0.5829282	0.8367649

IIR NORMALIZING FACTOR: $C_0 = 0.02205$

STAGE 1 NORMALIZING FACTOR: $C_1 = 0.06756$

STAGE 2 NORMALIZING FACTOR: $C_2 = 0.31930$

STAGE 3 NORMALIZING FACTOR: $C_3 = 1.02221$

FREQUENCY RESPONSE OUTPUT

Frequency	Magnitude	Frequency	Magnitude
0.000	− 90.000	10.000	− 59.793
20.000	− 54.455	30.000	− 52.355
40.000	− 52.853	50.000	− 60.229
60.000	− 52.185	70.000	− 40.039
80.000	− 31.767	90.000	− 24.328
100.000	− 16.651	110.000	− 7.928
120.000	− 0.493	122.400	− 0.042
124.800	− 0.017	127.200	− 0.170
129.600	− 0.339	132.000	− 0.454
134.400	− 0.493	136.800	− 0.462
139.200	− 0.378	141.600	− 0.264
144.000	− 0.148	146.400	− 0.053
148.800	− 0.000	151.200	− 0.000
153.600	− 0.053	156.000	− 0.148
158.400	− 0.264	160.800	− 0.378
163.200	− 0.462	165.600	− 0.493
168.000	− 0.454	170.400	− 0.339
172.800	− 0.170	175.200	− 0.017
177.600	− 0.042	180.000	− 0.493
189.231	− 7.214	198.462	− 15.393
207.692	− 22.609	216.923	− 29.449
226.154	− 36.659	235.385	− 45.622
244.615	− 65.675	253.846	− 55.595
263.077	− 52.350	272.308	− 52.629
281.538	− 55.008	290.769	− 60.456

EXAMPLE 4.A.10
BUTTERWORTH BANDSTOP (ORDER = 8)
BANDSTOP FILTER FORMAT

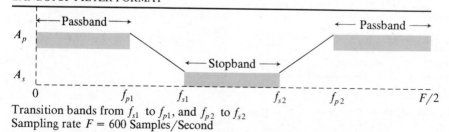

Transition bands from f_{s1} to f_{p1}, and f_{p2} to f_{s2}
Sampling rate $F = 600$ Samples/Second

	Passband		Stopband
f_{p1}	= 60.00 HZ	f_{s1}	= 120.00 HZ
f_{p2}	= 240.00 HZ	f_{s2}	= 180.00 HZ
Ripple =	0.50 dB	Ripple =	35.00 dB

s-PLANE POLES and ZEROS → LOWPASS

No.	Zeros		Poles	
	Real	**Imaginary**	**Real**	**Imaginary**
1	0.0000000	0.0000000	− 0.3826835	0.9238796
2	0.0000000	0.0000000	− 0.9238796	0.3826833

s-PLANE POLES AND ZEROS → BANDSTOP

No.	Zeros		Poles	
	Real	**Imaginary**	**Real**	**Imaginary**
1	0.0000000	1.0000002	− 0.6940045	2.3470006
2	0.0000000	1.0000002	− 1.5300284	1.1214997
3	0.0000000	1.0000002	− 0.1158595	− 0.3918164
4	0.0000000	1.0000002	− 0.4251559	− 0.3116362

z-PLANE POLES AND ZEROS → BANDSTOP

No.	Zeros		Poles	
	Real	**Imaginary**	**Real**	**Imaginary**
1	− 0.0000002	± 1.0000000	− 0.5956095	± 0.5602728
2	− 0.0000002	± 1.0000000	− 0.3393151	± 0.2928655
3	− 0.0000002	± 1.0000000	+ 0.5956093	± 0.5602730
4	− 0.0000002	± 1.0000000	+ 0.3393149	± 0.2928655

SECOND-ORDER SECTION COEFFICIENTS

Stage	Numerator coefficients		Denominator Coefficients	
	A_1	A_2	B_1	B_2
1	0.0000005	1.0000000	1.1912190	0.6686563
2	0.0000005	1.0000000	0.6786303	0.2009049
3	0.0000005	1.0000000	− 1.1912186	0.6686563
4	0.0000005	1.0000000	− 0.6786298	0.2009048

IIR NORMALIZING FACTOR: $C_0 = 0.08377$

STAGE 1 NORMALIZING FACTOR: $C_1 = 0.15031$

STAGE 2 NORMALIZING FACTOR: $C_2 = 0.35345$

STAGE 3 NORMALIZING FACTOR: $C_3 = 1.55668$

STAGE 4 NORMALIZING FACTOR: $C_4 = 1.01293$

FREQUENCY RESPONSE OUTPUT

Frequency	Magnitude	Frequency	Magnitude
0.000	−0.000	5.000	0.000
10.000	−0.000	15.000	−0.000
20.000	−0.000	25.000	−0.000
30.000	−0.001	35.000	−0.003
40.000	−0.011	45.000	−0.031
50.000	−0.083	55.000	−0.211
60.000	−0.500	67.200	−1.532
74.400	−3.840	81.600	−7.609
88.800	−12.405	96.000	−17.834
103.200	−23.811	110.400	−30.474
117.600	−38.148	124.800	−47.429
132.000	−59.520	139.200	−77.299
146.400	−89.989	153.600	−89.989
160.800	−77.299	168.000	−59.520
175.200	−47.429	182.400	−38.148
189.600	−30.474	196.800	−23.811
204.000	−17.834	211.200	−12.405
218.400	−7.609	225.600	−3.840
232.800	−1.532	240.000	−0.500
244.615	−0.226	249.231	−0.097
253.846	−0.039	258.462	−0.015
263.077	−0.005	267.692	−0.002
272.308	−0.000	276.923	−0.000
281.538	−0.000	286.154	−0.000
290.769	−0.000	295.385	−0.000

EXAMPLE 4.A.11
CHEBYSHEV BANDSTOP (ORDER = 6)
BANDSTOP FILTER FORMAT

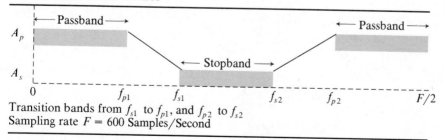

Transition bands from f_{s1} to f_{p1}, and f_{p2} to f_{s2}
Sampling rate F = 600 Samples/Second

Passband		Stopband	
f_{p1}	= 60.00 HZ	f_{s1}	=120.00 HZ
f_{p2}	= 240.00 H Z	f_{s2}	=180.00 HZ
Ripple =	0.50 dB	Ripple =	35.00 dB

s-PLANE POLES AND ZEROS → LOWPASS

	Zeros		Poles	
No.	Real	Imaginary	Real	Imaginary
1	0.0000000	0.0000000	− 0.3132283	1.0219275
2	0.0000000	0.0000000	− 0.6264565	0.0000000

s-PLANE POLES AND ZEROS → BANDSTOP

	Zeros		Poles	
No.	Real	Imaginary	Real	Imaginary
1	0.0000000	1.0000002	− 0.6735228	2.7999786
2	0.0000000	1.0000002	− 0.2407656	− 4.1534181
3	0.0000000	1.0000002	− 0.0812110	− 0.3376109

z-PLANE POLES AND ZEROS → BANDSTOP

	Zeros		Poles	
No.	Real	Imaginary	Real	Imaginary
1	− 0.0000002	± 1.0000000	− 0.6854445	± 0.5262842
2	− 0.0000002	± 1.0000000	+ 0.6119079	± 0.6119081
3	− 0.0000002	± 1.0000000	+ 0.6854441	± 0.5262842

SECOND-ORDER SECTION COEFFICIENTS

	Numerator Coefficients		Denominator Coefficients	
Stage	A_1	A_2	B_1	B_2
1	0.0000005	1.0000000	1.3708891	0.7468092
2	0.0000005	1.0000000	0.0000001	− 0.3744314
3	0.0000005	1.0000000	− 1.3708882	0.7468087

IIR NORMALIZING FACTOR: $C_0 = 0.09165$

STAGE 1 NORMALIZING FACTOR: $C_1 = 0.09562$

STAGE 2 NORMALIZING FACTOR: $C_2 = 0.56727$

STAGE 3 NORMALIZING FACTOR: $C_3 = 1.68955$

FREQUENCY RESPONSE OUTPUT

Frequency	Magnitude	Frequency	Magnitude
0.000	0.000	5.000	− 0.024
10.000	− 0.093	15.000	− 0.194
20.000	− 0.309	25.000	− 0.414
30.000	− 0.485	35.000	− 0.495
40.000	− 0.425	45.000	− 0.269
50.000	− 0.075	55.000	− 0.016
60.000	− 0.500	67.200	− 3.022

Frequency	Magnitude	Frequency	Magnitude
74.400	− 7.278	81.600	− 12.026
88.800	− 16.799	96.000	− 21.600
103.200	− 26.573	110.400	− 31.919
117.600	− 37.929	124.800	− 45.074
132.000	− 54.273	139.200	− 67.840
146.400	− 89.147	153.600	− 89.147
160.800	− 67.840	168.000	− 54.273
175.200	− 45.074	182.929	− 37.400
189.600	− 31.919	196.800	− 26.573
204.000	− 21.600	211.200	− 16.799
218.400	− 12.026	225.600	− 7.278
232.800	− 3.022	240.000	− 0.500
244.615	− 0.027	249.231	− 0.050
253.846	− 0.224	258.462	− 0.386
263.077	− 0.479	267.692	− 0.499
272.308	− 0.458	276.923	− 0.377
281.538	− 0.274	286.154	− 0.169
290.769	− 0.080	295.385	− 0.021

EXAMPLE 4.A.12
ELLIPTIC BANDSTOP (ORDER = 6)
BANDSTOP FILTER FORMAT

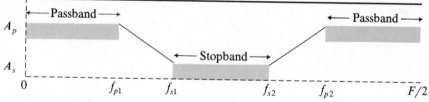

Transition bands from f_{s1} to f_{p1}, and f_{p2} to f_{s2}
Sampling rate $F = 600$ Samples/Second

	Passband		Stopband
f_{p1}	$=$ 60.00 HZ	f_{s1}	$=120.00$ HZ
f_{p2}	$=240.00$ HZ	f_{s2}	$=180.00$ HZ
Ripple $=$	0.50 dB	Ripple $=$	35.00 dB

s-PLANE POLES AND ZEROS → LOWPASS

No.	Zeros		Poles	
	Real	Imaginary	Real	Imaginary
1	0.0000000	2.3681457	− 0.1477767	0.4982435
2	0.0000000	0.0000000	− 0.3105073	0.0000000

s-PLANE POLES AND ZEROS → BANDSTOP

	Zeros		Poles	
No.	Real	Imaginary	Real	Imaginary
1	0.0000000	− 0.7567176	− 0.6530900	2.8054663
2	0.0000000	1.0000002	− 0.2462344	− 4.0611711*
3	0.0000000	1.3214975	− 0.0787125	− 0.3381235

*Two real poles from odd-order lowpass *s*-plane # = 2.

z-PLANE POLES AND ZEROS → BANDSTOP

	Zeros		Poles	
No.	Real	Imaginary	Real	Imaginary
1	+ 0.2717618	± 0.9623646	− 0.6881948	± 0.5291661
2	− 0.0000002	± 1.0000000	+ 0.6048345	− 0.6048345*
3	− 0.2717622	± 0.9623645	+ 0.6881944	± 0.5291662

*Two real poles from odd-order lowpass *s*-plane # = 2.

SECOND-ORDER SECTION COEFFICIENTS

	Numerator Coefficients		Denominator Coefficients	
Stage	A_1	A_2	B_1	B_2
1	− 0.5435236	1.0000000	1.3763895	0.7536287
2	0.0000005	1.0000000	0.0000001	− 0.3658248
3	0.5435243	1.0000000	− 1.3763888	0.7536284

IIR NORMALIZING FACTOR: $C_0 = 0.10107$

STAGE 1 NORMALIZING FACTOR: $C_1 = 0.06980$

STAGE 2 NORMALIZING FACTOR: $C_2 = 0.57609$

STAGE 3 NORMALIZING FACTOR: $C_3 = 2.51345$

FREQUENCY RESPONSE OUTPUT

Frequency	Magnitude	Frequency	Magnitude
0.000	0.000	5.000	− 0.024
10.000	− 0.091	15.000	− 0.190
20.000	− 0.303	25.000	− 0.409
30.000	− 0.482	35.000	− 0.497
40.000	− 0.432	45.000	− 0.280
50.000	− 0.082	55.000	− 0.013
60.000	− 0.500	67.200	− 3.151
74.400	− 7.668	81.600	− 12.736
88.800	− 17.918	96.000	− 23.310
103.200	− 29.230	110.400	− 36.331

Frequency	Magnitude	Frequency	Magnitude
117.600	− 46.617	124.800	− 65.567
132.000	− 52.566	139.200	− 53.391
146.400	− 61.624	153.600	− 61.624
160.800	− 53.391	168.000	− 52.566
175.200	− 65.567	182.400	− 46.617
189.600	− 36.331	196.800	− 29.230
204.000	− 23.310	211.200	− 17.918
218.400	− 12.736	225.600	− 7.668
232.800	− 3.151	240.000	− 0.500
244.615	− 0.024	249.231	− 0.056
253.846	− 0.235	258.462	− 0.394
263.077	− 0.482	267.692	− 0.498
272.308	− 0.454	276.923	− 0.371
281.538	− 0.268	286.154	− 0.165
290.769	− 0.078	295.385	− 0.020

References

1. A. Antoniou, *Digital Filters: Analysis and Design*, McGraw–Hill, New York, 1979.

2. R. W. Daniels, *Approximation Methods for the Design of Passive, Active, and Digital Filters*, McGraw–Hill, New York, 1974.

3. A. G. Constantinides, "Spectral Transformations for Digital Filters," *Proc. IEE*, 117(8):1585–1590, 1970.

4. B. Gold and C. M. Rader, *Digital Processing of Signals*, McGraw–Hill, New York, 1969.

5. M. T. Jong, *Methods of Discrete Signal and System Analysis*, McGraw–Hill, New York, 1982.

6. S. D. Stearns, *Digital Signal Analysis*, Hayden, New York, 1975.

7. A. V. Oppenheim and R. W. Schafer, *Digital Signal Processing*, Prentice–Hall, Englewood Cliffs, N.J., 1975.

8. L. R. Rabiner and B. Gold, *Theory and Application of Digital Signal Processing*, Prentice–Hall, Englewood Cliffs, N.J., 1975.

9. T. H. Crystal and L. Ehrman, "The Design and Application of Digital Filters with Complex Coefficients," *IEEE Trans. Audio Electroacoustics*, 16(3):315–320, 1968.

10. J. K. Kaiser, "Nonrecursive Digital Filter Design Using the I_0-sinh Window Function," IEEE Press, Digital Signal Processing II, Wiley, New York, 1976.

11. A. Peled and B. Liu, *Digital Signal Processing*, Wiley, New York, 1976.

12. R. Fletcher and M. J. D. Powell, "A Rapidly Convergent Descent Method for Minimization," *Computer J.*, No. 2, 163–168, 1963.

13. A. G. Deczky, "Synthesis of Recursive Digital Filters Using the Minimum P-Error Criterion," *IEEE Trans. Audio Electroacoustics*, AU-20(4):257–263, 1972.

14. *Programs for Digital Signal Processing*, IEEE Press, Wiley, New York, 1979.

15. C. S. Burrus and T. W. Parks, "Time Domain Design of Recursive Digital Filters," *IEEE Trans. Audio*, 18:137–141, 1970.

16. F. Brophy and A. C. Salazar, "Recursive Digital Filter Synthesis in the Time Domain," *IEEE Trans. ASSP*, ASSP-22(1) 1974.

Additional Reading

T. W. Parks, and C. S. Burrus; Digital Filter Design; Wiley, N.Y., 1987. This book provides an excellent supplement to the present text for the theory and design of IIR and FIR digital filters. A design example for the implementation of an IIR elliptic filter is discussed in detail, together with a listing of the assembly language program for implementing the filter on a signal processing chip (the TMS32010 from TI).

PROBLEMS

4.2.1   Determine the first 10 terms of Eq. 4.17.

4.2.2   Derive Eq. 4.20

4.2.3   Derive Eq. 4.33

4.4.1   From Eq. 4.48 show that

$$|z|^2 = \frac{(1 + \sigma)^2 + \omega^2}{(1 - \sigma)^2 + \omega^2}$$

4.4.2   From Eq. 4.48 obtain the z-plane complex poles from the s-plane complex poles given in Table 4.2.

4.4.3   Given the following s-plane points, obtain the z-plane points from Eq. 4.48. Is this mapping consistent with the property given in Table 4.2, that is, points on the $j\omega$ axis map onto the unit circle in the z-plane: $s = j0.1, j0.2, j0.3, j0.4, j0.5, j0.6, j07, j0.8, j0.9$ and $j0.9999$.

4.4.4   Given the following points in the s-plane, obtain the z-plane points from Eq. 4.48 and plot the results into the unit circle.

$$s = -0.1 + j0, -0.1 + j0.2, -0.1 + j0.4, -0.1 + j0.6,$$
$$-0.1 + j0.8, -0.1 + j.$$

Is the result consistent with one of the properties in Table 4.2?

4.5.1 Design a Butterworth lowpass digital filter satisfying the following specifications.

$$f_p = 800 \text{ Hz} \qquad A_p = 0.5 \text{ dB}$$
$$f_s = 1600 \text{ Hz} \qquad A_s = 45 \text{ dB}$$
$$F = 5000 \text{ Hz}$$

4.5.2 Repeat Problem 4.5.1 for a Chebyshev digital filter.

4.5.3 Repeat Problem 4.5.1 for an elliptic digital filter.

4.5.4 Design a Butterworth lowpass digital filter satisfying the following specifications.

$$f_p = 0.1 \text{ Hz} \qquad A_p = 1 \text{ dB}$$
$$f_s = 0.15 \text{ Hz} \qquad A_s = 15 \text{ dB}$$
$$F = 1 \text{ Hz}$$

4.5.5 Repeat Problem 4.5.4 for a Chebyshev digital filter.

4.5.6 Repeat Problem 4.5.4 for an elliptic digital filter.

4.5.7 The elliptic analog lowpass poles given by Eq. 4.35 are a function of the infinite series given by Eqs. 4.36 and 4.37. Perform an error analysis to determine the number of terms that would be required to implement Eqs. 4.36 and 4.37 into a computer program. The analysis should show the sensitivity the error has on Eq. 4.35. Use the specifications given by Prob. 4.5.3 for the analysis.

4.5.8 Design a Butterworth highpass digital filter satisfying the following specifications.

$$f_p = 0.32 \text{ Hz} \qquad A_p = 0.5 \text{ dB}$$
$$f_s = 0.16 \text{ Hz} \qquad A_s = 45 \text{ dB}$$
$$F = 1.0 \text{ Hz}$$

4.5.9 Repeat Problem 4.5.8 for a Chebyshev digital filter.

4.5.10 Repeat Problem 4.5.8 for an elliptic digital filter.

4.5.11 Design a Butterworth bandpass digital filter satisfying the following specifications.

$$f_{p1} = 40 \text{ Hz} \qquad f_{p2} = 60 \text{ Hz} \qquad A_p = 0.5 \text{ dB}$$
$$f_{s1} = 20 \text{ Hz} \qquad f_{s2} = 80 \text{ Hz} \qquad A_s = 35 \text{ Hz}$$
$$F = 200 \text{ Hz}$$

4.5.12 Repeat Problem 4.5.11 for a Chebyshev digital filter.

4.5.13 Repeat Problem 4.5.11 for an elliptic digital filter.

4.5.14 Design a Butterworth bandstop digital filter satisfying the following specifications.

$$f_{p1} = 20 \text{ Hz} \qquad f_{p2} = 80 \text{ Hz} \qquad A_p = 0.5 \text{ dB}$$

$$f_{s1} = 40 \text{ Hz} \qquad f_{s2} = 60 \text{ Hz} \qquad A_s = 35 \text{ dB}$$

$$F = 200 \text{ Hz}$$

4.5.15 Repeat Problem 4.5.14 for a Chebyshev digital filter.

4.5.16 Repeat Problem 4.5.14 for an elliptic digital filter.

4.5.17 Design a Butterworth lowpass digital filter satisfying the following specifications:

$F = 1 \text{ Hz}$	Sampling frequency
$f_p = 0.03125 \text{ Hz}$	Passband frequency
$f_s = 0.375 \text{ Hz}$	Stopband frequency
$A_p = 6.0206 \text{ dB}$	Passband attenuation
$A_s = 20 \text{ dB}$	Stopband attenuation

and determine the following.

    a. The digital filter transfer function.

    b. The difference equation.

    c. The normalization factor.

    d. The magnitude response at 0.03125 Hz.

    e. The realization (digital network).

4.5.18 Derive Eqs. 4.56 and 4.57.

4.5.19 Derive Eqs. 4.72 and 4.76.

4.5.20 Derive Eqs. 4.93 and 4.94 .

4.5.21 Derive Eq. 4.100.

4.5.22 Derive Eq. 4.103.

4.5.23 Derive Eq. 4.104.

4.5.24 From Eqs. 4.85a and b derive Eq. 4.102.

4.9.1 Derive Eqs. 4.116, and 4.118.

4.9.2 Given the following transfer function,

$$H(z) = \frac{1 - z_0 z^{-1}}{1 - p_0 z^{-1}}$$

where the zero z_0 and the pole p_0 are complex. Determine the difference equation, of the complex digital filter. Also sketch the realization of the filter.

5

Finite Impulse Response Digital Filter Design

5.0 INTRODUCTION

In Chapter 4 we considered the design of digital filters whose impulse response was of infinite duration. This property was described in Section 3.4; where we considered the response of linear systems to discrete signals, that is, if $H(z)$ is in the form of a rational function, and is both stable and causal, then the impulse response of the system decays exponentially to zero as the time index n approaches infinity.

We now consider digital filters whose impulse response is of finite duration. These filters are appropriately referred to as finite impulse response (FIR) digital filters. For the FIR filter the response of the filter depends only on the present and past input samples, whereas for the IIR filter if nT is taken as the present time, the present response is a function of the present and past N values of the excitation as well as past values of the response.

In this chapter we consider design techniques for FIR digital filters. The approximation problem in FIR filters (also called nonrecursive, moving average, transversal, or tapped delay line filters) is usually solved by using Fourier series or numerical analysis techniques. In this chapter we consider the time domain implementation of FIR filters. However, in Chapter 6 we will show that these filters can also be efficiently implemented in the frequency domain.

5.1 CHARACTERISTICS OF FIR DIGITAL FILTERS

Some advantages and disadvantages of FIR filters compared to their IIR counterparts are as follows.

1. FIR filters can be designed with exactly linear phase. Linear phase is important for applications where phase distortion due to nonlinear phase

can degrade performance, for example, speech processing, data transmission, and correlation processing.

2. FIR filters realized nonrecursively are inherently stable, that is, the filter impulse response is of finite length and therefore bounded.

3. Quantization noise due to finite precision arithmetic can be made negligible for nonrecursive realizations.

4. Coefficient accuracy problems inherent in sharp cutoff IIR filters can be made less severe for realizations of equally sharp FIR filters.

5. FIR filters can be efficiently implemented in multirate DSP systems (see Chapter 7).

6. A disadvantage of FIR filters compared to IIR filters is that an appreciably higher order filter is required to achieve a specified magnitude response, thereby requiring more filter coefficient storage.

In Chapter 7 we consider the design of efficient multirate systems implemented with FIR filters. Chapter 9 examines the finite arithmetic effects, and Chapter 10 provides a detection application example implemented with FIR filters.

5.2 PROPERTIES OF FIR FILTERS

5.2.1 Linear Phase Filters

The transfer function of a FIR causal filter is given by[1]

$$H(z) = \sum_{n=0}^{N-1} h(n)z^{-n} \tag{5.1}$$

where $h(n)$ is the impulse response of the filter. The difference equation of a FIR filter is obtained by taking the inverse Z-transform of Eq. 5.1, that is,

$$y(iT) = \sum_{n=0}^{N-1} h(n)x(iT - nT) \tag{5.2}$$

which can be recognized as the convolution summation. Notice that the finite impulse response is identical to the filter coefficients.

From Eq. 5.1 the Fourier transform of the finite sequence $h(n)$ is given by

$$H(e^{j\omega T}) = \sum_{n=0}^{N-1} h(n)e^{-j\omega nT} = |H(e^{j\omega T})|e^{j\theta(\omega)} \tag{5.3}$$

and the magnitude and phase response are defined as

$$M(\omega) = |H(e^{j\omega T})|$$

$$\theta(\omega) = \tan^{-1}\frac{-\operatorname{Im} H(e^{j\omega T})}{\operatorname{Re} H(e^{j\omega T})} \tag{5.4}$$

We now define phase delay and group (time) delay functions of a filter as follows.

$$\tau_p = -\frac{\theta(\omega)}{\omega} \quad \text{and} \quad \tau_g = -\frac{d\theta(\omega)}{d\omega} \tag{5.5}$$

where group delay is defined as the delayed response of the filter as a function of ω to a signal. Filters for which τ_p and τ_g are constant, that is, independent of frequency, are referred to as constant time delay or linear phase filters. Therefore, for the phase response to be linear we require

$$\theta(\omega) = -\tau\omega \qquad -\pi < \omega < \pi \tag{5.6}$$

where τ is a constant phase delay in samples. From Eqs. 5.3, 5.4, and 5.6 the phase response can be expressed as

$$\theta(\omega) = -\tau\omega = \tan^{-1}\frac{-\sum\limits_{n=0}^{N-1} h(n)\sin\omega nT}{\sum\limits_{n=0}^{N-1} h(n)\cos\omega nT} \tag{5.7}$$

or

$$\tan\omega\tau = \frac{\sum\limits_{n=0}^{N-1} h(n)\sin\omega nT}{\sum\limits_{n=0}^{N-1} h(n)\cos\omega nT} \tag{5.8}$$

Finally, we obtain

$$\sum_{n=0}^{N-1} h(n)\sin(\omega\tau - \omega nT) = 0 \tag{5.9}$$

It can be shown that a solution to Eq. 5.9 is given by[1]

$$\tau = \frac{(N-1)T}{2} \tag{5.10}$$

and

$$h(n) = h[(N - 1 - n)] \qquad \text{for } 0 < n < N - 1 \tag{5.11}$$

Therefore, FIR filters will have constant phase and group delays if the conditions of Eqs. 5.10 and 5.11 are satisfied. From the symmetry property of Eq. 5.11, we can see that Eq. 5.1 will result in a mirror-image polynomial. We will show later that efficient FIR filter realizations are achieved by taking advantage of this property.

Example 5.1:
For FIR filters to have linear phase, the properties described by Eqs. 5.10 and 5.11 must be satisfied. Given $N = 11$, show that Eq. 5.9 is satisfied.

Solution:

$$\sum_{n=0}^{10} h(n) \sin(5 - n)\omega T$$

$$= h(0) \sin(5\omega T) + h(1) \sin(4\omega T) + h(2) \sin(3\omega T)$$

$$+ h(3) \sin(2\omega T) + h(4) \sin(\omega T) + h(5) \sin(0)$$

$$+ h(6) \sin(-\omega T) + h(7) \sin(-2\omega T) + h(8) \sin(-3\omega T)$$

$$+ h(9) \sin(-4\omega T) + h(10) \sin(-5\omega T)$$

Since $h(n) = h(N - 1 - n)$, we have

$$h(0) = h(10) \qquad h(3) = h(7)$$
$$h(1) = h(9) \qquad h(4) = h(6)$$
$$h(2) = h(8) \qquad h(5)$$

Finally, Eq. 5.9 is satisfied.   ■

Consider the symmetry conditions of the impulse response for N-odd and even. From Eq. 5.10 τ is an integer for N-odd and the impulse response is symmetrical about the sample point $(N - 1)/2$. This implies that the filter delay is an integer number of samples. For N-even, we can see that τ will be a noninteger number; therefore, the filter delay is a noninteger number of samples. Typical filter impulse sequences for N-even and odd are shown in Figure 5.1. From Figure 5.1 we can see that the center of symmetry of the impulse response for N-even lies in the center between coefficients $(N - 2)/2$ and $N/2$, and for N-odd at $(N - 1)/2$. Functions that demonstrate this kind of coefficient symmetry are generally referred to as mirror-image polynomials.

In applications where only constant group delay is desired, it can be shown[1] that the impulse response is of the form

$$h(n) = -h(N - 1 - n) \tag{5.12}$$

The impulse response given by Eq. 5.12 is antisymmetrical about the center of the impulse response sequences just described. Applications of FIR filters with antisymmetrical impulse response sequences are for wideband differentiators and Hilbert transformers.[1]

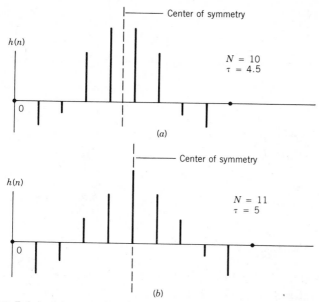

Figure 5.1 Impulse response for linear phase FIR filters: (a) N-even; (b) N-odd.

Example 5.2

A FIR digital filter is characterized by the following transfer function ($N = 11$). Determine the magnitude response, and show that the phase and group delay of the causal filter are constant.

$$H(z) = \sum_{n=0}^{N-1} h(n)z^{-n}$$

Solution:
Expanding $H(z)$, we obtain

$$H(z) = h(0) + h(1)z^{-1} + h(2)z^{-2} + h(3)z^{-3} + h(4)z^{-4} + h(5)z^{-5}$$
$$+ h(6)z^{-6} + h(7)z^{-7} + h(8)z^{-8} + h(9)z^{-9} + h(10)z^{-10}$$

Since $\tau = 5$, the transfer function $H(z)$ can be expressed in terms of the phase delay and noncausal response, that is,

$$H(z) = z^{-5}\big[h(0)z^5 + h(1)z^4 + h(2)z^3 + h(3)z^2 + h(4)z^1 + h(5)$$
$$+ h(6)z^{-1} + h(7)z^{-2} + h(8)z^{-3} + h(9)z^{-4} + h(10)z^{-5}\big]$$

Since $h(n) = h(N - 1 - n)$, we have

$$H(z) = z^{-5}\left[h(0)(z^5 + z^{-5}) + h(1)(z^4 + z^{-4}) + h(2)(z^3 + z^{-3})\right.$$
$$\left. + h(3)(z^2 + z^{-2}) + h(4)(z^1 + z^{-1}) + h(5)\right]$$

where z^{-n} implies a delay of nT seconds from the time nT = 0. The frequency response is obtained by substituting $z = e^{j\omega T}$. We then obtain

$$H(e^{j\omega T}) = e^{-j5\omega T}\left[2h(0)\cos 5\omega T + 2h(1)\cos 4\omega T + 2h(2)\cos 3\omega T\right.$$
$$\left. + 2h(3)\cos 2\omega T + 2h(4)\cos \omega T + h(5)\right]$$
$$= e^{-j5\omega T}\left[h(5) + 2\sum_{n=1}^{5} h(n)\cos n\omega T\right] = e^{-j5\omega T}M(\omega) \quad \blacksquare$$

which is the expected result from Eqs. 5.3 and 5.4, where $M(\omega)$ is the magnitude response and $\theta(\omega) = -5\omega T$ is the phase response. Notice that the number of multiplies (coefficients) is equal to $(N + 1)/2 = 6$. Finally, the phase and group delay are obtained from Eq. 5.5, that is,

$$\tau_p = \tau_g = 5T$$

5.2.2 Frequency Response of Linear Phase FIR Filters

The frequency response for a causal linear phase filter with N-odd can be obtained by expressing Eq. 5.3 as

$$H(e^{j\omega T}) = \sum_{n=0}^{(N-3)/2} h(n)e^{-j\omega nT} + h\left(\frac{N-1}{2}\right)e^{-j\omega(N-1)T/2}$$
$$+ \sum_{n=(N+1)/2}^{N-1} h(n)e^{-j\omega nT} \tag{5.13}$$

From Eq. 5.11, substituting $m = N - 1 - n$, changing the limits of summation, and finally letting $m = n$, the last summation becomes

$$\sum_{n=(N+1)/2}^{N-1} h(n)e^{-j\omega nT} = \sum_{n=0}^{(N-3)/2} h(n)e^{-j\omega(N-1-n)T} \tag{5.14}$$

Substituting Eq. 5.14 into Eq. 5.13, we obtain

$$H(e^{j\omega T}) = \sum_{n=0}^{(N-3)/2} h(n)\left[e^{-j\omega nT} + e^{-j\omega(N-1-n)T}\right]$$
$$+ h\left(\frac{N-1}{2}\right)e^{-j\omega(N-1)T/2} \tag{5.15}$$

Factoring $e^{-j\omega(N-1)T/2}$ in Eq. 5.15 and letting $k = (N-1)/2 - n$ we obtain

$$H(e^{j\omega T}) = e^{-j\omega(N-1)T/2}$$

$$\times \left[\sum_{k=1}^{(N-1)/2} h\left(\frac{N-1}{2} - k\right)(e^{j\omega kT} + e^{-j\omega kT}) + h\left(\frac{N-1}{2}\right)\right] \quad (5.16)$$

Equation 5.16 can now be simplified to the form

$$H(e^{j\omega T}) = e^{-j\omega(N-1)T/2}\left[\sum_{k=0}^{(N-1)/2} a(k)\cos\omega kT \right]$$

$$= e^{-j\omega(N-1)T/2}M(\omega) \quad (5.17)$$

$$a(0) = h\left(\frac{N-1}{2}\right) \qquad a(k) = 2h\left(\frac{N-1}{2} - k\right)$$

which is the result obtained in Example 5.2. Equation 5.17 defines the frequency response of a causal FIR filter with linear phase shift. It should be noted the the cosine function is real and represents the frequency response. The phase response function $\theta(\omega) = -\omega(N-1)T/2$ represents the constant delay of $(N-1)/2$ units in sampling time. It should also be emphasized that when the frequency response function $M(\omega)$ becomes negative, π radians should be added to the phase response function $\theta(\omega)$.

Using a similar derivation, the frequency response, functions for the symmetrical case with N-even and the two antisymmetrical cases result in the frequency response functions given in Table 5.1.

It is instructive to consider the frequency response of the noncausal impulse response. Again consider the case for a symmetrical impulse response with N-odd, where for the noncausal case $h(n) = h(-n)$, which corresponds to a zero-phase filter. The frequency response is given by

$$H(z) = \sum_{n=-M}^{M} h(n)z^{-n}$$

$$= \sum_{n=-M}^{-1} h(n)z^{-n} + h(0) + \sum_{n=1}^{M} h(n)z^{-n}$$

$$= \sum_{n=1}^{M} h(-n)z^{n} + h(0) + \sum_{n=1}^{M} h(n)z^{-n}$$

$$= h(0) + \sum_{n=1}^{M} h(n)[z^{n} + z^{-n}]$$

If we evaluate $H(z)$ for $z = e^{j\omega T}$, we obtain

$$H(e^{j\omega T}) = h(0) + \sum_{n=1}^{M} 2h(n)\cos\omega nT = \sum_{n=0}^{M} a(n)\cos\omega nT \quad (5.18)$$

TABLE 5.1
FREQUENCY RESPONSE OF CONSTANT-DELAY NONRECURSIVE FILTERS

h(nT)	N	$H(e^{j\omega T})$
Symmetrical	Odd	$e^{-j\omega(N-1)T/2} \displaystyle\sum_{k=0}^{(N-1)/2} a_k \cos \omega k T$
	Even	$e^{-j\omega(N-1)T/2} \displaystyle\sum_{k=1}^{N/2} b_k \cos[\omega(k-1/2)T]$
Antisymmetrical	Odd	$e^{-j[\omega(N-1)T/2-\pi/2]} \displaystyle\sum_{k=1}^{(N-1)/2} a_k \sin \omega k T$
	Even	$e^{-j[\omega(N-1)T/2-\pi/2]} \displaystyle\sum_{k=1}^{N/2} b_k \sin[\omega(k-1/2)T]$
where		$a_0 = h\left[\dfrac{(N-1)T}{2}\right] \quad a_k = 2h\left[\left(\dfrac{N-1}{2}-k\right)T\right] \quad b_k = 2h\left[\left(\dfrac{N}{2}-k\right)T\right]$

where

$$a(0) = h(0)$$
$$a(n) = 2h(n) \qquad n = 1, 2, \ldots, (N-1)/2$$
$$M = (N-1)/2$$

Equation 5.18 represents the noncausal frequency response. To obtain the desired causal frequency response, it is necessary to delay the noncausal impulse response by $(N-1)/2$ samples. This delay corresponds to a phase shift of $e^{-j\omega(N-1)T/2}$, which is equivalent to Eq. 5.17. Notice that for this case the frequency response is complex, whereas for the noncausal zero-phase filter the frequency response is purely real.

5.2.3 Locations of the Zeros of Linear Phase FIR Filters

The roots of the FIR mirror-image polynomial exhibit typical symmetry conditions in the z-plane. This property of linear phase FIR filters can be shown by writing Eq. 5.16 in the form

$$H(z) = z^{-(N-1)/2} \sum_{k=0}^{(N-1)/2} \frac{a(k)}{2}(z^k + z^{-k}) \qquad (5.19)$$

where a_0 and a_k are given by Eq. 5.17. On substituting z^{-1} for z in Eq. 5.19 we obtain

$$H(z^{-1}) = z^{(N-1)}H(z) \qquad (5.20)$$

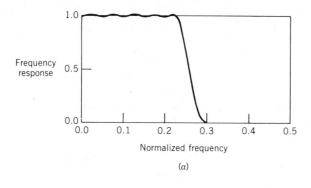

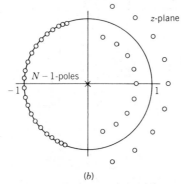

Figure 5.2 Typical zero-pole plot for a linear phase FIR filter: (a) 41-point equiripple low-pass filter; (b) plot of z-plane positions of poles and zeros.

We can see from Eq. 5.20 that $H(z^{-1})$ and $H(z)$ are identical except for the delay of $(N - 1)$ samples. The implication of Eq. 5.20 is that complex zeros of a mirror-image polynomial occur in complex conjugate pairs. Given that zero $z_i = r_i e^{j\theta_i}$, then the following zero locations result.

1. If $r_i \neq 1$, $\theta_i \neq 0, \pi$, then the zeros occur in quadruplets, that is, complex conjugate reciprocal pairs off the unit circle in the z-plane.

2. If $r_i = 1$, $\theta_i \neq 0, \pi$, then the zeros occur in complex conjugate pairs on the unit circle.

3. If $r_i \neq 1$, $\theta_i = 0$ or π, then the zeros are real and occur in reciprocal pairs off the unit circle.

4. If $r_i = 1$, $\theta_i = 0$, or π, then the zeros lie at either $+1$ or -1 on the unit circle.

It can be shown that the same relationships hold for N-even. A typical z-plane zero plot for a linear phase FIR filter is given in Fig. 5.2, where the out-of-band zeros are all on the unit circle and the inband zeros lie in quadruplets.

5.3 THE FOURIER SERIES METHOD OF DESIGNING FIR FILTERS

As shown in Sections 3.4.3 and 3.4.5, the frequency response of a digital filter is periodic with period equal to the sampling frequency F. From Fourier series analysis we know that any periodic function can be expressed as a linear combination of complex exponentials. Therefore, the desired frequency response of an FIR digital filter can be represented by the Fourier series.

$$H(e^{j2\pi fT}) = \sum_{n=-\infty}^{\infty} h_d(n)e^{-j2\pi fnT} \tag{5.21}$$

where the Fourier coefficients $h_d(n)$ are the desired impulse response sequence of the filter, which can be determined from

$$h_d(n) = \frac{1}{F}\int_{-F/2}^{F/2} H(e^{j2\pi fT})e^{j2\pi fnT}\,df \tag{5.22}$$

If in Eq. 5.21 we substitute $e^{j\omega T} = z$, we obtain the transfer function of the digital filter, that is,

$$H(z) = \sum_{n=-\infty}^{\infty} h_d(n)z^{-n} \tag{5.23}$$

Observe that there are two implementation problems with Eq. 5.23; that is, the transfer function represents a noncausal digital filter of infinite duration. A finite duration causal filter can be obtained by truncating the infinite duration impulse response and multiplying the resultant finite duration impulse response by $z^{-(N-1)/2}$. For N-odd we obtain

$$H(z) = z^{-(N-1)/2} \sum_{n=-(N-1)/2}^{(N-1)/2} h_d(n)z^{-n}$$

$$= z^{-(N-1)/2}\left[h_d(0) + \sum_{n=1}^{(N-1)/2} h_d(n)(z^n + z^{-n}) \right] \tag{5.24}$$

We can see from Eq. 5.24 causality was brought about by multiplying the transfer function by the delay factor (see Eq. 5.10). This modification does not effect the amplitude response of the filter; however, the abrupt truncation of the Fourier series results in oscillations in the passband and stopband. These oscillations are due to slow convergence of the Fourier series, particularly near points of discontinuity. This effect is known as the **Gibbs phenomenon**.[1] It will be shown that the undesirable oscillations can be reduced by multiplying the desired impulse response coefficients by an appropriate window function.

Example 5.3

a. Use the Fourier series method to design a lowpass digital filter to approximate the ideal specifications given by

$$H(e^{j2\pi fT}) = \begin{cases} 1 & \text{for } |f| \le f_p \\ 0 & \text{for } f_p < |f| \le F/2 \end{cases}$$

where f_p is the passband frequency and F is the sampling frequency.

b. Assume $N = 11$ and show that the transfer function given by Eq. 5.24 is causal.

Solution:

a. From Eq. 5.22,

$$h_d(n) = \frac{1}{F} \int_{-F/2}^{F/2} e^{j2\pi fnT} df = \frac{1}{n\pi} \sin 2\pi f_p nT$$

$$= \left(\frac{2f_p}{F}\right) \frac{\sin 2\pi nf_p/F}{2\pi nf_p/F} \tag{5.25}$$

where we note that for $n = 0$ Eq. 5.25 has the indeterminate form $0/0$, which may be evaluated by use of l'Hospital's rule. For this case, $h_d(0) = 2f_p/F$.

b. Multiplying Eq. 5.24 through by the delay factor $\tau_p = 5$, we obtain the causal transfer function shown here,

$$H(z) = h_d(0)z^{-5} + z^{-5} \sum_{n=1}^{5} h_d(n)[z^n + z^{-n}] \tag{5.26}$$

Expanding Eq. 5.26 yields

$$H(z) = h_d(0)z^{-5} + h_d(1)(z^{-4} + z^{-6}) + h_d(2)(z^{-3} + z^{-7})$$
$$+ h_d(3)(z^{-2} + z^{-8}) + h_d(4)(z^{-1} + z^{-9}) + h_d(5)(z^{-0} + z^{-10})$$
$$\tag{5.27} \blacksquare$$

Example 5.3 shows that the causality can be brought about by multiplying $H(z)$ in Eq. 5.23 by $z^{-(N-1)/2}$. Negative powers of z in Eq. 5.27 indicate that the filter generates an output that is a function of the past input samples. Figure 5.3 shows the relationship between the causal and noncausal impulse response for $N = 11$. It will be shown in Section 5.5 that efficient FIR realizations take advantage of the symmetry property of the impulse response; for example, for this case only, $(N + 1)/2 = 6$ multiplies are required to be implemented.

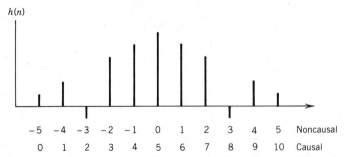

Figure 5.3 Time index relationship between causal and noncausal impulse response.

The magnitude (amplitude) response of an FIR filter can be obtained from Eq. 24 by making the substitution $z = e^{j\omega T}$, that is

$$M(\omega) = h_d(0) + 2 \sum_{n=1}^{(N-1)/2} h_d(n) \cos \omega nT \tag{5.28}$$

where $M(\omega)$ is also given in Table 5.1 for the case N-odd symmetrical; however, the impulse response given by Eq. 5.28 is noncausal, as described by Eq. 5.26. The relationship between the causal and noncausal impulse response (filter coefficients) is demonstrated in Figure 5.3.

Evaluating the frequency response of Eq. 5.28 as a function of a given passband and sampling frequency, for a family of increasing N demonstrates the behavior of the passband and stopband oscillations; that is, as the length of $h_d(n)$ is increased, the amplitude of the oscillations decrease. However, the amplitude of the ripple near the discontinuity tends to remain unchanged (Gibbs' oscillations). This type of filter design performance is objectionable; a method to reduce the undesired oscillations will be described in the next section by application of a particular class of window functions.

5.3.1 Window Functions

The previous section showed that truncating the Fourier series results in FIR filters with undesirable oscillations in the passband and stopband, which result from slow convergence of the Fourier series. To reduce these oscillations, a particular class of functions are used to modify the Fourier coefficients (impulse response). These time-limited weighting functions are generally referred to as window functions. Since truncation of the infinite Fourier series is equivalent to multiplying it with the rectangular window

$$a_R(n) = \begin{cases} 1 & \text{for } |n| \leq \dfrac{N-1}{2} \\ 0 & \text{otherwise} \end{cases} \tag{5.29}$$

we obtain

$$h(n) = h_d(n) a_R(n) \qquad (5.30)$$

Since multiplication in the time domain corresponds to convolution in the frequency domain, a design criterion for FIR filters is to find a finite window function whose Fourier transform has low level sidelobes with respect to the mainlobe peak.

From the complex convolution theorem developed in Section 3.3.4 we have

$$H_A(e^{j\omega T}) = \frac{1}{2\pi F} \int_0^{2\pi F} H(e^{j\omega T}) A(e^{j(\omega - \Omega)T}) \, d\Omega \qquad (5.31)$$

Thus, $H_A(e^{j\omega T})$ is a circular convolution of the desired frequency response with the Fourier transform of the window function. As a result, the discontinuities in the desired frequency response become transition bands of $H_A(e^{j\omega T})$. Further, it can be shown that the width of the transition band is a function of the length N of the window function.

As shown above, a technique for the design of FIR digital filters is to multiply the desired impulse response h_d by a class of time-domain functions known as window functions; therefore, in general we have

$$h(n) = h_d(n) a(n) \qquad (5.32)$$

Some of the most frequently used window functions, which yield in some sense optimum designs, are described in the following sections.

RECTANGULAR WINDOW FUNCTION

The rectangular window function is given by Eq. 5.29. The spectrum of $\omega_R(n)$ can be obtained from Eq. 5.21 as

$$
\begin{aligned}
A_R(e^{j\omega T}) &= \sum_{n=-(N-1)/2}^{(N-1)/2} e^{-j\omega nT} = e^{j\omega(N-1)T/2} \sum_{n=0}^{N-1} e^{-j\omega nT} \\
&= e^{j\omega(N-1)T/2} \frac{1 - e^{j\omega NT}}{1 - e^{j\omega T}} = \frac{e^{j\omega NT/2} - e^{j\omega NT/2}}{e^{j\omega T/2} - e^{-j\omega T/2}} \\
&= \frac{\sin(\omega NT/2)}{\sin(\omega T/2)} \qquad (5.33)
\end{aligned}
$$

The highest (first) sidelobe level is approximately 13 dB down from the mainlobe peak and rolloff is at 6 dB per octave. It is instructive to consider the frequency response for the causal rectangular window, that is,

$$A_R(e^{j\omega NT}) = \sum_{n=0}^{N-1} e^{-j\omega nT} = e^{-j\omega(N-1)T/2} \frac{\sin(\omega NT/2)}{\sin(\omega T/2)} \qquad (5.34)$$

As expected, the linear phase response is given by $\theta(\omega) = -\omega(N-1)T/2$, whereas the noncausal impulse response results in a zero phase filter.

HAMMING WINDOW FUNCTION

The Hamming window function is given by

$$a_H(n) = \begin{cases} 0.54 + 0.46\cos\dfrac{2\pi n}{N-1} & \text{for } |n| \le \dfrac{N-1}{2} \\ 0 & \text{otherwise} \end{cases} \tag{5.35}$$

The spectrum of the Hamming window can be obtained by noting that

$$a_H(n) = a_R(n)\left[0.54 + 0.46\cos\dfrac{2\pi n}{N-1}\right] \tag{5.36}$$

From Eq. 5.36 we then obtain

$$A_H(e^{j\omega T}) = 0.54\frac{\sin(\omega NT/2)}{\sin(\omega T/2)} + 0.46\frac{\sin[\omega NT/2 - N\pi/(N-1)]}{\sin[\omega T/2 - \pi/(N-1)]}$$
$$+ 0.46\frac{\sin[\omega NT/2 + N\pi/(N-1)]}{\sin[\omega T/2 + \pi/(N-1)]} \tag{5.37}$$

where the highest sidelobe level is approximately -43 dB and the sidelobe rolloff is 6 dB/octave. Note that since the window function is noncausal, the terms in Eq. 5.37 are all positive. For the causal window the second and third terms are negative, which is the form used for discrete Fourier transform processing (see Chapter 6). Equation 5.37 is also recognized as a three-point convolution in the frequency domain, where the convolution coefficients are 0.46, 0.54, 0.46, respectively.

BLACKMAN WINDOW FUNCTION

The noncausal Blackman window function is given by

$$a_B(n) = \begin{cases} 0.42 + 0.5\cos\dfrac{2\pi n}{N-1} + 0.08\cos\dfrac{4\pi n}{N-1}, & \text{for } |n| < \dfrac{N-1}{2} \\ 0 & \text{otherwise} \end{cases} \tag{5.38}$$

It can be shown that the spectrum of the Blackman window has its highest sidelobe level down -58 dB from the mainlobe peak.

KAISER WINDOW FUNCTION

For the foregoing windows the width of the mainlobe is inversely proportional to N, that is, increasing the window length decreases the mainlobe width, which results in a decrease in the transition band of the filter. However, the minimum stopband attenuation is independent of the window length and is a function of the selected window. Therefore, to achieve a desired stopband attenuation the designer must find a window that meets the design specifications. It should also be emphasized that windows with low sidelobe levels have broader mainlobe widths requiring an increase in the order of the filter N to achieve the desired transition width.

Kaiser windows have a variable parameter, α, which can be varied to control the sidelobe level with respect to the mainlobe peak. Similar to the other windows the mainlobe width can be varied by adjusting the length of the window, which in turn adjusts the transition width of the filter. Therefore, FIR digital filters can be efficiently designed using the Kaiser window function.[2]

From the preceding discussion it can be seen that a desirable property of a window function is that the function be of limited duration in the time domain and that the Fourier transform best approximates a bandlimited function, that is, has maximum energy in the mainlobe for a given peak sidelobe amplitude. The prolate spheroidal functions[3] have this desirable property in an optimal sense; however, the form of these functions are complicated and therefore difficult to compute. A simple approximation to these functions have been developed by Kaiser in terms of zero-order modified Bessel functions of the first kind, that is, $I_0(x)$. This window function is given by

$$
a_K(n) = \begin{cases} \dfrac{I_0(\beta)}{I_0(\alpha)} & \text{for } |n| \leq \dfrac{N-1}{2} \\ 0 & \text{otherwise} \end{cases} \tag{5.39}
$$

where α is an independent variable empirically determined by Kaiser. The parameter β is expressed by

$$
\beta = \alpha \left[1 - \left(\frac{2n}{N-1} \right)^2 \right]^{0.5} \tag{5.40}
$$

The modified Bessel function of the first kind, $I_0(x)$, can be computed from its power series expansion given by

$$
I_0(x) = 1 + \sum_{k=1}^{\infty} \left[\frac{1}{k!} \left(\frac{x}{2} \right)^k \right]^2
$$

$$
= 1 + \frac{0.25x^2}{(1!)^2} + \frac{(0.25x^2)^2}{(2!)^2} + \frac{(0.25x^2)^3}{(3!)^2} + \cdots \tag{5.41}
$$

This series converges rapidly and can be computed to any desired accuracy, where 25 terms are sufficient for most purposes.

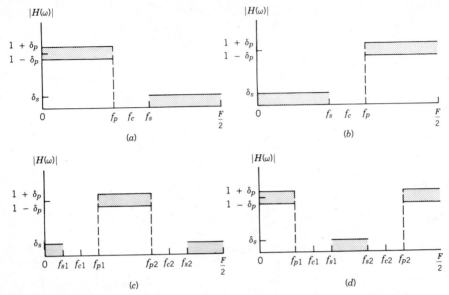

Figure 5.4 Idealized frequency responses: (*a*) lowpass filter; (*b*) highpass filter; (*c*) bandpass filter; (*d*) bandstop filter.

The spectrum of the Kaiser window can be obtained from

$$\sum_{-(N-1)/2}^{(N-1)/2} a_K(n) e^{-j\omega n T} = a_K(0) + 2 \sum_{n=1}^{(N-1)/2} a_K(n) \cos \omega n T \quad (5.42)$$

Consider the lowpass design specifications of Fig. 5.4*a*, where the actual passband ripple (A_p) and minimum stopband attenuation (A_s) in decibels are given by

$$A_p = 20 \log_{10} \frac{1 + \delta_p}{1 - \delta_p} \quad (5.43)$$

and

$$A_s = -20 \log_{10} \delta_s \quad (5.44)$$

The transition bandwidth is given by

$$\Delta F = f_s - f_p \quad (5.45)$$

For convenience, we define the specified passband ripple and minimum stopband attenuation by A_p' and A_s', respectively. We now proceed to outline a design procedure with the specifications

$$A_p \leq A_p'$$

$$A_s \geq A_s' \quad (5.46)$$

where A_p and A_s are the actual passband peak-to-peak ripple and minimum stopband attenuation, respectively.

5.3.2 FIR Filter Design Using the Kaiser Window Function

The following design procedure derived from the work done by Kaiser[2] and Antoniou[4] can be used to design FIR digital filters using the Fourier series (window) method. It will be shown that the procedure can be used to design lowpass (LP), highpass (HP), bandpass (BP), and bandstop (BS) filters for the impulse response duration of N-odd coefficients.

DESIGN SPECIFICATIONS

1. Filter type: LP, HP, BP, BS.

2. Filter critical passband and stopband frequencies in Hz:
LP/HP: f_p and f_s
BP/BS: f_{p1}, f_{p2}, f_{s1}, and f_{s2} (see Figures 5.4 and 4.11)

3. Filter passband ripple and minimum stopband attenuation in positive decibels: A'_p and A'_s.

4. Filter sampling frequency in Hz: F.

5. Filter order (N)-odd.

DESIGN PROCEDURE

1. Determine δ according to Eqs. 5.43, 5.44, and 5.46, where the actual design parameter can be determined from

$$\delta = \min\left(\delta_p, \delta_s\right) \tag{5.47}$$

where from Eqs. 5.43 and 5.44 we obtain

$$\delta_s = 10^{-0.05A'_s} \qquad \delta_p = \frac{10^{0.05A'_p} - 1}{10^{0.05A'_p} + 1} \tag{5.48}$$

Notice that the passband and stopband ripple amplitudes are equal, that is, $\delta = \delta_p = \delta_s$. We will return to this requirement later when we consider the design of half-band filters.

2. Calculate A_s using Eq. 5.44.

3. Determine the parameter α from the empirical design equation.

$$\alpha = \begin{cases} 0 & \text{for } A_s \leq 21 \\ 0.5842(A_s - 21)^{0.4} + 0.07886(A_s - 21) & \text{for } 21 < A_s \leq 50 \\ 0.1102(A_s - 8.7) & \text{for } A_s > 50 \end{cases} \tag{5.49}$$

4. Determine the parameter D from the empirical design equation.

$$D = \begin{cases} 0.9222 & \text{for } A_s \leq 21 \\ \dfrac{A_s - 7.95}{14.36} & \text{for } A_s > 21 \end{cases} \qquad (5.50)$$

5. Calculate the filter order for the lowest odd value of N.

$$N \geq \frac{FD}{\Delta F} + 1 \qquad (5.51)$$

6. Compute the modified impulse response using Eqs. 5.32, 5.39, and 5.40, that is,

$$h(n) = a_K(n)h_d(n) \qquad \text{for } |n| \leq \frac{N-1}{2} \qquad (5.52)$$

7. The transfer function is given by

$$H(z) = z^{-(N-1)/2}\left[h(0) + 2 \sum_{n=0}^{(N-1)/2} h(n)(z^n + z^{-n}) \right] \qquad (5.53)$$

where

$$\begin{aligned} h(0) &= a_K(0)h_d(0) \\ h(n) &= a_K(n)h_d(n) \end{aligned} \qquad (5.54)$$

The magnitude response can be obtained from Eq. 5.53, that is,

$$M(\omega) = h(0) + 2 \sum_{n=1}^{(N-1)/2} h(n) \cos 2\pi f n T \qquad (5.55)$$

which is equivalent to Eq. 5.28; however, for this case $h(n)$ represents the modified impulse response.

5.3.3 General FIR Filter Design Equations

In this section we consider the design equations for the lowpass, highpass, bandpass, and bandstop FIR digital filters. It will also be shown that the foregoing design equations are applicable to these filter types.

LOWPASS FIR FILTER

$$h_d(n) = \begin{cases} \left(\dfrac{2f_c}{F}\right)\dfrac{\sin 2\pi n f_c/F}{2\pi n f_c/F} & \text{for } n > 0 \\ \dfrac{2f_c}{F} & \text{for } n = 0 \end{cases} \qquad (5.56)$$

where

$$f_c = 0.5(f_p + f_s) \qquad \Delta F = f_s - f_p \tag{5.57}$$

HIGHPASS FIR FILTER

$$h_d(n) = \begin{cases} -\left(\dfrac{2f_c}{F}\right)\dfrac{\sin 2\pi n f_c/F}{2\pi n f_c/F} & \text{for } n > 0 \\ 1 - 2f_c/F & \text{for } n = 0 \end{cases} \tag{5.58}$$

where

$$f_c = 0.5(f_p + f_s) \qquad \Delta F = f_p - f_s \tag{5.59}$$

BANDPASS FIR FILTER

$$h_d(n) = \begin{cases} \dfrac{1}{n\pi}\left[\sin\left(2\pi n f_{c2}/F\right) - \sin\left(2\pi n f_{c1}/F\right)\right] & \text{for } n > 0 \\ \dfrac{2}{F}(f_{c2} - f_{c1}) & \text{for } n = 0 \end{cases} \tag{5.60}$$

where

$$
\begin{aligned}
f_{c1} &= f_{p1} - \frac{\Delta F}{2} & f_{c2} &= f_{p2} + \frac{\Delta F}{2} \\
\Delta F_1 &= f_{p1} - f_{s1} & \Delta F_h &= f_{s2} - f_{p2} \\
\Delta F &= \min\left[\Delta F_1, \Delta F_h\right]
\end{aligned}
\tag{5.61}
$$

BANDSTOP FIR FILTER

$$h_d(n) = \begin{cases} \dfrac{1}{n\pi}\left[\sin\left(2\pi n f_{c1}/F\right) - \sin\left(2\pi n f_{c2}/F\right)\right] & \text{for } n > 0 \\ \dfrac{2}{F}(f_{c1} - f_{c2}) + 1 & \text{for } n = 0 \end{cases} \tag{5.62}$$

where

$$
\begin{aligned}
f_{c1} &= f_{p1} + \frac{\Delta F}{2} & f_{c2} &= f_{p2} - \frac{\Delta F}{2} \\
\Delta F_1 &= f_{s1} - f_{p1} & \Delta F_h &= f_{p2} - f_{s2} \\
\Delta F &= \min\left[\Delta F_l, \Delta F_h\right]
\end{aligned}
\tag{5.63}
$$

The BASIC program shown in Figure 5.5 can be used to compute the Kaiser window FIR lowpass, highpass, bandpass, and bandstop filter coefficients.

```
15000 '**************************************************************
15001 '*          KAISER ORDER (NK) CALCULATION                     *
15002 '*   TYPE (1=LOWPASS,2=HIGHPASS,3=BANDPASS,4=BANDSTOP)        *
15003 '*   PASSBANDS (FP1,FP2), STOPBANDS (FA1,FA2), SAMPLING RATE (F)*
15004 '*          (AP in dB)                (AA in dB)              *
15005 '*   SEE DESIGN PROCEDURE SECTION 5.3.2 STEPS 1-5             *
15006 '**************************************************************
15010 BT1=ABS(FP2-FA2)
15020 BT2=ABS(FP1-FA1)
15030 IF TYPE=1 THEN BT=BT1
15040 IF TYPE=2 THEN BT=BT2
15050 IF TYPE=3 OR TYPE=4 THEN IF BT1<BT2 THEN BT=BT1 ELSE BT=BT2
15060 D2=10^(-.05*AA)
15070 D1=((10^(.05*AP))-1)/((10^(.05*AP))+1)
15080 IF D1<D2 THEN DEL=D1 ELSE DEL=D2
15090 AAP=-20*(LOG(DEL))/LOG(10)
15100 IF AAP<=21 THEN PARD=.9222 ELSE PARD=(AAP-7.95)/14.36
15110 NK=INT(2+PARD*FS/BT)
15120 IF (NK/2)=INT(NK/2) THEN NK=NK+1
15130 N=NK
15199 RETURN
```

(a)

```
15200 '**************************************************************
15201 '*          COMPUTE KAISER COEFFICIENTS WK=AK EQ. 5.52        *
15202 '*   AAP FROM 15090, ALP USING EQ. 5.49                      *
15203 '*   IOBE & IOALP USING EQ. 5.41                             *
15204 '*   WK USING EQ. 5.39                                       *
15205 '*   SEE DESIGN PROCEDURE SECTION 5.3.2 STEPS 3,6            *
15206 '*   TO COMPLETE EQ. 5.52 BRANCH TO SUBROUTINE BASED ON "TYPE" *
15207 '*       EQS. GIVEN IN SECTION 5.3.3                         *
15208 '**************************************************************
15209 N=NK
15210 IF AAP<=21 THEN ALP=0 ELSE ALP=.1102*(AAP-8.7)
15220 IF AAP>21 AND AAP<=50 THEN ALP=(.5842*(AAP-21)^.4)+.07886*(AAP-21)
15221 KFAC(1)=1
15222 FOR K%=2 TO 30
15223     KFAC(K%)=KFAC(K%-1)*K%
15224 NEXT K%
15225 LOCATE 22,10:PRINT "COMPUTING KAISER COEFFICIENTS          "
15230 FOR I%=0 TO (NK-1)/2
15235 LOCATE 22,45:PRINT I%;" OUT OF ";(NK-1)/2
15240     BE=ALP*SQR(1-(2*I%/(NK-1))^2)
15250     IOBE=1:IOALP=1
15260     FOR K%=1 TO 30
15270         IOBE=IOBE+(((BE/2)^K%)/KFAC(K%))^2
15280         IOALP=IOALP+(((ALP/2)^K%)/KFAC(K%))^2
15290     NEXT K%
15300     WK(I%+1)=IOBE/IOALP
15310 NEXT I%
15399 RETURN
```

(b)

Figure 5.5 Kaiser FIR filter coefficient BASIC program listing: (*a*) filter order; (*b*) Kaiser coefficients $a_k(n)(c)$ LP, HP, BP, and BS filter coefficients.

```
15400 '*******************************************************************
15401 '*          KAISER LOWPASS SUBROUTINE EQS. 5.52, 5.56, & 5.57     *
15402 '*******************************************************************
15410 WC=.5*(FP2+FA2)
15420 DEF FNSX(X,Y)=(SIN(X*Y*2*PI/FS))/(X*Y*2*PI/FS)
15430 H1=2*WC/FS
15440 H(1)=H1*WK(1)
15450 FOR I%=1 TO (NK-1)/2
15460     H(I%+1)=H1*FNSX(WC,I%)*WK(I%+1)
15461 'LPRINT "I= ";I%;" H(I)= ";H(I%+1)
15470 NEXT I%
15499 RETURN
15500 '*******************************************************************
15501 '*          KAISER HIGHPASS SUBROUTINE EQS. 5.52, 5.58, & 5.59     *
15502 '*******************************************************************
15510 WC=.5*(FP1+FA1)
15520 DEF FNSX(X,Y)=(SIN(X*Y*2*PI/FS))/(X*Y*2*PI/FS)
15530 H1=-2*WC/FS
15540 H(1)=(1+H1)*WK(1)
15550 FOR I%=1 TO (NK-1)/2
15560     H(I%+1)=H1*FNSX(WC,I%)*WK(I%+1)
15570 NEXT I%
15599 RETURN
15600 '*******************************************************************
15601 '*          KAISER BANDPASS SUBROUTINE EQS. 5.52, 5.60, & 5.61     *
15602 '*******************************************************************
15610 WC1=FP1-BT/2:WC2=FP2+BT/2
15620 H(1)=(2/FS)*(WC2-WC1)*WK(1)
15630 FOR I%=1 TO (NK-1)/2
15640     ARG=I%*2*PI/FS
15650     H(I%+1)=(1/(PI*I%))*(SIN(WC2*ARG)-SIN(WC1*ARG))*WK(I%+1)
15651     'LPRINT "I= ";I%;" H(I)= ";H(I%+1)
15660 NEXT I%
15799 RETURN
15800 '*******************************************************************
15801 '*          KAISER BANDSTOP SUBROUTINE EQS. 5.52, 5.62, & 5.63     *
15802 '*******************************************************************
15810 WC1=FP1+BT/2:WC2=FP2-BT/2
15820 H(1)=(2*(WC1-WC2)/FS)+1)*WK(1)
15830 FOR I%=1 TO (NK-1)/2
15840     ARG=I%*2*PI/FS
15850     H(I%+1)=(1/(PI*I%))*(SIN(WC1*ARG)-SIN(WC2*ARG))*WK(I%+1)
15860 NEXT I%
15999 RETURN
```

(c)

Figure 5.5 Continued

Example 5.4

Using the Kaiser window method, design an FIR lowpass digital filter satisfying the following specifications.

$$A_p \leq 0.1 \text{ dB}$$

$$A_s \geq 44 \text{ dB}$$

$$f_p = 500 \text{ Hz}$$

$$f_s = 750 \text{ Hz}$$

$$F = 2500 \text{ Hz}$$

Solution:

1. Compute ripple parameter. From Eqs. 5.48 and 5.49, $\delta = 0.005756$.
2. Compute actual stopband attenuation. From Eq. 5.44, $A_s = 44.796948$ dB.
3. Determine parameter D. From Eq. 5.50, $D = 2.565943$.
4. Compute filter order. From Eq. 5.51, $N = 25.7$ (round up to 27).
5. Determine parameter α. From Eq. 5.49, $\alpha = 3.952354$.
6. Compute modified impulse response. From Eqs. 5.56, 5.39, and 5.52 we obtain the following noncausal impulse response for $h_d(n)$, $a_K(n)$, and $h(n)$:

Normalized frequencies:

$f_p = 0.2$ Hz

$f_s = 0.3$ Hz

$F = 1$ Hz

Passband ripple: $A_p = 0.1$ dB
Stopband attenuation: $A_s = 44$ dB

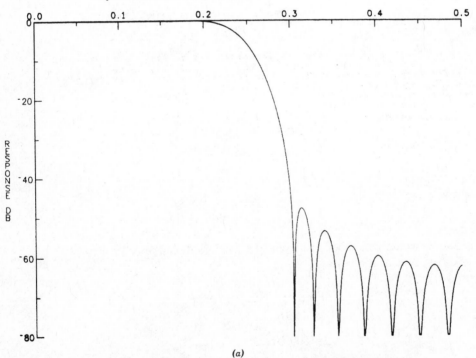

(a)

Figure 5.6 Kaiser window lowpass filter design (Example 5.4), (*a*) frequency response, (*b*) numerical values of frequency, (*c*) numerical values q impulse response (filter coefficients).

FREQUENCY	RESPONSE
.000	-.02970940
.005	-.03045296
.010	-.03252172
.015	-.03546384
.020	-.03863273
.025	-.04132445
.030	-.04292941
.035	-.04306688
.040	-.04167317
.045	-.03902362
.050	-.03568168
.055	-.03238425
.060	-.02988559
.065	-.02879168
.070	-.02941947
.075	-.03171157
.080	-.03522718
.085	-.03921602
.090	-.04276563
.095	-.04499651
.100	-.04526710
.105	-.04334421
.110	-.03949779
.115	-.03449088
.120	-.02945573
.125	-.02567065
.130	-.02427407
.135	-.02596864
.140	-.03077659
.145	-.03790669
.150	-.04578331
.155	-.05226939
.160	-.05508618
.165	-.05239660
.170	-.04347906
.175	-.02938694
.180	-.01347714
.185	-.00170919
.190	-.00266366
.195	-.02729157
.200	-.08846380
.205	-.20042379
.210	-.37825064
.215	-.63741845
.220	-.99350570
.225	-1.46207675
.230	-2.05873695
.235	-2.79935414
.240	-3.70044413
.245	-4.77973370
.250	-6.05694375
.255	-7.55488299
.260	-9.30102484
.265	-11.32989407
.270	-13.68690352
.275	-16.43498405
.280	-19.66710874
.285	-23.53286642
.290	-28.30486777
.295	-34.59343008
.300	-44.49859476
.305	-59.67296196
.310	-48.24829775
.315	-48.00509192
.320	-51.61712811
.325	-61.96084509
.330	-61.74804318
.335	-54.70965688
.340	-53.36552075
.345	-54.77870061
.350	-59.31838854
.355	-74.93505774
.360	-63.89704031
.365	-58.48751649
.370	-57.17262165
.375	-58.24141519
.380	-62.04718564
.385	-73.45439763
.390	-68.83709365
.395	-61.91370817
.400	-59.80661826
.405	-60.07394459
.410	-62.66271146
.415	-69.64980523
.420	-77.85368842
.425	-65.49443843
.430	-62.00156577
.435	-61.24743446
.440	-62.60711230
.445	-66.85766626
.450	-80.00000000
.455	-70.25198518
.460	-64.28183296
.465	-62.26467560
.470	-62.44404520
.475	-64.86945683
.480	-71.62124090
.485	-79.92257191
.490	-67.16464059
.495	-63.40609153

(b)

FIR Coefficients ($N = 27$)

Causal n	Noncausal ±n	$h_d(n)$	$a_K(n)$	$a_K(n) h_d(n)$
13 13	0	.500000000	1.000000000	.500000000
14 12	1	.318309886	.989962711	.315114918
15 11	2	.000000000	.960308906	.000000000
16 10	3	-.106103295	.912385341	-.096807091
17 9	4	.000000000	.848347050	.000000000
18 8	5	.063661977	.771029432	.049085258
19 7	6	.000000000	.683781532	.000000000
20 6	7	-.045472841	.590271779	-.026841335
21 5	8	.000000000	.494279190	.000000000
22 4	9	.035367765	.399483855	.014128851
23 3	10	.000000000	.309270295	.000000000
24 2	11	-.028937262	.226556162	-.006555915
25 1	12	.000000000	.153656683	.000000000
26 0	13	.024485376	.092192517	.002257369

(c)

Figure 5.6 Continued

$\pm n$	$h_d(n)$	$a_K(n)$	$h(n)$
0	0.5	1.0	0.5
1	0.318310	0.989963	0.315115
2	0	0.960309	0
3	-0.106103	0.912385	-0.096807
4	0	0.848347	0
5	0.063662	0.771029	0.049085
6	0	0.683782	0
7	-0.045473	0.590272	-0.026841
8	0	0.494279	0
9	0.035368	0.399484	0.014129
10	0	0.309270	0
11	-0.028937	0.226556	-0.006556
12	0	0.153657	0
13	0.024485	0.092193	0.002257

As shown in Fig. 5.6, the magnitude frequency response for the FIR lowpass filter can be obtained from Eq. 5.55. ■

Example 5.5

Derive the bandpass impulse response given by Eq. 5.60.

Solution:

From Eq. 5.22 and Fig. 5.4 we obtain

$$h_d(n) = \frac{1}{F} \int_{-F/2}^{F/2} H(e^{j2\pi fT}) e^{j2\pi fnT} \, df$$

$$= \frac{1}{F} \int_0^{F/2} \left[H(e^{j2\pi fT}) e^{j2\pi fnT} + H(e^{-j2\pi fT}) e^{-j2\pi fnT} \right] df$$

$$= \frac{1}{F} \int_{f_{c1}}^{f_{c2}} \left[e^{j2\pi fnT} + e^{-j2\pi fnT} \right] df$$

$$= \frac{1}{F} \int_{f_{c1}}^{f_{c2}} 2 \cos \left(2\pi fnT \right) \, df$$

$$= \frac{1}{\pi n} \left(\sin 2\pi f_{c2} nT - \sin 2\pi f_{c1} nT \right)$$

which is the result given by Eq. 5.60. Notice that for $n = 0$ $h_d(n)$ is determined by application of l'Hospital's rule. (See Eq. 5.60). ∎

Example 5.6

Design a bandpass filter satisfying the following specifications.

Passband	Stopband
$f_{p1} = 120$ Hz	$f_{s1} = 60$ Hz
$f_{p2} = 180$ Hz	$f_{s2} = 240$ Hz
$A_p = 0.5$ dB	$A_s = 35$ dB

Sampling frequency $(F) = 600$ Hz.

Solution:

1. Compute ripple parameter. From Eqs. 5.48 and 5.49, $\delta = 0.0177828$.
2. Compute actual stopband attenuation. From Eq. 5.44, $A_s = 35$ dB.
3. Determine parameter D. From Eq. 5.50, $D = 1.883705$.
4. Compute filter order. From Eq. 5.51, $N = 21$.
5. From Eqs. 5.60, 5.61, 5.39, and 5.52 we obtain the following noncausal impulse response:

$\pm n$	$h_d(n)$	$a_K(n)$	$h(n)$
0	0.4	1.0	0.4
1	0	0.989014	0
2	−0.302731	0.956550	−0.289577
3	0	0.904066	0
4	0.093549	0.833904	0.078011
5	0	0.749153	0
6	0.062366	0.653488	0.040755
7	0	0.550961	0
8	−0.075683	0.445786	−0.033738
9	0	0.342117	0
10	0	0.243827	0

Normalized frequencies:

$$f_{p1} = 0.2 \text{ Hz} \qquad f_{s1} = 0.1 \text{ Hz}$$

$$f_{p2} = 0.3 \text{ Hz} \qquad f_{s2} = 0.4 \text{ Hz}$$

$$F = 1 \text{ Hz}$$

Passband ripple: $A_p = 0.5$ dB
Stopband attenuation: $A_s = 35$ dB

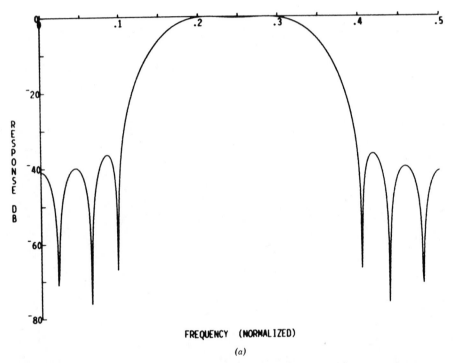

Figure 5.7 Kaiser window bandpass filter design (Example 5.6): (*a*) frequency response; (*b*) numerical values of frequency response; (*c*) numerical values of impulse response.

As shown in Fig. 5.7, the magnitude frequency response for the bandpass filter can be obtained from Eq. 5.55. ∎

It has been shown that the Kaiser window design technique provides a closed-form solution for designing linear phase FIR filters. In Section 5.6, a computer-aided design technique will be described, which is more efficient than the Kaiser window technique, in that it requires less filter coefficients to achieve the passband ripple and stopband attenuation performance specifications.

FREQUENCY	RESPONSE				
.000	-40.95871194	.160	-4.03122748	.330	-2.43294818
.005	-41.53972527	.165	-3.16781809	.335	-3.16781809
.010	-43.47408288	.170	-2.43294818	.340	-4.03122748
.015	-47.67955806	.175	-1.81575053	.345	-5.03564985
.020	-60.37714136	.180	-1.30642476	.350	-6.19588906
.025	-52.26415913	.185	-.89581703	.355	-7.53009991
.030	-44.98973102	.190	-.57508097	.360	-9.06130297
.035	-41.72196862	.195	-.33539938	.365	-10.81976064
.040	-40.19650153	.200	-.16775908	.370	-12.84694166
.045	-39.93992605	.205	-.06278179	.375	-15.20263872
.050	-40.98592055	.210	-.01062274	.380	-17.97897906
.055	-43.93314474	.215	-.00095600	.385	-21.33161805
.060	-51.84050200	.220	-.02306906	.390	-25.56272301
.065	-53.62988040	.225	-.06608794	.395	-31.41634109
.070	-43.12179137	.230	-.11934344	.400	-42.06227736
.075	-38.87616856	.235	-.17286909	.405	-46.28836755
.080	-36.84238683	.240	-.21799247	.410	-38.30447119
.085	-36.48371486	.245	-.24794964	.415	-36.48371486
.090	-38.30447119	.250	-.25842937	.420	-36.84238683
.095	-46.28836755	.255	-.24794964	.425	-38.87616856
.100	-42.06227736	.260	-.21799247	.430	-43.12179137
.105	-31.41634109	.265	-.17286909	.435	-53.62988040
.110	-25.56272301	.270	-.11934344	.440	-51.84050200
.115	-21.33161805	.275	-.06608794	.445	-43.93314474
.120	-17.97897906	.280	-.02306906	.450	-40.98592055
.125	-15.20263872	.285	-.00095600	.455	-39.93992605
.130	-12.84694166	.290	-.01062274	.460	-40.19650153
.135	-10.81976064	.295	-.06278179	.465	-41.72196862
.140	-9.06130297	.300	-.16775908	.470	-44.98973102
.145	-7.53009991	.305	-.33539938	.475	-52.26415913
.150	-6.19588906	.310	-.57508097	.480	-60.37714136
.155	-5.03564985	.315	-.89581703	.485	-47.67955806
		.320	-1.30642476	.490	-43.47408288
		.325	-1.81575053	.495	-41.53972527

(b)

FIR Coefficients (N = 21)

Causal n	Noncausal ±n	$h_d(n)$	$a_K(n)$	$a_K(n)\, h_d(n)$
10	10	0	.400000000	1.000000000	.400000000
11	9	1	.000000000	.989013547	.000000000
12	8	2	-.302730691	.956549514	-.289576896
13	7	3	-.000000000	.904066210	.000000000
14	6	4	.093548928	.833903592	.078010787
15	5	5	.000000000	.749153304	.000000000
16	4	6	.062365952	.653488069	.040755406
17	3	7	.000000000	.550960881	.000000000
18	2	8	-.075682673	.445786231	-.033738293
19	1	9	-.000000000	.342116577	.000000000
20	0	10	.000000000	.243827427	.000000000

(c)

Figure 5.7 Continued

5.4 FIR HALF-BAND DIGITAL FILTERS

Consider a lowpass filter with a desired frequency response given by

$$H(e^{j2\pi fT}) = \begin{cases} 1 & \text{for } 0 < f < f_p \\ 0 & \text{for } \dfrac{F}{2} - f_p < f < \dfrac{F}{2} \end{cases} \tag{5.64}$$

where the transition band is not specified since it does not effect the band of interest. The desired filter response is shown in Fig. 5.8.

Filters designed with the symmetry illustrated in Fig. 5.8 are called **half-band filters.** Half-band filters are odd-order filters used to decimate or interpolate the sampling rate by a factor of 2 (see Chapter 7). As shown in Fig. 5.8, the frequency response has odd symmetry around $F/4$, which results in the even filter coefficients being zero. As a result, the realization requires half as many multiplies as standard FIR filters.

Half-band FIR filters have the following general properties.

1. Equal amplitude ripples in the passband and stopband, that is,

$$\delta = \delta_s = \delta_p \tag{5.65}$$

2. The stopband frequency is given by

$$f_s = \frac{F}{2} - f_p \tag{5.66}$$

3. The transition band is then determined as

$$\Delta F = \frac{F}{2} - 2f_p \tag{5.67}$$

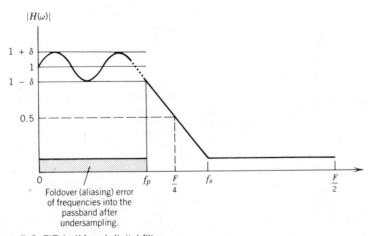

Figure 5.8 FIR half-band digital filter.

4. The value of the impulse response at $n = 0$ is 0.5, that is,

$$h(0) = 0.5 \qquad (5.68)$$

5. The frequency response is antisymmetric around $f = F/4$, that is, $H(F/4 + f) = 1 - H(F/4 - f)$. Also, at $f = F/4$ we obtain

$$H(e^{j2\pi fT}) = H(e^{j\pi/2}) = 0.5 \qquad (5.69)$$

6. The impulse response (filter coefficients) for values of N-odd given by

$$N = 4n + 3 \qquad \text{for } n = 0, 1, 2, \ldots \qquad (5.70)$$

results in the following general coefficient sequence

$$h(n) = \{x\, 0\, x\, 0\, x\, 0 \,\cdots\, 0\, x\, 0.5\, x\, 0 \,\cdots\, 0\, x\, 0\, x\, 0\, x\} \qquad (5.71)$$

where every other coefficient is exactly zero, eliminating the need for multiplies. Note that for this case the end coefficients are actual multiplies, where x denotes actual multiply, for example, see Example 5.4. The number of nonzero coefficients is given by

$$N' = \frac{N + 5}{4} \qquad (5.72)$$

For the case where N-odd is given by

$$N = 4n + 5 \qquad \text{for } n = 0, 1, 2, \ldots \qquad (5.73)$$

results in a coefficient sequence with the endpoints equal to zero, that is,

$$h(n) = \{0\, x\, 0\, x \,\cdots\, 0\, x\, 0.5\, x\, 0 \,\cdots\, x\, 0\, x\, 0\, x\, 0\} \qquad (5.74)$$

The number of nonzero coefficients for this case is given by

$$N' = \frac{N + 3}{4} \qquad (5.75)$$

An example of a half-band FIR filter using the window design technique is given in Example 5.4, where it can be seen that for the case of $N = 27$ the end coefficient is an actual multiply equal to 0.002257. Also, there are 8 nonzero multiplies, as determined from Eq. 5.72.

5.5 DIGITAL NETWORKS FOR LINEAR PHASE FIR DIGITAL FILTERS

In the previous sections we were concerned with the approximation of FIR digital filters, that is, obtaining a magnitude response satisfying the desired specifications. We now investigate the realization of FIR filters, which is the process of transforming the FIR difference equation into a digital network.

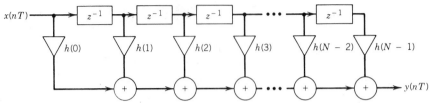

Figure 5.9 Direct-form realization of an FIR digital filter.

The finite arithmetic effects and implementation considerations will be considered in Chapters 7 and 9.

Consider the causal FIR transfer function

$$H(z) = \sum_{n=0}^{N-1} h(n) z^{-n} \tag{5.76}$$

where the difference equation obtained by taking the inverse Z-transform Eq. 5.76 is given by

$$y(iT) = \sum_{n=0}^{N-1} h(n) x[(i-n)T] \tag{5.77}$$

which is recognized as the convolution-summation equation. The direct-form realization of Eq. 5.77 is shown in Figure 5.9.

As shown in Section 5.2.1, the impulse response for a causal FIR filter with linear phase has the property

$$h(n) = h(N - 1 - n) \tag{5.78}$$

Using this property, efficient realizations of FIR filters can be obtained. Consider Eq. 5.15 for the case N-odd, where we make the substitution $z = e^{j\omega T}$

$$H(z) = \sum_{n=0}^{(N-3)/2} h(n)[z^{-n} + z^{-(N-1-n)}] + h\left(\frac{N-1}{2}\right) z^{-[(N-1)/2]} \tag{5.79}$$

Similarly, it can be shown that the transfer function for N-even is given by

$$H(z) = \sum_{n=0}^{N/2-1} h(n)[z^{-n} + z^{-(N-1-n)}] \tag{5.80}$$

We can now determine the difference equations of Eqs. 5.79 and 5.80 by taking the inverse Z-transform, that is, for the case of N-odd we obtain

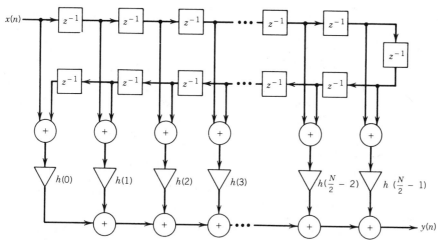

Figure 5.10 Direct-form realization of FIR digital filter exploiting impulse response symmetry for *N*-even.

$$y(iT) = \sum_{n=0}^{(N-3)/2} h(n)\{x[(i-n)T] + x[(i+n+1-N)T]\} \quad (5.81)$$
$$+ h\left(\frac{N-1}{2}\right)x\left[\left(i - \frac{N-1}{2}\right)T\right]$$

Similarly, for the case *N*-even we obtain

$$y(iT) = \sum_{n=0}^{N/2-1} h(n)\{x[(i-n)T] + x[(i+n+1-N)T]\} \quad (5.82)$$

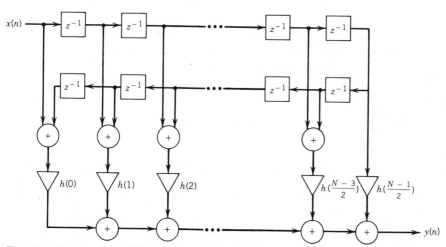

Figure 5.11 Direct-form realization of FIR digital filter exploiting impulse response symmetry for *N*-odd.

The digital networks for the difference equations given by Eqs. 5.81 and 5.82 are shown in Figures 5.10 and 5.11, respectively.

5.6 COMPUTER-AIDED DESIGN OF FIR DIGITAL FILTERS

The window method described in Section 5.3 is an analytic approach to the design of FIR digital filters. A disadvantage of the window technique is that the passband ripple δ_p and the stopband ripple δ_s are required to be equal. The filter is also suboptimal in the sense that the number of coefficients N required to meet the design specifications is not minimized. However, the window technique can be efficiently used when ease of design is important, especially for the case of real-time coefficient generation. It should also be noted that FIR filters designed using the window technique can compare reasonably well with computer-aided design techniques; for example, the designer can compare the range of values required for the filter order N to achieve the given design specifications.

In this section we present a technique for the design of FIR digital filters where the passband and stopband ripples are not required to be equal and the filter coefficient length N required to meet the design specifications is minimized.

The design procedure to be described can be applied to lowpass, highpass, bandpass, and bandstop filters. However, for simplicity we focus on the case of lowpass filters with symmetric impulse response, assuming N-odd. For Table 5.1 we obtain

$$H(e^{j2\pi fT}) = h(0) + \sum_{n=0}^{M} 2h(n) \cos 2\pi fT \tag{5.83}$$

where $M = (N-1)/2$.

Consider the design a lowpass filter according to the tolerance specifications in Figure 5.12. As shown in Figure 5.12, we desire to approximate unity in the passband $0 \leq f \leq f_p$ with maximum error δ_p and zero in the band $f_s \leq f \leq F/2$ with maximum error δ_s, that is,

- δ_p = allowed passband deviation (ripple)
- δ_s = allowed stopband deviation (ripple)
- f_p = desired passband frequency
- f_s = desired stopband frequency

From Figure 5.12 the filter approximation problem can be expressed as

$$1 - K\delta_s \leq H(e^{j2\pi fT}) \leq 1 + K\delta_s \qquad 0 \leq f \leq f_p$$

$$-\delta_s \leq H(e^{j2\pi fT}) \leq \delta_s \qquad f_s \leq f \leq \frac{F}{2} \tag{5.84}$$

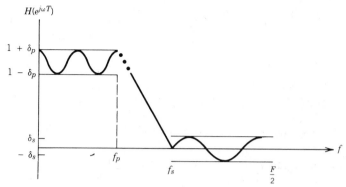

Figure 5.12 Idealized response for FIR lowpass approximation showing error tolerance.

where we define the parameter K as the ripple ratio, which determines the relative relationship between the approximation error parameters δ_p and δ_s, that is,

$$\delta_p = K\delta_s \tag{5.85}$$

From Eq. 5.84 we define the desired weighting function as

$$W(f) = \begin{cases} \dfrac{1}{K} = \dfrac{\delta_s}{\delta_p} & 0 \le f \le f_p \\[2ex] 1 & f_s \le f \le \dfrac{F}{2} \end{cases} \tag{5.86}$$

Also, from Figure 5.11 the desired response is given by

$$H_d(e^{j2\pi fT}) = \begin{cases} 1 & 0 \le f \le f_p \\[1ex] 0 & f_s \le f \le \dfrac{F}{2} \end{cases} \tag{5.87}$$

A computer-aided iterative design procedure has been developed by Parks and McCellan,[1] whereby the parameters M, f_p, f_s are fixed and δ_s is allowed to vary. The procedure is basically an iterative procedure for estimating frequencies at which extrema occur in the passband and stopband. The Lagrange interpolation formula is then used to obtain a polynomial that has prescribed extremal values, $1 \pm \delta_p$ in the passband and $\pm \delta_s$ in the stopband, at these frequencies. The iterative procedure is terminated when the maximum and minimum values of the polynomial are within the specified tolerance limits. This problem, therefore, is a Chebyshev approximation over disjoint sets and L_∞ is the norm considered for the error function.

The Chebyshev (minimax) approximation solution to Eq. 5.84 requires an algorithm for minimizing the stopband ripple, that is,

$$\delta_s = \min_{\{h(n)\}} \; \text{maximum}_{\substack{0 \le f \le f \\ f \le f \le F/2}} |E(\omega)| \qquad (5.88)$$

where for the lowpass design the frequency bands, given in Eq. 5.88, are disjoint in the interval $0 \le f \le F/2$. A solution to Eq. 5.88 is realized when the polynomial error function just meets the error tolerance shown in Figure 5.11.

For a given set of extremal frequencies $\{\omega_k\}$ the error function may be written as

$$E(\omega_k) = -(-1)^k \delta = W(\omega_k)\left[H(e^{j\omega T}) - H(e^{j\omega T})\right]$$
$$k = 0, 1, \ldots, r \qquad (5.89)$$

where $r = M + 1$ and δ is a constant. This relation can be put in matrix form as

$$
\begin{bmatrix}
1 \cos(\omega_0) \cos(2\omega_0) \ldots \cos[(r-1)\omega_0] & \dfrac{1}{W(\omega_0)} \\[2ex]
1 \cos(\omega_1) \cos(2\omega_1) \ldots \cos[(r-1)\omega_1] & \dfrac{-1}{W(\omega_1)} \\[2ex]
\vdots \quad \vdots \quad \vdots \quad\quad \vdots & \vdots \\[2ex]
1 \cos(\omega_r) \cos(2\omega_r) \ldots \cos[(r-1)\omega_r] & \dfrac{(-1)^r}{W(\omega_r)}
\end{bmatrix}
$$

$$
\times
\begin{bmatrix}
h(0) \\
h(1) \\
\vdots \\
h(r-1) \\
\delta
\end{bmatrix}
=
\begin{bmatrix}
H_d(\omega_0) \\
H_d(\omega_1) \\
\vdots \\
H_d(\omega_r)
\end{bmatrix}
\qquad (5.90)
$$

Therefore, if the extremal frequencies are known, the coefficients $\{h(n)\}$, the peak error δ, and hence the frequency response of the filter, can be computed by inverting the matrix. A more efficient approach, however, is to only solve for δ using the Remez exchange algorithm. For the details of this algorithm and a FORTRAN computer program see Rabiner and Gold.[1]

5.6.1 Optimum FIR Filter Order Estimate

LOWPASS FILTER ORDER EQUATION

The optimal (minimax) FIR filter requires the filter order N as an input parameter. An empiric relationship among the specification parameters has been established,[1] and is given by the following algorithm.

$$\hat{N} = \frac{D_{\infty}(\delta_p, \delta_s) F}{(f_s - f_p)} - f(\delta_p, \delta_s) \frac{(f_s - f_p)}{F} + 1$$

$$D_{\infty}(\delta_p, \delta_s) = \log_{10} \delta_s \left[a_1 (\log_{10} \delta_p)^2 + a_2 \log_{10} \delta_p + a_3 \right]$$

$$+ \left[a_4 (\log_{10} \delta_p)^2 + a_5 \log_{10} \delta_p + a_6 \right]$$

$$f(\delta_p, \delta_s) = 11.01217 + 0.51244(\log_{10} \delta_p - \log_{10} \delta_s), \quad \text{for } |\delta_s| \leq |\delta_p|$$

$$a_1 = 0.005309$$

$$a_2 = 0.07114$$

$$a_3 = -0.4761$$

$$a_4 = -0.00266$$

$$a_5 = -0.5941$$

$$a_6 = -0.4278 \tag{5.91}$$

Another approximation that can be used as a rough estimate of the lowpass filter order is given by

$$\hat{N} = \frac{-10 \log_{10}(\delta_p \delta_s) - 15}{14(f_s - f_p)/F} + 1 \tag{5.92}$$

This equation is similar to Eq. 5.51, which was used to calculate the filter order of FIR filters using the Kaiser window method.

BANDPASS FILTER ORDER EQUATION

An empiric relationship for the estimation of the filter order for the FIR bandpass filter has also been developed for the optimum FIR bandpass digital filter.[6] The equation is a refinement of Eq. 5.91 where we have

$$N = \frac{C_{\infty}(\delta_p, \delta_s) F}{\Delta F} + g(\delta_p, \delta_s) \frac{\Delta F}{F} + 1 \tag{5.93}$$

$$C_{\infty}(\delta_p, \delta_s) = \log_{10} \delta_s \left[b_1 (\log_{10} \delta_p)^2 + b_2 \log_{10} \delta_p + b_3 \right]$$

$$+ \left[b_4 (\log_{10} \delta_p)^2 + b_5 \log_{10} \delta_p + b_6 \right]$$

$$b_1 = 0.01201 \qquad b_2 = 0.09664$$

$$b_3 = -0.51325 \qquad b_4 = 0.00203$$

$$b_5 = -0.57054 \qquad b_6 = -0.44314$$

$$g(\delta_p, \delta_s) = -14.6 \log_{10}(\delta_p/\delta_s) - 16.9$$

where ΔF is given by Eq. 5.57. Further, the order of the filter is nearly

```
              FINITE IMPULSE RESPONSE
     LINEAR PHASE DIGITAL FILTER DESIGN
             REMEZ EXCHANGE ALGORITHM

              BANDPASS FILTER

     FILTER LENGTH = 23

     *****    IMPULSE RESPONSE    *****
     H(  1) =    ‾.00721048331 = H( 23)
     H(  2) =     .00000000000 = H( 22)
     H(  3) =     .01362148952 = H( 21)
     H(  4) =     .00000000000 = H( 20)
     H(  5) =    ‾.02629043395 = H( 19)
     H(  6) =     .00000000000 = H( 18)
     H(  7) =     .04859472811 = H( 17)
     H(  8) =     .00000000000 = H( 16)
     H(  9) =    ‾.09647795372 = H( 15)
     H( 10) =     .00000000000 = H( 14)
     H( 11) =     .31500080461 = H( 13)
     H( 12) =     .50000000000 = H( 12)


                         BAND 1      BAND 2

     LOWER BAND EDGE     .000000     .300000
     UPPER BAND EDGE     .200000     .500000
     DESIRED VALUE      1.000000     .000000
     WEIGHTING          1.000000    1.000000
     DEVIATION           .005524     .005524
     DEVIATION IN DB     .095957   ‾45.155405
```

(a)

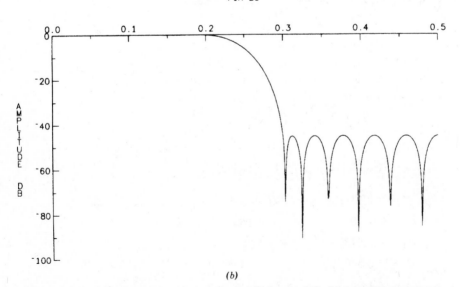

(b)

Figure 5.13 Optimal lowpass FIR filter design (Example 5.7): (*a*) impulse response; (*b*) frequency response.

```
              FINITE IMPULSE RESPONSE
       LINEAR PHASE DIGITAL FILTER DESIGN
            REMEZ EXCHANGE ALGORITHM

                BANDPASS FILTER

     FILTER LENGTH = 17

     *****    IMPULSE RESPONSE   *****
     H(  1) =    ˉ.03115459811 = H( 17)
     H(  2) =     .00000000000 = H( 16)
     H(  3) =     .03360974696 = H( 15)
     H(  4) =     .00000000000 = H( 14)
     H(  5) =     .08222434251 = H( 13)
     H(  6) =     .00000000000 = H( 12)
     H(  7) =    ˉ.28764822101 = H( 11)
     H(  8) =     .00000000000 = H( 10)
     H(  9) =     .39519596566 = H(  9)
```

	BAND 1	BAND 2	BAND 3
LOWER BAND EDGE	.000000	.200000	.400000
UPPER BAND EDGE	.100000	.300000	.500000
DESIRED VALUE	.000000	1.000000	.000000
WEIGHTING	1.618102	1.000000	1.618102
DEVIATION	.010741	.017381	.010741
DEVIATION IN DB	ˉ39.378707	.301966	ˉ39.378707

(a)

MAGNITUDE OF TRANSFER FUNCTION [IN DB]
VS
NORMALIZED FREQUENCY
FIR 17

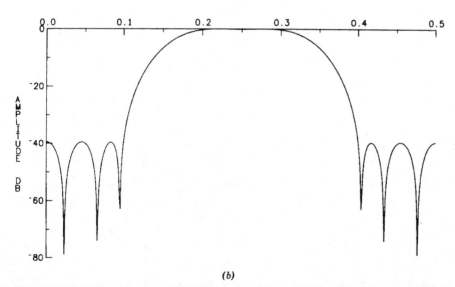

(b)

Figure 5.14 Optimal bandpass FIR filter design (Example 5.8): (a) impulse response; (b) frequency response.

proportional to ΔF. Therefore, the approximate filter order can be expressed by $\hat{N} = C_\infty(\delta_p, \delta_s)/\Delta F$.

Example 5.7
Using the design specifications given in Example 5.4, design an FIR lowpass digital filter using the optimal (minimax) design program.[1]

Solution:
The solution to this example is shown in Figure 5.13. Included in Figure 5.13 are the frequency response, filter coefficients, and system design specification parameters, where the normalized passband frequencies range from 0 to 0.2 Hz and stopband frequencies range from 0.3 to 0.5 Hz. Notice that the actual peak-to-peak ripple in the passband is 0.095957 dB and maximum stopband attenuation is -45.1554 dB, which are well within the required specifications. ∎

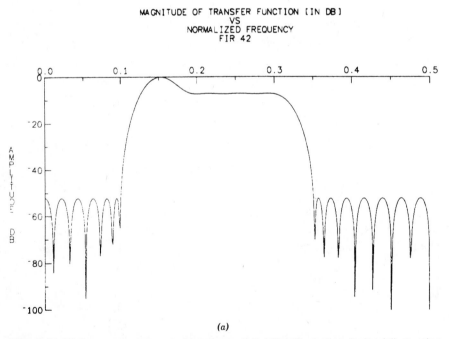

(a)

Figure 5.15 Frequency response of bandpass digital filter illustrating design failure using optimal design techniques: (a) frequency response showing overshoot in transition band; (b) impulse response.

Example 5.8

Using the design specifications given in Example 5.6, design an FIR bandpass digital filter using the optimal (minimax) design program.[1]

Solution:

The solution to this example is shown in Figure 5.14. Included in Figure 5.14 are the frequency response, filter coefficients, and system design specification parameters, where the normalized passband frequencies range from 0.2 to 0.3 Hz. The stopband frequencies range from 0 to 0.1 Hz and 0.4 to 0.5 Hz,

```
            FINITE IMPULSE RESPONSE
       LINEAR PHASE DIGITAL FILTER DESIGN
            REMEZ EXCHANGE ALGORITHM


                  BANDPASS FILTER


      FILTER LENGTH = 42


      *****      IMPULSE RESPONSE     *****
      H(  1) =    .004201134667 = H( 42)
      H(  2) =    .015646505170 = H( 41)
      H(  3) =    .000552624930 = H( 40)
      H(  4) =   ¯.022184020840 = H( 39)
      H(  5) =   ¯.020738669671 = H( 38)
      H(  6) =   ¯.014415261336 = H( 37)
      H(  7) =    .010127097834 = H( 36)
      H(  8) =    .062801850960 = H( 35)
      H(  9) =    .049450416584 = H( 34)
      H( 10) =   ¯.026971192099 = H( 33)
      H( 11) =   ¯.058666433208 = H( 32)
      H( 12) =   ¯.076866748277 = H( 31)
      H( 13) =   ¯.060733306687 = H( 30)
      H( 14) =    .083189944737 = H( 29)
      H( 15) =    .153843100183 = H( 28)
      H( 16) =    .043854179326 = H( 27)
      H( 17) =    .003487338778 = H( 26)
      H( 18) =   ¯.068590316456 = H( 25)
      H( 19) =   ¯.308196174446 = H( 24)
      H( 20) =   ¯.160284618381 = H( 23)
      H( 21) =    .393288682215 = H( 22)
```

	BAND 1	BAND 2	BAND 3
LOWER BAND EDGE	.000000	.200000	.350000
UPPER BAND EDGE	.100000	.300000	.500000
DESIRED VALUE	.000000	1.000000	.000000
WEIGHTING	2.047229	1.000000	2.047229
DEVIATION	.005592	.011449	.005592
DEVIATION IN DB	¯45.048238	.198892	¯45.048238

(b)

Figure 5.15 Continued

respectively. Notice that the peak-to-peak passband ripple is 0.301966 dB and the maximum stopband attenuation is − 39.378707 dB, which are well within the required specifications. ∎

5.6.2 Design Comments on Optimum FIR Digital Filters

As shown above, the Parks-McClellan algorithm provides an efficient method for the design of FIR linear-phase digital filters. However, since the design procedure only considers the specified passbands and stopbands, the transition bands are not considered in the numeric solution. In fact, the transition

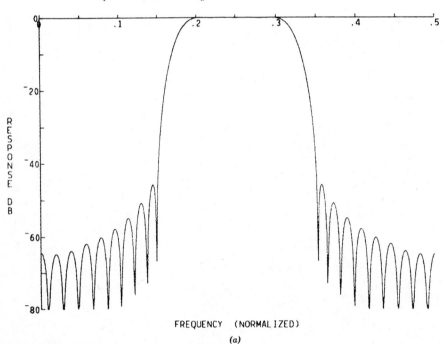

Normalized frequencies

$f_{p1} = 0.2$ Hz $f_{s1} = 0.1$ Hz
$f_{p2} = 0.3$ Hz $f_{s2} = 0.35$ Hz
$F = 1$ Hz

Passband ripple: $A_p = 0.2$ dB
Stopband attenuation: $A_B = 45$ dB

FREQUENCY (NORMALIZED)

(a)

Figure 5.16 Frequency response of bandpass digital filter shown in Figure 5.15 using Kaiser-window design techniques: (a) frequency response; (b) impulse response.

FIR Coefficients $(N = 53)$

Causal n	Noncausal ±n	$h_d(n)$	$a_K(n)$	$a_K(n) h_d(n)$
26	26	0	.300000000	1.000000000	.300003052
27	25	1	_.000000000	.997466133	_.000000000
28	24	2	_.257518107	.989893824	_.254913330
29	23	3	.000000000	.977370495	.000000000
30	22	4	.151365346	.960040381	.145309448
31	21	5	.000000000	.938102335	_.000000000
32	20	6	_.032787721	.911806822	_.029891968
33	19	7	_.000000000	.881452144	_.000000000
34	18	8	_.046774464	.847379957	_.039642334
35	17	9	.000000000	.809970162	.000000000
36	16	10	.063661977	.769635258	.048995972
37	15	11	_.000000000	.726814254	_.000000000
38	14	12	_.031182976	.681966236	_.021270752
39	13	13	_.000000000	.635563715	_.000000000
40	12	14	_.014051881	.588085846	_.008270264
41	11	15	_.000000000	.540011653	_.000000000
42	10	16	.037841336	.491813351	.018615723
43	9	17	_.000000000	.443949898	_.000000000
44	8	18	_.028613123	.396860858	_.011352539
45	7	19	_.000000000	.350960688	.000000000
46	6	20	_.000000000	.306633533	.000000000
47	5	21	_.000000000	.264228610	.000000000
48	4	22	.023410737	.224056242	.005249023
49	3	23	_.000000000	.186384606	_.000000000
50	2	24	_.025227558	.151437230	_.003814697
51	1	25	_.000000000	.119391272	.000000000
52	0	26	.007566397	.090376586	.000686646

(b)

Figure 5.16 Continued

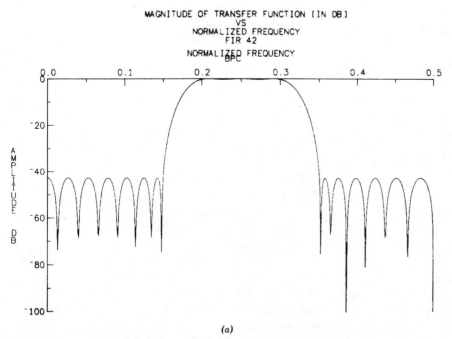

(a)

Figure 5.17 Frequency response of bandpass digital filter demonstrating correction to transition band overshoot: (*a*) frequency response; (*b*) impulse response.

```
              FINITE IMPULSE RESPONSE
         LINEAR PHASE DIGITAL FILTER DESIGN
              REMEZ EXCHANGE ALGORITHM

                   BANDPASS FILTER

     FILTER LENGTH = 42

     *****      IMPULSE RESPONSE      *****
     H(  1) =      .000201364048  =  H( 42)
     H(  2) =      .002186041325  =  H( 41)
     H(  3) =     -.006633010693  =  H( 40)
     H(  4) =     -.010704745073  =  H( 39)
     H(  5) =      .011465479154  =  H( 38)
     H(  6) =      .009401301853  =  H( 37)
     H(  7) =     -.005653346423  =  H( 36)
     H(  8) =      .000934309326  =  H( 35)
     H(  9) =     -.010826231912  =  H( 34)
     H( 10) =     -.021743043792  =  H( 33)
     H( 11) =      .030406014528  =  H( 32)
     H( 12) =      .033917461056  =  H( 31)
     H( 13) =     -.029853371438  =  H( 30)
     H( 14) =     -.015921343118  =  H( 29)
     H( 15) =     -.008598659653  =  H( 28)
     H( 16) =     -.042425529100  =  H( 27)
     H( 17) =      .082402644213  =  H( 26)
     H( 18) =      .123888595495  =  H( 25)
     H( 19) =     -.161673124414  =  H( 24)
     H( 20) =     -.190364312846  =  H( 23)
     H( 21) =      .205875332933  =  H( 22)
```

	BAND 1	BAND 2	BAND 3
LOWER BAND EDGE	.000000	.200000	.350000
UPPER BAND EDGE	.150000	.300000	.500000
DESIRED VALUE	.000000	1.000000	.000000
WEIGHTING	2.047229	1.000000	2.047229
DEVIATION	.007436	.015224	.007436
DEVIATION IN DB	-42.572804	.264487	-42.572804

(b)

Figure 5.17 Continued

regions are considered "don't care" regions in the design procedure. As a result, the numeric solution may fail, especially in the transition region. One type of failure, for a bandpass filter with one transition band larger than the other is illustrated in Figure 5.15.

For the optimum design procedure, the filter length is determined by the narrower transition band. If the transition band is too wide, the algorithm will fail in this region, resulting in the frequency response shown in Figure 5.15. Figure 5.16 shows the same filter design using the Kaiser window technique. We can see that a filter design failure does not occur in the transition region since the filter coefficients are determined analytically using the Fourier series method. Notice, however, the Kaiser window design is less efficient since it requires more coefficients to implement the filter, that is $N = 53$ for the Kaiser filter design compared to $N = 42$ for the optimal filter design.

Therefore, when using the optimum design technique, the frequency response should be examined for design failure. It should be noted, however, that it may be possible to eliminate the response overshoot by varying the transition bandwidths. Figure 5.17 demonstrates that the transition band overshoot, shown in Figure 5.15, has been eliminated by allowing the larger transition bandwidth to be equal to the narrower transition bandwidth. This can be seen by referring to Figure 5.15b where the lower stopband bandwidth (BAND 1) is $\Delta f_1 = 0.1 - 0.0 = 0.1$ Hz, and the upper stopband bandwidth (BAND 3) is $\Delta f_3 = 0.5 - 0.35 = 0.15$ Hz, therefore $\Delta f_1 \neq \Delta f_3$. Now consider Figure 5.17b where $\Delta f_1 = 0.15$ Hz and $\Delta f_3 = 0.15$ Hz, therefore, when $\Delta f_1 = \Delta f_3$ the overshoot of the frequency response in the lower transition band was corrected (see Figure 5.17a).

References

1 L. R. Rabiner and B. Gold, *Theory and Application of Digital Signal Processing*, Prentice–Hall, Englewood Cliffs, N.J., 1975.

2 J. F. Kaiser, *Nonrecursive Digital Filter Design Using The I-sinh Window Function*, Selected Papers in Digital Signal Processing, II, IEEE Press, New York, 1976.

3 H. Landau and H. Pollak, "Prolate Spheroidal Wave Functions, Fourier Analysis and Uncertainty," Bell Tel. Syst. J., 40:43–64, January 1961.

4 A. Antoniou, *Digital Filters: Analysis and Design*, McGraw-Hill, New York, 1979.

5 E. Remez, *General Computational Methods of Chebyshev Approximation*, Kiev, 1957, Atomic Energy Translation 4491, pp. 1–85.

6 F. Mitzer and B. Liu, "Practical Design Rules for Optimum FIR Bandpass Digital Filters," IEEE Trans. ASSP, pp. 204–206, April 1979.

PROBLEMS

5.2.1 Determine the intermediate steps in the derivation of Eq. 5.17.

5.2.2 Derive the symmetrical N-even frequency response given in Table 5.1.

5.2.3 Derive the antisymmetrical N-odd frequency response given in Table 5.1.

5.2.4 Determine the magnitude and phase response of the causal filter whose transfer function is

$$H(z) = \sum_{m=0}^{2M} h(m) z^{-m}$$

where $h(m) = h(2M - m), \quad 0 \le m \le M - 1$

5.2.5 Given the FIR transfer function

$$H(z) = \sum_{n=0}^{N-1} h(n) z^{-n}$$

show that

$$H(z^{-1}) = z^{(N-1)} H(z)$$

5.3.1 Derive an expression for the frequency response of the Blackman window given in Eq. 5.38.

5.3.2 Derive the highpass impulse response given by Eq. 5.58.

5.3.3 Derive the bandstop impulse response given by Eq. 5.62.

5.3.4 Write a computer program to determine the frequency response of Example 5.4. Verify that the design specifications are satisfied.

5.3.5 Write a computer program to determine the frequency response of Example 5.6. Verify that the design specifications are satisfied.

5.3.6 Write a computer program to design a Kaiser window FIR lowpass digital filter satisfying the following specifications.

$$f_p = 150 \text{ Hz} \qquad A_p = 0.1 \text{ dB}$$
$$f_s = 250 \text{ Hz} \qquad A_s = 40 \text{ db}$$
$$F = 1000 \text{ Hz}$$

5.3.7 Write a computer program to design a Kaiser window FIR highpass digital filter satisfying the following specifications.

$$f_p = 3200 \text{ Hz} \qquad A_p = 0.1 \text{ dB}$$
$$f_s = 1600 \text{ Hz} \qquad A_s = 40 \text{ dB}$$
$$F = 10{,}000 \text{ Hz}$$

5.3.8 Write a computer program to design a Kaiser window FIR bandpass digital filter satisfying the following specifications.

$$.f_{p1} = 40 \text{ Hz} \qquad A_p = 0.5 \text{ dB}$$
$$f_{p2} = 60 \text{ Hz} \qquad A_s = 35 \text{ dB}$$
$$f_{s1} = 20 \text{ Hz}$$
$$f_{s2} = 80 \text{ Hz}$$
$$F = 200 \text{ Hz}$$

5.3.9 Write a computer program to design a Kaiser window FIR bandstop digital filter satisfying the following specifications.

$$f_{p1} = 100 \text{ Hz} \qquad A_p = 0.3 \text{ db}$$
$$f_{p2} = 700 \text{ Hz} \qquad A_s = 35 \text{ dB}$$
$$f_{s1} = 200 \text{ Hz}$$
$$f_{s2} = 400 \text{ Hz}$$
$$F = 3000 \text{ Hz}$$

5.3.10 Write a generalized computer program to design a lowpass, highpass, bandpass, and bandstop FIR digital filter satisfying the specifications of Problems 5.3.6 through 5.3.9.

5.6.1 Write a computer program to determine the optimum lowpass filter order given by Eq. 5.91.

6

The Discrete Fourier Transform and Fast Fourier Transform Algorithms

The discrete Fourier transform (DFT) is one of the most important digital signal processing operations. Its implementation using the fast Fourier transform (FFT) algorithm has made it widely used for real-time signal processing applications.

6.0 INTRODUCTION

Detection and estimation of signals are fundamental signal processing functions required by many applications. **Detection** is the process of determining the presence or absence of specific signals. **Estimation** is the process of extracting information (e.g., amplitude, frequency, phase, and time delay) from the noise-corrupted signals. A broad class of applications required the analysis of periodic signals by determining their sinusoidal components. The discrete Fourier transform (DFT) provides the decomposition of the input into N periodic components.

The DFT is used extensively in digital signal processing for the analysis of discrete signals. The concept of signal frequency spectrums for both continuous signal representations and the discrete signal representation was discussed in Chapter 2. Discrete system Fourier transform relationships were also introduced in Chapter 2. Alternative approaches to the discussion of the DFT are possible. Brigham[1] derives the DFT as a special case of the continuous Fourier transform. Oppenheim and Schafer[2] based their discussion of the DFT on the Fourier series representation of periodic sequences and applied this result to the representation of finite-length sequences.

The DFT is presented and an example is given to demonstrate the transform process. Short-length DFTs are easily programmed and provide insight into the properties of the DFT via observing the output for various input signals. The frequency response resulting from the DFT is completely defined, and the use of weighting functions is described. DFT properties and filtering applications are developed.

Computation of the DFT, although possible with digital processors, requires complex multiplications on the order of the square of the number of samples being processed. For example, a DFT of 1000 samples would require 1 million complex multiplies to obtain 1000 complex frequency output points. This computational load prevented widespread use of the DFT as an analysis tool. In 1965, the fast Fourier transform (FFT) algorithm was published.[3] It significantly reduced the computational requirements of the DFT and initiated work that has led to several FFT algorithms that exploit various machine-processing capabilities. Several FFT hardware processors have been developed based on these algorithms, and many software programs have been written to implement the DFT efficiently on standard digital computers.[4]

This chapter develops both the decimation-in-time (DIT) and decimation-in-frequency (DIF) algorithms. These algorithms are easily programmed on a personal computer. A set of efficient primitive operations for FFT implementation is discussed. Algorithms for addressing data and coefficients are presented. In addition, computer programs that can be implemented on a personal computer are given.

6.1 THE DISCRETE FOURIER TRANSFORM (DFT)

The discrete Fourier transform (DFT) of a discrete-time signal $x(n)$ is defined as

$$X(k) = \frac{1}{N} \sum_{n=0}^{N-1} x(n) W_N^{nk} \qquad k = 0, 1, \ldots, N-1 \qquad (6.1)$$

where $W_N^{nk} = e^{-j2\pi nk/N}$ constitute the complex basis functions, or **twiddle factors** of the DFT. The twiddle factors are periodic and define points on the unit circle in the complex plane. Figure 6.1 illustrates the cyclic property of the twiddle factors for an eight-point DFT. The twiddle factors are equally spaced around the circle at frequency increments of F/N, where F is the sampling rate of the input signal sequence. Therefore, the set of frequency samples,

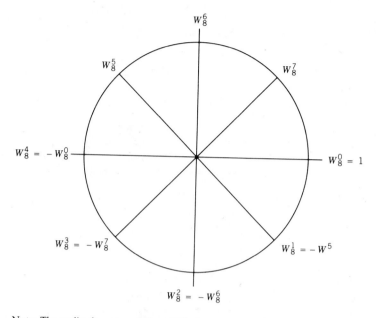

Note: The cyclic character of the twiddle factors are illustrated as follows.

$$w_8^0 = w_8^8 = w_8^{16} = w_8^{24} = \cdots$$

$$w_8^1 = w_8^9 = w_8^{17} = w_8^{25} = \cdots$$

$$w_8^2 = w_8^{10} = w_8^{18} = w_8^{26} = \cdots$$

$$\vdots \qquad \vdots \qquad \vdots \qquad \vdots$$

$$w_8^7 = w_8^{15} = w_8^{23} = w_8^{31} = \cdots$$

or in general the twiddle factors can be expressed as

$$w_8^{p \text{ MODULO } 8} = w_8^{8(\text{RESIDUE } p/8)}$$

where $w_8 = \text{EXP}(-j2\pi/8)$

Figure 6.1 Twiddle factors for $N = 8$-point DFT.

which defines the spectrum $X(k)$, are given on a frequency axis whose discrete frequency locations are given by

$$f_k = k\frac{F}{N} \qquad k = 0, 1, \ldots, N - 1 \tag{6.2}$$

The **frequency resolution** of the DFT is equal to the frequency increment F/N and is sometimes referred to as the *bin spacing* of the DFT outputs. The frequency response of any DFT bin output is determined by applying a complex exponential input signal and evaluating the DFT bin output response as the frequency is varied. Therefore, let $x(n) = e^{j2\pi fn/F}$; then the DFT of $x(n)$ can be expressed as a function of the arbitrary frequency variable f by

$$X(k) = \frac{1}{N}\sum_{n=0}^{N-1} x(n)W_N^{nk} = \frac{1}{N}\sum_{n=0}^{N-1} e^{j2\pi fn/F}W_N^{nk}$$

Combining exponential terms and using the geometric series summation, $X(k)$ becomes

$$X(k) = \frac{1}{N}\frac{1 - e^{-j2\pi(k/N - f/F)N}}{1 - e^{-j2\pi(k/N - f/F)}}$$

Defining $\Omega = 2\pi[(k/N) - (f/F)]$ and rewriting $X(k)$, the DFT of $x(nT)$ is given by

$$X(k) = e^{-j\Omega(N-1)/2}\frac{\sin(\Omega N/2)}{N\sin(\Omega/2)} \tag{6.3}$$

The exponential term is the phase of the response, and the ratio of sines is the amplitude response. If k is an integer, for example, k_0, then $X(k) = 1$ at $k = k_0$ on the frequency axis and zero elsewhere. If k is not an integer, then none of the DFT values are zero. This phenomenon is called **spectral**

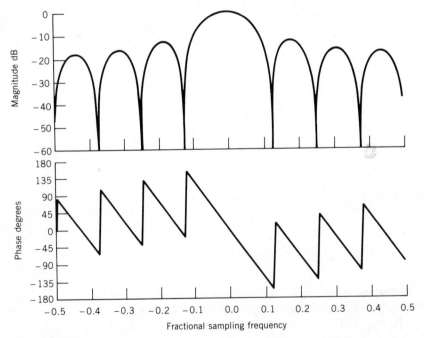

Figure 6.2

leakage. Thus, we only obtain a unit impulse at the kth frequency location on the DFT frequency axis when $f_k = kF/N$, k integer valued.

Figure 6.2 presents the frequency response of a DFT bin output for an eight-point DFT. The high sidelobe levels of the response are a direct result of truncation; that is, the DFT approximates the Fourier transform by forming the product $x(n)$ with a rectangular window. A detailed discussion of the frequency response is given in Section 6.6, where spectral leakage, scalloping loss, and effective noise bandwidth will be defined and analyzed with respect to window functions.

6.2 THE INVERSE DISCRETE FOURIER TRANSFORM (IDFT)

Given the frequency values $X(k)$, the inverse discrete Fourier transform (IDFT) gives the time sequence

$$x(n) = \sum_{k=0}^{N-1} X(k) W_N^{-nk} \tag{6.4}$$

The proof of this transform relationship is easily developed by substituting Eq. 6.1 into Eq. 6.4 and evaluating the expression

$$x(n) = \sum_{k=0}^{N-1} \left[\frac{1}{N} \sum_{m=0}^{N-1} x(m) W_N^{mk} \right] W_N^{-nk}$$

Interchanging the order of summation and summing over k first,

$$x(n) = \sum_{m=0}^{N-1} \frac{1}{N} x(m) \sum_{k=0}^{N-1} W_N^{k(m-n)}$$

where (see Problem 6.2.1)

$$\sum_{k=0}^{N-1} W_N^{k(m-n)} = \begin{cases} N & \text{for } m = n \\ 0 & \text{otherwise} \end{cases}$$

The transform relationship is developed since the expression is equal to $x(n)$, which is given by Eq. 6.4.

The form of the equation for the IDFT is identical to the DFT with the exception of the normalizing factor $1/N$ and the sign of the exponent of the twiddle factors. This is an important observation and provides a method for computing the IDFT without any changes in the DFT algorithm. Equation 6.1 is called the DFT or **analysis transform**, and Eq. 6.4 is referred to as the inverse DFT or the **synthesis transform**. It is of interest to note that the DFT is equal to the Z-transform of a sequence, $x(n)$, evaluated at equally spaced inputs on the unit circle in the z-plane.

6.3 DFT PROPERTIES

The DFT is important to the analysis and design of digital signal processing systems. The properties of the DFT are defined, and discussed in this section. Similarities with the Fourier transform are noted, and the unique properties due to the finite length transform are discussed in detail.

6.3.1 Linearity

Linearity is a key property that allows us to compute the DFTs of several different signals and determine the combined DFT via the summation of the individuals DFTs. This result is used extensively in digital signal processing analysis. The system output frequency response can be easily evaluated for specific frequency components and then combined to determine the total system frequency response.

If $X_i(k)$ represents the DFTs of time sequences $x_i(n)$ for $i = 1, 2, \ldots, L$, then if the time sequences are multiplied by arbitrary constants and summed, the DFT is given by the sum of the individual DFTs multiplied by the arbitrary constants. The **linearity** property is given by

$$\text{DFT}\left[\sum_{i=1}^{L} \alpha_i x_i(n) \right] = \sum_{i=1}^{L} \alpha_i X_i(k) \tag{6.5}$$

where

$$X_i(k) = \text{DFT}[x_i(n)] \quad \text{and} \quad \alpha_i = \text{arbitrary constants}$$

The DFTs of $x_i(n)$ each must be computed over N samples, where

$$N = \text{Max}[N_i] \quad i = 1, 2, \ldots, L$$

Each of the input sequences with less than N samples must be augmented with $N - N_i$ zeros before performing the DFT.

6.3.2 Circular Shift of DFT Input

The circular shift is shown in Figure 6.3 as a shift in a finite-length sequence viewed as a periodic extension of the finite sequence. If $X(k)$ is the DFT of $x(n)$, then the DFT of $x(n + m)$ is given by

$$\text{DFT}[x(n + m)] = W_N^{-km} X(k)$$

This result is derived from the IDFT expression for $x(n)$ by substituting $n + m$ in place of n.

$$x(n) = \sum_{k=0}^{N-1} X(k) W_N^{-nk}$$

$$x(n + m) =$$

$$\sum_{k=0}^{N-1} X(k) W_N^{-(n+m)k} = \sum_{k=0}^{N-1} X(k) W_N^{-mk} W_N^{-nk}$$

$$= \text{IDFT}\left[X(k) W_N^{-mk} \right]$$

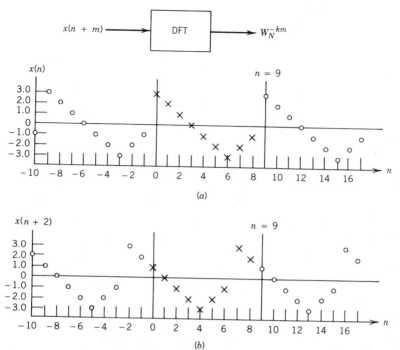

Figure 6.3 Circular shift: (*a*) periodic sequence $x(n)$; (*b*) circular shift of sequence $x(n)$ by two samples.

The DFT of $x(n + m)$ is the DFT of $x(n)$ multiplied by the complex exponential

$$\mathrm{DFT}[x(n + m)] = X(k)W_N^{-mk} \tag{6.6}$$

When using the DFT to evaluate a linear convolution, zero padding is required to negate the circular wrap-around effect. The method of zero padding two signals of different lengths in order to obtain a linear convolution from a circular convolution is discussed in Section 6.4. A similar result is obtained when the frequency is shifted by m samples.

6.3.3 Circular Shift of DFT Frequency Output

If the frequency variable k is shifted by an integer value m, then the resulting time sequence from the IDFT of the unshifted frequency sequence is modulated by a complex exponential given by

$$X(k + m) = \mathrm{DFT}[x(n)W_N^{mn}]$$

This can be rewritten to show the output in terms of the IDFT of $X(k + m)$

$$\mathrm{IDFT}[X(k + m)] = x(n)W_N^{mn} \tag{6.7}$$

Figure 6.4 illustrates the circular frequency shift for $m = -3$.

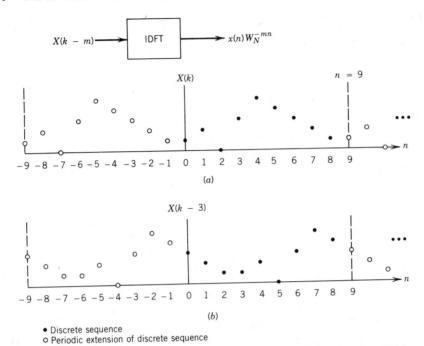

Figure 6.4 Circular shift of DFT output sequence: (*a*) frequency sequence $X(k)$; (*b*) frequency sequence $X(k)$ right-shifted by three samples.

Equation 6.7 defines a modulation process of the input signal by a complex exponential with frequency mF/N, where F is the input sampling rate. This process is often used in signal processing applications to translate the center of a frequency band to zero Hz. Note that the output of this modulation process is complex, thereby requiring processing of each of the quadrature channels.

6.3.4 Periodicity

The DFT and IDFT both produce periodic results, with period N (transform length). The periodicity is the result of the complex exponential periodic property shown in Figure 6.1. Therefore, the time and frequency samples are periodic with period N, that is,

$$x(n) = x(n + N) \qquad \text{for } n = 0, 1, 2, \ldots$$

and

$$X(k) = X(k + N) \qquad \text{for } k = 0, 1, 2, \ldots$$

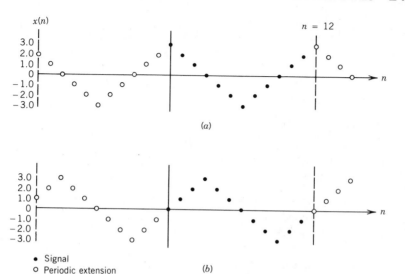

Figure 6.5 Periodicity of DFT. The signal is viewed as a periodic extension of a finite length DFT input sample sequence ($N = 12$): (*a*) even sequence; (*b*) odd sequence.

This periodic relation is demonstrated by taking the transform of a signal at frequency $(k + N)$

$$X(k + N) = \frac{1}{N} \sum_{n=0}^{N-1} x(n) W_N^{n(k+N)} = \frac{1}{N} \sum_{n=0}^{N-1} x(n) W_N^{nk} W_N^{nN}$$

where

$$W_N^{nN} = e^{-j2\pi n} = 1 \qquad \text{for all } n$$

therefore $X(k) = X(k + N)$.

An interpretation of this periodicity is that the DFT of a finite-length sequence results in a periodic sequence as shown in Figure 6.5. Figure 6.5 illustrates the periodic extension of the finite length sequences. Note that an **even** function is symmetric about its midpoint and open on the right-hand side. The missing endpoint is the start of the periodic extension of the sequences. Harris[5] distinguishes this symmetry from the conventional even symmetry (that is, a sequence with matching endpoints) by referring to it as DFT-even. The endpoint symmetry becomes important when developing windows from the DFT, which will be used to reduce spectral leakage inherent in the sidelobes of the DFT frequency response. An **odd** function is antisymmetric about the midpoint of the sequence with the right-hand side open.

6.3.5 DFT of Even and ODD Sequences

The DFT of an even sequence is purely real. The DFT of an odd sequence is purely imaginary. We consider these sequences to be even and odd, respectively, in the DFT sense discussed in the preceding section. Therefore, the

DFT can be evaluated using the cosine transform for even sequences and the sine transform for odd sequences.

$$X(k) = \frac{1}{N} \sum_{n=0}^{N-1} x(n) \cos(2\pi nk/N) \qquad \text{Even transform}$$

$$X(k) = \frac{1}{N} \sum_{n=0}^{N-1} x(n) \sin(2\pi nk/N) \qquad \text{Odd transform}$$

6.3.6 DFT of Real Sequence

The DFT of a real sequence produces symmetric real frequency components and antisymmetric imaginary frequency components about the $N/2$ DFT output point. For a real input sequence, the real part of the DFT output is an even function and the imaginary part of the DFT output is an odd function. Thus, for a real sequence the frequency components are related by

$$X(k) = X^*(N - k) \qquad k = 0, 1, 2, \ldots, N - 1 \qquad (6.8)$$

This equation is interpreted to mean that only the frequency components from zero to $N/2$ need to be computed in order to define the output completely. This agrees with the sampling result which showed that frequencies alias above half the sampling rate (i.e., $N/2$ output represents the frequency output at $F/2$ where F is the sampling rate).

This property is easily shown by taking the transform of a real sinusoidal input sequence given by

$$x(n) = \sin(2\pi nf/F) = \frac{e^{j2\pi nf/F} - e^{-j2\pi nf/F}}{2j}$$

The DFT of $x(n)$ is taken for each of the complex exponential terms

$$X(k) = \frac{1}{j2N} \sum_{k=0}^{N-1} e^{j2\pi nf/F} W_N^{nk} - \frac{1}{j2N} \sum_{k=0}^{N-1} e^{-j2\pi nf/F} W_N^{nk}$$

If $f = kF/N$ (i.e., a frequency at bin k), then

$$X(k) = \frac{1}{j2N} \sum_{k=0}^{N-1} W_N^{-nk} W_N^{nk} - \frac{1}{j2N} \sum_{k=0}^{N-1} W_N^{nk} W_N^{nk}$$

$$X(k) = \frac{1}{2j}$$

The second term is zero since the summation is over an even number of cycles (also see Problem 6.2.1). Taking the DFT at frequency bin $N - k$ yields

$$X(N - k) = \frac{1}{j2N} \sum_{k=0}^{N-1} W_N^{nk} W_N^{n(N-k)} - \frac{1}{j2N} \sum_{k=0}^{N-1} W_N^{-nk} W_N^{n(N-k)}$$

$$X(N - k) = \frac{-1}{2j}$$

Hence, Eq. 6.8 is shown for a sinusoidal signal at any bin k of the DFT. This result can be generalized to apply to any input signal. Note that the signal used was real-odd and the result is imaginary with amplitude equal to half the input. You should repeat this result for a cosine (real-even) signal and observe the output is real (Problem 6.2.3).

6.3.7 Complex-Conjugate Sequence / DFT Relationships

The inverse discrete Fourier transform is given in Eq. 6.4. Taking the complex conjugate of both sides yields

$$x^*(n) = \sum_{k=0}^{N-1} X^*(k) W_N^{nk} = N\,\text{DFT}[X^*(k)]$$

Therefore, the IDFT can be computed using the DFT via the following relationship

$$x(n) = N(\text{DFT}[X^*(k)])^*$$

The DFT of a complex-conjugate sequence can be obtained by taking the complex conjugate of both sides of the DFT given by Eq. 6.1

$$X^*(k) = \frac{1}{N} \sum_{n=0}^{N-1} x^*(n) W_N^{-nk} = \text{IDFT}[x^*(n)/N]$$

Using the periodicity property discussed in Section 6.3.4 and replacing k by $-k$, the DFT of the complex-conjugate sequence is

$$\text{DFT}[x^*(n)] = X^*(N - k)$$

Therefore, the DFT of the complex conjugate of a sequence is equal to the complex conjugate of the frequency output of the sequence reversed in order. For future preference the basic DFT properties are summarized in Table 6.1.

6.4 CIRCULAR CONVOLUTION

The process of convolution was defined in Chapter 2 for a discrete-time, linear, time-invariant system. In Section 2.4 the product relationship for the convolution sum was developed for the frequency responses of the input signal

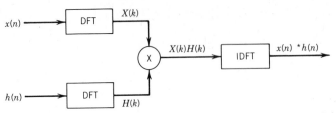

Figure 6.6 Frequency domain convolution process (digital filtering).

and system impulse response. This relationship was based on a Fourier transform for a discrete signal from $-\infty$ to $+\infty$. The convolution sum process defined in Chapter 2 is a linear convolution. Figure 6.6 illustrates the process of convolution performed as the IDFT of the product of the DFTs of two input signals. This process results in a circular convolution due to the periodicity of the DFT operation. That is, the DFT of a finite-length signal results in a periodic sequence in the frequency domain. To eliminate the circular effect and ensure the DFT method of evaluating the convolution results in a linear convolution, the signals must be zero-padded.

The circular convolution of two periodic discrete signals with period N is given by

$$y(n) = \sum_{m=0}^{N-1} x(m)h(n-m) = \sum_{m=0}^{N-1} h(m)x(n-m) \qquad (6.9)$$

where $y(n)$ is also periodic with period N. This expression is derived by performing the operation shown in Figure 6.6; that is, consider the IDFT $y(n)$ of the product of two DFTs:

$$y(n) = \sum_{k=0}^{N-1} X(k)H(k)W_N^{-kn}$$

where

$$X(k) = \frac{1}{N} \sum_{i=0}^{N-1} x(i)W_N^{ik}$$

$$H(k) = \frac{1}{N} \sum_{m=0}^{N-1} h(m)W_N^{mk}$$

substituting the expressions for $X(k)$ and $H(k)$ into the expression for $y(n)$ gives

$$y(n) = \sum_{k=0}^{N-1} \left[\frac{1}{N} \sum_{i=0}^{N-1} x(i)W_N^{ik} \right]\left[\frac{1}{N} \sum_{m=0}^{N-1} h(m)W_N^{mk} \right]W_N^{-kn}$$

Rewriting by combining the twiddle factor terms

$$y(n) = \frac{1}{N * N} \sum_{i=0}^{N-1} \sum_{m=0}^{N-1} x(i)h(m) \sum_{k=0}^{N-1} W_N^{(i+m-n)k}$$

The summation over k equals N for $i = n - m$ and zero otherwise. Therefore,

$$y(n) = \frac{1}{N} \sum_{q=0}^{N-1} x(n - m)h(m)$$

where the quantity $n - m$ is modulo N.

To demonstrate the cyclic property of the samples, several authors describe the circular convolution via the use of two concentric rotating circles. To perform circular (or cyclic) convolution, N samples of one signal are displayed equally spaced around an outer circle in a clockwise direction and N

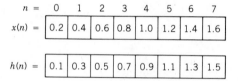

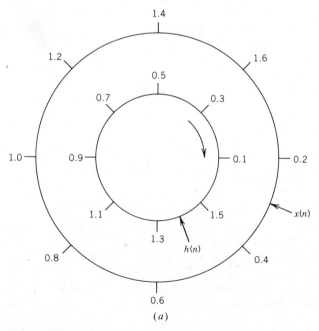

(a)

Figure 6.7 Circular convolution of two sequences: (a) concentric circle approach; (b) equivalent linear convolution approach of two periodic sequences.

$n =$	-1	-2	-3	-4	-5	-6	-7	0	1	2	3	4	5	6	7	...
$x(n)$	0.4	0.6	0.8	1.0	1.2	1.4	1.6	0.2	0.4	0.6	0.8	1.0	1.2	1.4	1.6	...

$h(-m)$	1.5	1.3	1.1	0.9	0.7	0.5	0.3	0.1				Rotate inner circle			$y(0)$
$h(1-m)$		1.5	1.3	1.1	0.9	0.7	0.5	0.3	0.1			one sample clockwise			$y(1)$
$h(2-m)$			1.5	1.3	1.1	0.9	0.7	0.5	0.3	0.1		"			$y(2)$
$h(3-m)$				1.5	1.3	1.1	0.9	0.7	0.5	0.3	0.1	"			$y(3)$
$h(4-m)$					1.5	1.3	1.1	0.9	0.7	0.5	0.3	0.1	"		$y(4)$
$h(5-m)$						1.5	1.3	1.1	0.9	0.7	0.5	0.3	0.1	"	$y(5)$
$h(6-m)$							1.5	1.3	1.1	0.9	0.7	0.5	0.3	0.1	"	$y(6)$
$h(7-m)$								1.5	1.3	1.1	0.9	0.7	0.5	0.3	0.1	$y(7)$

Process repeats with period $N = 8$

$$y(n) = \sum_{m=0}^{N-1} x(m)h(n-m)$$

$$y(0) = \sum_{m=0}^{7} x(m)h(0-m) = 5.20 \qquad y(4) = \sum_{m=0}^{7} x(m)h(4-m) = 6.48$$

$$y(1) = \sum_{m=0}^{7} x(m)b(1-m) = 6.00 \qquad y(5) = \sum_{m=0}^{7} x(m)h(5-m) = 6.00$$

$$y(2) = \sum_{m=0}^{7} x(m)h(2-m) = 6.48 \qquad y(6) = \sum_{m=0}^{7} x(m)h(6-m) = 5.20$$

$$y(3) = \sum_{m=0}^{7} x(m)h(3-m) = 6.64 \qquad y(7) = \sum_{m=0}^{7} x(m)h(7-m) = 4.08$$

$$h(-1) = h(7) \qquad h(-5) = h(3)$$
$$h(-2) = h(6) \qquad h(-6) = h(2)$$
$$h(-3) = h(5) \qquad h(-7) = h(1)$$
$$h(-4) = h(4)$$

(b)

Figure 6.7 Continued

samples of the signal are displayed on the inner circle in a counterclockwise direction starting at the same point. Next, corresponding samples on the two circles are multiplied, and the resultant products are summed to produce an output. Successive values of the circular convolution are obtained by rotating the inner circle one sample at a time in a clockwise direction; the outputs are computed via the summation of the corresponding products. The process is iterated until the inner circle first sample lines up with the first sample of the exterior circle again.

This process is illustrated in Figure 6.7, where it is noted that an equal number of samples are required to perform circular convolution.

The linear convolution of two sequences of lengths M and N will result in a sequence of length $M + N - 1$. For the circular convolution to yield the same result as the linear convolution, the sequences must be appended with zeros, as follows.

$$N_{zi} = N + M - 1 - i \qquad \text{for } i = N, M \tag{6.10}$$

where N_{zi} is the number of zeros to be padded to the end of the sequence of length M or N.

$$
\begin{bmatrix} y(0) \\ y(1) \\ y(2) \\ \cdot \\ \cdot \\ \cdot \\ y(N-2) \\ y(N-1) \end{bmatrix}
=
\begin{bmatrix}
h(0) & h(N-1) & h(N-2) & \cdots & h(2) & h(1) \\
h(1) & h(0) & h(N-1) & \cdots & h(3) & h(2) \\
h(2) & h(1) & h(0) & \cdots & h(4) & h(3) \\
\cdot & \cdot & \cdot & \cdots & \cdot & \cdot \\
\cdot & \cdot & \cdot & \cdots & \cdot & \cdot \\
\cdot & \cdot & \cdot & \cdots & \cdot & \cdot \\
h(N-2) & h(N-3) & h(N-4) & \cdots & h(2) & h(1) \\
h(N-1) & h(N-2) & h(N-3) & \cdots & h(1) & h(0)
\end{bmatrix}
\begin{bmatrix} x(0) \\ x(1) \\ x(2) \\ \cdot \\ \cdot \\ \cdot \\ x(N-2) \\ x(N-1) \end{bmatrix}
$$

$$
\begin{bmatrix} 5.20 \\ 6.00 \\ 6.48 \\ 6.64 \\ 6.48 \\ 6.00 \\ 5.20 \\ 4.08 \end{bmatrix}
=
\begin{bmatrix}
0.1 & 1.5 & 1.3 & 1.1 & 9.9 & 0.7 & 0.5 & 0.3 \\
0.3 & 0.1 & 1.5 & 1.3 & 1.1 & 0.9 & 0.7 & 0.5 \\
0.5 & 0.3 & 0.1 & 1.5 & 1.3 & 1.1 & 0.9 & 0.7 \\
0.7 & 0.5 & 0.3 & 0.1 & 1.5 & 1.3 & 1.1 & 0.9 \\
0.9 & 0.7 & 0.5 & 0.3 & 0.1 & 1.5 & 1.3 & 1.1 \\
1.1 & 0.9 & 0.7 & 0.5 & 0.3 & 0.1 & 1.5 & 1.3 \\
1.3 & 1.1 & 0.9 & 0.7 & 0.5 & 0.3 & 0.1 & 1.5 \\
1.5 & 1.3 & 1.1 & 0.9 & 0.7 & 0.5 & 0.3 & 0.1
\end{bmatrix}
\begin{bmatrix} 0.2 \\ 0.4 \\ 0.6 \\ 0.8 \\ 1.0 \\ 1.2 \\ 1.4 \\ 1.6 \end{bmatrix}
$$

(a)

(b)

Figure 6.8 Circular convolution process: (a) matrix representation of Figure 6.7; (b) plot of circular convolution process.

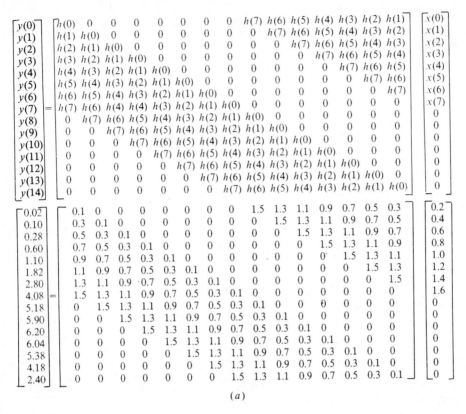

$$
\begin{bmatrix} y(0) \\ y(1) \\ y(2) \\ y(3) \\ y(4) \\ y(5) \\ y(6) \\ y(7) \\ y(8) \\ y(9) \\ y(10) \\ y(11) \\ y(12) \\ y(13) \\ y(14) \end{bmatrix} =
\begin{bmatrix}
h(0) & 0 & 0 & 0 & 0 & 0 & 0 & 0 & h(7) & h(6) & h(5) & h(4) & h(3) & h(2) & h(1) \\
h(1) & h(0) & 0 & 0 & 0 & 0 & 0 & 0 & 0 & h(7) & h(6) & h(5) & h(4) & h(3) & h(2) \\
h(2) & h(1) & h(0) & 0 & 0 & 0 & 0 & 0 & 0 & 0 & h(7) & h(6) & h(5) & h(4) & h(3) \\
h(3) & h(2) & h(1) & h(0) & 0 & 0 & 0 & 0 & 0 & 0 & 0 & h(7) & h(6) & h(5) & h(4) \\
h(4) & h(3) & h(2) & h(1) & h(0) & 0 & 0 & 0 & 0 & 0 & 0 & 0 & h(7) & h(6) & h(5) \\
h(5) & h(4) & h(3) & h(2) & h(1) & h(0) & 0 & 0 & 0 & 0 & 0 & 0 & 0 & h(7) & h(6) \\
h(6) & h(5) & h(4) & h(3) & h(2) & h(1) & h(0) & 0 & 0 & 0 & 0 & 0 & 0 & 0 & h(7) \\
h(7) & h(6) & h(4) & h(4) & h(3) & h(2) & h(1) & h(0) & 0 & 0 & 0 & 0 & 0 & 0 & 0 \\
0 & h(7) & h(6) & h(5) & h(4) & h(3) & h(2) & h(1) & h(0) & 0 & 0 & 0 & 0 & 0 & 0 \\
0 & 0 & h(7) & h(6) & h(5) & h(4) & h(3) & h(2) & h(1) & h(0) & 0 & 0 & 0 & 0 & 0 \\
0 & 0 & 0 & h(7) & h(6) & h(5) & h(4) & h(3) & h(2) & h(1) & h(0) & 0 & 0 & 0 & 0 \\
0 & 0 & 0 & 0 & h(7) & h(6) & h(5) & h(4) & h(3) & h(2) & h(1) & h(0) & 0 & 0 & 0 \\
0 & 0 & 0 & 0 & 0 & h(7) & h(6) & h(5) & h(4) & h(3) & h(2) & h(1) & h(0) & 0 & 0 \\
0 & 0 & 0 & 0 & 0 & 0 & h(7) & h(6) & h(5) & h(4) & h(3) & h(2) & h(1) & h(0) & 0 \\
0 & 0 & 0 & 0 & 0 & 0 & 0 & h(7) & h(6) & h(5) & h(4) & h(3) & h(2) & h(1) & h(0)
\end{bmatrix}
\begin{bmatrix} x(0) \\ x(1) \\ x(2) \\ x(3) \\ x(4) \\ x(5) \\ x(6) \\ x(7) \\ 0 \\ 0 \\ 0 \\ 0 \\ 0 \\ 0 \\ 0 \end{bmatrix}
$$

$$
\begin{bmatrix} 0.02 \\ 0.10 \\ 0.28 \\ 0.60 \\ 1.10 \\ 1.82 \\ 2.80 \\ 4.08 \\ 5.18 \\ 5.90 \\ 6.20 \\ 6.04 \\ 5.38 \\ 4.18 \\ 2.40 \end{bmatrix} =
\begin{bmatrix}
0.1 & 0 & 0 & 0 & 0 & 0 & 0 & 0 & 1.5 & 1.3 & 1.1 & 0.9 & 0.7 & 0.5 & 0.3 \\
0.3 & 0.1 & 0 & 0 & 0 & 0 & 0 & 0 & 0 & 1.5 & 1.3 & 1.1 & 0.9 & 0.7 & 0.5 \\
0.5 & 0.3 & 0.1 & 0 & 0 & 0 & 0 & 0 & 0 & 0 & 1.5 & 1.3 & 1.1 & 0.9 & 0.7 \\
0.7 & 0.5 & 0.3 & 0.1 & 0 & 0 & 0 & 0 & 0 & 0 & 0 & 1.5 & 1.3 & 1.1 & 0.9 \\
0.9 & 0.7 & 0.5 & 0.3 & 0.1 & 0 & 0 & 0 & 0 & 0 & 0 & 0 & 1.5 & 1.3 & 1.1 \\
1.1 & 0.9 & 0.7 & 0.5 & 0.3 & 0.1 & 0 & 0 & 0 & 0 & 0 & 0 & 0 & 1.5 & 1.3 \\
1.3 & 1.1 & 0.9 & 0.7 & 0.5 & 0.3 & 0.1 & 0 & 0 & 0 & 0 & 0 & 0 & 0 & 1.5 \\
1.5 & 1.3 & 1.1 & 0.9 & 0.7 & 0.5 & 0.3 & 0.1 & 0 & 0 & 0 & 0 & 0 & 0 & 0 \\
0 & 1.5 & 1.3 & 1.1 & 0.9 & 0.7 & 0.5 & 0.3 & 0.1 & 0 & 0 & 0 & 0 & 0 & 0 \\
0 & 0 & 1.5 & 1.3 & 1.1 & 0.9 & 0.7 & 0.5 & 0.3 & 0.1 & 0 & 0 & 0 & 0 & 0 \\
0 & 0 & 0 & 1.5 & 1.3 & 1.1 & 0.9 & 0.7 & 0.5 & 0.3 & 0.1 & 0 & 0 & 0 & 0 \\
0 & 0 & 0 & 0 & 1.5 & 1.3 & 1.1 & 0.9 & 0.7 & 0.5 & 0.3 & 0.1 & 0 & 0 & 0 \\
0 & 0 & 0 & 0 & 0 & 1.5 & 1.3 & 1.1 & 0.9 & 0.7 & 0.5 & 0.3 & 0.1 & 0 & 0 \\
0 & 0 & 0 & 0 & 0 & 0 & 1.5 & 1.3 & 1.1 & 0.9 & 0.7 & 0.5 & 0.3 & 0.1 & 0 \\
0 & 0 & 0 & 0 & 0 & 0 & 0 & 1.5 & 1.3 & 1.1 & 0.9 & 0.7 & 0.5 & 0.3 & 0.1
\end{bmatrix}
\begin{bmatrix} 0.2 \\ 0.4 \\ 0.6 \\ 0.8 \\ 1.0 \\ 1.2 \\ 1.4 \\ 1.6 \\ 0 \\ 0 \\ 0 \\ 0 \\ 0 \\ 0 \\ 0 \end{bmatrix}
$$

(a)

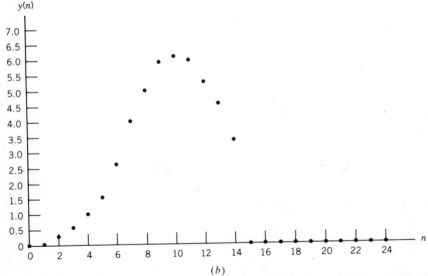

(b)

Figure 6.9 Linear convolution process: (a) matrix approach; (b) plot of y(n).

For example, consider two sequences of lengths $M = 8$ and $N = 16$. The linear convolution would produce an output sequence of length 23. To construct two sequences of length 23, the sequence of length 8 is padded with 15 zeros at the end of the sequence and the sequence of length 16 is padded with 7 zeros. For two sequences of equal length N, the number of zeros to be appended is

$$N_z = N - 1$$

While the concentric circle approach is useful conceptually it is cumbersome for constructing specific numerical solutions. A more convenient method is the **matrix approach**, whereby solutions to extensive problems can be obtained by matrix algebra. In the matrix representation, one sequence repeated via circular shifting of the samples is shown in Figure 6.8a for the same two sequences given in Figure 6.7. A linear convolution is produced from the circular convolution of Figure 6.8 by appending zeros (Eq. 6.10) to the sequences. The resulting linear convolution is shown in Figure 6.9.

The preceding results are limited to a convolution between two finite-length sequences. Generally, applications require real-time continuous

TABLE 6.1
DFT PROPERTIES

Time Domain Sequence $x(n)$	Frequency Domain Sequence $X(k)$
6.3.1 Linearity	
$\displaystyle\sum_{i=1}^{I} \alpha_i x_i(n)$	$\displaystyle\sum_{i=1}^{I} \alpha_i X_i(k)$
6.3.2 Circular shift of $x(n)$ (time shifting)	
$\quad x(n+m)$	$W_N^{-km} X(k)$
$\quad x(n-m)$	$W_N^{km} X(k)$
6.3.3 Circular shift of $X(k)$ (quadrature modulation)	
$\quad x(n)W_N^{mn}$	$X(k+m)$ (*Left shift*)
$\quad x(n)W_N^{-mn}$	$X(k-m)$ (*Right shift*)
6.3.4 Periodicity	
$\quad x(n) = x(n+N)$	$X(k) = X(k+N)$
6.3.5 DFT of even and odd sequences	
$\quad x(n)$ even	$X(k) = \dfrac{1}{N} \displaystyle\sum_{n=0}^{N-1} x(n) \cos(2\pi nk/N)$
$\quad x(n)$ odd	$X(k) = \dfrac{-j}{N} \displaystyle\sum_{n=0}^{N-1} x(n) \sin(2\pi nk/N)$
6.3.6 DFT of real sequence	
$\quad x(n)$ real	$X(k) = X^*(N-k)$
6.3.7 Complex conjugate sequences	
$\quad x^*(n)$	$X^*(N-k) = X^*(-k)$

processing in performing a convolution sum between an FIR filter impulse response and long duration discrete-time signals. The output of the convolution is computed for each successive input sample. Since the input sequence is not limited to a small number of nonzero values, an approach is required that permits sections of input samples to be processed and the results combined to form an equivalent linear convolution. The two algorithms for performing linear convolution of long sequences are referred to as the **overlap-save** and **overlap-add methods**. These methods will be presented in Section 6.9.

For future reference, the basic DFT properties are summarized in Table 6.1.

6.5 EIGHT-POINT DFT EXAMPLE

Equation 6.1 defined the DFT of an arbitrary input signal $x(n)$. In order to gain some insight into the computation of the DFT, an eight-point DFT is performed on a real input signal consisting of a sine and cosine component. Figure 6.10 illustrates the periodic signal components. Each component was selected with frequency exactly at the center of the DFT bin frequency,

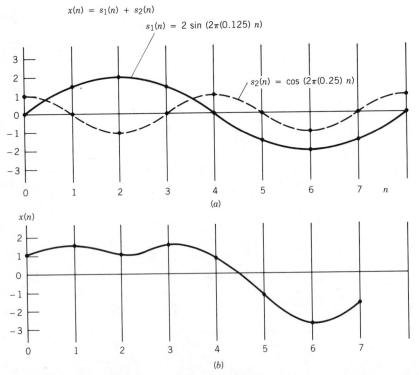

Figure 6.10 Input signal for eight-point DFT example: (a) signal components s_1 and s_2; (b) composite sum signal $x(n)$.

$$X(0) = \frac{1}{8} \sum_{n=0}^{7} x(n) \cos[0] - j\frac{1}{8} \sum_{n=0}^{7} x(n) \sin[0] = 0 + j0$$

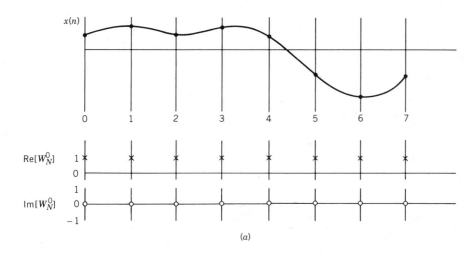

(a)

$$X(1) = \frac{1}{8} \sum_{n=0}^{7} x(n) \cos\left[\frac{2\pi}{8}n\right] - j\frac{1}{8} \sum_{n=0}^{7} x(n) \sin\left[\frac{2\pi}{8}n\right] = 0 + j1$$

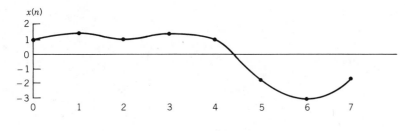

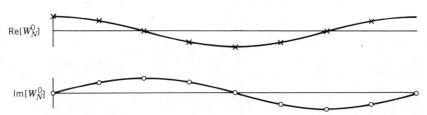

Note: The real and imaginary parts of W_N^n cover one period over the DFT length for $k = 1$.

Figure 6.11 DFT process for input signal $x(n)$: (*a*) DFT output for bin 0; (*b*) DFT output for bin 1; (*c*) DFT output for bin 2.

$$X(2) = \frac{1}{8} \sum_{n=0}^{7} x(n) \cos\left[\frac{4\pi}{8}n\right] - j\frac{1}{8} \sum_{n=0}^{7} x(n) \sin\left[\frac{4\pi}{8}n\right] = 0.5 - j0$$

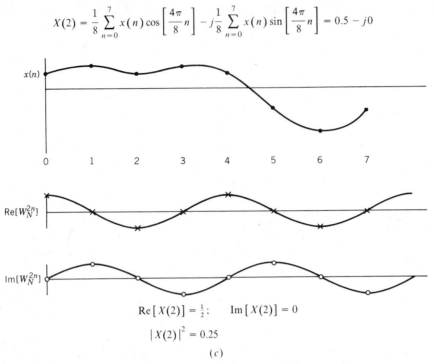

$$\text{Re}\left[X(2)\right] = \tfrac{1}{2}; \qquad \text{Im}\left[X(2)\right] = 0$$

$$\left|X(2)\right|^2 = 0.25$$

(c)

Note: The real and imaginary parts of W_N^{2n} cover two periods over the DFT length for $k = 2$.

Figure 6.11 Continued

assuming an input sampling rate of 1 Hz. The input signal is given by

$$x(n) = s_1(n) + s_2(n)$$
$$s_1(n) = 2\sin\left[2\pi(1)n/8\right]$$
$$s_2(n) = \cos\left[2\pi(2)n/8\right]$$

The DFT of $x(n)$ is computed for each of the eight output points. For an assumed sampling frequency of 1 Hz, the frequency resolution is obtained from Eq. 6.2, that is, $F/N = 1/8 = 0.125$ Hz. The s_1 signal frequency component is at bin 1 (0.125 Hz) and the s_2 signal frequency component is at bin 2 (0.25 Hz). From the properties for even and odd sequences (6.3.5) and real sequences (6.3.6), the sine component should appear as an imaginary output as bin 1 with its complex conjugate at bin 7, and the cosine component should appear as a real output at bin 2 and bin 6.

The DFT process is shown in Figure 6.11. Figure 6.11a illustrates the DFT calculation for bin 0. The twiddle factors are given by the amplitudes of the real and imaginary terms equal to unity and zero, respectively. The DFT calculation for bin 1 is shown in Figure 6.11b. Note that the twiddle factors represent one cycle of the unit circle given in Figure 6.1. Therefore, components with frequencies that exactly match the bin will be maximized. Frequencies at other bin frequencies will be zero. This is because the nulls of the frequency response of the bin occur at the bin centers. Frequency components that are between bin centers will smear across bins based on the frequency response. This will be discussed in detail in the next section. Likewise, the calculation for bin 2 is shown in Figure 6.11c. Bins 4 through 7 were computed

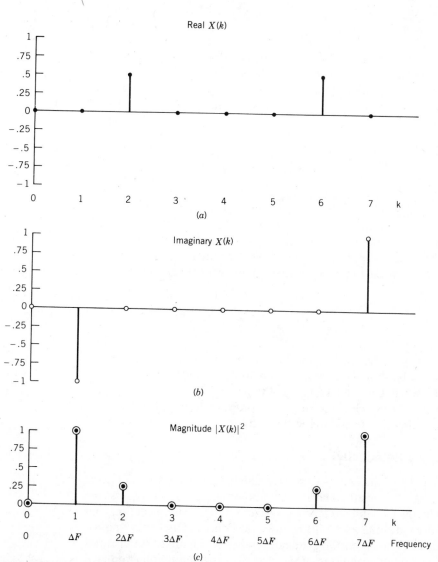

Figure 6.12 Composite frequency domain output of eight-point DFT example: (*a*) real part of frequency response; (*b*) imaginary part of frequency response; and (*c*) magnitude-squared frequency response.

using Eq. 6.8. Finally, the DFT output for frequency bin indices $k = 0, 1, \ldots, 7$ are presented in Figure 6.12, where it is noted that the cosine component is an even function and the sine component is an odd function. Also, since $x(n)$ is a real function, the spectrum repeats at $N/2$.

6.6 DFT FREQUENCY RESPONSE CHARACTERISTICS

The DFT frequency response of Eq. 6.3 is shown in Figure 6.2. The DFT frequency response is characterized by the following properties: frequency selectivity, spectral leakage, scalloping loss, and equivalent noise bandwidth. These inherent properties of the DFT directly relate to the spectral analysis performance. A technique for modifying the frequency response resulting in improved spectral analysis performance is to use a weighting (window) function. If the weighting function, $a(n)$, is applied to the input signal, then the DFT output is given by

$$Y(k) = \frac{1}{N} \sum_{n=0}^{N-1} a(n)x(n)W_N^{nk} \tag{6.11}$$

The unweighted DFT output is obtained by setting $a(n) = 1$ for all n, and is referred to as uniform or rectangular weighting. For a given weighting function, the frequency response is obtained by substituting a complex exponential for $x(n)$ and computing the output $Y(k)$ as the frequency is varied. For $a(n) = 1$, the frequency response is defined as (from Eq. 6.3)

$$X(k) = e^{-j\Omega(N-1)/2} \frac{\sin(\Omega N/2)}{N \sin(\Omega/2)}, \qquad \Omega = 2\pi[(k/N) - (f/F)]$$

This is recognized as the Dirichlet kernel[5] (i.e., the transform of the rectangular window). It will be shown that many of the useful windows are derived as summations of shifted Dirichlet kernels multiplied by coefficients to obtain desired frequency response characteristics. The properties of the DFT weighting functions are discussed in the following sections.

6.6.1 Frequency Selectivity

Frequency selectivity is the ability to resolve different frequency components of the input signal. Several common definitions are used to discuss the frequency selectivity including frequency resolution, analysis resolution, bin resolution, 3-dB bandwidth, and 6-dB bandwidth. Components within the bin crossover points of the adjacent bins are not resolvable. Figure 6.13 illustrates the frequency selectivity definitions. The use of a weighting function will broaden the mainlobe of the DFT bin frequency response, resulting in loss of frequency selectivity. This is necessary in order to reduce losses resulting from

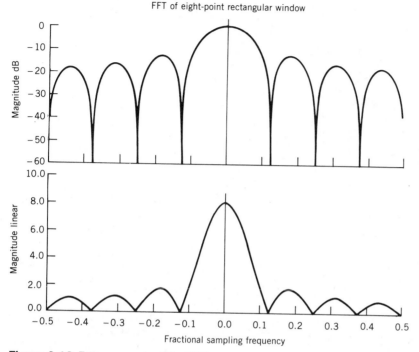

Figure 6.13 Frequency selectivity: DFT magnitude frequency response.

the spectral leakage and scalloping loss characteristics of the rectangular weighted DFT output.

6.6.2 Spectral Leakage

Harris[5] shows that a discontinuity results at the boundary of the finite length sequence due to the periodicity property of the DFT. This occurs for all signal components that do not exactly fall on one of the bin centers. Figure 6.14 depicts this phenomena, showing the samples that result if they were taken directly from the signal versus the discontinuity resulting from the periodic extension of the signal. This discontinuity gives rise to the leakage (or spectral contributions) across the entire frequency set. Weighting functions reduce the contribution of the samples near the endpoints, and therefore reduce the discontinuity and its effect on the frequency response.

The high sidelobe levels of the DFT response result in false frequency detections. Therefore it is desirable to reduce the sidelobe levels and the resultant false detections via use of a weighting function. Figure 6.15 presents an illustration of adjacent bin outputs for an input signal with frequency at the center between two bins. The output is shown for several bins on each side of the frequency. Note that the levels for the adjacent bins are only 13 dB

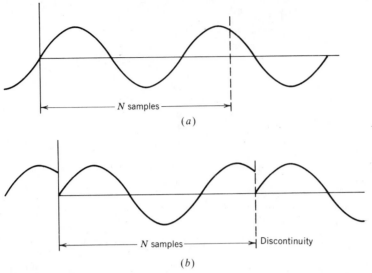

Figure 6.14 Discontinuity resulting from periodic extension of signal by DFT process: (*a*) continuous signal; (*b*) signal showing discontinuity due to DFT process.

down and that subsequent bins fall off at approximately 4 dB per bin farther away from the mainlobe.

6.6.3 Scalloping Loss

Figure 6.16 illustrates the bin crossovers that result in a signal detection loss due to the reduced signal levels at frequency points off the DFT bin centers. This is commonly referred to as **scalloping loss (SL)** or the **picket-fence effect**. Since the weighting function modifies the frequency responses, it can be used to reduce the bin crossover losses. The maximum loss occurs at the bin crossover given by evaluating the frequency response at the frequency midway between adjacent bin centers, that is, at frequencies given by $(k + 0.5)F/N$, k integer valued. Therefore, SL represents the maximum reduction in processing gain due to signal frequency and is given by

$$\text{SL} = \frac{\left| \sum\limits_{n=0}^{N-1} a(n)e^{-j\pi n/N} \right|}{\sum\limits_{n=0}^{N-1} a(n)} = \frac{A(\pi/N)}{NA(0)} \tag{6.12}$$

If the signal is equally likely to occur at any frequency across the bin, an average scalloping loss is more appropriate to use in performance calculations.

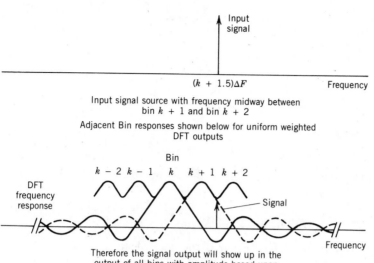

Input signal source with frequency midway between
bin $k + 1$ and bin $k + 2$

Adjacent Bin responses shown below for uniform weighted
DFT outputs

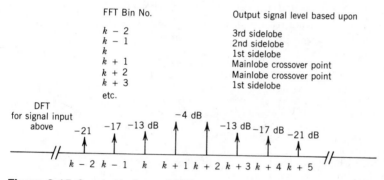

Therefore the signal output will show up in the
output of all bins with amplitude based upon
the bin responses

FFT Bin No.	Output signal level based upon
$k - 2$	3rd sidelobe
$k - 1$	2nd sidelobe
k	1st sidelobe
$k + 1$	Mainlobe crossover point
$k + 2$	Mainlobe crossover point
$k + 3$	1st sidelobe
etc.	

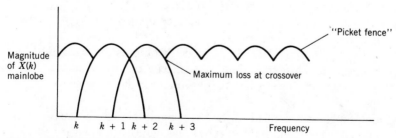

Figure 6.15 Spectral leakage of DFT.

Figure 6.16 Scalloping loss of DFT.

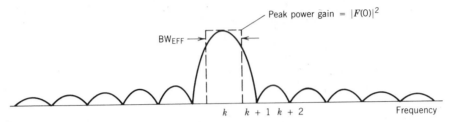

Figure 6.17 Equivalent noise bandwidth.

This loss is computed by integrating over the weighting function between the bin crossover points.

6.6.4 Equivalent Noise Bandwidth (ENBW)

Another key characteristic of the DFT output is the equivalent noise bandwidth. The **ENBW** is interpreted as the bandwidth of a hypothetical rectangular filter, which passes a white noise* signal with a mean square value equal to the actual filter output, as illustrated in Figure 6.17. ENBW is used as a measure for comparing the desirability of various weighting functions. The equivalent noise bandwidth is given by

$$ENBW = \frac{N \displaystyle\sum_{n=0}^{N-1} a^2(n)}{\left[\displaystyle\sum_{n=0}^{N-1} a(n) \right]^2} \qquad (6.13)$$

The ENBW is used to normalize the power spectrum measurements to power spectral density (power per unit frequency).

6.6.5 Overlap Correlation

It will be shown in Chapter 8 that in order to obtain good DFT spectral analysis results many contiguous DFT outputs must be computed and averaged. This results in a reduction of the variance of the DFT bin outputs containing noise-only by a factor equal to the number of outputs averaged. The details will be shown in Chapter 8, but the concept of redundancy processing and overlap correlation is presented here owing to its dependencies on weighting functions.

When a weighting function is applied to the input data, the data points near each end are reduced. This reduces the spectral leakage as discussed in Section 6.5.2, but is also creates voids in the input data being analyzed, as

*White noise refers to a random input signal having a flat power spectrum for all frequencies. We consider signals to be white provided they have a flat spectrum over a specified range of frequencies.

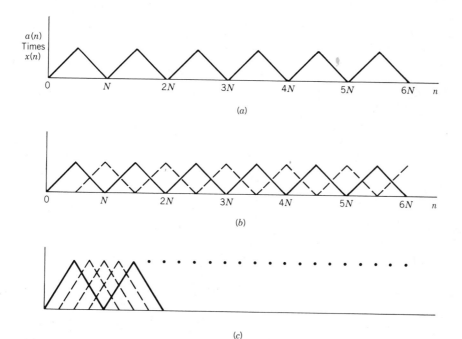

Figure 6.18 Triangular weighted data sections showing: (*a*) nonoverlapped data sections; (*b*) 50% overlapped data sections; (*c*) 75% overlapped data sections.

shown in Figure 6.18*a* using a triangular weighting function. It has been shown[5,6] that the performance is improved if additional DFTs are inserted to fill the voids, as shown in Figure 6.18*b*. The process of filling the voids via additional DFTs is referred to as **redundancy (overlap) processing**.

The performance enhancement achieved by redundancy processing is dependent on the weighting function and the amount of overlap used. Figures 6.18*b*, and *c* illustrate 50% and 75% overlaps of the data used for successive DFT computations. To quantify the performance benefits due to the additional processing performed, the correlation coefficient between successive DFTs for a given fractional overlap r, and bin is computed

$$c(r) = \frac{\displaystyle\sum_{n=0}^{rN-1} a(n)a[n + (1 - r)N]}{\displaystyle\sum_{n=0}^{N-1} a^2(n)} \qquad (6.14)$$

Generally, 50% to 75% overlap processing provides 90% of the possible performance improvement for most weighting functions (see Chapter 8).

6.7 WEIGHTING FUNCTIONS

Weighting functions (i.e., windows)[5] are used to modify the DFT frequency output response characteristics. Weighting functions are used in conjunction with DFT spectral analysis to reduce spectral leakage, reduce scalloping loss, provide variable resolution, and change DFT output bin centers. The weighted DFT output represents the product of two sequences; therefore the output can also be computed as the convolution of the DFTs of each of the sequences

$$Y(k) = \sum_{n=0}^{N-1} A(n) X(k-n) \qquad (6.15)$$

Equation 6.15 is the form generally used to implement weighting functions for spectral processing. Very good results can be achieved with weighting functions with five or fewer nonzero $A(n)$ values, which was shown by Harris.[5]

Table 6.2 presents a summary of the key characteristics of several weighting functions taken from Harris.[5] Each of these weighting functions is described in the following sections based on the results of Harris. For more detail, you should refer to the referenced paper, which provides one of the most comprehensive treatments of weighting functions for harmonic analysis available.

6.7.1 Triangular Weighting Function

The triangular weighting function[7] for the DFT is defined as

$$a(n) = \begin{cases} 2n/N & n = 0, 1, \ldots, N/2 \\ a(N-n) & n = (N/2)+1, (N/2)+2, \ldots, N-1 \end{cases} \qquad (6.16)$$

and is illustrated in Figure 6.19 for a 16-point window. Note that the weighting function is symmetrical with the endpoint missing at N. The frequency response is computed by evaluating Eq. 6.11 with the input signal $x(n)$ equal to a complex exponential as the frequency is varied. This results in an expression that is given by

$$A(f) = 2e^{-j2\pi f((N/2)-1)/F} \left[\frac{\sin(\pi f N/2F)}{N \sin(\pi f/F)} \right]^2$$

which is seen to be equal to the square of the rectangular weighted DFT frequency response of length $N/2$. This result can be explained by viewing the triangular window of length N as a convolution of two $N/2$-point rectangular windows. Actually, $N+1$ points would result if the zero endpoints were counted. Since convolution in the time domain is equivalent to multiplication in the frequency domain, the expected result is the product of the spectrums of the two rectangular windows. Therefore, the mainlobe width is doubled, the first sidelobe is approximately -26 dB down from the peak of the mainlobe, and the maximum scalloping loss that occurs at the bin crossover point is

TABLE 6.2
WEIGHTING FUNCTION CHARACTERISTICS (from Harris[5])

Window		Highest Side Lobe Level (dB)	Side Lobe Fall Off (dB / OCT)	Coherent Gain	Equiv. Noise BW (BINS)	3.0-dB BW (BINS)	Scallop Loss (dB)	Worst Case Process Loss (dB)	6.0-dB BW (BINS)	Overlap Correlation (PCNT) 75% OL	50% OL
Rectangle		−13	−6	1.00	1.00	0.89	3.92	3.92	1.21	75.0	50.0
Triangle		−27	−12	0.50	1.33	1.28	1.82	3.07	1.78	71.9	25.0
cos$^\alpha$ (x)	α = 1.0	−23	−12	0.64	1.23	1.20	2.10	3.01	1.65	75.5	31.8
Hanning	α = 2.0	−32	−18	0.50	1.50	1.44	1.42	3.18	2.00	65.9	16.7
	α = 3.0	−39	−24	0.42	1.73	1.66	1.08	3.47	2.32	56.7	8.5
	α = 4.0	−47	−30	0.38	1.94	1.86	0.86	3.75	2.59	48.6	4.3
Hamming		−43	−6	0.54	1.36	1.30	1.78	3.10	1.81	70.7	23.5
Riesz		−21	−12	0.67	1.20	1.16	2.22	3.01	1.59	76.5	34.4
Riemann		−26	−12	0.59	1.30	1.26	1.89	3.03	1.74	73.4	27.4
De La Valle Poussin		−53	−24	0.38	1.92	1.82	0.90	3.72	2.55	49.3	5.0
Tukey	α = 0.25	−14	−18	0.88	1.10	1.01	2.96	3.39	1.38	74.1	44.4
	α = 0.50	−15	−18	0.75	1.22	1.15	2.24	3.11	1.57	72.7	36.4
	α = 0.75	−19	−18	0.63	1.36	1.31	1.73	3.07	1.80	70.5	25.1
Bohman		−46	−24	0.41	1.79	1.71	1.02	3.54	2.38	54.5	7.4
Poisson	α = 2.0	−19	−6	0.44	1.30	1.21	2.09	3.23	1.69	69.9	27.8
	α = 3.0	−24	−6	0.32	1.65	1.45	1.46	3.64	2.08	54.8	15.1
	α = 4.0	−31	−6	0.25	2.08	1.75	1.03	4.21	2.58	40.4	7.4
Hanning Poisson	α = 0.5	−35	−18	0.43	1.61	1.54	1.26	3.33	2.14	61.3	12.6
	α = 1.0	−39	−18	0.38	1.73	1.64	1.11	3.50	2.30	56.0	9.2
	α = 2.0	NONE	−18	0.29	2.02	1.87	0.87	3.94	2.65	44.6	4.7
Cauchy	α = 3.0	−31	−6	0.42	1.48	1.34	1.71	3.40	1.90	61.6	20.2
	α = 4.0	−35	−6	0.33	1.76	1.50	1.36	3.83	2.20	48.8	13.2
	α = 5.0	−30	−6	0.28	2.06	1.68	1.13	4.28	2.53	38.3	9.0
Gaussian	α = 2.5	−42	−6	0.51	1.39	1.33	1.69	3.14	1.86	67.7	20.0
	α = 3.0	−55	−6	0.43	1.64	1.55	1.25	3.40	2.18	57.5	10.6
	α = 3.5	−69	−6	0.37	1.90	1.79	0.94	3.73	2.52	47.2	4.9

TABLE 6.2 (*Continued*)

Window	Highest Side Lobe Level (dB)	Side Lobe Fall Off (dB/OCT)	Coherent Gain	Equiv. Noise BW (BINS)	3.0-dB BW (BINS)	Scallop Loss (dB)	Worst Case Process Loss (dB)	6.0-dB BW (BINS)	Overlap Correlation (PCNT) 75% OL	50% OL
Dolph Chebyshev $\alpha = 2.5$	−50	0	0.53	1.39	1.33	1.70	3.12	1.85	69.6	22.3
$\alpha = 3.0$	−60	0	0.48	1.51	1.44	1.44	3.23	2.01	64.7	16.3
$\alpha = 3.5$	−70	0	0.45	1.62	1.55	1.25	3.35	2.17	60.2	11.9
$\alpha = 4.0$	−80	0	0.42	1.73	1.65	1.10	3.48	2.31	55.9	8.7
Kaiser Bessel $\alpha = 2.0$	−46	−6	0.49	1.50	1.43	1.46	3.20	1.99	65.7	16.9
$\alpha = 2.5$	−57	−6	0.44	1.65	1.57	1.20	3.38	2.20	59.5	11.2
$\alpha = 3.0$	−69	−6	0.40	1.80	1.71	1.02	3.56	2.39	53.9	7.4
$\alpha = 3.5$	−82	−6	0.37	1.93	1.83	0.89	3.74	2.57	48.8	4.8
Barcilon Temes $\alpha = 3.0$	−53	−6	0.47	1.56	1.49	1.34	3.27	2.07	63.0	14.2
$\alpha = 3.5$	−58	−6	0.43	1.67	1.59	1.18	3.40	2.23	58.6	10.4
$\alpha = 4.0$	−68	−6	0.41	1.77	1.69	1.05	3.52	2.36	54.4	7.6
Exact Blackman	−51	−6	0.46	1.57	1.52	1.33	3.29	2.13	62.7	14.0
Blackman	−58	−18	0.42	1.73	1.68	1.10	3.47	2.35	56.7	9.0
Minimum 3-Sample Blackman-Harris	−67	−6	0.42	1.71	1.66	1.13	3.45	1.81	57.2	9.6
Minimum 4-Sample Blackman-Harris	−92	−6	0.36	2.00	1.90	0.83	3.85	2.72	46.0	3.8
61 dB 3-Sample Blackman-Harris	−61	−6	0.45	1.61	1.56	1.27	3.34	2.19	61.0	12.6
74 dB 4-Sample Blackman-Harris	−74	−6	0.40	1.79	1.74	1.03	3.56	2.44	53.9	7.4
4-Sample Kaiser-Bessel 3.0	−69	−6	0.40	1.80	1.74	1.02	3.56	2.44	53.9	7.4

Source: Reference 5.

266

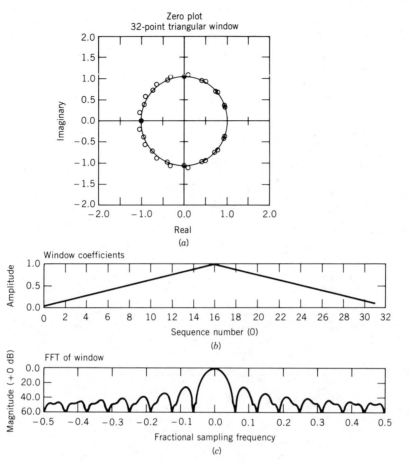

Figure 6.19 Triangular weighting function.

approximately -1.82 dB. The sidelobes falloff at a rate of approximately -12 dB per octave. An **octave** is defined as a successive doubling in frequency away from the reference point (e.g., 100, 200, 400, 800, . . .). As shown in Table 6.2, the ENBW is 1.33.

Harris[5] demonstrates that the triangular window is the simplest window that exhibits a positive transform. Any weighting function (of half-extent) can be convolved with itself to obtain a positive transform that is the square of the original window's transform.

6.7.2 Hann (Hanning) Weighting Function

The Hann window, more commonly known as the Hanning window,[7] is a member of the $\cos^{\alpha}(\theta)$ family of weighting functions with the parameter

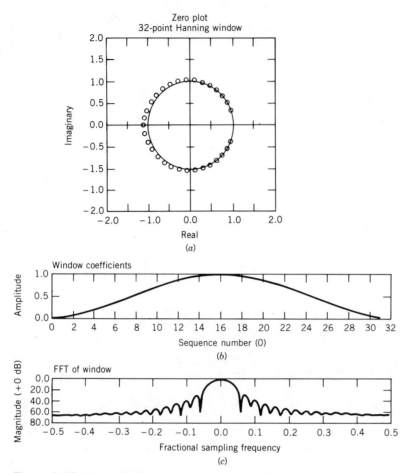

Figure 6.20 Hann weighting function.

$\alpha = 2$. The Hann weighting function is defined as

$$a(n) = 0.5[1 - \cos(2\pi n/N)], \qquad n = 0, 1, 2, \ldots, N - 1 \quad (6.17)$$

and is illustrated in Figure 6.20a. The frequency response is computed by evaluating Eq. 6.11 with a complex exponential input. Rewriting $a(n)$ in complex exponential form and substituting into Eq. 6.11 results in

$$Y(k) = \frac{0.5}{N} \sum_{n=0}^{N-1} x(n) W_N^{nk} - \frac{0.25}{N} \sum_{n=0}^{N-1} x(n) W_N^{n(k+1)}$$

$$- \frac{0.25}{N} \sum_{n=0}^{N-1} x(n) W_N^{n(k-1)}$$

where the first term is 0.5 times the DFT of $x(n)$ for bin k, the second and third terms represent 0.25 times the DFTs of $x(n)$ for bins $k - 1$ and $k + 1$, respectively.

$$Y(k) = -0.25X(k - 1) + 0.5X(k) - 0.25X(k + 1)$$

From this expression the frequency response is easily computed as the summation of the appropriate unweighted DFT bin frequency responses.

$$A(f) = 0.5e^{j\Omega(N-1)/2} \frac{\sin(\Omega_1 N/2)}{N \sin(\Omega_1/2)} - 0.25e^{j\Omega(N-1)/2} \frac{\sin(\Omega_2 N/2)}{N \sin(\Omega_2/2)}$$

$$- 0.25e^{j\Omega(N-1)/2} \frac{\sin(\Omega_3 N/2)}{N \sin(\Omega_3/2)}$$

$$\Omega_1 = 2\pi[(f/F) - (k/N)]$$
$$\Omega_2 = 2\pi[(f/F) - ((k - 1)/N)]$$
$$\Omega_3 = 2\pi[(f/F) - ((k + 1)/N)] \qquad (6.18)$$

The frequency response is shown in Figure 6.20*b*. Note that the first sidelobe is approximately -32 dB, the bin crossover point is approximately -1.42 dB, and the ENBW is 1.5.

Computationally, the Hanning weighting function can be computed by the convolution summation of the three unweighted outputs multiplied by -0.25, 0.5, and -0.25, respectively. The DFT unweighted output is complex, requiring that the operation be performed on the real and imaginary outputs. Therefore, each bin output can be obtained via 10 shift operations and 4 adds.

6.7.3 Hamming Weighting Function

The Hamming weighting function[7] is very much like the Hann window, but with different coefficients for the cosine terms

$$a(n) = 0.54 - 0.46 \cos(2\pi n/N) \qquad n = 0, 1, 2, \dots, N - 1 \quad (6.19)$$

The frequency response of the Hamming window is of the same form as the Hann window with the coefficients replaced by 0.54 and 0.23. The coefficients were computed by adjusting the contribution of each of the three kernels of the Hann frequency response to achieve cancellation of the first sidelobe. Since the coefficients are not simple (i.e., 0, 1, or 2^i), implementation via convolution in the frequency domain requires six multiplications and four adds for each weighted output bin. The Hamming window characteristics are given in Table 6.2. Note that the first sidelobe is reduced to approximately -43 dB, the sidelobes fall off at -6 dB per octave, the scalloping loss is 1.78 dB, and the ENBW is equal to 1.36.

6.7.4 Blackman-Harris Weighting Function

The Hann and Hamming weighting functions[5] are examples of time domain raised cosine bell-shaped functions that result in frequency responses constructed as summations of shifted Dirichlet kernals given by

$$D(f) = e^{j\Omega(N-1)/2} \frac{\sin(\Omega N/2)}{N \sin(\Omega/2)}, \qquad \Omega = 2\pi[(f/F) - (k/N)] \quad (6.20)$$

$D(f)$ is the basic frequency response function for the rectangular weighted input signal. This concept can be generalized by expanding the weighting function as a summation of cosine terms

$$a(n) = \sum_{m=0}^{M} (-1)^m b(m) \cos(2\pi nm/N), \qquad n = 0, 1, \ldots, M$$

where the summation of the $b(m)$ coefficients is constrained to unity and the general frequency response function is given by

$$A(f) = \sum_{m=0}^{M} (-1)^m \frac{b(m)}{2} [D(f - 2\pi m/N) + D(f + 2\pi m/N)]$$

The Blackman[5,7] weighting function is equal to a three-term expression with $b(0) = 0.42$, $b(1) = 0.50$, and $b(3) = 0.08$. Harris uses a gradient search technique to find a set of windows with three and four nonzero terms that result in minimum sidelobe levels. He extends this set by trading-off mainlobe width for sidelobe level. Harris calls this family of weighting functions the Blackman-Harris windows. The four windows summarized in Table 6.2 are itemized in Table 6.3.

TABLE 6.3
BLACKMAN-HARRIS WINDOW COEFFICIENTS

Coefficients	3-Term (−67 dB)	3-Term (−61 dB)	4-Term (−92 dB)	4-Term (−74 dB)
a_0	0.42323	0.44959	0.35875	0.40217
a_1	0.49755	0.49364	0.48829	0.49703
a_2	0.07922	0.05677	0.14128	0.09392
a_3	—	—	0.01168	0.00183

Blackman-Harris Window Function

$$a(n) = a_0 - a_1 \cos\left(\frac{2\pi}{N-1}n\right) + a_2 \cos\left(\frac{2\pi}{N-1}2n\right) - a_3 \cos\left(\frac{2\pi}{N-1}3n\right)$$

$$n = 0, 1, \ldots, N-1$$

Source: Reference 5.

6.7.5 Kaiser-Bessel Weighting Function

Kaiser[8] developed an approximation to the zero-order prolate-spheroidal wave functions in terms of the zero-order modified Bessel function of the first kind, I_0. The coefficients are given by

$$a(n) = \frac{I_0(\pi\beta)}{I_0(\pi\alpha)} \qquad 0 \le |n| \le N/2$$

$$\beta = \alpha\left[1 - (2n/N)^2\right]^{0.5}$$

$$I_0(x) = \sum_{k=0}^{\infty}\left[\frac{(x/2)^k}{k!}\right]^2$$

where the parameter $\pi\alpha$ is half the time-bandwidth product, TW. This provides a measure of the mean square time and frequency concentration of the weighting function, which are inversely related. Harris computed the characteristics of these weighting functions for four values of α, as listed in Table 6.2. The functions were computed by restricting the frequency and determining as a function of restricted time T the maximum energy in the frequency band W.

6.7.6 Weighting Function Performance

The characteristics of some popular weighting functions have been presented in the preceding sections. To gain insight into their relative effectiveness, you are encouraged to experiment with them. Harris chose an harmonic analysis sample problem to discuss the relative merits of the windows. The analysis considered two signals separated by six bins. One signal was placed half way between bins, and the smaller signal (-40 dB) was placed on the center of the bin. Using the various weighting functions, a comparative spectral analysis was performed to determine the relative ability of the windows to provide frequency selectivity between the two signals. It was concluded that the Kaiser-Bessel and Blackman-Harris windows were top performers. The Kaiser-Bessel window was preferred because the coefficients are easily generated by a computer and sidelobe versus time-bandwidth product can be varied. For applications where the implementation is performed as a spectral comvolution, it is important to have few coefficients, and therefore the Blackman-Harris weighting functions provide excellent performance and computational efficiency.

6.8 FAST FOURIER TRANSFORM (FFT)

Spectral analysis applications often require DFTs in realtime on contiguous sets of input samples. Computation of the DFT (Eq. 6.1) for N input sample points requires N^2 complex multiplies and $N^2 - N$ complex additions for N

frequency output points. This assumes that all twiddle factor coefficients require complex multiplications, even those that have real or imaginary parts equal to 1 or 0. Accounting for these simple coefficients is addressed in the following sections. In addition to multiplication and addition, storage must be provided for data. Also, the twiddle factors W_N^p must be either computed or stored for use in the computation.

The realtime computational requirements of the DFT are expressed as a function of the DFT input data rate, F.

$$\text{DFT CMPS} = NF \qquad \text{(Complex multiplies per second)}$$

$$\text{DFT CAPS} = (N - 1)F \qquad \text{(Complex additions per second)}$$

Typically, the requirements are shown as raw multiplies and additions required to implement the transform. We have chosen the CMPS and CAPS representations since the rate at which the processing is performed is key for a hardware/software perspective. Sampling rates and DFT lengths in the thousands are frequently encountered in applications resulting in CMPS and CAPS requirements in the millions. These requirements made the application of the DFT prohibitive.

The FFT is a fast algorithm for efficient implementation of the DFT where the number of time samples of the input signal N are transformed into N frequency points. The computational requirements of the FFT are expressed as

$$\text{FFT CMPS} = \frac{F}{2} \log_2 N$$

$$\text{FFT CAPS} = F \log_2 N$$

TABLE 6.4
DFT / FFT COMPUTATION REQUIREMENTS ($F = 1000$ s/s)

Number of Samples	DFT CMPS	CAPS	FFT CMPS	CAPS	Savings (%) CMPS	CAPS
8	8000	7000	1500	3000	81.25	57.14
16	16000	15000	2000	4000	87.50	73.33
32	32000	31000	2500	5000	92.19	83.87
64	64000	63000	3000	6000	95.31	90.48
128	128000	127000	3500	7000	97.27	94.49
256	256000	255000	4000	8000	98.44	96.86
512	512000	511000	4500	9000	99.12	98.24
1024	1024000	1023000	5000	10000	99.51	99.02
2048	2048000	2047000	5500	11000	99.73	99.46
4096	4096000	4095000	6000	12000	99.85	99.71
8192	8192000	8191000	6500	13000	99.92	99.84
16384	16384000	16383000	7000	14000	99.96	99.91

These expressions are developed in the following paragraphs. Table 6.4 presents a comparison of the DFT and FFT computational requirements for several values of N with $F = 1000$ samples per second (s/s). Note the significance of the FFT reduction in processing requirements, which is the source of the popularity of the use of DFT analysis via an FFT in many signal processing applications.

Chapter 10 discusses the implementation of signal processing algorithms from the perspective of hardware and software. For the purpose of comparisons, in this chapter the number of multiplications and additions will be used as a measure of the processing capacity required. You are cautioned that this is a rough-order method of comparing algorithms and that the actual estimates of processing time should be made based on the planned hardware and software implementation. This will be discussed in Chapter 10.

6.8.1 FFT Decimation-in-Time (DIT) Algorithm

Several approaches can be used to develop the FFT algorithm. Bellanger[9] uses matrix notation to show the successive decomposition of the DFT algorithm. Brigham[1] transforms the indices for frequency and time into binary form in order to do the decomposition. We start with the DFT expression and factor it into two DFTs of length $N/2$ by splitting the input samples into even and odd samples

$$X(k) = \frac{1}{N} \sum_{n=0}^{N-1} x(n) W_N^{nk} \qquad k = 0, 1, 2, \ldots, N$$

$$X(k) = \frac{1}{N} \sum_{m=0}^{(N/2)-1} x(2m) W_N^{2mk} + \frac{1}{N} \sum_{m=0}^{(N/2)-1} x(2m+1) W_N^{(2m+1)k}$$

The remainder of the development assumes that the normalizing factor is handled by dividing the input signal by $1/N$. Since hardware implementations of the FFT are often limited by cost and size (e.g., for signal processing on-board a military aircraft), it is important to consider wordlength requirements while maintaining dynamic range of the input and output. Normalization techniques will be discussed in Chapter 9, along with finite arithmetic effects of implementing the FFT algorithm. Let x_1 and x_2 equal the even and odd input sample components respectively; that is, $x_1(m) = x(2m)$ and $x_2(m) = x(2m+1)$, $m = 0, 1, \ldots, (N/2) - 1$, the foregoing summation can be written as

$$X(k) = \sum_{m=0}^{(N/2)-1} x_1(m) W_{N/2}^{mk} + W_N^k \sum_{m=0}^{(N/2)-1} x_2(m) W_{N/2}^{mk}$$

where $W_N^{2n} = W_{N/2}^n$, and each of the summation terms is reduced to an $N/2$ point DFT. Graphically, the DFT computation has been decomposed as

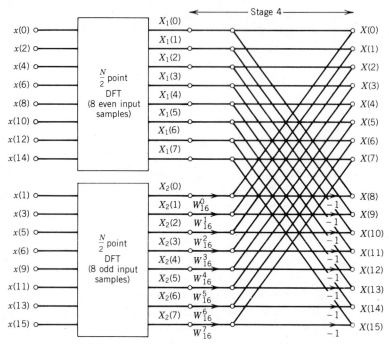

Figure 6.21 First step in developing 16-point decimation-in-time FFT signal flow graph.

shown in Figure 6.21 for $N = 16$ and the general form is written as

$$X(k) = X_1(k) + W_N^k X_2(k)$$

$$X(k + N/2) = X_1(k) + W_N^{k+N/2} X_2(k)$$

$$= X_1(k) - W_N^k X_2(k) \tag{6.21}$$

where $W_N^{k+N/2} = -W_N^k$, and $W_{N/2}^{m(k+N/2)} = W_{N/2}^{mk}$.

Since the DFT output is periodic, $X_1(k) = X_1(k + N/2)$ and $X_2(k) = X_2(k + N/2)$. Equation 6.21 is commonly referred to as the **FFT DIT butterfly**, and is shown in Figure 6.22 in three different forms. For forms 1 and 2, the intersection nodes of the X_1 and X_2 values represent complex additions. The form 2 butterfly is derived from the symmetry of the twiddle factors as shown in the figure and is the form implemented since it requires only one complex multiply operation and two complex additions. Form 3 is an alternative method sometimes used[10] to depict the second form of the butterfly.

For this first decomposition (Fig. 6.21), the twiddle factors are indexed consecutively and the butterfly values are separated by $N/2$ samples. This observation is important for developing general expressions using the basic butterfly operations. Also, the order of the input samples has been rearranged, which will be important for selecting the desired FFT implementation.

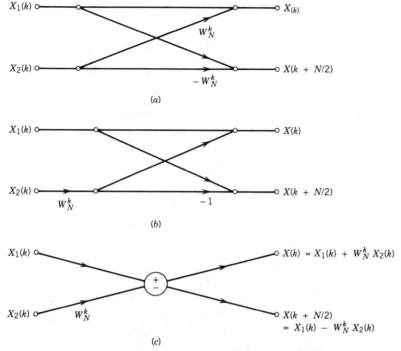

Figure 6.22 FFT decimation-in-time radix-2 butterfly representations: (a) normal form 1; (b) normal form 2; (c) alternative form 2.

Now we apply the same approach to each of the $N/2$ sample DFTs, where $x_{11}(m) = x_1(2m)$ and $x_{12}(m) = x_1(2m + 1)$, $m = 0, 1, \ldots, (N/4) - 1$. Then

$$X_1(k) = \sum_{m=0}^{(N/4)-1} x_1(2m) W_{N/2}^{2mk} + \sum_{m=0}^{(N/4)-1} x_1(2m + 1) W_{N/2}^{(2m+1)k}$$

$$X_1(k) = \sum_{m=0}^{(N/4)-1} x_{11}(m) W_{N/4}^{mk} + W_N^{2k} \sum_{m=0}^{(N/4)-1} x_{12}(m) W_{N/4}^{mk}$$

Resulting in the butterfly expressions for the $N/4$ decomposition of x_1 given by

$$X_1(k) = X_{11}(k) + W_N^{2k} X_{12}(k)$$

$$X_1(k + N/4) = X_{11}(k) - W_N^{2k} X_{12}(k)$$

Therefore, sequence x_1 has been decomposed into two DFTs of length $N/4$. The same process is followed for sequence x_2. The resulting decomposition of the two $N/2$ point DFTs is shown in Figure 6.23 for $N = 16$, and the $N/4$

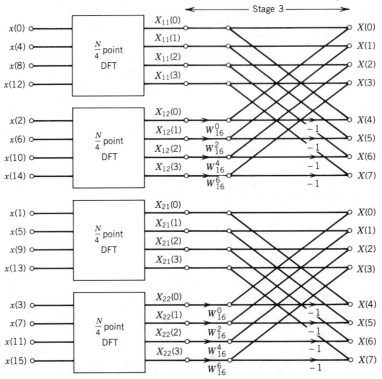

Figure 6.23 Second step in developing 16-point decimination-in-time FFT signal flow graph.

point butterflies are expressed as

$$X_2(k) = X_{21}(k) + W_N^{2k}X_{22}(k)$$

$$X_2(k + N/4) = X_{21}(k) - W_N^{2k}X_{22}(k)$$

Note that the input sequence is again reordered, the input samples to each butterfly are separated by $N/4$ samples, and the twiddle factor exponents are separated by a factor of two.

The decomposition process is repeated until two-point DFTs are generated. Each decomposition is called a stage, and the total number of stages is given by

$$M = \log_2 N \qquad (6.22)$$

The 16-point DFT requires four stages. The decompositions for stage 2 and stage 1 are shown in Figure 6.24, and the total flow graph for the 16-point DIT radix-2 FFT is shown in Figure 6.25. The twiddle factors for the two-point

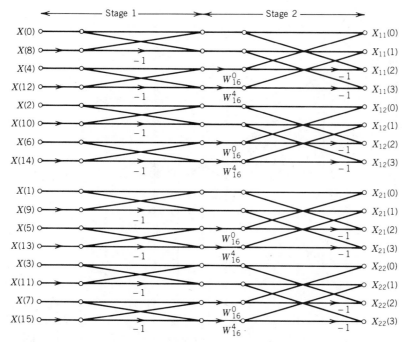

Figure 6.24 FFT 16-point decimation-in-time stage 1 and stage 2 decompositions.

DFTs are equal to 1 and -1, and therefore the first stage can be implemented without multiplies; that is, only additions and subtractions are required.

Figure 6.25 describes an in-place radix-2 DIT FFT algorithm. From Figure 6.25 we note that the in-place algorithm requires that the input data must be stored in bit-reversed order in consecutive memory locations. The in-place version of the algorithm allows the input data storage registers to be reused for intermediate storage for subsequent stage computations. Alternative in-place structures can be developed that preserve the order of the input and result in a bit-reversed output.

BIT REVERSAL*

The input sample sequence has been reordered, as seen from Figure 6.25, in a bit-reversed order. The **bit reversal** process is defined in Figure 6.26. Each of the decimal time sample indices is converted to its binary representation. Then the binary bit streams are reversed. Converting the reversed binary numbers to decimal values gives the reordered time indices.

Radar[11] presented a bit-reversal algorithm that iteratively generates the reversed time indices. The algorithm can be initialized at any point in the

---

*Bit is used to represent the positions of the binary number. The number 12 requires four bits for its binary representation (1100).

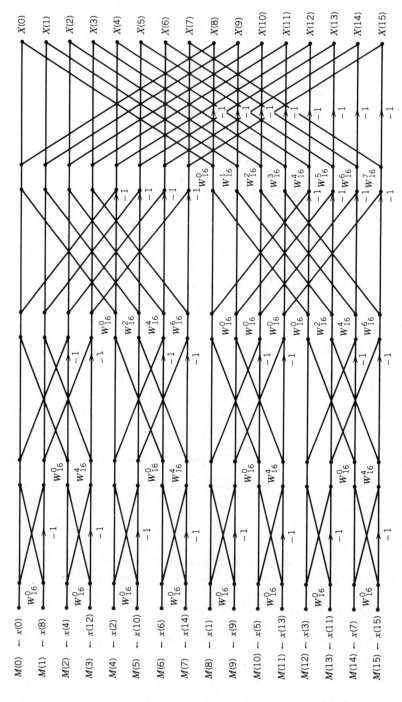

Figure 6.25 Flow graph of 16-point DIT FFT.

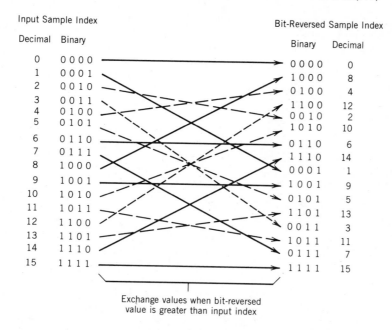

Exchange values when bit-reversed
value is greater than input index

Number System	Input Sample Index		Bit-Reversed Sample Index	
	$2^L \ 2^{L-1} \ \ldots 2^3 \ 2^2 \ 2^1 \ 2^0$		$2^L \ 2^{L-1} \ 2^{L-2} \ 2^{L-3} \ \ldots \ 2^1 \ \ 2^0$	
Binary	$B_L \ B_{L-1} \ \ldots \ B_3 \ B_2 \ B_1 \ B_0$		$B_0 \ B_1 \quad B_2 \quad B_3 \quad \ldots \ B_{L-1} B_L$	
Decimal	$\displaystyle\sum_{r=0}^{L} B_r 2^r$		$\displaystyle\sum_{r=0}^{L} B_r 2^{L-r}$	

Figure 6.26 Bit-reversal process.

reversed sequence and only requires the preceding reversed indices. A flowchart of the algorithm is presented in Figure 6.27a. Another algorithm commonly used reverses any given sample index into its bit-reversed index. This procedure is shown in Figure 6.27b along with a BASIC subroutine. The process works by iteratively determining if the input index binary bit positions are unity and building up the bit-reversed index decimal value.

Example 6.1
Use the procedures in Figures 6.27a and b to determine the fourth input sample to the 16-point FFT shown in Figure 6.25. The fourth input is the twelfth time-ordered sample of input x. Since we are using a 16-point FFT, four bits are required to represent the binary sample indices.

Using the first bit-reversal procedure shown in Figure 6.27a, the preceding bit-reversed value, 0100 (4 decimal), is needed to generate the next value.

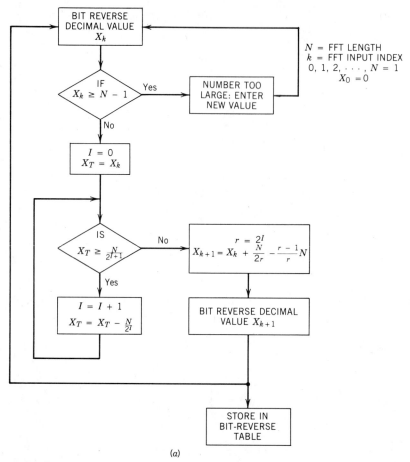

N = FFT LENGTH
k = FFT INPUT INDEX
$0, 1, 2, \cdots, N = 1$
$X_0 = 0$

(a)

Figure 6.27 Bit-reversal algorithm: (*a*) bit-reverse sequence generation algorithm; (*b*) algorithm to convert one value to bit-reverse value; (*c*) Basic code to convert one value to bit-reverse value.

The first test is applied, and since 4 is not greater than or equal to 8, the result is obtained as $4 + 8 = 12$.

The second procedure (Fig. 6.27*b*) uses the sample index of the input directly. Therefore, we apply the sample index 3 to the bit-reversal algorithm. Since $N = 16$, the number of bits is 4 as before. From the BASIC program the following calculations are made:

a. The number to be bit-reversed is set equal to B.

b. The variable $B1$ is initialized to 3 (B) and will be used to hold the results of successive integer divisions by 2.

c. The variable BR is initialized to zero and will be used to develop the bit-reversed output value.

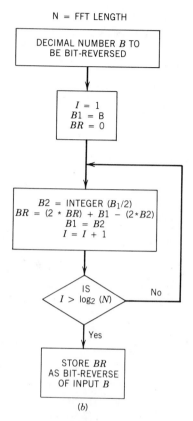

N = FFT LENGTH

DECIMAL NUMBER *B* TO
BE BIT-REVERSED

$I = 1$
$B1 = B$
$BR = 0$

$B2 = \text{INTEGER } (B_1/2)$
$BR = (2 * BR) + B1 - (2*B2)$
$B1 = B2$
$I = I + 1$

IS
$I > \log_2 (N)$ No

Yes

STORE *BR*
AS BIT-REVERSE
OF INPUT *B*

(b)   **Figure 6.27** Continued

d. The remaining calculations are iterated based on the number of bits four
times.

1. $B2 = 3 \setminus 2 = 1$: $BR = 0*2 + (3 - 2*1) = 1$: $B1 = 1$.
2. $B2 = 1 \setminus 2 = 0$: $BR = 1*2 + (1 - 2*0) = 3$: $B1 = 0$.
3. $B2 = 0 \setminus 2 = 0$: $BR = 3*2 + (0 - 2*0) = 6$: $B1 = 0$.
4. $B2 = 0 \setminus 2 = 0$: $BR = 6*2 + (0 - 2*0) = 12$: $B1 = 1$.
and the result, BR, equals 12, which agrees with Figure 6.25 and the
procedure given in Figure 6.27. In the above equations the backslash
symbol $(\setminus)$ denotes integer division, for example, $10 \setminus 4 = 2$, $25.68 \setminus$
$6.99 = 3$. ■

Figure 6.26 shows the input/output data shuffling that results for a
16-point transform. The operation can be performed without another storage
array to hold the bit-reversed index sample values. As shown in the figure,
when the input index and bit-reversed index are equal, no operation is
required. When the bit-reversed index is different than the input index, then
the input sample values are swapped. This swapping operation must be
performed only once, and therefore the algorithm must exercise control by

```
1000 '*****************************************************************
1001 '*              Bit Reverse Data Vector                         *
1002 '*     N=FFT LENGTH                                             *
1003 '*     XR=REAL INPUT VECTOR: XI=IMAGINARY INPUT VECTOR          *
1004 '*****************************************************************
1005 DIM XR(1024),XI(1024)
1010 CLS:LOCATE 5,10:PRINT "BIT REVERSAL FOR DATA VECTOR IN PROCESS"
1020 NLOG2%=LOG(N)/LOG(2)
1030 FOR J%=0 TO N-1
1040 B%=J%:GOSUB 1500
1050 LOCATE 10,15:PRINT "BIT REVERSAL FOR J%=";J%;" BR%=";BR%
1070     K%=BR%
1080     IF K%<=J% THEN GOTO 1150
1090     REAL=XR(J%)
1100     IMAG=XI(J%)
1110     XR(J%)=XR(K%)
1120     XI(J%)=XI(K%)
1130     XR(K%)=REAL
1140     XI(K%)=IMAG
1150 NEXT J%
1500 '*****************************************************************
1501 '*                BIT REVERSE SUBROUTINE                        *
1510 '*     NLOG2%=# OF BITS FOR B%                                  *
1520 '*     B%=INPUT # TO BE BIT REVERSED                            *
1521 '*****************************************************************
1530 B1%=B%
1540 BR%=0
1550 FOR I%=1 TO NLOG2%
1560     B2%=B1%\2
1570     BR%=BR%*2+(B1%-2*B2%)
1580     B1%=B2%
1590 NEXT I%
1599 RETURN
```

(c)

Figure 6.27 Continued

performing the swap only when the bit-reversed index is greater than the input index or vice versa. This process of performing the bit-reversal without requiring an additional storage array is referred to as an **in-place** operation.

For implementation of the bit reversal, one of the preceding approaches could be used, another algorithm could be used, or the bit-reversed addresses could be stored for the sets of FFT lengths desired. The primary factors in selecting a suitable bit-reversal approach are the time required to perform the operation and the amount of hardware and software required.

IN-PLACE COMPUTATIONS

The DIT FFT algorithm presented in Figure 6.25 results in an FFT with the intermediate data storage in-place. The storage requirements for intermediate FFT stage outputs are satisfied by the input data storage requirements since sample values are used only once in a butterfly operation.

NOT-IN-PLACE COMPUTATIONS

The FFT could be restructured to provide an ordered input and output. Such FFTs are defined as **not-in-place**, require an additional N element complex storage array for intermediate computations, and are described in several of the references.[2,10] These algorithms preserve the data ordering, eliminating the processing time required for bit-reversal algorithms of the input or output data. The twiddle factor ordering for each stage will be modified and must be assessed in analyzing the suitability of the various FFT design choices. The symmetry of the butterfly input/output addresses is also complicated and must be assessed when selecting an FFT for implementation.

RADIX-2 DIT IN-PLACE FFT SUMMARY

The following list summarizes the radix-2 DIT in-place FFT algorithm characteristics.

1. $N = 2^M$, where M is an integer, and
   N equals the number of input and output samples.

2. The input sequence is reordered via bit-reversal.

3. The number of stages is given by
   $M = \log_2 N$.

4. Each stage requires $N/2$ butterfly computations.

5. The symmetry of the butterfly input/output addresses for the mth stage is equal to 2^{m-1}. This is the separation between sample addresses $A1$ and $A2$ for each butterfly of stage m.

6. The number of complex multiplies is approximated by

$$\text{CMULT} \approx \frac{N}{2} \log_2 N$$

where no advantage is taken for simple (i.e., zero, unity or j) twiddle factor complex coefficients.

7. The number of complex additions is given by

$$\text{CADD} = N \log_2 N$$

8. The powers of W_N involved in computing the mth stage from the $(m-1)$th stage are given by

$$W(N : m : Nk/r^m; t) = W_N^{Nk/r^m}, \qquad k = 0, 1, 2, \ldots, 2^{m-1} - 1$$
$$t = 0, 1, 2, \ldots, (N/2) - 1$$

where t is the butterfly index for each stage. There are 2^{m-1} distinct twiddle factors per stage, which are repeated according to

$$R = r^{M-m} \text{ MOD } r^{m-1}$$

Also, if the implementation is restructured to provide sequential input data to the first stage, the powers will be in bit-reversed order.

PRIMITIVE OPERATION

The process for each butterfly operation can be defined as a **primitive operation**, as shown in Figure 6.28. In order to perform the operation, the input data addresses for the butterfly are required along with the twiddle factor. The output is stored back into the input addresses since this is an in-place operation.

Table 6.5 lists the twiddle factor exponents ($Nk/2^m$) for $N = 16$, 64, 256, and 1024. As shown in the table, the exponents are a function of the stage index m and the exponent repeat factor (ERF), which is the number of times the exponent sequence associated with m is repeated. For example, consider stage $m = 3$ for $N = 16$. For this case, the ERF = 2; therefore, the exponent sequence (0, 2, 4, 6) is repeated two times. The utility of Table 6.5 as a twiddle factor look-up table is illustrated in the following example.

Example 6.2

For a 16-point radix-2 DIT FFT, use the foregoing summary statements to verify the 16-point implementation shown in Figure 6.25.

Solution:

Each of the 8 summary statements are computed for the 16-point FFT.

1. The number of input samples, $N = 16$.

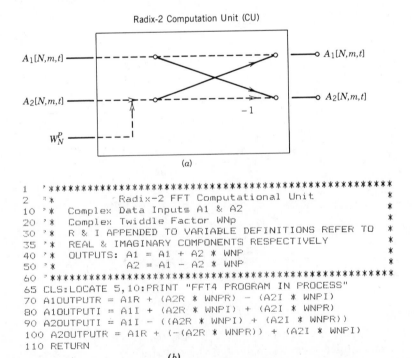

Radix-2 Computation Unit (CU)

(a)

```
1    ' *********************************************************
2    ' *          Radix-2 FFT Computational Unit             *
10   ' *   Complex Data Inputs A1 & A2                        *
20   ' *   Complex Twiddle Factor WNp                         *
30   ' *   R & I APPENDED TO VARIABLE DEFINITIONS REFER TO    *
35   ' *   REAL & IMAGINARY COMPONENTS RESPECTIVELY           *
40   ' *   OUTPUTS: A1 = A1 + A2 * WNP                        *
50   ' *            A2 = A1 - A2 * WNP                        *
60   ' *********************************************************
65   CLS:LOCATE 5,10:PRINT "FFT4 PROGRAM IN PROCESS"
70   A1OUTPUTR = A1R + (A2R * WNPR)  -  (A2I * WNPI)
80   A1OUTPUTI = A1I + (A2R * WNPI)  +  (A2I * WNPR)
90   A2OUTPUTI = A1I - ((A2R * WNPI)  +  (A2I * WNPR))
100  A2OUTPUTR = A1R + (-(A2R * WNPR))  +  (A2I * WNPI)
110  RETURN
```

(b)

Figure 6.28 (*a*) FFT2 CU primitive operation; (*b*) radix-2 CU BASIC program.

TABLE 6.5
FFT2 TWIDDLE FACTOR EXPONENTS FOR $N = 16, 64, 256,$ AND 1024 POINT FFTS.

N	m	ERF	Radix-2 Twiddle Factor Exponents
16	1	8	0
	2	4	0 4
	3	2	0 2 4 6
	4	1	0 1 2 3 4 5 6 7
64	1	32	0
	2	16	0 16
	3	8	0 8 16 24
	4	4	0 4 8 12 16 20 24 28
	5	2	0 2 4 6 8 10 12 14 16 18 20 22 24 26 28 30
	6	1	0 1 2 3 4 5 6 7 8 9 10 11 12 13 14 15 16 … 31
256	1	128	0
	2	64	0 64
	3	32	0 32 64 96
	4	16	0 16 32 48 64 80 96 112
	5	8	0 8 16 24 32 40 48 56 64 72 80 88 96 104 112 120
	6	4	0 4 8 12 16 20 24 28 32 36 40 44 48 52 56 60 … 124
	7	2	0 2 4 6 8 10 12 14 16 18 20 22 24 26 28 30 … 126
	8	1	0 1 2 3 4 5 6 7 8 9 10 11 12 13 14 15 … 127
1024	1	512	0
	2	256	0 256
	3	128	0 128 256 384
	4	64	0 64 128 192 256 320 384 448
	5	32	0 32 64 96 128 160 192 224 256 288 320 352 384 … 480
	6	16	0 16 32 48 64 80 96 112 128 144 160 176 192 … 496
	7	8	0 8 16 24 32 40 48 56 64 72 80 88 96 … 504
	8	4	0 4 8 12 16 20 24 28 32 36 40 44 48 … 508
	9	2	0 2 4 6 8 10 12 14 16 18 20 22 24 … 510
	10	1	0 1 2 3 4 5 6 7 8 9 10 11 12 … 511

2. The bit-reversal was shown in Example 6.1.

3. The number of stages, $M = 4$.

4. The number of butterflies per stage is $N/2 = 8$.

5. The butterfly input/output memory addresses, that is, the separation between the addresses of two complex input samples to a butterfly is $2^{(m-1)}$, that is,

Stage 1 Inputs/outputs for each butterfly are 1 sample apart.
Stage 2 Inputs/outputs for each butterfly are 2 samples apart.
Stage 3 Inputs/outputs for each butterfly are 4 samples apart.
Stage 4 Inputs/outputs for each butterfly are 8 samples apart.

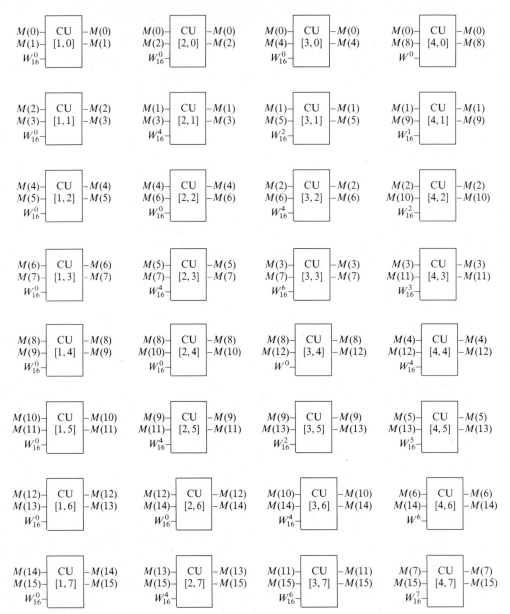

Figure 6.29 FFT2 16-point DIT FFT representation using primitive operations.

6. The number of complex multiplies is approximated by 32 compared to 256, the square of 16, for the DFT implementation. Count the number of multiplications required in Figure 6.25, eliminating the simple twiddle factors, which are zero or unity, for further reductions.

7. The number of additions required is 64 compared to the DFT requirements of $N^2 - N = 240$.

8. The twiddle factors for each stage are computed as follows:

Stage 1   One unique coefficient with power equal to zero, therefore,

$$W(16:1:0; t) = 1 \text{ for all}$$
$$t = 0, 1, \dots, 7$$

Stage 2   Two unique coefficients with powers equal to zero and four, therefore,

$$W(16:2:0; 0, 2, 4, 6) = 1$$
$$W(16:2:4; 1, 3, 5, 7) = -j$$

Stage 3   Four unique coefficients with powers equal to 0, 2, 4, and 6, therefore,

$$W(16:3:0; 0, 4) = 1$$
$$W(16:3:2; 1, 5) = 0.7071068 - j0.7071068$$
$$W(16:3:4; 2, 6) = -j$$
$$W(16:3:6; 3, 7) = -0.7071068 - j0.7071068$$

Stage 4   Eight unique coefficients with powers equal to 0, 1, 2, 3, 4, 5, 6, and 7, therefore,

$$W(16:4:0; 0) = 1$$
$$W(16:4:1; 1) = 0.9238795 - j0.3826834$$
$$W(16:4:2; 2) = 0.7071068 - j0.7071068$$
$$W(16:4:3; 3) = 0.3826834 - j0.9238795$$
$$W(16:4:4; 4) = -j$$
$$W(16:4:5; 5) = -0.3826834 - j0.9238795$$
$$W(16:4:6; 6) = -0.7071068 - j0.7071068$$
$$W(16:4:7; 7) = -0.9238795 - j0.3826834.$$

Note that the twiddle factors for stages 1, 2, 3 and 4 could have easily been obtained from Table 6.5.

The primitive operations are shown in Figure 6.29 for the 16-point FFT. ■

IMPLEMENTATIONS WITH COMPUTATIONAL SAVINGS

If the hardware and/or software control provides the capability to avoid multiplications by simple twiddle factors (real or imaginary components equal to 0 or 1); then the number of complex multiplies is given by

$$\text{CMULT} = \frac{N}{2}(\log_2 N/2) - \left(N \sum_{i=2}^{M} 2^{-i} \right)$$

Additional savings are possible if the implementation is designed to take advantage of the twiddle factors at 45, 135, 225, and 315 degrees on the unit circle. The real and imaginary components are equal and therefore only half of a complex multiply is actually required. We have assumed that a complex multiply requires four real multiplications and two real additions. A simple example of a fast algorithm for complex multiplication is[12]

$$E + jF = (A + jB)(C + jD) = AC - BD + j(AD + BC)$$

$$E = AC - BD \text{ or } E = A(C - D) + D(A - B)$$

$$F = AD + BC \text{ or } F = B(C + D) + D(A - B)$$

The normal form of the complex multiplication requires four real multiplications and two real additions. The modified form requires three real multiplications and five real additions. Therefore, if additions are much easier

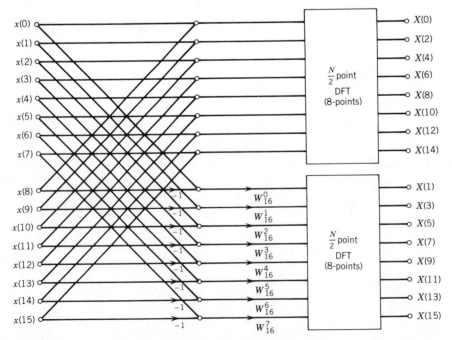

Figure 6.30 DFT decomposition for decimation-in-frequency FFT.

to implement than multiplications on the hardware/software being used, the modified form would be better. Also, if the twiddle factors are stored as the C and D coefficients, the $C + D$ and $C - D$ terms could be stored for the modified form, reducing the number of additions required to three. A trade-off between control and computational requirements must be performed to determine the best design approach.

6.8.2 FFT Decimation-In-Frequency (DIF) Algorithm

The development of the radix-2 DIF FFT algorithm is very similar to the DIT development. For the DIF decomposition, the data are split into batches of $N/2$ samples of contiguous input data. Therefore, let $X(k)$ be expressed as

$$X(k) = \sum_{D=0}^{(N/2)-1} x(n)W_N^{nk} + \sum_{n=0}^{(N/2)-1} x(n + (N/2))W_N^{(n+(N/2))k}$$

Now we can factor W^{nk} out of the summations, and combining terms in one summation results in

$$X(k) = \sum_{n=0}^{(N/2)-1} \left[x(n) + W_N^{(N/2)k}x(n + (N/2))\right]W_N^{nk}$$

The next step is to separate the frequency terms $X(k)$ into even and odd samples of k; the resulting expression provides the first decomposition into $N/2$ point DFTs shown in Figure 6.30 and expressed as

$$X(2h) = \sum_{n=0}^{(N/2)-1} \left[x(n) + x(n + (N/2))\right]W_{N/2}^{nh}$$

$$X(2h + 1) = \sum_{n=0}^{(N/2)-1} \left[x(n) - x(n + (N/2))\right]W_N^n W_{N/2}^{nh} \qquad (6.23)$$

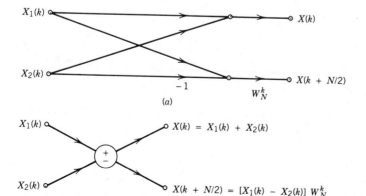

Figure 6.31 FFT decimation-in-frequency (DIF) two-point butterfly representations: (a) form 2; (b) alternative form 2.

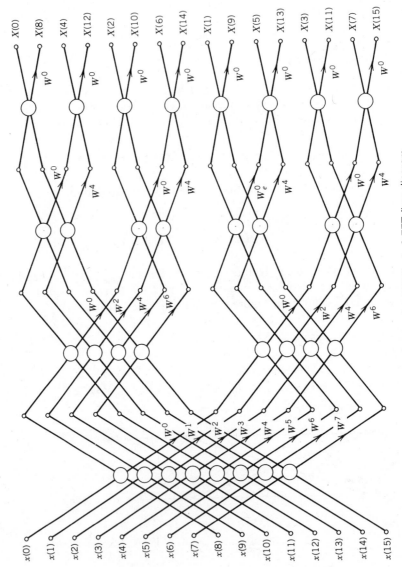

Figure 6.32 16-point DIF Raoix-2 FFT flow diagram.

Again, the process of decomposition is continued until the last stage is made up of two-point DFTs. The butterfly representation for the DIF FFT implementation is shown in Figure 6.31. A 16-point DIF FFT is illustrated in Figure 6.32. Note that the decomposition proceeds from left to right for the DIF development and the symmetry relationships are reversed from the DIT algorithm (e.g., the bit reversal occurs at the output instead of the input). Similar to the DIT algorithm, alternative structures can be developed that preserve the in-place computations and provide an ordered output and a bit-reversed input. Similarly, a DIF not-in-place structure provides an ordered input and output but requires an additional N-complex storage array for intermediate calculations.

6.8.3 FFT Radix-4 (FFT4) DIT Implementation

The radix-2 FFT algorithms produce the lowest possible level of DFT decomposition. Alternative decompositions are possible. Radix-4 and radix-8 FFT algorithms are popular because they are better adapted to pipeline* hardware implementations.[13,14] This section presents a detailed approach to the implementation of a radix-4 DIT FFT algorithm. A decomposition of the DFT into factors of four results in the FFT4 algorithm, where for this case the DFT length N is a power of 4. The radix-8 FFT used in combination with the FFT4 algorithm provides implementation of any power of two.

FFT4 COMPUTATIONAL UNIT (CU)

This development uses the radix-2 FFT structure and builds four-point primitive operations from the two-point structures. Figure 6.33 illustrates a four-point computational unit (CU). Note that the computational unit requires four input data memory addresses (A_i, $i = 1, 2, 3, 4$), three twiddle factor complex coefficients (TW_i, $i = 1, 2, 3$), and four output memory addresses (C_i, $i = 1, 2, 3, 4$). Similar to the FFT2 algorithm, the computations will be performed in-place. Therefore, $A_i = C_i$ and the input data will be read into consecutive memory locations in bit-reversed order.

Consider the FFT4 CU given in Figure 6.33. The general expressions for the FFT4 primitive operation are easily derived by writing the equations in complex form. From the signal flow graph, A_i are the complex input values obtained from a read memory location, B_i are the intermediate CU values, and C_i are the complex output values. Since the computation is performed in-place, the output values from stage m are read back into the input memory address locations to be used for the next ($m + 1$) FFT4 stage computation.

---

*A *pipeline* architecture provides multiple hardware stages, each performing a basic operation (e.g., multiplication, addition) simultaneously, through which the data progress at the hardware cycle rate. The pipeline is discussed in Chapter 10.

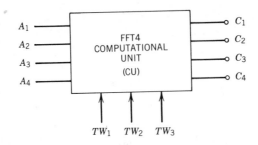

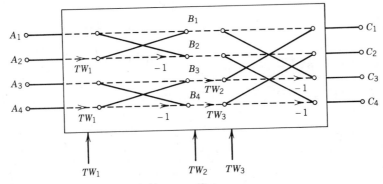

Figure 6.33 FFT4 CU primitive operation.

The basic CU primitive algorithm is developed by assuming complex input data, and twiddle factors are read in from memory, that is

$$A_1 = AR_1 + jAI_1 \qquad TW_1 = TWR_1 + jTWI_1$$
$$A_2 = AR_2 + jAI_2 \qquad TW_2 = TWR_2 + jTWI_2$$
$$A_3 = AR_3 + jAI_3 \qquad TW_3 = TWR_3 + jTWI_3$$
$$A_4 = AR_4 + jAI_4$$

where the real and imaginary components of the complex values are represented by appending R and I, respectively, to the complex values. On substitution, the CU intermediate complex values are

$$B_1 = A_1 + A_2 TW_1$$
$$= AR_1 + jAI_1 + (AR_2 + jAI_2)(TWR_1 + jTWI_1)$$
$$= BR_1 + jBI_1$$

where the real and imaginary parts are

$$BR_1 = AR_1 + (AR_2 TWR_1) - (AI_2 TWI_1)$$
$$BI_1 = AI_1 + (AI_2 TWR_1) + (AR_2 TWI_1)$$

Similarly, the complex intermediate values B_2, B_3, and B_4 are given by

$$BR_2 = AR_1 - AR_2 TWR_1 + AI_2 TWI_1$$
$$BI_2 = AI_1 - AI_2 TWR_1 - AR_2 TWI_1$$
$$BR_3 = AR_3 + (AR_4 TWR_1) - (AI_4 TWI_1)$$
$$BI_3 = AI_3 + (AI_4 TWR_1) + (AR_4 TWI_1)$$
$$BR_4 = AR_3 - AR_4 TWR_1 + AI_4 TWI_1$$
$$BI_4 = AI_3 - AI_4 TWR_1 - AR_4 TWR_1$$

Finally, the complex values at the output of the radix-4 CU are determined to be

$$C_1 = B_1 + B_3 TW_2$$
$$= BR_1 + jBI_1 + (BR_3 + jBI_3)(TWR_2 + jTWI_2)$$
$$= CR_1 + jCI_1 \tag{6.24}$$

where

$$CR_1 = BR_1 + (BR_3 TWR_2) - (BI_3 TWI_2)$$
$$CI_1 = BI_1 + (BR_3 TWI_2) + (BI_3 TWR_2)$$

Similarly, the complex CU output values C_2, C_3, and C_4 are given by

$$CR_2 = BR_2 + (BR_4 TWR_3) - (BI_4 TWI_3)$$
$$CI_2 = BI_2 + (BR_4 TWI_3) + (BI_4 TWR_3)$$
$$CR_3 = BR_1 - BR_3 TWR_2 + BI_3 TWI_2$$
$$CI_3 = BI_1 - BR_3 TWI_2 - BI_3 TWR_2$$
$$CR_4 = BR_2 - BR_4 TWR_3 + BI_4 TWI_3$$
$$CI_4 = BI_2 - BR_4 TWI_3 - BI_4 TWR_3$$

The BASIC expressions implementing the FFT4 computational unit are given in Figure 6.34. To implement the FFT4 algorithm it is required to develop a method for determining the read and write memory addresses as well as a method for obtaining the respective complex twiddle factors. This method is developed in the following section.

For the FFT4 algorithm developed, there are 16 unique real multiplies (indicated by parentheses) and 32 unique real adds per FFT4 CU. The radix-4 FFT processing load can be determined by computing the number of real multiplies and adds per second, that is, RMPS = $R4^{M+1}M$ and the number of

```
2000 '***********************************************************************
2001 '*            FFT4 COMPUTATIONAL UNIT  IMPLEMENTATION                  *
2002 '*   GENERATE REAL=WR & IMAGINARY=WI TWIDDLE FACTORS GOSUB 3000 *
2003 '*   GENERATE READ/WRITE CU ADDRESSES = RW GOSUB 4000            *
2004 '*    GENERATE TWIDDLE FACTOR ADDRESSES = TA GOSUB 5000          *
2005 '***********************************************************************
2010 GOSUB 3000
2020 FOR M%=1 TO NLOG4 'M=LOOP ON STAGE INDEX
2030   GOSUB 4000 ' GENERATE READ/WRITE CU ADDRESSES = RW
2040   GOSUB 5000 ' GENERATE TWIDDLE FACTOR ADDRESSES = TA
2050   FOR I%=1 TO N/4 'I=LOOP OF ALL CUs FOR STAGE M
2060     M1=XR(RW(I%,2))*WR(TA(I%,1))
2061     M2=XI(RW(I%,2))*WR(TA(I%,1))
2062     M3=XI(RW(I%,2))*WI(TA(I%,1))
2063     M4=XR(RW(I%,2))*WI(TA(I%,1))
2070     M5=XR(RW(I%,4))*WR(TA(I%,1))
2071     M6=XI(RW(I%,4))*WR(TA(I%,1))
2072     M7=XI(RW(I%,4))*WI(TA(I%,1))
2073     M8=XR(RW(I%,4))*WI(TA(I%,1))
2080     B1R=XR(RW(I%,1))+M1-M3
2081     B1I=XI(RW(I%,1))+M2+M4
2082     B2R=XR(RW(I%,1))-M1+M3
2083     B2I=XI(RW(I%,1))-M2-M4
2090     B3R=XR(RW(I%,3))+M5-M7
2091     B3I=XI(RW(I%,3))+M6+M8
2092     B4R=XR(RW(I%,3))-M5+M7
2093     B4I=XI(RW(I%,3))-M6-M8
2100     M1=B3R*WR(TA(I%,2))
2101     M2=B3R*WI(TA(I%,2))
2102     M3=B3I*WI(TA(I%,2))
2103     M4=B3I*WR(TA(I%,2))
2104     M5=B4R*WR(TA(I%,3))
2105     M6=B4R*WI(TA(I%,3))
2106     M7=B4I*WI(TA(I%,3))
2107     M8=B4I*WR(TA(I%,3))
2110     XR(RW(I%,1))=B1R+M1-M3
2111     XI(RW(I%,1))=B1I+M2+M4
2112     XR(RW(I%,2))=B2R+M5-M7
2113     XI(RW(I%,2))=B2I+M6+M8
2120     XR(RW(I%,3))=B1R-M1+M3
2121     XI(RW(I%,3))=B1I-M2-M4
2122     XR(RW(I%,4))=B2R-M5+M7
2123     XI(RW(I%,4))=B2I-M6-M8
2200   NEXT I%
2300 NEXT M%
3000 '***********************************************************************
3001 '*                  TWIDDLE FACTOR GENERATION                         *
3010 '* REAL & IMAGINARY TERMS GENERATED FOR ENTIRE 360 DEGREES   *
3020 '* N=NUMBER OF POINTS                                                 *
3030 '***********************************************************************
3040 FOR I%=0 TO N
3050   WR(I%)=COS(PI2N*I%)
3060   WI(I%)=-SIN(PI2N*I%)
3070 NEXT I%
3080 RETURN
```

(a)

Figure 6.34 BASIC programs for the implementation of the FFT4 CU: (*a*) FFT4 CU implementation and twiddle factor coefficients; (*b*) FFT4 CU read / write memory address locations: (*c*) FFT4 CU twiddle factor addresses.

```
4000 '*******************************************************************
4001 '*         GENERATE READ/WRITE MEMORY ADDRESSES FOR FFT4          *
4010 '*    M% IS THE STAGE NUMBER OF FFT4                              *
4020 '*    NLOG4 IS NUMBER OF STAGES REQUIRED                         *
4025 '*******************************************************************
4030 NLOG4=LOG(N)/LOG(4):Q=4^(M%-1)
4040 IF M%=1 THEN J%=0 ELSE J%=1
4050 KK%=1:II%=1:JJ%=4^(NLOG4-M%)
4055 QQ=(N/JJ%)-J%
4060 FOR P%=0 TO JJ%-1
4070    FOR I%=0 TO (N/JJ%)-1
4076       IMOD#=((I%*Q) MOD QQ)
4077 IF I%>0 AND IMOD#=0 THEN IMOD#=QQ
4080       RW(KK%,II%)=IMOD#+(P%*N/JJ%)
4090       II%=II%+1
4100       IF II%>4 THEN II%=1:KK%=KK%+1
4110    NEXT I%
4120 NEXT P%
4125 RW(N/4,4)=N-1
4130 RETURN
```

(b)

```
5000 '*******************************************************************
5005 '*          GENERATE TWIDDLE ADDRESSES FOR FFT4                  *
5010 '*    N IS FFT LENGTH                                            *
5020 '*    M% IS FFT4 STAGE #; NLOG4 IS # OF FFT4 STAGES              *
5025 '*******************************************************************
5030 TA(1,1)=0:TA(1,2)=0:TA(1,3)=N/4
5040 TRF=N/(4^M%) ' TWIDDLE REPEAT FACTOR
5045 VRF=N/(4*TRF) ' REPEAT FACTOR MULTIPLYER
5050 IF M%=1 THEN 5200
5055 Q=N/(4^(NLOG4+1-M%))
5056    ID%=4^(NLOG4-M%)
5060 FOR I%=2 TO Q
5065       TA(I%,1)=TA(I%-1,1)+2*ID%
5070       TA(I%,2)=TA(I%-1,2)+ID%
5080       TA(I%,3)=TA(I%-1,3)+ID%
5090 NEXT I%
5100 FOR I%=1 TO Q
5110    FOR II%=2 TO TRF
5125       FOR L%=1 TO 3
5130          TA(I%+((II%-1)*VRF),L%)=TA(I%,L%)
5140       NEXT L%
5150    NEXT II%
5160 NEXT I%
5170 GOTO 5300
5200 FOR I%=2 TO N/4
5210    FOR L%=1 TO 3
5220       TA(I%,L%)=TA(1,L%)
5230    NEXT L%
5240 NEXT I%
5300 RETURN
```

(c)

Figure 6.34 Continued

real adds per second is RAPS $= 2R4^{M+1}M$. The parameter R is the FFT rate (FFTs/second).

FFT4 CU READ / WRITE MEMORY ADDRESSES

The address symmetry of the radix-2 FFT butterfly structure allows the FFT4 addresses to be obtained by inspection. For example, a 16-point radix-4 flow graph using FFT4 CUs can readily be developed by referring to Figure 6.25. Figure 6.35 presents the 16-point FFT4 implementation using the CUs from Figure 6.33. The read/write memory addresses for stage m are determined by

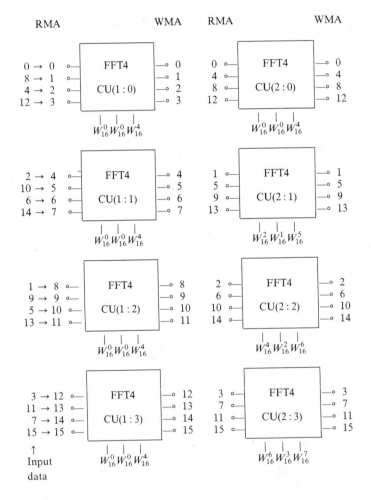

Notes: 1. RMA: read memory address
WMA: write memory address
2. Input data read into consecutive memory locations in bit-reversed order.

Figure 6.35 FFT4 16-point DIT FFT representation using primitive operations.

the following algorithm.

$$RW(m\colon i\colon l) = iL + (l4^{m-1}) \, \text{MOD} \, (L - 1) \qquad (6.25)$$

$$i = 0, 1, \ldots, I - 1 \qquad \text{Outer loop index}$$

$$l = 0, 1, \ldots, L - 1 \qquad \text{Inner loop index}$$

$$I = 4^{M-m}$$

$$L = \frac{N}{4^{M-m}}$$

$$M = \text{Log}_4 \, N$$

$$m = 1, 2, \ldots, M$$

Note: 1. $X \, \text{MOD} \, Y = Y \, \text{RESIDUE} \, (X/Y)$, where $\text{RESIDUE} \, (X/Y)$ denotes the fractional part of (X/Y).

2. For $l = (N/4^{M-m}) - 1$, $\text{RES}(X/Y) = 1$

FFT4 CU TWIDDLE FACTOR EXPONENT ALGORITHM

The twiddle factor *TW* exponents $p(m\colon t; i)$, $i = 1, 2, 3$ for CU $(m\colon t)$ can be found by inspection from the FFT2 requirements, where $m =$ stage number; $t =$ computational unit number: that is,

$$p(1\colon t; 1) = 0$$

$$p(1\colon t; 2) = 0 \qquad\qquad\qquad\quad m = 1$$

$$p(1\colon t; 3) = N/4 \qquad\qquad\qquad t = 0, 1, \ldots, (N/4) - 1 \quad (6.26a)$$

$$p(m\colon 0; 1) = 0$$

$$p(m\colon 0; 2) = 0 \qquad\qquad\qquad\quad m > 1$$

$$p(m\colon 0; 3) = N/4 \qquad\qquad\qquad t = 0 \qquad\qquad\qquad\qquad (6.26b)$$

$$p(m\colon t; 1) = p(m\colon t - 1; 1) + 2I$$

$$p(m\colon t; 2) = p(m\colon t - 1; 2) + I \qquad m > 1$$

$$p(m\colon t; 3) = p(m\colon t - 1; 3) + I \qquad t = 1, 2, \ldots, (N/4) - 1 \quad (6.26c)$$

Equation 6.26a defines the exponents for all stage 1 CUs. For subsequent stages the successive exponents for each CU are computed iteratively starting with the initial exponent values defined by Eq. 6.26 b and using Eq. 6.26c to compute the successive CU twiddle factor exponents for stage m. Note that the iteration is repeated based on the value of $N/4^m$.

Another approach for obtaining the twiddle factor exponents for the FFT4 CUs is given by addressing Table 6.6. For example, consider the case for $N = 16$. For stage $m = 1$ we obtain the three exponents $(0, 0, 4)$, which are

TABLE 6.6
FFT4 TWIDDLE FACTOR EXPONENTS FOR N = 16, 64, 256, AND 1024 POINT FFTS.

$N = 16$

m	1	2			
ERF	4	1			
t	0	0	1	2	3
$p(1)$	0	0	2	4	6
$p(2)$	0	0	1	2	3
$p(3)$	4	4	5	6	7

$N = 64$

m	1	2				3															
ERF	16	4				1															
t	0	0	1	2	3	0	1	2	3	4	5	6	7	8	9	10	11	12	13	14	15
$p(1)$	0	0	8	16	24	0	2	4	6	8	10	12	14	16	18	20	22	24	26	28	30
$p(2)$	0	0	4	8	12	0	1	2	3	4	5	6	7	8	9	10	11	12	13	14	15
$p(3)$	16	16	20	24	28	16	17	18	19	20	21	22	23	24	25	26	27	28	29	30	31

$N = 256$

m	1	2				3															
ERF	64	16				4															
t	0	0	1	2	3	0	1	2	3	4	5	6	7	8	9	10	11	12	13	14	15
$p(1)$	0	0	32	64	96	0	8	16	24	32	40	48	56	64	72	80	88	96	104	112	120
$p(2)$	0	0	16	32	48	0	4	8	12	16	20	24	28	32	36	40	44	48	52	56	60
$p(3)$	64	64	80	96	112	64	68	72	76	80	84	88	92	96	100	104	108	112	116	120	124

m	4																							
ERF	1																							
t	0	1	2	3	4	5	6	7	8	9	10	11	12	13	14	15	16	17	18	19	20	21	...	64
$p(1)$	0	2	4	6	8	10	12	14	16	18	20	22	24	26	28	30	32	34	36	38	40	42	...	126
$p(2)$	0	1	2	3	4	5	6	7	8	9	10	11	12	13	14	15	16	17	18	19	20	21	...	63
$p(3)$	64	65	66	67	68	69	70	71	72	73	74	75	76	77	78	79	80	81	82	83	84	85	...	127

$N = 1024$

m	1	2			
ERF	256	64			
t	0	0	1	2	3
$p(1)$	0	0	128	256	384
$p(2)$	0	0	64	128	192
$p(3)$	256	256	320	384	448

TABLE 6.6 (*Continued*)

m	3														
ERF	16														
t	0	1	2	3	4	5	6	7	8	9	10	11	12	13	14	15
$p(1)$	0	32	64	96	128	160	192	224	256	288	320	352	384	416	448	480
$p(2)$	0	16	32	48	64	80	96	112	128	144	160	176	192	208	224	240
$p(3)$	256	272	288	304	320	336	352	368	384	400	416	432	448	464	480	496

m	4																
ERF	4																
t	0	1	2	3	4	5	6	7	8	9	10	11	12	13	14	15	...	63
$p(1)$	0	8	16	24	32	40	48	56	64	72	80	88	96	104	216	224	...	504
$p(2)$	0	4	8	12	16	20	24	28	32	36	40	44	48	52	56	60	...	252
$p(3)$	256	260	264	268	272	276	280	284	288	292	296	300	304	308	312	316	...	508

m	5																
ERF	1																
t	0	1	2	3	4	5	6	7	8	9	10	11	12	13	14	15	...	255
$p(1)$	0	2	4	6	8	10	12	14	16	18	20	22	24	26	28	30	...	510
$p(2)$	0	1	2	3	4	5	6	7	8	9	10	11	12	13	14	15	...	255
$p(3)$	256	257	258	259	260	261	262	263	264	265	266	267	268	269	270	271	...	511

associated with twiddle factors TW_1, TW_2, and TW_3, respectively. The indicated exponent repeat factor (ERF) denotes the number of times the exponents are to be repeated for successive CUs. From Table 6.6 for stage $m = 1$ the ERF = 4, which indicates the twiddle factor exponents $(0, 0, 4)$ are to be repeated four times, that is, the twiddle factors $TW_1 = W_{16}^0$, $TW_2 = W_{16}^0$, and $TW_3 = W_{16}^4$ are read into the computational units CU$(1, t)$ ($t = 0, 1, 2, 3$). For the final stage ($m = 2$), the FFT4 CU twiddle factor exponent sequence $(0, 0, 4)$, $(2, 1, 5)$, $(4, 2, 6)$, and $(6, 3, 7)$ is associated with CU $(2: 0)$, CU $(2: 1)$, CU $(2: 2)$, and CU $(2: 3)$, respectively. Since the ERF = 1, the sequence is not repeated.

As another example, consider the FFT4 twiddle factor exponents for $N = 64$, where for stage $m = 1$ the ERF = 16. Therefore, the twiddle factor exponents $(0, 0, 16)$ are repeated for all 16 FFT4 CUs; that is, CU $(1: t)$, $t = 0, 1, \ldots, 15$. For $m = 2$ the ERF = 4. For this case the exponent sequence $(0, 0, 16)$, $(8, 4, 20)$, $(16, 8, 24)$, and $(24, 12, 28)$ is associated with the FFT4 computational units CU $(2: 0)$, CU $(2: 1)$, CU $(2: 2)$, and CU $(2: 3)$, respectively. Likewise, since ERF = 4 this exponent sequence is repeated for CU $(2: t)$, $t = 4, 5, 6, 7$; CU $(2: t)$, $t = 8, 9, 10, 11$, and finally CU $(2: t)$,

$t = 12, 13, 14, 15$. For the final stage $m = 3$ the ERF = 1; therefore the exponent sequence given in the table is not repeated; that is, the exponent sequence $(0, 0, 16)$, $(2, 1, 17)$, $(4, 2, 18), \ldots, (30, 15, 31)$ are associated with the FFT4 computational units CU $(3\!: t)$, $t = 0, 1, \ldots, 15$.

Example 6.4

Generate the read/write memory addresses and twiddle factor exponents that are computed for the FFT4 16-point transform.

Solution:

For a 16-point FFT4 implementation the number of stages is $\log_4 16 = 2$. From Eq. 6.25 the twiddle factors are computed for each stage. For stage 1 the

```
1 '***************************************************************
2 '*                      FFT4 PROGRAM                          *
3 '*    N=FFT LENGTH                                            *
5 '*    PROGRAM USES FFT4 COMPUTATIONAL UNIT SUBROUTINE 2000    *
6 '*    TWIDDLE FACTOR GENERATION SUBROUTINE 3000              *
7 '*    READ WRITE MEMORY ADDRESSES SUBROUTINE 4000            *
8 '*    GENERATE TWIDDLE FACTOR ADDRESSES SUBROUTINE 5000      *
9 '***************************************************************
10 DIM WR(1025),WI(1025),RW(257,4)
20 DIM TA(257,3),MAG(1025)
30 DIM XR(1025),XI(1025),Y(513),ZR(1025),ZI(1025)
35 CLS:LOCATE 5,10:PRINT "FFT4 PROGRAM IN PROCESS"
40 LOCATE 10,10:
50 INPUT "ENTER THE FFT LENGTH N=";N
60 NLOG4=LOG(N)/LOG(4)
70 PI=3.14159265#:PI2N=2*PI/N
100 GOSUB 1000
110 FOR I%=0 TO N
120    MAG(I%)=0
130 NEXT I%
200 GOSUB 2000 'FFT4 COMPUTATIONAL UNIT IMPLEMENTATION FIG. 6.34
210 GOSUB 3000 ' GENERATE REAL & IMAGINARY TWIDDLE FACTORS
400 MAX=0
410 FOR I%=0 TO N-1
420    MAG(I%)=(XR(I%)^2)+(XI(I%)^2)
430    IF MAG(I%)>MAX THEN MAX=MAG(I%)
440 NEXT I%
445 IF MAX=0 THEN MAX=1
450 FOR I%=0 TO N-1
455    MAG(I%)=(MAG(I%)/MAX)+1E-08
460    MAG(I%)=10*LOG(MAG(I%))/LOG(10)
470 NEXT I%
500 'PRINTOUT OF MAGNITUDE FFT OUTPUT
505 LPRINT "MAGNITUDE OUTPUT N= ";N;" F= ";F
510 FOR I%=0 TO N/2-1
515    K%=I%
520    LPRINT USING "  #### ";K%;:LPRINT USING " ###.## ";MAG(K%);
530    IF K%=I% THEN K%=I%+N/2:GOTO 520 ELSE LPRINT ""
540 NEXT I%
900 FOR I%=0 TO N/2-1
910    LPRINT "I%= ";I%;XR(I%);XI(I%);" I%= ";I%+N/2;XR(I%+N/2);XI(I%+N/
920 NEXT I%
999 END
```

Figure 6.36 FFT4 DIT in-place BASIC program.

following RW values result:

$$\text{Stage 1 } (m = 1)$$

$$I = 4^{2-1} = 4$$

$$L = \frac{16}{4^{2-1}} = 4$$

$$M = \text{Log}_4 16 = 2$$

$$RW(1: i: l) = 4i + l \text{ MOD } 3 \qquad i = 0, 1, 2, 3$$
$$l = 0, 1, 2, 3$$

CU: $t = 0$ CU: $t = 2$

$RW(1: 0: 0) = 0$ $RW(1: 2: 0) = 8$

$RW(1: 0: 1) = 1$ $RW(1: 2: 1) = 9$

$RW(1: 0: 2) = 2$ $RW(1: 2: 2) = 10$

$RW(1: 0: 3) = 3$ $RW(1: 2: 3) = 11$

CU: $t = 1$ CU: $t = 3$

$RW(1: 1: 0) = 4$ $RW(1: 3: 0) = 12$

$RW(1: 1: 1) = 5$ $RW(1: 3: 1) = 13$

$RW(1: 1: 2) = 6$ $RW(1: 3: 2) = 14$

$RW(1: 1: 3) = 7$ $RW(1: 3: 3) = 15$

For stage 2 the RW addresses are given by

$$\text{Stage 2 } (m = 2)$$

$$I = 1$$

$$L = 16$$

$$RW(2: i: l) = 4l \text{ MOD } 15 \qquad i = 0$$
$$l = 0, 1, \dots, 15$$

CU: $t = 0$ CU: $t = 2$

$RW(2: 0: 0) = 0$ $RW(2: 0: 8) = 2$

$RW(2: 0: 1) = 4$ $RW(2: 0: 9) = 6$

$RW(2: 0: 2) = 8$ $RW(2: 0: 10) = 10$

$RW(2: 0: 3) = 12$ $RW(2: 0: 11) = 14$

CU: $t = 1$ CU: $t = 3$

$RW(2: 0: 4) = 1$ $RW(2: 0: 12) = 3$

$RW(2: 0: 5) = 5$ $RW(2: 0: 13) = 7$

$RW(2: 0: 6) = 9$ $RW(2: 0: 14) = 11$

$RW(2: 0: 7) = 13$ $RW(2: 0: 15) = 15$

The twiddle factors are computed using Eq. 6.26. For stage 1, all CUs have the same TW factors given by Eq. 6.26a

					TWR	*TWI*	*TW* **Index**
CU:							
$p(1: t; 1) = 0$					1	0	1
$p(1: t; 2) = 0$					1	0	2
$p(1: t; 3) = 16/4 = 4$					0	-1	3
$t = 0, 1, 2, 3$							

Next the TW factors for stage 2 are computed from Eqs. 6.26b and 6.26c

$$\text{CU: } t = 0$$
$$p(2: 0; 1) = 0$$
$$p(2: 0; 2) = 0$$
$$p(2: 0; 3) = 4$$
$$\text{CU: } t > 0$$
$$p(2: t; 1) = p(m: t - 1; 1) + 2$$
$$p(2: t; 2) = p(m: t - 1; 2) + 1$$
$$p(2: t; 3) = p(m: t - 1; 3) + 1$$

		TWR	*TWI*	*TW* **Index**
CU: $t = 0$				
$p(2: 0; 1) = 0$	$\rightarrow$	1	0	1
$p(2: 0; 2) = 0$	$\rightarrow$	1	0	2
$p(2: 0; 3) = 4$	$\rightarrow$	0	-1	3
CU: $t = 1$				
$p(2: 1; 1) = 0 + 2 = 2$	$\rightarrow$	0.7071068	-0.7071068	1
$p(2: 1; 2) = 0 + 1 = 1$	$\rightarrow$	0.9238795	-0.3826834	2
$p(2: 1; 3) = 4 + 1 = 5$	$\rightarrow$	-0.3826834	-0.9238795	3
CU: $t = 2$				
$p(2: 2; 1) = 2 + 2 = 4$	$\rightarrow$	0	-1	1
$p(2: 2; 2) = 1 + 1 = 2$	$\rightarrow$	0.7071068	-0.7071068	2
$p(2: 2; 3) = 5 + 1 = 6$	$\rightarrow$	-0.7071068	-0.7071068	3
CU: $t = 3$				
$p(2: 3; 1) = 4 + 2 = 6$	$\rightarrow$	-0.7071068	-0.7071068	1
$p(2: 3; 2) = 2 + 1 = 3$	$\rightarrow$	0.3826834	-0.9238795	2
$p(2: 3; 3) = 6 + 1 = 7$	$\rightarrow$	-0.9238795	-0.3826834	3

The twiddle factor exponents can also be obtained directly from Table 6.6. ∎

Given the foregoing read/write addresses and the twiddle factor coefficients, the 16-point FFT4 implementation is completed as previously shown in Figure 6.35. A BASIC program for implementing the FFT4 DIT in-place algorithm is presented in Figure 6.36. The BASIC program is provided to illustrate the implementation of the FFT4 algorithm; it is not intended as an efficient personal computer program. Several computer programs are available for use or you can write your own to gain experience with the FFT algorithm.

6.8.4 FFT Algorithm for *N*, a Product of Factors

We have considered FFT algorithms for N equal to a power of two and four in the preceding sections. If N is made up of a set of a factors

$$N = n_1 n_2 n_3 \ldots n_a$$

then we can perform a decomposition of the DFT based on n_1. This provides the designer with added flexibility in the design of the system. For systems with fixed sampling rates and rigid spectral resolution requirements, the value of N must be chosen to meet the resolution requirement.

Starting with the DFT expression and N as specified above, let $N = n_1 p_1$, then

$$p_1 = n_2 n_3 \ldots n_a$$

and the DFT can be decomposed into a set of n_1 summations of length p_1

$$X(k) = \sum_{r=0}^{P} x(n_1 r) W_N^{n_1 rk} + \sum_{r=0}^{P} x(n_1 r + 1) W_N^{k} W_N^{n_1 rk} + \cdots$$

$$\cdots + \sum_{r=0}^{P} x(n_1 r + n_1 - 1) W_N^{(n_1 - 1)k} W_N^{n_1 rk}$$

$$P = p_1 - 1$$

Rewriting the expression by factoring out the W terms, which are now summed on r, and rewriting the W_N summation terms we obtain

$$X(k) = \sum_{s=0}^{n-1} W_N^{sk} \sum_{r=0}^{P} x(n_1 r + s) W_{p_1}^{rk} \qquad (6.27)$$

$$W_{p_1}^{rk} = W_N^{n_1 rk}$$

The summation over r represents a DFT of length p_1. We can proceed with this decomposition by further factoring the inner summation. The results of this process is a set of computational units based on the factors chosen. A general treatment of FFT algorithms is found in several texts.[1,2,9]

Example 6.5
Develop the FFT algorithm for three-point factors, FFT3, and implement a nine-point transform.

Solution:

From Eq. 6.27 the expression for $X(k)$ is given by

$$X(k) = \sum_{s=0}^{2} W_9^{sk} \sum_{r=0}^{2} x(3r + s)W_3^{rk}$$

$$= W_9^0 \sum_{r=0}^{2} x(3r + 0)W_3^{rk} + W_9^k \sum_{r=0}^{2} x(3r + 1)W_3^{rk}$$

$$+ W_9^{2k} \sum_{r=0}^{2} x(3r + 2)W_3^{rk}, \qquad k = 0, 1, 2$$

Redefining the input sequences for each of the three summations, each summation is written as a three-point DFT.

$$x_0(r) = x(3r); \; x_1(r) = x(3r + 1); \; x_2(r) = x(3r + 2);$$
$$r = 0, 1, 2$$

$$X_0(k) = \sum_{r=0}^{2} x_0(r)W_3^{rk}$$

$$X_1(k) = \sum_{r=0}^{2} x_1(r)W_3^{rk}$$

$$X_2(k) = \sum_{r=0}^{2} x_2(r)W_3^{rk}$$

Figure 6.37 presents the three-point DFT flow diagram. The flow diagram applies to each of the three input signal sets (x_1, $i = 0, 1, 2$), resulting in the intermediate DFT outputs (X_1, $i = 0, 1, 2$). Combining the intermediate three-point DFTs results in the nine-point FFT output.

$$X(k) = \sum_{r=0}^{2} x_0(r)W_3^{rk} + W_9^k \sum_{r=0}^{2} x_1(r)W_3^{rk} + W_9^{2k} \sum_{r=0}^{2} x_2(r)W_3^{rk}$$

$$= X_0(k) + W_9^k X_1(k) + W_9^{2k} X_2(k)$$

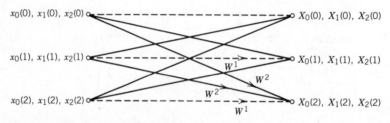

Figure 6.37 Three-point butterfly.

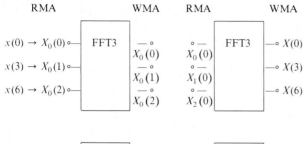

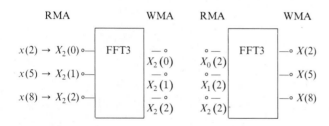

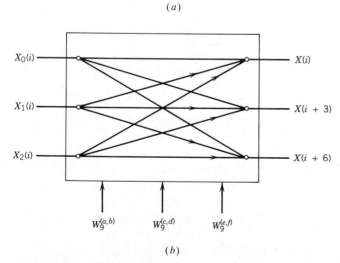

(a)

(b)

Figure 6.38 Nine-point FFT (a) flow graph showing three-point computational units (b) three-point computational unit.

The radix-3 FFT output is obtained by combining the three-point DFTs and using the periodic property of the X_i and twiddle factor terms

$$X_0(k) = X_0(k + 3) = X_0(k + 6) \qquad k = 0, 1, 2$$

$$X_1(k) = X_1(k + 3) = X_1(k + 6) \qquad k = 0, 1, 2$$

$$X_2(k) = X_2(k + 3) = X_2(k + 6) \qquad k = 0, 1, 2$$

$$X(0) = X_0(0) + W_9^0 X_1(0) + W_9^0 X_2(0)$$

$$X(1) = X_0(1) + W_9^1 X_1(1) + W_9^2 X_2(1)$$

$$X(2) = X_0(2) + W_9^2 X_1(2) + W_9^4 X_2(2)$$

$$X(3) = X_0(0) + W_9^3 X_1(0) + W_9^6 X_2(0)$$

$$X(4) = X_0(1) + W_9^4 X_1(1) + W_9^8 X_2(1)$$

$$X(5) = X_0(2) + W_9^5 X_1(2) + W_9^1 X_2(2)$$

$$X(6) = X_0(0) + W_9^6 X_1(0) + W_9^3 X_2(0)$$

$$X(7) = X_0(1) + W_9^7 X_1(1) + W_9^5 X_2(1)$$

$$X(8) = X_0(2) + W_9^8 X_1(2) + W_9^7 X_2(2)$$

The nine-point FFT implementation is shown in Figure 6.38 using the results from Figure 6.37. ∎

6.9 FAST CONVOLUTION

In Section 6.4 we discussed the process of circular convolution obtained by taking the IDFT of the product of the DFTs of two signals. In order to produce a result equal to a linear convolution, it was necessary to append zeros to the signals negating the circular effect. A FIR digital filter operation is a linear convolution of the finite duration filter impulse response with the input signal sequence. For many applications the input signal sequence is of long duration compared to the filter impulse response duration. Therefore, in order to perform the filtering operation in the frequency domain using the FFT/IFFT algorithms, the input sequence must be grouped into batches of samples. The selection of the sample batch size is a function of the filter impulse response duration and the hardware/software available for implementing the FFT/IFFT. Since the filter impulse response is known a priori, the FFT of the impulse response can be precalculated and stored. The process of fast convolution is shown in Figure 6.39. The overlap-save and overlap-add techniques for performing filtering in the frequency domain are presented in the following paragraphs.

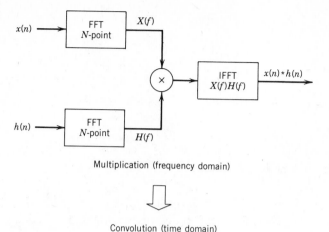

Multiplication (frequency domain)

Convolution (time domain)

Figure 6.39 Fast convolution process.

6.9.1 Overlap-Save Fast Convolution Algorithm

The **overlap-save** method makes use of the convolution property of the DFT, which can be computed rapidly using the FFT algorithm. It can be shown that multiplication of two DFTs corresponds to a circular convolution of their time-domain sequences. However, in order to implement a FIR filter, a linear convolution is required. A procedure for converting a circular convolution into a linear convolution is the overlap-save algorithm. The overlap-save process is carried out by overlapping long-duration signal sections by $(Q - 1)$ samples, where Q is the length of the FIR filter impulse response $h(n)$. If the overlap is less than $(Q - 1)$, wrap-around errors resulting from the circular convolution are introduced. As a result of these errors, the filter is changed from being time-invariant into a time-varying periodic linear filter with period $(N - Q - 1)$, where N is the FFT length.[13]

SELECT-SAVE (OVERLAP-SAVE) PROCEDURE

The process for select-save is shown in Figure 6.40 and described by the following procedure:

1. Perform N-point FFT of the FIR impulse response $h(n)$.

$$h(n) = \begin{cases} h(n) & n = 0, 1,, \ldots, Q - 1 \\ 0 & n = Q, Q + 1, \ldots, N - 1 \end{cases}$$

Store the FFT output for subsequent use.

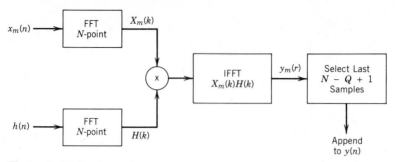

Figure 6.40 Overlap-save convolution process (digital filtering in the frequency domain).

2. Select N points from the input signal sequence based on the following expression

$$x_m(n) = x[n + (m-1)(N-Q+1)] \qquad (6.28)$$

$$n = 0, 1, 2, \ldots, N-1$$

$$m = 1, 2, 3, 4, \ldots$$

$$N = \text{FFT length}$$

$$Q = \text{impulse response length}$$

Figure 6.41 shows the overlap resulting from Eq. 6.28 for successive FFTs of the input signal.

3. Multiply the stored frequency response of the filter obtained in step 1 by the FFT of input signal batch m.

4. Perform an N-point IFFT on the product obtained in step 3.

5. Discard the first $(Q-1)$ points from each successive output of step 4, and append the remaining outputs to $y(n)$, that is,

$y_1(n)$	$n = Q-1, \ldots, N-1$
$y_2[n - (N-Q+1)]$	$n = N, \ldots, 2N-Q$
$y_3[n - 2(N-Q+1)]$	$n = 2N-Q+1, \ldots, 3N-2Q+1$
$\vdots$	$\vdots$
$y_m[n - (m-1)(N-Q+1)]$	$n = (m-1)(N-Q+1) + (Q-1),$
	$\ldots, (m-1)(N-Q+1) + (N-1)$
$\vdots$	

Finally,

$$y(n) = y_1(n), \; y_2(n), \; y_3(n), \ldots, \; y_m(n), \ldots$$

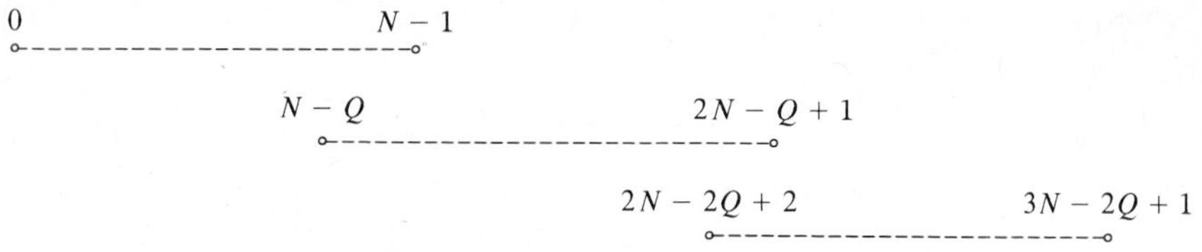

Figure 6.41 Overlap data records for the overlap-save fast convolution processing.

Example 6.6

The impulse response and the signal to be filtered is given by

$$h(n) = \begin{cases} 1 & \text{for } 0 \le n \le 3 \\ 0 & \text{elsewhere} \end{cases}$$

$$x(n) = \begin{cases} 0 & \text{for } n < 0 \\ 1 & \text{for } 0 \le n \le 3 \\ -1 & \text{for } 4 \le n \le 7 \end{cases}$$

$$x(n) = x(n+8) \qquad \text{for } n \ge 0$$

Assuming the FFT length $N = 8$, determine the linear convolution of the sequence $x(n)$ with the impulse response $h(n)$ using the overlap-save method.

Solution:

Sectioning the sequence $x(n)$ into overlapping subsections we obtain the overlapped sequences of Figure 6.42. Notice in Figure 6.42 that the first $(Q-1)$ samples of $x_1(n)$ are padded with zeros since the preceding section does not exist; that is, the initial conditions of the system are set to zero. As a result, we can rewrite Eq. 6.28 in the form

$$x_m(n) = x[n + (m-1)(N-Q+1) - (Q-1)]$$
$$n = 0, 1, \ldots, N-1$$
$$m = 1, 2, \ldots \qquad (6.29)$$

$n =$	0	1	2	3	4	5	6	7	8	9	10	11	12	13	14	15	16	17	18	19	...
$x(n) =$	1	1	1	1	−1	−1	−1	−1	1	1	1	1	−1	−1	−1	−1	1	1	1	1	...

$n =$	−3	−2	−1	0	1	2	3	4	
$x_1(n) =$	0	0	0	1	1	1	1	−1	$-(Q-1), \ldots, N-Q$

$n =$	2	3	4	5	6	7	8	9	
$x_2(n) =$	1	1	−1	−1	−1	−1	1	1	$N-2(Q-1), \ldots, 2(N-Q)+1$

$n =$	7	8	9	10	11	12	13	14	
$x_3(n) =$	−1	1	1	1	1	−1	−1	−1	$2N-3(Q-1), \ldots, 3(N-Q)+2$

$\ddots$

Figure 6.42 Overlap-save method of sectioning input data sequence $x(n)$ (Example 6.6).

The circular convolution of sequence $x_1(n)$ with $h(n)$ yields

$$
\begin{bmatrix} y_1(0) \\ y_1(1) \\ y_1(2) \\ y_1(3) \\ y_1(4) \\ y_1(5) \\ y_1(6) \\ y_1(7) \end{bmatrix}
=
\begin{bmatrix}
0 & -1 & 1 & 1 & 1 & 1 & 0 & 0 \\
0 & 0 & -1 & 1 & 1 & 1 & 1 & 0 \\
0 & 0 & 0 & -1 & 1 & 1 & 1 & 1 \\
1 & 0 & 0 & 0 & -1 & 1 & 1 & 1 \\
1 & 1 & 0 & 0 & 0 & -1 & 1 & 1 \\
1 & 1 & 1 & 0 & 0 & 0 & -1 & 1 \\
1 & 1 & 1 & 1 & 0 & 0 & 0 & -1 \\
-1 & 1 & 1 & 1 & 1 & 0 & 0 & 0
\end{bmatrix}
\begin{bmatrix} 1 \\ 1 \\ 1 \\ 1 \\ 0 \\ 0 \\ 0 \\ 0 \end{bmatrix}
$$

$$
=
\begin{bmatrix} 1 \\ 0 \\ -1 \\ 1 \\ 2 \\ 3 \\ 4 \\ 2 \end{bmatrix}
\begin{array}{l} \\ \text{Discard} \\ \\ \text{------} \\ \\ \\ \text{Save} \\ \\ \end{array}
$$

where $h(n)$ was padded with $(N - Q)$ zeros. The circular convolution of sequence $x_2(n)$ with $h(n)$ yields

$$
\begin{bmatrix} y_2(0) \\ y_2(1) \\ y_2(2) \\ y_2(3) \\ y_2(4) \\ y_2(5) \\ y_2(6) \\ y_2(7) \end{bmatrix}
=
\begin{bmatrix}
1 & 1 & 1 & -1 & -1 & -1 & -1 & 1 \\
1 & 1 & 1 & 1 & -1 & -1 & -1 & -1 \\
-1 & 1 & 1 & 1 & 1 & -1 & -1 & -1 \\
-1 & -1 & 1 & 1 & 1 & 1 & -1 & -1 \\
-1 & -1 & -1 & 1 & 1 & 1 & 1 & -1 \\
-1 & -1 & -1 & -1 & 1 & 1 & 1 & 1 \\
1 & -1 & -1 & -1 & -1 & 1 & 1 & 1 \\
1 & 1 & -1 & -1 & -1 & -1 & 1 & 1
\end{bmatrix}
\begin{bmatrix} 1 \\ 1 \\ 1 \\ 1 \\ 0 \\ 0 \\ 0 \\ 0 \end{bmatrix}
$$

$$
=
\begin{bmatrix} 2 \\ 4 \\ 2 \\ 0 \\ -2 \\ -4 \\ -2 \\ 0 \end{bmatrix}
\begin{array}{l} \\ \text{Discard} \\ \\ \text{--------} \\ \text{Save} \\ \text{Append} \\ \text{to} \\ \text{previous} \\ \text{set} \end{array}
$$

Since circular convolutions are performed on overlapping sections of data, the last $(N - Q + 1)$ samples of output sequences $y_1(n)$ with $y_2(n)$ are saved from each circular convolution and concatenated to form the filtered output, that is

$$y(n) = y_1(n), y_2(n) = \{1 \quad 2 \quad 3 \quad 4 \quad 2 \quad 0 \quad -2 \quad -4 \quad -2 \quad 0\}$$

Continuation of the overlap-save process results in the indefinitely long filtered sequence given by

$$y(n) = y_1(n), y_2(n), y_3(n), y_4(n), \ldots \quad \blacksquare$$

6.9.2 Overlap-Add Fast Convolution Algorithm

The procedure for the **overlap-add** process modifies the input by making the last $Q - 1$ samples of each N-point FFT zero and requires an addition of the first $Q - 1$ IFFT points of the mth iteration with the last $Q - 1$ IFFT points of the $m - 1$ iteration. The process is shown in Figure 6.43 and the procedure is itemized.

OVERLAP-ADD PROCEDURE

1. Perform N-point FFT of FIR impulse response, where zeros are appended to $h(n)$ to obtain an N-point sequence. Store the values for subsequent use.

2. Select $N - Q + 1$ points from the input signal sequence based on the following expression:

$$x_m(n) = \begin{matrix} x[r + (m-1)(N - Q + 1)] & 0 \le r \le N - Q \\ 0 & N - Q < r \le N - 1 \end{matrix}$$

$$\begin{aligned} &n = 0, 1, 2, \ldots, N - 1 \\ &m = 1, 2, 3, 4, \ldots \\ &N = \text{FFT length} \\ &Q = \text{impulse response length} \end{aligned} \qquad (6.30)$$

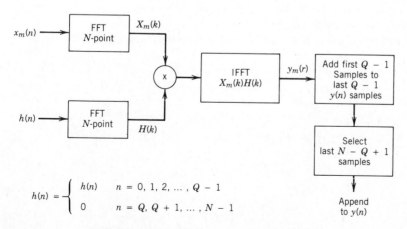

$$h(n) = \begin{cases} h(n) & n = 0, 1, 2, \ldots, Q - 1 \\ 0 & n = Q, Q + 1, \ldots, N - 1 \end{cases}$$

Figure 6.43 Overlap-add fast convolution.

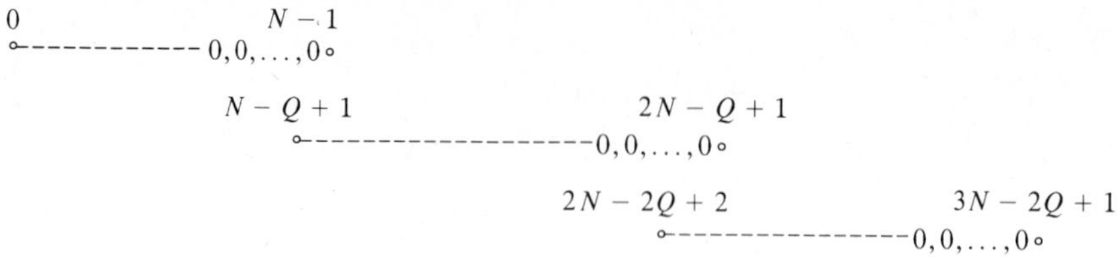

$$0 \qquad\qquad N - 1$$
$$\circ\text{-----------}\,0,0,\dots,0\,\circ$$

$$N - Q + 1 \qquad\qquad 2N - Q + 1$$
$$\circ\text{----------------}\,0,0,\dots,0\,\circ$$

$$2N - 2Q + 2 \qquad\qquad 3N - 2Q + 1$$
$$\circ\text{--------------}\,0,0,\dots,0\,\circ$$

Figure 6.44 Overlap data records for the overlap-add fast convolution processing.

Figure 6.44 shows the overlap resulting from Eq. 6.30 for successive FFTs of the input signal.

3. Multiply the stored frequency response of the filter obtained in step 1 by the FFT on input signal batch m.

4. Perform an N-point IFFT on the product obtained in step 3.

5. Add the first $Q - 1$ points from each successive output of step 5 with the last $Q - 1$ outputs of the preceding output $y(n)$ given by

$$y(n) = \begin{cases} y(n) + y_m(r) & r = 0, 1, 2, \dots, Q - 2 \\ y_m(r) & Q - 1 \le r \le N - 1 \\ & n = (m - 1)(N - Q + 1) + r \end{cases}$$

6.9.3 Fast Convolution Computational Considerations

Two approaches to fast convolution were presented in the preceding sections. The value of these procedures is that they provide a technique for performing the convolution-sum operation (i.e., digital filtering) in the frequency domain using any of the numerous FFT algorithms available. This is useful in instances where the signal is already transformed to the frequency domain and one or more band definition processes are desired. Also, the hardware may be efficient in performing FFTs rather than FIR or IIR difference equations. Therefore, the computational advantages of either the time-domain filtering or frequency domain filtering approaches must be evaluated based on the specific application. The result will be dependent on the hardware/software solutions available. From a pure computational comparison (i.e., multiplications and additions) the expressions are easily developed from the preceding digital filter and FFT computational requirements discussed in Sections 2.3.2 and 6.8, respectively. The multiplications and additions for the FIR filter implementations assuming symmetry is taken advantage of in the implementation were given by

$$\text{MPS} = F(Q + 1)/2$$
$$\text{ADS} = FQ$$

where F = filter input sampling rate
Q = filter impulse response length

Using the overlap-save procedure, the number of multiplications and additions per second is approximately

$$MPS = 4FN(1 + \log_2 N)/(N - Q + 1)$$
$$ADS = 2FN(1 + 3\log_2 N)/(N - Q + 1)$$

where the FFT length is N, F and Q are as above, and the FFT rate is given by

$$FFT\ Rate = F/(N - Q + 1)$$

It was assumed that the FFT of the impulse response is precomputed. For a fixed-length filter Q and a specified sampling rate F, the value of N can be selected to minimize the number of multiplications, as shown in Figure 6.45. Note that N has been constrained to a power of two (i.e., FFT2 implementation) and we have not attempted to take advantage of simple twiddle factors. A comparison between the multiplications required for the symmetric FIR implementation and the optimum FFT2 implementation is presented in Table 6.7. Quadrature data (i.e., real and imaginary channels) are assumed, which double the MPS and APS FIR numbers. Three sets of values for N, the FFT length, are given in the three tables provided. The minimum MPS FFT requirements were given in Figure 6.45.

The comparison is meant to make you aware of the trade-offs that must be considered when evaluating various approaches for designing signal processing systems. As mentioned earlier, the specific requirements and the available hardware/software can have significant affects on the results and must be included in the analysis. We address system design alternatives in Chapter 10.

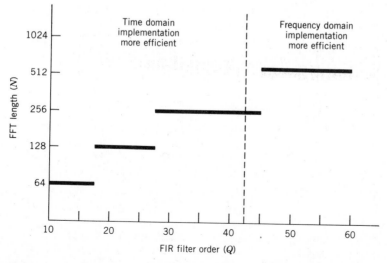

Figure 6.45 Optimum FFT length versus FIR filter length for fast convolution.

TABLE 6.7
FAST CONVOLUTION COMPUTATIONAL REQUIREMENTS: FIR TIME DOMAIN
VERSUS FREQUENCY DOMAIN (FAST CONVOLUTION) APPROACHES
(IN-PHASE AND QUADRATURE INPUT DATA ASSUMED).

$F = 1000$ Hz Filter Q	FFT N	FIR MPS	FIR APS	Fast Convolution MPS	Fast Convolution APS
10	64	11000	20000	32581.82	55854.55
11	64	12000	22000	33185.19	56888.89
12	64	13000	24000	33811.32	57962.26
13	64	14000	26000	34461.54	59076.92
14	64	15000	28000	35137.25	60235.29
15	64	16000	30000	35840.00	61440.00
16	64	17000	32000	36571.43	62693.88
17	64	18000	34000	37333.33	64000.00
18	64	19000	36000	38127.66	65361.70
19	64	20000	38000	38956.52	66782.61
20	128	21000	40000	37577.98	65761.47
21	128	22000	42000	37925.93	66370.37
22	128	23000	44000	38280.37	66990.65
23	128	24000	46000	38641.51	67622.64
24	128	25000	48000	39009.52	68266.67
25	128	26000	50000	39384.62	68923.08
26	128	27000	52000	39766.99	69592.23
27	128	28000	54000	40156.86	70274.51
28	128	29000	56000	40554.46	70970.30
29	128	30000	58000	40960.00	71680.00
30	256	31000	60000	40599.12	72176.21
31	256	32000	62000	40778.76	72495.58
32	256	33000	64000	40960.00	72817.78
33	256	34000	66000	41142.86	73142.86
34	256	35000	68000	41327.35	73470.85
35	256	36000	70000	41513.51	73801.80
36	256	37000	72000	41701.36	74135.75
37	256	38000	74000	41890.91	74472.73
38	256	39000	76000	42082.19	74812.79
39	256	40000	78000	42275.23	75155.96
40	256	41000	80000	42470.05	75502.30
41	256	42000	82000	42666.67	75851.85
42	256	43000	84000	42865.12	76204.65
43	256	44000	86000	43065.42	76560.75
44	256	45000	88000	43267.61	76920.19
45	256	46000	90000	43471.70	77283.02
46	256	47000	92000	43677.73	77649.29
47	256	48000	94000	43885.71	78019.05
48	256	49000	96000	44095.69	78392.34
49	256	50000	98000	44307.69	78769.23
50	512	51000	100000	44233.26	79619.87
51	512	52000	102000	44329.00	79792.21
52	512	53000	104000	44425.16	79965.29
53	512	54000	106000	44521.74	80139.13
54	512	55000	108000	44618.74	80313.73
55	512	56000	110000	44716.16	80489.08
56	512	57000	112000	44814.00	80665.21
57	512	58000	114000	44912.28	80842.11
58	512	59000	116000	45010.99	81019.78
59	512	60000	118000	45110.13	81198.24

6.10 SUMMARY

We presented the DFT pair and developed important properties of the DFT. The frequency response of the DFT was derived, and characteristics including frequency selectivity, spectral leakage, scalloping loss, equivalent noise bandwidth, and overlap correlation were presented. Several weighting functions were introduced, and their impact on the characteristics were presented.

The FFT DIT and DIF radix-2 algorithms were developed. The computational efficiency of the FFT over the DFT was developed. Computational units for FFT2, FFT3, and FFT4 point elements were introduced. Procedures for implementing FFTs using the computational units were developed, which provide general expressions for data-addressing and twiddle factors. Bit-reversal algorithms were defined. BASIC computer programs were provided to illustrate the algorithms.

The use of the FFT to perform a filtering operation (i.e., convolution) was presented. Methods for avoiding the circular effect of the FFT were developed. Overlap-add and overlap-save procedures for performing the convolution were presented. A comparison between the time domain and frequency domain filtering approaches was performed.

References

1. E. O. Brigham, *The Fast Fourier Transform*, Prentice–Hall, Englewood Cliffs, N.J., 1974.

2. A. V. Oppenheim and R. W. Schafer, *Digital Signal Processing*, Prentice–Hall, Englewood Cliffs, N.J., 1975.

3. J. W. Cooley and J. W. Tukey, *An Algorithm for the Machine Calculation of Complex Fourier Series*, *Mathematics of Computation*, Vol. 19, No. 90, The American Mathematical Society, Providence, 1965.

4. Digital Signal Processing Committee, IEEE ASSPS Society, *Programs for Digital Signal Processing*, IEEE Press, New York, 1979.

5. F. J. Harris, "On the Use of Windows for Harmonic Analysis with the Discrete Fourier Transform," *Proc. IEEE*, 66(1): January 1978.

6. J. R. Williams and G. G. Ricker, "Spectrum Analyzer Overlap Requirements," *J. Acoustical Society of America*, 64(3): September 1978.

7. R. B. Blackman and J. W. Tukey, *The Measurement of Power Spectra*, Dover, New York, 1958.

8. F. F. Kuo and J. F. Kaiser, *System Analysis by Digital Computer*, Wiley, New York, 1966.

9. M. Bellanger, *Digital Processing of Signals, Theory and Practice*, Wiley, New York, 1984.

10. L. R. Rabiner and B. Gold, *Theory and Applications of Digital Signal Processing*, Prentice–Hall, Englewood Cliffs, N.J., 1975.

11. B. Gold and C. M. Rader, *Digital Processing of Signals*, McGraw–Hill, New York, 1969.

12. R. E. Blahut, *Fast Algorithms for Digital Signal Processing*, Addison–Wesley, Reading, Mass., 1984.

13. L. Pelkowitz, "Frequency Domain Analysis of Wraparound Error in Fast Convolution Algorithms," *IEEE Trans. ASSP* June 1981.

14. B. Gold and T. Bailly, "Parallelism in Fast Fourier Transform Hardware," *IEEE Trans. Audio Electroacoustics*, February 1973.

Additional Readings

ANTONIOU, A. *Digital Filters Analysis and Design*. McGraw–Hill, New York, 1979.

Selected Papers in Digital Signal Processing, II, Part 1: General. Digital Signal Processing Committee, IEEE ASSP Society, IEEE Press, New York, 1975.

DOETSCH, G. *Guide to Applications of Laplace Transform*, (translated). Van Nostrand, Princeton, N.J., 1963.

JUNG, M. T. *Methods of Discrete Signal and System Analysis*. McGraw–Hill, New York, 1982.

Oppenheim, A. V. (ed.). *Applications of Digital Signal Processing*, Prentice–Hall, Englewood Cliffs, N.J., 1978.

Oppenheim, A. V. (ed.). *Papers on Digital Signal Processing*, MIT Press, Cambridge, Mass., 1969.

Rabiner, L. R., and Rader, C. M. (eds.). *Digital Signal Processing, Part 2—The Fast Fourier Transform*. IEEE Press, New York, 1972.

S. D. STEARNS, *Digital Signal Analysis*. Hayden, Rochelle Park, N.J., 1975.

S. A. TRETTER, *Introduction to Discrete-Time Signal Processing*. Wiley, New York, 1976.

C. LANCZOS, Discourse on Fourier Series, New York, Hafner Publishing Co., 1966.

PROBLEMS

6.2.1. Prove the following relationship.

$$\sum_{n=0}^{N-1} e^{j(2\pi/N)(r-k)n} = \begin{cases} N & \text{for } r = k \\ 0 & \text{for } r \neq k \end{cases}$$

6.2.2. Find the DFT of the Hann window given by

$$x(n) = 0.5\left(1 - \cos\frac{2\pi n}{N}\right) \qquad n = 0, 1, \ldots, N - 1$$

6.2.3. Find the DFT of the following sequences.

a. $x(n) = e^{j(2\pi/N)kn} \qquad n = 0, 1, \ldots, N - 1$

b. $x(n) = \cos\left(\frac{2\pi kn}{N}\right) \qquad n = 0, 1, \ldots, N - 1$

c. $x(n) = \sin\left(\frac{2\pi kn}{N}\right) \qquad n = 0, 1, \ldots, N - 1$

where k is an integer value frequency (bin) index.

6.4.1. a. Determine the circular convolution of the following sequences.

$$x(n) = \{1,1,1,1,-1,-1,-1,-1\}$$
$$h(n) = \{0,1,2,3,4,3,2,1\}$$

b. Compare the result of part a with the linear convolution of the sequences.

6.4.2. Use the matrix approach of the circular convolution to obtain a linear convolution (see Section 7.4) of the two sequences:

$$h(n) = \{1,2,3,4,3,2,1\}$$
$$x(n) = \{1,1,1,1,-1,-1,-1,-1\}$$

6.6.1. Consider the following sequence:

$$x(nT) = e^{j2\pi f_0 nT}$$

a. Show that the DFT of $x(nT)$ is given by

$$X_k = e^{-j[2\pi(k-k_0)(N-1)/2N}\frac{\sin \pi(k-k_0)}{\sin \pi(k-k_0)/N}$$

b. If k and k_0 are integers show that

$$X_k = \begin{cases} N & \text{for } k = k_0 \\ 0 & \text{otherwise} \end{cases}$$

c. If f_0 is not one of the frequency sampling points $k_0 F/N$ show that none of the DFT values are zero, that is, spectral leakage results.

6.6.2. a. Find a closed-form expression for the DFT of the following discrete sequences.

$$x_1(n) = \begin{cases} 1 & \text{for } 0 \le n \le 2 \\ 0 & \text{for } 3 \le n \le 8 \end{cases}$$

$$x_2(n) = \begin{cases} e^{j(2\pi/9)3n} & \text{for } 0 \le n \le 2 \\ 0 & \text{for } 3 \le n \le 8 \end{cases}$$

$$x_3(n) = \begin{cases} -e^{j(2\pi/9)3n} & \text{for } 0 \le n \le 2 \\ 0 & \text{for } 3 \le n \le 8 \end{cases}$$

b. Show that the DFT of sequences $x_2(n)$ and $x_3(n)$ could have been deduced from the DFT of sequence $x_1(n)$. *Hint:* See modulation property.

c. Plot the magnitude response of sequences $X_1(k)$, $X_2(k)$, and $X_3(k)$. Comment on the resultant frequency responses.

6.6.3. Determine the scallop loss (SL) in decibels for the following spectral weighting functions:

a. Rectangular

$$A_R(\theta) = \frac{\sin N\theta/2}{N \sin \theta/2}$$

b. Hann (set $\alpha = 0.5$ in $A_H(\theta)$)

c. Hamming (set $\alpha = 0.54$ in $A_H(\theta)$)

$$A_H(\theta) = \frac{1}{N}\left[\frac{\alpha \sin \theta N/2}{\sin \theta/2} + \frac{(1-\alpha)}{2} \frac{\sin\left[(N/2)(\theta - 2\pi/N)\right]}{\sin\left[(\theta - 2\pi/N)/2\right]} \right.$$
$$\left. + \frac{(1-\alpha)}{2} \frac{\sin\left[(N/2)(\theta - 2\pi/N)\right]}{\sin\left[(\theta + 2\pi/N)/2\right]} \right]$$

where

$$SL_{dB} = 20\log_{10} \frac{A(\pi/N)}{NA(0)}$$

Let $N = 512$ and $\theta = \pi/N$ (center of two frequency bins).

6.6.4. Determine the equivalent noise bandwidth (ENBW) for the Hanning data window (let $N = 4$).

6.8.1. a. It is desired to use the IFFT as a waveform generator to synthesize a composite waveform described by the following discrete function.

$$x(nT) = \sin(2\pi f_1 nT) + 0.5 \sin(2\pi f_2 nT)$$
$$+ 0.25 \sin(2\pi f_3 nT) + 0.125 \sin(2\pi f_4 nT)$$

where

$$F = 1024 \text{ Hz (sampling frequency)}$$
$$N = 2048 \text{ (FFT size)}$$
$$f_1 = 128.5 \text{ Hz}$$
$$f_2 = 129.5 \text{ Hz}$$
$$f_3 = 130.5 \text{ Hz}$$
$$f_4 = 131.5 \text{ Hz}$$

Determine the input spectral sequence to the IFFT that will generate the 2048 point sequence $x(nT)$.

b. The generation of arbitrary frequency sinusoids can be obtained by using the IFFT. Given the discrete complex sinusoidal function

$$x(nT) = e^{j(2\pi f_m nT + \theta)}$$

where $f_m = (k_0 + d)F/N$, $0 \le f_m \le F/2$ and $m = k_0 + d$.

$0 \le d < 1$, fraction of bin spacing above $k_0 F/N$

k_0 is the bin index (integer valued)

θ is the phase angle

N is the FFT size

Determine the following:

  i. $X(k) = \text{DFT}[x(nT)]$

  ii. The spectrum of $X(k)$ convolved with the spectrum of the Hanned window given by

$$W_H(k) = -0.25\delta(k+1) + 0.5\delta(k) - 0.25\delta(k-1)$$

that is, determine

$$X_H(k) = X(k) * W_H(k), \text{ where } * \text{ denotes convolution.}$$

  c. Plot the magnitude of the spectral sequences.

6.8.2. Two digital signal processing channels of a spectral analysis system are shown in Figure P6.8.2. The highest frequency component at the output of LPF1 is 1280 Hz. It is desired to use a radix-2 FFT algorithm to determine the power spectrum of the input signal with a frequency resolution no worse than 0.5 Hz for channel 1 and 0.1 Hz for channel 2. (*Note*: Assume all filter responses are idealized). Determine the following.

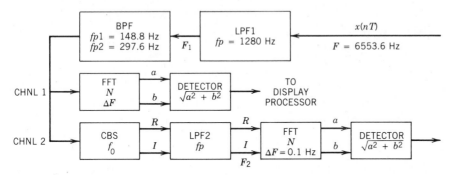

Figure P6.8.2

Channel 1

  a. The sampling rate F_1, where $F_1 = F/M$, M integer-valued.
  b. The number of samples N to be processed by the FFT.
  c. The actual frequency resolution, ΔF.
  d. The FFT bin numbers associated with the BPF passband frequencies.
  e. The time duration of the signal, that is, the FFT rate (sec/FFT).

Channel 2

For channel 2, the bandpass signal is complex band-shifted (CBS), that is, complex is demodulated by $e^{-j2\pi f_0 kT}$ to form a lowpass signal, where the center of the band is shifted to dc. Determine the following.

a. The center frequency of the band, f_0, that is, the modulation frequency.

b. The passband frequency, f_P, of LPF2.

c. The FFT size, N.

d. The sampling frequency, F_2, at the output of LPF2.

e. The reduction in sampling frequency, that is, F_1/F_2.

f. The FFT rate (sec/FFT).

g. Sketch the idealized spectrum at the output of the detector.

6.8.3. Efficient evaluation of the FFT for real discrete signals. This problem describes a procedure for calculating the FFTs of two real signals by applying one FFT. Let $x(n)$ and $y(n)$ be two real discrete signals of length N. Then

a. Form the complex signal

$$z(n) = x(n) + jy(n)$$

b. Compute the FFT of $z(n)$

$$Z(k) = \text{FFT}[z(n)]$$

c. Sort $Z(k)$ to obtain $X(k)$ and $Y(k)$

$$X(0) = \text{Re}\,[Z(0)], \qquad X(N/2) = \text{Re}\,[Z(N/2)]$$
$$Y(0) = \text{Im}\,[Z(0)], \qquad Y(N/2) = \text{Im}\,[Z(N/2)]$$
$$X(k) = 0.5[Z(k) + Z^*(N - k)]$$
$$Y(k) = -0.5j[Z(k) - Z^*(N - k)], \qquad k = 1, 2, \ldots, (N/2) - 1$$

Given the sequences

$$x(0) = 1,\ x(1) = 1,\ x(2) = 1,\ x(3) = 1$$
$$y(0) = 0.5,\ y(1) = 1,\ y(2) = 1,\ y(3) = 0.5$$

Find the FFT of the complex signal

$$z(n) = x(n) + jy(n)$$

and sort $Z(k)$ using the foregoing procedure to obtain $X(k)$ and $Y(k)$. Check results by finding the FFT of $x(n)$ and $y(n)$.

6.8.4. It is desired to implement an in-place DIT 64-point FFT algorithm using FFT4 computational units (CU).

a. Determine the bit-reverse order input sequence.

b. Determine the CU read/write memory address locations and twiddle factors for each stage.

c. Construct a 64-point DIT FFT representation using primitive operations.

6.8.5. a. Determine an efficient algorithm to implement an in-place DIT FFT8 computational unit. The implementation equations should be in terms of real and imaginary parts (the input array and intermediate sum node outputs (see Figure P6.8.5).

    b. Using the FFT8 algorithm developed in part a, implement a 32-point FFT by using an FFT8 CU for the first stage and an FFT4 CU for the second stage.

    c. Show that for $N = 8 \times 4^n$, where n is an integer, the algorithm developed in part b can be extended to implement larger size FFTs. As an example, determine the read/write memory locations and twiddle factors for $N = 8 \times 4 \times 4 = 128$. It should be noted that the twiddle factors for the first stage FFT8 CU are independent of the FFT size.

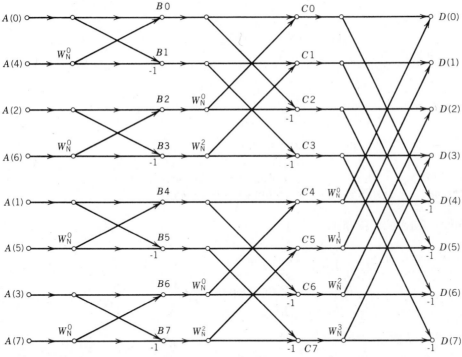

Figure P6.8.5

6.8.6. It is desired to implement the not-in-place version of the DIT N-point FFT algorithm using FFT4 computational units (CUs).

From the four-point signal flow graph in Figure P6.8.6, and using the method of Section 6.8.3, determine the general expressions for the FFT4 primitive operation in terms of real and imaginary parts. It should be noted that for this algorithm, bit reversal is not required and therefore the complex input data are in sequential order. Let the consecutive memory locations $A1$, $A2$, $A3$, and $A4$ be represented by the complex

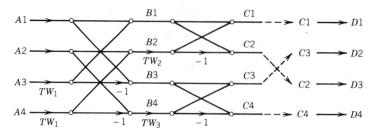

Figure P6.8.6

input data given by

$$A1 = X1R + jX1I$$
$$A2 = X2R + jX2I$$
$$A3 = X3R + JX3I$$
$$A4 = X4R + jX4I$$

Also define the complex twiddle factors in terms of their real and imaginary parts as follows.

$$TW_1 = TWR_1 + jTWI_1$$
$$TW_2 = TWR_2 + jTWI_2$$
$$TW_3 = TWR_3 + jTWI_3$$

6.8.7. It can be shown that the read/write memory addresses and twiddle factors for the not-in-place DIT FFT4 algorithm can be expressed as

Read Memory Address (RMA)

$$\text{RMA}(i, k) = \left[iI \bmod (L - 1)\right] + kL = \left[(L-1)\text{res}\frac{iI}{L-1}\right] + kL$$

$$k = 0, 1, \ldots, R - 1 \quad \text{(Outer loop index)}$$
$$i = 0, 1, \ldots, L - 1 \quad \text{(Inner loop index)}$$

Where the number of FFT4 stages is given by

$$M = \log_4 N, \quad N = 4, 16, 64, 256, 1024, \ldots$$

The FFT4 stage index is given by

$$1 \le m \le M$$

and

$$L = 4^{(M-m+1)}$$
$$I = 4^{(M-m)}$$
$$R = 4^{(m-1)}$$

Write Memory Address (WMA)

$$\text{WMA} = (N/4)i \bmod (N-1) = \left[\text{res} \frac{iN}{4(N-1)} \right](N-1)$$

$$i = 0, 1, \ldots, N-1$$

FFT4 Twiddle Factor Exponent Algorithm

The FFT4 twiddle factor factor is defined as

$$W_N^p = \cos 2\pi p/N - j \sin 2\pi p/N$$

The recursive procedure for generating the exponents of W for the FFT4 butterflies is given by

$$TW_1 = W^{p_1}$$
$$TW_2 = W^{p_2}$$
$$TW_3 = W^{p_3}$$

We define a three element vector as

$$\mathbf{p}(i) = \begin{bmatrix} p(1) \\ p(2) \\ p(3) \end{bmatrix}$$

The FFT4 twiddle factor exponent recursive algorithm is

$$\mathbf{p}(i) = \mathbf{p}(i-1) + I(m-1)$$

where

$$i = 2, 3, \ldots, \frac{N}{4^{(M+1-m)}}$$

The initial three-element vector $\mathbf{p}(1)$ is defined for all FFT4 stages as

$$\mathbf{p}(1) = \begin{bmatrix} 0 \\ 0 \\ N/4 \end{bmatrix} \quad \text{for } m = 1, 2, \ldots$$

For the increment index $\mathbf{I}$ is given by

$$\mathbf{I}(0) = \begin{bmatrix} 0 \\ 0 \\ 0 \end{bmatrix} \quad \text{for } m = 1$$

$$\mathbf{I}(m-1) = \begin{bmatrix} 2 \cdot 4^{(M-m)} \\ 4^{(M-m)} \\ 4^{(M-m)} \end{bmatrix} \cdot \quad \text{for } m > 1$$

Finally, the $\mathbf{p}(i)$ three-element vectors are repeated according to the vector repeat factor (VRF).

$$\text{VRF} = N4^{-m} \qquad m = 1, 2, \ldots$$

Determine the read memory addresses (RMA), write memory addresses (WMA), and twiddle factor exponents for $N = 16$ and 64.

6.8.8. Using the Cooley-Tukey formulation of the FFT algorithm,[1] determine the FFT stage equations for (a) $N = 8$, (b) $N = 16$, (c) $N = 32$, (d) $N = 64$.

Cooley-Tukey Formulation of the FFT Algorithm:

The FFT algorithm involves evaluating the expression

$$X(k) = \sum_{n=0}^{N-1} x_0(n) W_N^{nk}$$

where $W_N = e^{-j2\pi/N}$

$$N = 2^m, \ m \text{ integer valued}$$
$$k = \text{frequency index} = 0, 1, \ldots, N - 1$$
$$n = \text{time-sample index} = 0, 1, \ldots, N - 1$$

The Cooley-Tukey algorithm is performed by expressing the indices n and k in m-bit binary numbers as,

$$n = 2^{m-1} n_{m-1} + 2^{m-2} n_{m-2} + \cdots + 2n_1 + n_0$$
$$k = 2^{m-1} k_{m-1} + 2^{m-2} n_{m-2} + \cdots + 2k_1 + k_0$$

where n_i and k_i are equal to 0 or 1 and are contents of the respective bit positions in the binary representation of n and k. All arrays (stages) will now be written as functions of the bits of their indices. With this convention, $X(k)$ is written as

$$X(k_{m-1}, k_{m-2}, \ldots, k_0) = \sum_{n_0=0}^{1} \sum_{n_1=0}^{1} \cdots \sum_{n_{m-1}=0}^{1} x(n_{m-1}, n_{m-2}, \ldots, n_0) W_N^{nk}$$

where n and k are expressed as m-bit binary numbers. Performing each of the summations separately, and labeling the intermediate results, we obtain the stage equations,

$$x_1(k_0, n_{m-2}, \ldots, n_0) = \sum_{n_{m-1}=0}^{1} x_0(n_{m-1}, n_{m-2}, \ldots, n_0) W_N^{(N/2)k_0 n_{m-1}}$$

$$x_2(k_0, k_1, n_{m-3}, \ldots, n_0) = \sum_{n_{m-2}=0}^{1} x_1(k_0, n_{m-2}, \ldots, n_0) W_N^{(N/4)(2k_1+k_0)n_{m-2}}$$

$$\vdots$$

$$x_m(k_0, k_1, \ldots, k_{m-1}) = \sum_{n_0=0}^{1} x_{m-1}(k_0, k_1, \ldots, n_0)$$

$$\times W^{[(N/2)k_{m-1}+(N/4)k_{m-2}+\cdots+k_0]n_0}$$

Finally,

$$X(k_{m-1}, k_{m-2}, \ldots, k_0) = x_m(k_0, k_1, \ldots, k_{m-1})$$

That is, the final result is in bit-reversed order. To demonstrate the foregoing equations, consider the case for $N = 2^2 = 4$. For this case we represent n and k as $m = 2$-bit binary numbers,

$$n = 0, 1, 2, 3 \quad \text{or} \quad n = (n_1, n_0) = 00, 01, 10, 11$$
$$k = 0, 1, 2, 3 \quad \text{or} \quad k = (k_1, k_0) = 00, 01, 10, 11$$

That is, $n = 2n_1 + n_0$, $k = 2k_1 + k_0$ where n_0, n_1, k_0, and k_1 can take on values of 0 and 1 only. From the foregoing general equations the stage equations can be expressed by

$$x_1(k_0, n_0) = \sum_{n_1=0}^{1} x_0(n_1, n_0) W_N^{2k_0 n_1}$$

$$x_2(k_0, k_1) = \sum_{n_0=0}^{1} x_1(k_0, n_0) W_N^{(2k_1+k_0)n_0}$$

and

$$X(k_1, k_0) = x_2(k_0, k_1)$$

These recursive equations represent the Cooley-Tukey formulation of the FFT algorithm for $N = 4$.

Using the general recursive equations, determine the stage equations for $N = 8, 16, 32,$ and 64.

6.9.1. Use the overlap-save method to obtain the linear convolution of the impulse response and periodic sequence given below:

$$x(n) = \begin{cases} 0 & \text{for } n < 0 \\ 1 & \text{for } 0 \le n \le 4 \\ -1 & \text{for } 5 \le n \le 9 \end{cases} \qquad h(n) = \begin{cases} 1 & \text{for } 0 \le n \le 4 \\ 0 & \text{elsewhere} \end{cases}$$

Assume 16-point DFTs are used.

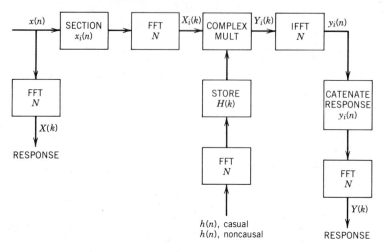

Figure P6.9.2

6.9.2. Use the overlap-save method to filter the indefinitely long sequence

$$x(n) = \cos(2\pi n/N) + \cos(38\pi n/N) + \cos(52\pi n/N) + \cos(64\pi n/N)$$

$$n = 0, 1, \ldots$$

with the causal impulse response obtained from the specifications of the following lowpass FIR filter.

$$F = 1.0 \text{ Hz}$$
$$f_p = 0.15 \text{ Hz}$$
$$f_s = 0.25 \text{ Hz}$$
$$A_p = 0.1 \text{ dB}$$
$$A_s = 40 \text{ dB}$$

Overlap save method (filtering in the frequency domain).

a. Determine the input sequences $x_1(n)$, $x_2(n)$, $x_3(n)$, $x_4(n)$ and the corresponding filtered output sequences $y_1(n)$, $y_2(n)$, $y_3(n)$, and $y_4(n)$.

b. Using the FFT subroutine, determine the magnitude and phase of the spectrum of $x_i(n)$ and $y_i(n)$. Comment on the resultant spectra and provide the following computer outputs (see above Figure P6.9.2):

$$h(n), H(k), X(k), \text{ and } Y(k).$$

c. Repeat the problem using the DFT of the noncausal impulse response.

6.9.3. The impulse response of a desired FIR digital filter is given by

$$h(n) = \begin{cases} a(n) & \text{for } 0 \le n \le 3 \\ 0 & \text{elsewhere} \end{cases}$$

The signal to be filtered is the indefinitely long sequence

$$x(n) = b(n) \qquad n = 0, 1, \ldots$$

Section the $\{x(n)\}$ sequence into eight-point segments, and, using the overlap-save method to perform digital filtering of a discrete signal, determine how to modify the overlap-save method so that the last $L - 1$ points of each output section, $y_i(n)$, are deleted rather than the first $L - 1$ points.

6.9.4. Using the matrix approach repeat Example 6.6 using the overlap-add fast convolution algorithm.

7

Multirate Digital Signal Processing

Multirate processing provides the method for changing the sampling rate and forms the basis for implementing complex signal processing applications with reduced computational requirements.

7.0 INTRODUCTION

Many papers on multirate digital signal processing were published starting in the 1970s. The papers addressed both the theoretical and practical aspects of multirate processing. This included a large volume of work by Crochiere and Rabiner[1-3] that developed design ground rules to minimize the computational requirements for multirate systems. This work addressed both IIR and FIR filter implementations and showed how the sampling rate could be changed within the system by any ratio of integers. A key result was a set of design curves optimizing the number of multiplications and the storage requirements for implementing digital filters when the sampling rate was being modified. Other works have concentrated on implementation structures and trade-offs between FIR and IIR implementations.[4,5] This has provided the signal processing system designer with powerful alternatives for efficient system design.

Multirate processing is defined as a signal processing implementation that uses more than one sampling rate to perform the desired digital system

328

operations. The two basic operations performed are decimation and interpolation. **Decimation** results in a sample rate reduction, and **interpolation** provides a sample rate increase. The processes can be used to implement any change in sampling rate that can be represented by a ratio of integers, I/D, as illustrated in Figure 7.1. The interpolation process is shown resulting in an increase in the input sampling frequency by factor I, followed by the decimation process that decreases the sampling rate by factor D. Using network concepts discussed in Appendix 7.A, the sampling rate changes can be brought within the filter structures such that the multiplications are performed at a lower rate to minimize computational requirements. We describe the multirate process and the resultant computational requirements in detail in this chapter.

Multirate processing results in several advantages including:

- Less computational requirements.
- Less storage for filter coefficients and histories.
- Lower order filter designs/implementations.
- Less finite arithmetic effects.
- Less sensitivity to filter coefficient lengths.

The advantages must be weighed against the additional complexity in the control portion of the hardware/software implementation required to handle the changes in the sampling rates and the selection of certain samples and the deletion of others. The designer must assure that the effects of aliasing for decimation and pseudoimages for interpolation are avoided.

A brief introduction to the concepts of multirate processing was given in Chapter 1, and a spectral interpretation of sampling rate conversion was presented in Section 2.5.4. In this chapter we summarize the primary concepts and implementation guidelines for developing multirate systems. In Chapter 10 we apply the results to a representative system design.

7.1 MULTIRATE DSP DECIMATION PROCESS

Decimation is the process of decreasing the sampling rate. Following the sampling rate decrease operation, frequencies above the Nyquist rate are

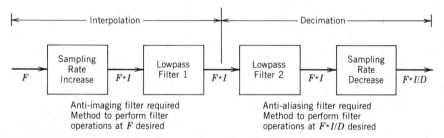

Figure 7.1 Multirate interpolation and decimation processes.

indistinguishable from frequencies below the Nyquist rate. The operation consists of band-limiting the input spectrum via a filter such that when the data rate is reduced the aliasing effects are constrained below a specified level. The filter is typically called an antialiasing filter. The spectral interpretation of the decimation process is shown in Figure 7.2 for a decimation factor of four ($D = 4$). The input spectrum shows two discrete frequency components above the broadband signal energy spectrum. One discrete signal is within the desired output passband, and one discrete signal is shown outside the passband.

The out-of-band signal is a potential source of interference due to aliasing after the sample rate reduction process. The antialiasing filter must attenuate all interfering signals that would alias (i.e., fold) into the passband after the sampling rate is reduced. The filter output spectrum shows the aliased

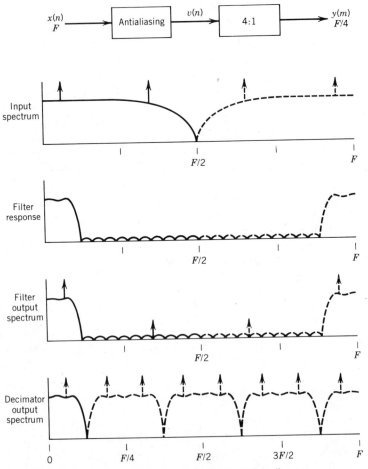

Figure 7.2 Spectral interpretation of decimation ($D = 4$).

spectrum within the passband. Two types of errors result due to the aliasing: interferring signals impersonate passband signals; and the out-of-band noise folds into the passband. The interferring signals can result in false detections, and the noise causes a degradation in the detection performance due to the higher noise level causing lower signal-to-noise ratios in the passband. The repeated spectrums are shown at multiples of the output sampling rate $F/4$.

Figure 7.3a presents the procedure for determining the filter specification requirements as a function of the desired decimation factor D. Since the input signal spectrum has been previously sampled at a rate of F samples per

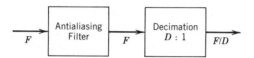

Define the following decimation filter requirements:

1. Passband frequency f_p 3. Passband ripple δ_p
2. Stopband frequency f_s 4. Stopband ripple δ_s

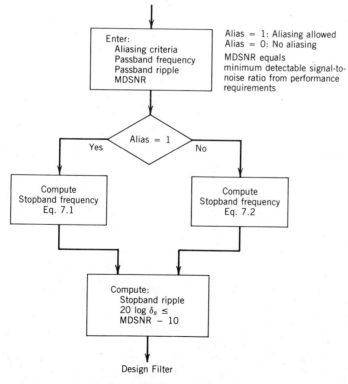

(a) Compute filter requirements.

Figure 7.3 Antialiasing decimator filter requirements (two aliasing criteria shown).

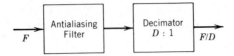

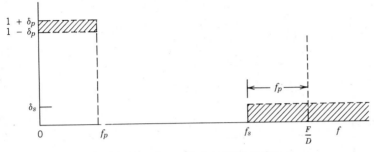

(b) Aliasing allowed in transition region.

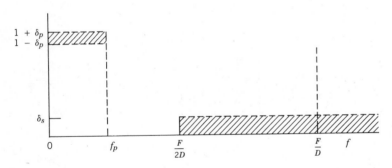

(c) Aliasing not allowed in transition region.

Figure 7.3 Continued

second, the input spectrum must satisfy the relationship given by

$$f_{in} \leq \frac{F}{2} \quad \text{(Nyquist rate)}$$

Now if we desire to reduce the sampling rate F by a factor of D, then the filter passband, f_p, must be constrained to satisfy the new Nyquist rate.

$$f_p < \frac{F}{2D}$$

The stopband cutoff frequency must be selected to limit aliasing in the frequency range of interest. Two stopband criteria are used in practice. One criterion allows aliasing above the stopband deviation level in the transition band (Fig. 7.3b). This criterion is for applications that are limited to frequencies within the passband. For this criterion, **aliasing allowed**, the stopband **critical frequency** is defined as the lowest frequency that will fold into the passband of the filter, which is given by

$$f_s = \frac{F}{D} - f_p \tag{7.1}$$

The second criterion, **no aliasing allowed**, sets the stopband at one half the output sampling rate $F/2D$.

$$f_s = \frac{F}{D} - \frac{F}{2D} = \frac{F}{2D} \tag{7.2}$$

No aliasing allowed means aliasing above the stopband deviation level is not allowed in the transition band (Fig. 7.3c). This places harder requirements on the filter (i.e., smaller transition width), resulting in higher filter orders and, consequently, higher computational and storage requirements. This criterion is used for applications that combine adjacent spectral bands, ensuring that the contribution from the adjacent bands is minimized in order to maintain the integrity of the signals.

The stopband critical frequency f_s, is determined by selecting the desired criterion, Eq. 7.1 or 7.2, as a function of the passband frequency, f_p, and the decimation factor, D. Now only the passband and stopband deviations (ripples) need to be determined in order to compute the filter order. Both ripple factors are based on system performance requirements that will be addressed in Chapter 10. The passband variation is generally in the tenths of decibels, whereas the stopband attenuation level is a function of the allowable interference levels for the application. Stopband attenuation levels of 40, 50, 60, and higher values are not uncommon. We consider a spectral detection application where the stopband attenuation is a function of the in-band probability-of-detection performance, the maximum interfering signal level, and the specified allowable interfering signal probability-of-detection in the passband. A good rule is that the interfering signal must be attenuated to a level of 10 dB below the minimum signal detection level in the passband, as shown in Figure 7.3. A discussion on detection performance calculations for systems, which is given in Chapter 8, and a system design example is presented in Chapter 10.

Example 7.1

For an input signal with a sampling rate of 10,000 Hz, determine the FIR filter order required as a function of D to satisfy the following filter characteristics.

$$f_p = 100 \qquad \delta_p = 0.1 \qquad \delta_s = 0.01$$

Use the criteria for aliasing allowed in the transition region. Use the filter order approximation given by Eq. 5.92,[6] which is an approximation to the

Parks-McClellan (PM) optimum FIR design

$$N = \frac{-10\log\left(\delta_p * \delta_s\right) - C_1}{C_2(f_s - f_p)/F} + 1$$

$C_1 = 15$ (13 given in reference)

$C_2 = 14$ (14.6 given in reference) $\qquad\qquad$ (7.3)

We have used the simpler approximation since it can be more easily used with a hand calculator and it does not effect the intent of the example. For a more accurate estimate of the order, Eq. 5.91 is available. This is the equation that is implemented when automating the procedure for trading off multirate design alternatives.

Solution:

For illustrative purposes, we show the effect of varying D on the filter order and the resultant computations required. In general, the decimation factor required will be determined based on the overall system processing requirements. Using the procedure defined in Figure 7.3, from Eq. 7.1 the stopband critical frequency is calculated

$$f_s = (10000/D) - 100$$

and substituting into the filter approximation results in a filter order equation as a function of D

$$N = D\frac{-10\log(0.1*0.01) - 15}{14(1 - D/50)} + 1 = D\left(\frac{15}{14(1 - (D/50))} + 1\right)$$

The filter order requirements as a function of D are listed in Table 7.1. Note the order required increases significantly with the increase in D because of the reduced transition width as discussed in Chapter 5. ■

Given the results of Example 7.1, the selection of D is dependent on an evaluation of the hardware and software complexity of implementation. For a rough order estimate, we assume that the complexity is proportional to the computations (MPS and APS) required and the storage requirements as defined by Eqs. 7.4a, b, c and presented in Table 7.1.

$$\text{MPS} = F(N + 1)/2D \qquad \text{(Multiplications per second)} \qquad (7.4a)$$

$$\text{APS} = F(N - 1)/D \qquad \text{(Additions per second)} \qquad (7.4b)$$

$$\text{Storage} = 2N + 1 \qquad\qquad\qquad (7.4c)$$

The expressions for the computations are based on the operations being performed at the sampling rate following the data decimation. This will be illustrated in the following section on filter flow graphs.

TABLE 7.1

FIR ORDER, COMPUTATION, AND STORAGE REQUIREMENTS VERSUS D

D	N	(MPS) $F(N + 1) / 2D$	(APS) $F(N - 1) / D$	Storage $2N + 1$
1	3	20,000	20,000	7
2	5	15,000	20,000	11
4	9	12,500	20,000	19
5	11	12,000	20,000	23
6	13	11,666 ← Min	20,000	27
7	17	12,875	22,857	35
10	25	13,000	24,000	51
15	39	13,333	25,333	79
20	57	14,500	28,000	115
25	79	16,000	31,200	159
30	111	18,666	36,666	223

From Table 7.1 the minimum multiplication requirement occurs at $D = 6$ for the filtering portion of the processing. The final selection of the decimation factor will depend on the overall system processing requirements. Therefore, a higher value of D may be desirable owing to reduced processing load for subsequent operations. Also, finite arithmetic effects that are more sensitive for the higher order filters must be considered in the overall design. In addition, the filter design programs do not always produce solutions for high-order filters and therefore a new implementation would be required.

7.1.1 Filter Flow Graphs for Integer Factor Decimators

We introduced the concepts of signal flow graphs for implementing filter difference equations in Chapter 3. Direct form implementations were shown for both FIR and IIR filters (Figs. 3.7 and 3.10). It was shown that for FIR filters the symmetric property of the coefficients could be exploited, resulting in a factor of 2 savings in the multiplications and additions required, as shown in Figure 7.4a. The sampling rate decrease operation is shown following the FIR filter. If we were to implement the structure as shown in the figure, it appears as if the filter must be computed at the input sampling rate. Since the FIR filter output is a function of the current input and past inputs only, the outputs that are ignored by the sampling rate decrease operation need not be computed. From a flow graph perspective, the sampling rate decrease operation is commuted within the filter structure as shown in Figure 7.4b. FIR designs for decimators are discussed in Section 7.1.3.

Commutation is the process of interchanging the order of two cascaded operations without affecting the input-to-output response of the system. Typical operations that are equivalent representations for DSP decimator signal flow graphs are presented in Appendix 7.A. As shown, interchanging the order of operations between scalars and the sampling rate decrease operation result

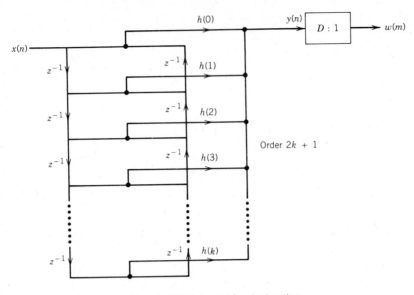

(a) Symmetric FIR followed by decimation.

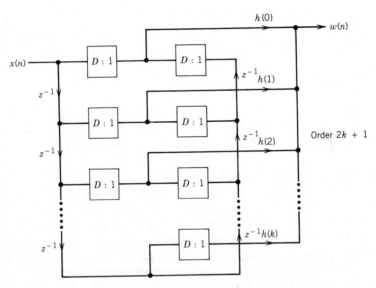

(b) Symmetric FIR with decimator commuted.

Figure 7.4 Multirate decimator FIR filter implementation.

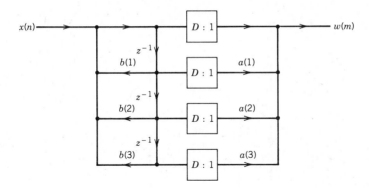

(a) IIR canonic form with decimator commuted in feed forward portion.

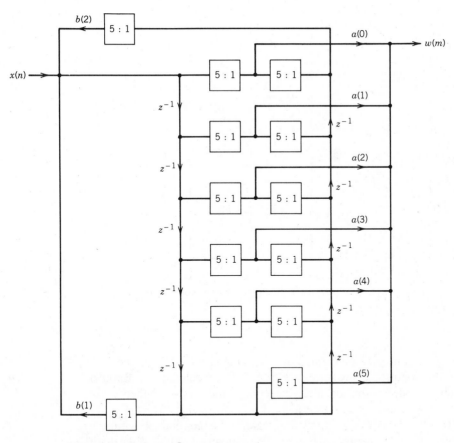

(b) Decimating IIR (z^5) with symmetric feed forward coefficients.

Figure 7.5 IIR multirate decimator structures.

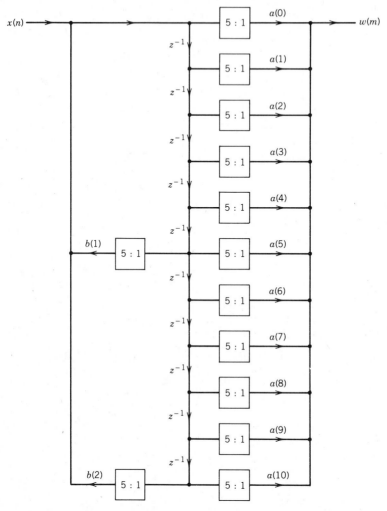

(c) Decimating IIR (z^5) with non-symmetric feed forward coefficients.

Figure 7.5 Continued

in the scalar operation being performed at the changed rate. This is the only rule used for developing the structure in Figure 7.4*b*. Crochiere and Rabiner present a detailed discussion of filter structures and network theory for multirate digital systems.[7]

For IIR filters the feedback portion cannot be commuted with the sampling rate decrease operation. Figure 7.5*a* shows the IIR filter with the decimation commuted in the feed-forward portion. This is identical to the process performed for the FIR filters. For IIR filters, a cascade of second-order stages is a typical method of implementing a higher order filter. This results in smaller finite arithmetic errors and shorter coefficient lengths. This complicates the multirate design, since a direct canonic form is required to commute the feed-forward portion. Another IIR implementation developed specifically for multirate systems is shown in Figure 7.5*b*. This requires that the denominator of the filter transfer function have only terms in z that are powers of D. This implementation results from transforming standard IIR transfer functions. Special design procedures have been developed for these decimating IIR structures that provide symmetry in the numerator coefficients.[4] The IIR designs for multirate systems are discussed in Section 7.1.3.

7.1.2 Multistage Decimators

It has been shown[1-3] that the computational requirements can be significantly reduced by implementing the antialiasing filter in a multistage form with intermediate decimations, as illustrated in Figure 7.6 for K stages. The

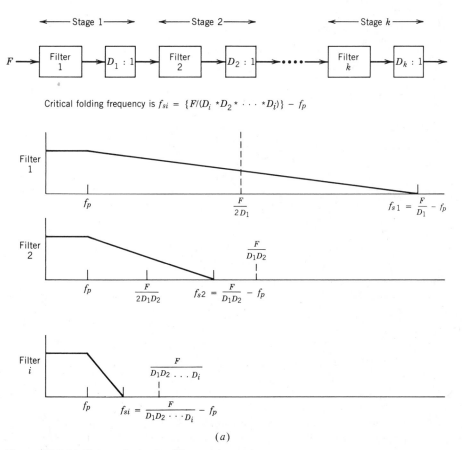

Figure 7.6 Multistage decimator filter requirements.

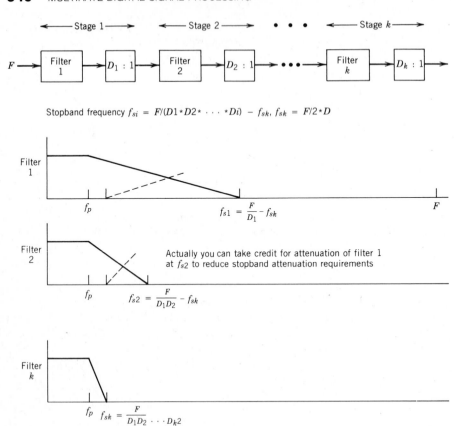

(b)

Figure 7.6 Continued

referenced work developed guidelines for determining the decimation factors $D(i)$ for each stage, i, of a multistage decimator filter implementation in order to minimize either the multiplications required or the storage required. The filter design requirements for each stage are based on the intermediate stopband critical frequencies. For stage k, the stopband frequency based on Eq. 7.1 is expressed as

$$f_s(k) = \left(F \bigg/ \prod_{i=1}^{k} D(i) \right) - f_p = F(k) - f_p \qquad (7.5a)$$

where the aliasing is allowed in the transition band (see Fig. 7.6a). For the more stringent requirement; that is, aliasing not allowed in the transition band, substitute in for f_p one half the final sampling rate resulting in (see Fig. 7.6b)

$$f_s(k) = \left(F \Big/ \prod_{i=1}^{k} D(i) \right) - F/2 \prod_{i=1}^{K} D(i) = F(k) - F(K)/2 \quad (7.5b)$$

For the multistage decimator implementation the passband ripple factor δ_p is divided by the number of stages K. This preserves the composite passband deviation at the output of the filter.

$$\delta_p(k) = \frac{\delta_p}{K} \quad k = 1, 2, \ldots, K \quad (7.5c)$$

A set of curves were developed[1,7] for both FIR PM and elliptic IIR filters for up to four cascaded stages showing the optimum $D(i)$ as a function of D for selected values of $(f_s - f_p)/f_s$. These curves can be used to provide a good starting point for each of the stage decimation factors, $D(i)$. We have found that the process of determining the optimum choice of decimation factors is well suited to the use of one of the spreadsheet programs available for personal computers in conjunction with a program for computing the filter order requirements. This approach is presented in Example 7.2.

The computation and storage requirements for stage k of a K multistage implementation are given by

$$MPS_K(k) = F(k) * (N_K(k) + 1)/2 \quad (7.6a)$$
$$APS_K(k) = F(k) * (N_K(k) - 1) \quad (7.6b)$$
$$Storage_K(k) = 2 * N_K(k) + 1 \quad (7.6c)$$

where $F(k) = F(k - 1)/D(k)$ and $F(0) = F$. Then the total computational and storage requirements are given by

$$MPS_K = \sum_{k=1}^{K} MPS_K(k) \quad (7.7a)$$

$$APS_K = \sum_{k=1}^{K} APS_K(k) \quad (7.7b)$$

$$Storage_K = \sum_{k=1}^{K} Storage_K(k) \quad (7.7c)$$

Example 7.2

For the filter specifications given by

$$f_p = 100 \quad \delta_p = 0.1 \quad \delta_s = 0.001 \quad F = 25{,}000 \quad D = 100$$

determine the optimum multistage implementation for a two-, three-, and four-stage approach. The results should be optimized for both storage and multiplication requirements. Assume a FIR PM design. Use the criteria where aliasing is allowed in the transition band.

Solution:
First we design a one-stage decimator to meet the filter specifications. We compute the one-stage design filter order using Eq. 7.3.

$$N = \frac{-10 \log (0.1*0.001) - 15}{14(150 - 100)/25000} + 1 = 893.9 \rightarrow 895$$

The computation and storage requirements are obtained using Eqs. 7.4.

$$\text{MPS} = 25000(895 + 1)/2(100) = 112{,}000$$

$$\text{APS} = 25000(895 - 1)/100 = 223{,}500$$

$$\text{Storage} = 2(895) + 1 = 1791$$

Next we design a set of two-stage implementations. We start with a matrix of seven sets of possible decimation factors $D_2(i)$, $i = 1, 2$ listed in Table 7.2 that provide the desired decimation, $D = 100$.

The design criterion for the two-stage options is shown in Figure 7.7. The passband ripple is computed using Eq. 7.5a. The stopband critical frequency is found using Eq. 7.5a. The filter order requirements are computed using Eq. 7.3. The stopband critical frequencies and the order calculations are summarized in Table 7.3. From Eq. 7.5c the passband ripple is given by

$$\delta_p(1) = \delta_p(2) = 0.1/2 = 0.05$$

TABLE 7.2
DECIMATION FACTORS FOR TWO-STAGE DESIGN (EXAMPLE 7.2)

| | **Decimation Factors** | |
| | Stage 1 | Stage 2 |
Option	$D_2(1)$	$D_2(2)$
1	50	2
2	25	4
3	20	5
4	10	10
5	5	20
6	4	25
7	2	50

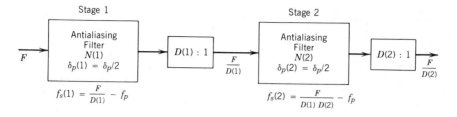

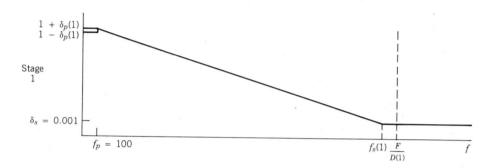

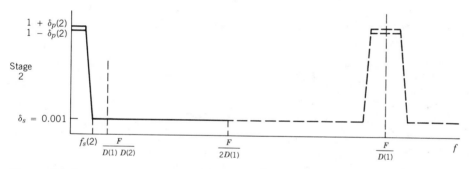

Figure 7.7 Two-stage decimator implementation aliasing allowed in final transition region.

TABLE 7.3
FILTER ORDERS REQUIRED FOR TWO-STAGE DESIGN (EXAMPLE 7.2)

| | Stage 1 | | Stage 2 | |
Option	$f_s(1)$	$N_2(1)$	$f_s(2)$	$N_2(2)$
1	400	169	150	23
2	900	65	150	43
3	1150	49	150	53
4	2400	23	150	103
5	4900	13	150	203
6	6150	11	150	253
7	12400	5	150	503

Now we compute the computation and storage requirements using Eqs. 7.6 and 7.7 for each of the possible implementations identified and present the results in Table 7.4. Selection of the optimum implementation is based on minimizing the resultant complexities of the hardware and software. Note that option 3 is the best choice for both storage and computations. Crochiere and Rabiner[7] developed closed-form expressions for the optimum decimation factors for the two-stage design. Expressions were developed for minimizing the number of multiplications and for minimizing the storage requirements. The optimization as a function of the number of multiplications required is given by

$$D_{opt}(1) = \frac{2 * D\left(1 - \sqrt{D * F_{sp}/(2 - F_{sp})}\right)}{2 - F_{sp} * (D + 1)} \tag{7.8a}$$

$$D_{opt}(2) = \frac{D}{D_{opt}(1)} \tag{7.8b}$$

$$F_{sp} = \frac{(f_s - f_p)}{f_s} \tag{7.8c}$$

For optimization of the storage requirements the expressions are given by

$$D_{opt}(1) = \frac{2 * D}{(2 - F_{sp}) + \sqrt{2 * D * F_{sp}}} \tag{7.9a}$$

$$D_{opt}(2) = \frac{D}{D_{opt}(1)} \tag{7.9b}$$

$$F_{sp} = \frac{(f_s - f_p)}{f_s} \tag{7.9c}$$

TABLE 7.4
TWO-STAGE COMPUTATION AND STORAGE REQUIREMENTS (EXAMPLE 7.2)

Option	Stage 1 MPS	APS	Storage	Stage 2 MPS	APS	Storage	Total MPS	APS	Storage
1	42500	84000	339	3000	5500	47	45000	89500	386
2	33000	64000	131	5500	10500	87	38500	74500	218
3	31250	60000	99	6750	13000	107	38000	73000	206 ←
4	30000	55000	47	13000	25500	207	43000	80500	254
5	35000	60000	27	25500	50500	407	60500	110500	434
6	37500	62500	23	31750	63000	507	69250	125500	530
7	50000	75000	15	63000	125500	1007	113000	200500	1022

← Best choice

Using these equations provides a starting point for choosing the integer values that should be used. Optimization to minimize the number of multiplications by evaluation of Eqs. 7.8 for our example results in

$$F_{sp} = (150 - 100)/150 = 0.333$$

$$D_{opt}(1) = 21.9$$

$$D_{opt}(2) = 4.56$$

Optimization for minimizing the storage by evaluating Eqs. 7.9 results in

$$F_{sp} = (150 - 100)/150 = 0.333$$

$$D_{opt}(1) = 20.34$$

$$D_{opt}(2) = 4.92$$

These values agree with our results from evaluation of the seven options. In cases where the optimum points are between two options, both options can be evaluated. A general result[7] that limits the options that need to be evaluated for stage k is given by

$$D_K(k) \geq D_K(k + 1) \qquad k = 1, 2, \ldots, K - 1 \qquad (7.10)$$

Using this result, options 5, 6, and 7 could have been eliminated from consideration for the two-stage alternatives. For low decimation factors (less than 10), the rule may not provide optimum results.

Now for three stages we repeat the same process. First we list the possible decimation factors (Table 7.5), then compute the stopband critical frequencies and filter orders (Table 7.6), and finally compute the computation and storage requirements (Table 7.7). The passband ripple is computed using Eq. 7.5c.

TABLE 7.5
DECIMATION FACTORS FOR THREE STAGE DESIGN (EXAMPLE 7.2)

	Decimation Factors		
Option	**Stage 1** $D_3(1)$	**Stage 2** $D_3(2)$	**Stage 3** $D_3(3)$
1	25	2	2
2	10	5	2
3	10	2	5
4	5	5	4
5	5	4	5
6	4	5	5
7	2	25	2
8	2	2	25

TABLE 7.6
FILTER ORDERS REQUIRED FOR THREE-STAGE DESIGN (EXAMPLE 7.2)

Option	Stage 1		Stage 2		Stage 3	
	$f_s(1)$	$N_3(1)$	$f_s(2)$	$N_3(2)$	$f_s(3)$	$N_3(3)$
1	900	69	400	9	150	23
2	2400	25	400	19	150	23
3	2400	25	1150	7	150	55
4	4900	13	900	15	150	45
5	4900	13	1150	13	150	55
6	6150	11	1150	15	150	55
7	12400	7	400	91	150	23
8	12400	7	6150	7	150	267

TABLE 7.7
THREE-STAGE COMPUTATION AND STORAGE REQUIREMENTS (EXAMPLE 7.2)

Option	Stage 1			Stage 2			Stage 3		
	MPS	APS	Storage	MPS	APS	Storage	MPS	APS	Storage
1	35000	68000	139	2500	4000	19	3000	5500	47
2	32500	60000	51	5000	9000	39	3000	5500	47
3	32500	60000	51	5000	7500	15	7000	13500	111
4	35000	60000	27	8000	14000	31	5750	11000	91
5	35000	60000	27	8750	15000	27	7000	13500	111
6	37500	62500	23	10000	17500	31	7000	13500	111
7	50000	75000	15	23000	45000	183	3000	5500	47
8	50000	75000	15	25000	37500	15	33500	66500	535

Option	Total		
	MPS	APS	Storage
1	40500	77500	205
2	40500	74500	137
3	44500	81000	177
4	48750	85000	149
5	50750	88500	165
6	54500	93500	165
7	76000	125500	245
8	108500	179000	565

← Best choice

As noted in Table 7.7, option 2 is the best choice for the three-stage design. Comparison with the two-stage optimum design shows that the storage requirements are significantly lower (33%) and the computations required are slightly higher (4%). This comparison is based on the use of the filter order approximation given by Eq. 7.3. A more accurate result can be obtained by performing the design and plotting the frequency response of the filters. This could change the results for cases where differences are less then 10%.

Now for four stages we repeat the same process. First we list the possible decimation factors (Table 7.8), then compute the stopband critical frequencies and filter orders (Table 7.9), and finally compute the computation and storage requirements (Table 7.10). For the four stage design, option 1 requires the fewest computations and storage. ■

The best choice for each of the different stage designs of Example 7.2 is summarized in Table 7.11. From this summary we would choose either the two-stage design or the three-stage design. The two-stage design requires the least number of computations. The three-stage design is very close in computations and requires less storage. Remember that we have not considered finite arithmetic effects or included the hardware and software complexity within the design selection. Also, we have used an approximation to the filter order required. These could have an impact on the final selection. Therefore, the frequency response of the filters should be computed using the planned coefficient precision to ensure that the filter order estimate is adequate. Hence, the results presented herein could change once the filter responses are checked

TABLE 7.8

DECIMATION FACTORS FOR FOUR-STAGE DESIGN (EXAMPLE 7.2)

Option	Decimation Factors			
	Stage 1 $D_4(1)$	Stage 2 $D_4(2)$	Stage 3 $D_4(3)$	Stage 4 $D_4(4)$
1	5	5	2	2
2	5	2	5	2
3	5	2	2	5
4	2	5	2	5
5	2	5	5	2
6	2	2	5	5

TABLE 7.9

FILTER ORDERS REQUIRED FOR FOUR-STAGE DESIGN (EXAMPLE 7.2)

Option	Stage 1		Stage 2		Stage 3		Stage 4	
	$f_s(1)$	$N_4(1)$	$f_s(2)$	$N_4(2)$	$f_s(3)$	$N_4(3)$	$f_s(4)$	$N_4(4)$
1	4900	13	900	15	400	9	150	25
2	4900	13	2400	7	400	21	150	25
3	4900	13	2400	7	1150	7	150	57
4	12400	7	2400	15	1150	7	150	57
5	12400	7	2400	15	400	21	150	25
6	12400	7	6150	7	1150	15	150	57

TABLE 7.10
FOUR-STAGE COMPUTATION AND STORAGE REQUIREMENTS (EXAMPLE 7.2)

	Stage 1			Stage 2			Stage 3		
Option	MPS	APS	Storage	MPS	APS	Storage	MPS	APS	Storage
1	35000	60000	27	8000	14000	31	2500	4000	19
2	35000	60000	27	10000	15000	15	5500	10000	43
3	35000	60000	27	10000	15000	15	5000	7500	15
4	50000	75000	15	20000	35000	31	5500	10000	15
5	50000	75000	15	20000	35000	31	5500	10000	43
6	50000	75000	15	25000	37500	15	10000	17500	31

	Stage 4			Total			
Option	MPS	APS	Storage	MPS	APS	Storage	
1	3250	6000	51	48750	84000	128	← Best choice
2	3250	6000	51	53750	91000	136	
3	7250	14000	115	57250	96500	172	
4	7250	14000	115	82250	131500	176	
5	3250	6000	51	78750	126000	140	
6	7250	14000	115	92250	144000	176	

TABLE 7.11
OPTIMUM DESIGNS FOR ONE, TWO, THREE, AND FOUR STAGES (EXAMPLE 7.2)

Stage Design	Option	MPS	APS	Storage
1	—	112000	223500	1791
2	3	38000	73000	206
3	2	40500	74500	137
4	1	48750	84000	128

since many of the options resulted in small differences in the computation and storage requirements.

Ideally, Crochiere and Rabiner found that the computational and storage requirements decrease with increasing K, the number of stages.[7] Since we have limited our choices to implementable integer decimation factors and used the filter order approximation in the calculations, these results do not contradict their general result. For example, if option 2 of the three-stage design resulted in a filter order for stage 1, $N_3(1) = 24$ instead of 25, then MPS = 38000, which equals the two-stage result. Therefore, small errors due to the inaccuracy in the estimates can effect the conclusions. The designs must be done in detail to determine the optimum result, but the additional computational savings will generally be small.

7.1.3 FIR and IIR Designs for Decimators

Any of the IIR and FIR design techniques presented in Chapter 4 and Chapter 5, respectively, are acceptable for use in the multirate implementations. You are reminded that the IIR filters have nonlinear phase and therefore should not be used without compensation for applications requiring linear phase. The PM optimum FIR designs and the elliptic IIR designs provide good performance versus computational requirements. For decimating filters, only a portion of the out-of-passband frequencies alias into the passband. A design procedure that provides multiple stopbands tuned to the frequencies that alias into the passband may provide computational efficiency over the single stopband design. These filters are referred to as **multiband designs**. Two special cases of FIR filter designs that are particularly useful for reducing the computational requirements in multirate implementations are *halfband* and *comb* filter designs. For implementing multirate filters using IIR filters, *decimating IIR* design techniques were developed.[4] Each of these design approaches is discussed in this section.

MULTIBAND FIR DESIGNS

Multiband designs[8] take advantage of regions that will not alias into the passband. These regions are called **don't care** regions. The multiband design is specified as one passband with multiple stopbands and don't care regions. The multiple stopbands are chosen to attenuate all frequencies that would fold into the passband. Figure 7.8 illustrates the multiband design requirements for a filter with passband f_p and data decimation factor $D = 10$. These multiband frequency response constraints require a design optimization technique that allows the multiple band constraints to be specified. The PM FIR design program[9] allows multiple band specifications and therefore is suited to the multiband design problem. The PM designs can result in unacceptable performance in the transition regions; therefore the frequency response of the filter should always be analyzed to ensure the design solution actually satisfies

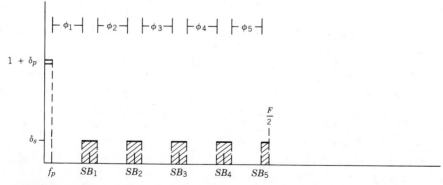

Figure 7.8 Multiband decimator design requirements ($D = 10$).

the requirements. The multiple stopband regions for the kth multirate stage are defined by

$$SB_n(k) = \frac{nF}{\displaystyle\prod_{i=1}^{k} D(i)} \pm f_p \qquad n = 1, 2, 3, \ldots, \mathrm{Int}\,(D(k)/2) \quad (7.11)$$

where the regions outside the passband and not defined by Eq. 7.11 represent the don't care bands (i.e., ϕ-bands).[7] Note that the number of ϕ-bands is equal to the integer portion of $(D(k) + 1) \div 2$. For D an even integer, the upper stopband region extends to one half the input sampling rate including only the negative portion of f_p in Eq. 7.11.

Example 7.3

For the two-stage multirate filter defined by option 3 in Example 7.2, determine the multiband filter specifications.

Solution:

The passband frequency, passband ripple, and stopband ripple specifications are given in Example 7.2. The multiple stopbands are given by Eq. 7.11. For the stage 1 filter, which precedes a decimation of 20, 10 stopbands are defined

$$SB_n(1) = 1250n \pm 100 \qquad n = 1, 2, 3 \ldots, 10$$

where the tenth stopband is stopped at one half the sampling rate. For the stage 2 filter, which precedes a decimation of five, two stopbands are given by

$$SB_n(2) = 250n \pm 100 \qquad n = 1, 2 \quad \blacksquare$$

There are no simple formulas for determining the order required to achieve the multiband requirements. We could turn to a trial-and-error process using the conventional PM filter order as a starting point and trying a 25% reduction. If a satisfactory design is achieved, then try 35%. If 25% does not work then try 15% and so on. By performing many multiband designs trends in the modified order requirements as a function of the passband and the decimation factor can be established.[7] The savings can be significant for cases where the *don't care* regions represent a high percentage of the out-of-passband frequency range. This occurs for cases where the passband is small compared to the sampling rate and moderate decimation factors are desired. Negative results sometimes occur for lower ϕ-band percents and/or high decimation factors. Apparently, more zeros are required to adapt to several stopbands than the conventional design in the negative cases.

The ϕ-band percent portion of the transition width and stopband for multiband stage k is given by

$$\phi\text{-band}(k) = \left| 1 - \frac{\dfrac{(D(k) - 1) * f_p}{F}}{2 \displaystyle\prod_{i=1}^{k} D(i - 1)} - f_p \right| * 100 \qquad D(0) = 1 \quad (7.12)$$

We designed the filters from Example 7.3 and found that they did not result in a better design. The stage 1 filter design required a 59 order filter versus a 55 order filter for the conventional PM lowpass design. Recall from Table 7.3 that the order estimate was 49 instead of the 55 actually required. This demonstrates the need to design the filters in order to obtain the most accurate estimate possible. Likewise for the stage 2 filter, a 55 order multiband design was required versus a 55 order conventional design. Actually the 55 order conventional design resulted in a stopband attenuation level of 59.87 dB versus the 60 dB specified. The designer would have to assess the impact of this on the system to determine if a higher order design was required.

The use of trend and/or design curves has been popular in DSP design and analysis. The introduction of programmable calculators, personal computers, and portable computers allows the DSP engineer more flexibility. For our multiband design problem, many designs can be tried and evaluated interactively until an acceptable design is achieved. The trend can be embedded into the program and used to select the initial filter order based on the ϕ-band percent.

Example 7.4
Demonstrate the effectiveness of using a multiband filter to decimate a speech signal.[10] The following design requirements were given

$$f_p = 3000 \text{ Hz} \qquad f_s = 5000 \text{ Hz} \qquad F = 64{,}000 \text{ Hz}$$
$$\delta_p = (0.1 \text{ dB}) \qquad \delta_s = (-50 \text{ dB}) \qquad D = 8$$

Solution:
Peled and Liu[10] found that the conventional lowpass design required a 160 order filter to meet the specifications. This results in the following computational and storage requirements (Eq. 7.4).

$$\text{MPS} = 64{,}000(160 + 1)/(2 \times 8) = 644{,}000$$
$$\text{APS} = 64{,}000(160 - 1)/8 = 1{,}272{,}000$$
$$\text{Storage} = 2(160) + 1 = 321$$

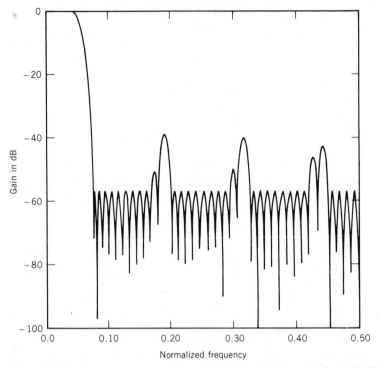

Figure 7.9 Frequency response of multiband filter discussed in Example 7.4.

They found that a multiband filter of order 80 met the requirements. The filter frequency response is shown in Figure 7.9. The computational and storage requirements are reduced by a factor of two.

$$\text{MPS} = 324{,}000 \rightarrow 49.7\% \text{ savings}$$
$$\text{APS} = 632{,}000 \rightarrow 50.3\% \text{ savings}$$
$$\text{Storage} = \quad 161 \rightarrow 49.8\% \text{ savings}$$

We compute the percentage portion of the ϕ-bands relative to the stopband for the filter design.

$$\phi\text{-band} = \left(1 - \frac{7*3000}{32{,}000 - 3000}\right)100 = 27.6\%$$

Therefore, a significant gain was realized for a moderate percentage and a relatively large passband. The designer must assess the significance of the savings for the application and decide whether a multiband design is necessary. ■

The foregoing results indicate that the conclusions reached in the previous section on the best multirate implementation may be altered if the multiband designs are taken into account. You are therefore cautioned that the approaches developed herein should be used as a guide only and that to reach a final design several factors must be considered.

The don't care regions can exhibit rather large amplitude levels compared to the passband response; therefore the overall output noise could be significantly effected. This effect must be analyzed to ensure that the system performance is not degraded.

Multiband IIR Designs. For an IIR multiband design, one of the design procedures that allow for multiple stopband specifications[11] is required. The filter requirements are derived as discussed in the preceding section for FIR multiband designs. Little interest has been given to the use of multiband IIR filters.

Halfband FIR Designs. Halfband FIR filters have the unique property that all the even coefficients of the filter are zero. Therefore, the number of computations and storage for stage k is expressed by

$$\text{Multiplications} = [N(k) + 5]/4 \qquad (7.13a)$$

$$\text{Additions} = [N(k) + 1]/2 \qquad (7.13b)$$

$$\text{Storage} = [3N(k) + 5]/2 \qquad (7.13c)$$

which represents a reduction by approximately a factor of two. The storage required for coefficients is given by Eq. 7.13c. The computation rates are given by

$$\text{MPS}(k) = F(k) * [N(k) + 5]/4 \qquad (7.14a)$$

$$\text{APS}(k) = F(k) * [N(k) + 1]/2 \qquad (7.14b)$$

Figure 7.10 illustrates the frequency response characteristics required for the

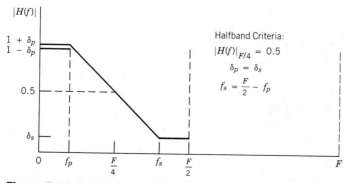

Figure 7.10 Halfband filter frequency response characteristics.

halfband filter. Note that the passband and stopband have equal ripple factors, that the stopband width is equivalent to the passband width, and that the symmetry point is at $0.25F$, where the power is at 0.5. Therefore the halfband constraints for stage k of a K-stage multiband filter are defined by

$$\delta_{hb}(k) = \delta_p(k) = \delta_s(k) = \text{MIN}\left(\frac{\delta_p}{K}, \delta_s\right) \qquad k = 1, 2, \ldots, K \quad (7.15a)$$

$$f_p(k) = f_p \qquad\qquad (7.15b)$$

$$f_s(k) = \left[\frac{F}{\displaystyle\prod_{i=1}^{k} D(i)}\right] - f_p(k) = F(k) - f_p \qquad (7.15c)$$

These criteria allow aliasing to occur in the final transition band. For the case where no aliasing is allowed in the final transition band, the passband is extended to meet the no aliasing requirement

$$f_p(k) = \begin{cases} \dfrac{F(K)}{2} & k = 1, 2, \ldots, K - 1 \\ f_p & k = K \end{cases} \qquad (7.15d)$$

The stopband critical frequency is expressed by

$$f_s(k) = \frac{F}{2^k} - f_p(k) \qquad k = 1, 2, \ldots, K \qquad (7.15e)$$

Note that for the case where aliasing is not permitted in the final transition band, $0.5F(k)$, the last stage does not meet the halfband constraints (i.e., the passband length is not equal to the stopband length). Any power of two can be obtained by cascading halfband filters. For cases that are not a power of two, only a portion of the multirate filter can be implemented using halfband filters. The number of stages, K, for D a power of two is given by

$$K = \log_2(D) \qquad (7.15f)$$

Halfband filters are very useful for decimating the data by a factor of 2. This is a result of the frequency response being constrained to less than $0.5F$ and the stopband limiting the aliasing level to δ_s in the passband. It can be shown that halfband filters are always odd-order filters. This is left as a problem. Also, since the even coefficients are zero, as shown in Fig. 7.11, the only orders that need to be implemented are expressed by

$$N_{hb} = 3, 7, 11, 15, 19, 23, \ldots \qquad (7.16)$$

Since the alternating odd values result in zero filter coefficients (i.e., $5, 9, 13, \ldots$),

n	-10	-9	-8	-7	-6	-5	-4	-3	-2	-1	-0	1	2	3	4	5	6	7	8	9	10
h_n	0	h_9	0	h_7	0	h_5	0	h_3	0	h_1	h_0	h_1	0	h_3	0	h_5	0	h_7	0	h_9	0

Order	Required Coefficient Multiplies									# Mult
3				x	x	x				2
7			x	x	x	x	x			3
13		x	x	x	x	x	x	x		4
15	x	x	x	x	x	x	x	x		5
19	x	x	x	x	x	x	x	x	x	6

Figure 7.11 Halfband filter orders versus multiplications.

the filter designs can specify these values, resulting in a better frequency response that can be implemented with the next lower odd-order halfband.

Since the reduction in computations and storage implies that only a portion of the data be processed, the hardware and software must have the intelligence to perform the operation. This will require special software code and/or hardware complexity to address the proper data and perform the halfband algorithm. This could offset the advantages gained by the reduced computations required and must be included in the overall design process.

Example 7.5

Design a multirate filter system using halfband filter designs resulting in a decimation of 32 for the following specifications.

$$f_p = 50 \qquad \delta_p = 0.01 \qquad \delta_s = 0.001 \qquad F = 4800 \qquad D = 32$$

Allow aliasing in the final transition band.

Solution:

First we determine the number of stages (Eq. 7.15f) and the filter specifications for each stage Eqs. 7.15a–c.

$$K = \log_2(32) = 5$$

$$\delta_{hb}(k) = \delta_p(k) = \delta_s(k) = \text{MIN}\left(\frac{0.01}{5}, 0.001\right) = 0.001$$

$$f_p(k) = 50$$

$$f_s(k) = \left\lceil \frac{4800}{\prod_{i=1}^{k} D(i)} \right\rceil - 50 = F(k) - 50$$

The results are presented in Table 7.12. Next we compute the filter orders

TABLE 7.12
EXAMPLE 7.5 HALFBAND FILTER REQUIREMENTS

Stage k	Stage Sampling Rate		Stopband Frequency $f_s(k)$	Transition Width	Filter Order $N(k)$	Halfband Filter Order
	Input $F(k-1)$	Output $F(k)$				
1	4800	2400	2350	0.479	7.7	9
2	2400	1200	1150	0.458	8.0	9
3	1200	600	550	0.417	8.7	9
4	600	300	250	0.333	10.6	11
5	300	150	100	0.167	20.3	21

required for each stage and list the results in Table 7.12. The filters orders were calculated using the simple form of the PM filter approximation. Next we compute the computational and storage requirements for the halfband design and compare it to the single-stage implementation. These results are presented in Table 7.13. The single-stage design results in

$$N = 243 \qquad MPS = 18300 \qquad APS = 36300 \qquad Storage = 487$$

which requires higher computations and significantly larger storage. Several other options are available for implementing this multirate filter, including two-stage and three-stage designs. Some of these options are summarized in Table 7.14. These designs in general result in less computational requirements than the total halfband implementation. Note that the lowest computational requirements result for the three-stage design, with the last two stages designed as halfband stages. Again, this example is provided to illustrate the importance of trading off potential implementations. We did not design the filters, and therefore the results could change slightly owing to inaccuracies in the filter order estimates. Also, the complexity in hardware and software must ultimately be assessed, including finite arithmetic effects, before a final selection is made. ∎

TABLE 7.13
HALFBAND COMPUTATION AND STORAGE REQUIREMENTS (EXAMPLE 7.5)

Stage k	Halfband Filter Order	Computations		
		MPS	APS	Storage
1	9	8400	12000	16
2	9	4200	6000	16
3	9	2100	3000	16
4	11	1200	1800	19
5	21	975	1650	34
Total		16875	24450	101

TABLE 7.14
ALTERNATIVE DESIGN IMPLEMENTATIONS (EXAMPLE 7.5)

Number of Stages	Decimation			Computational Requirements		
	$D(1)$	$D(2)$	$D(3)$	MPS	APS	Storage
2	16	2	—	11700	22500	174
2	8	4	—	11700	21900	130
2	16	2HB	—	11175	21450	169
3	8	2	2	12300	22500	121
3	4	4	2	14100	24900	109
3	8	2HB	2HB	11175	20250	121
3	4	4	2HB	13575	23850	104

HB = Halfband filter.

Comb FIR Designs. A simple FIR filter is constructed by averaging N samples (i.e., $h(n) = 1/N$ for $n = 1, 2, \ldots, N$). This class of FIR filter is called a **Comb** filter. We have seen the Fourier transform of this function before when we developed the frequency response of the DFT in Chapter 6. The transfer function for this filter is given by

$$H_N(z) = \frac{1}{N} \sum_{i=0}^{N-1} z^{-i}$$

and the frequency response is given by

$$H_N(f) = e^{-j2\pi f(N-1)/F} \frac{\sin(\pi f N/F)}{N \sin(\pi f/F)}$$

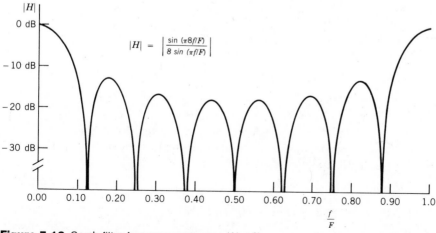

Figure 7.12 Comb filter frequency response ($N = 8$).

For $N = 8$ the frequency response is shown in Figure 7.12. Note that the response has high sidelobes, but if the desired passband is small compared to the input sampling frequency, the nulls in the response will reject the bands, which will fold into the passband for a decimator with $D = N$. There are $N - 1$ nulls in the response that occur at frequencies

$$f_n = nF/N \qquad n = 1, 2, \ldots, N - 1 \qquad (7.17)$$

The allowance for the passband width is based on the required passband and stopband deviations. The response is evaluated at each stopband edge frequency and at the passband edge to assure the requirements are met. The response constraints are expressed by

$$SB_n(k) = \frac{nF(k-1)}{N(k)} \pm f_p \qquad n = 1, 2, \ldots, \text{INT}\left(N(k)/2\right) \quad (7.18a)$$

$$\left| H_N\left(SB_n(k)\right) \right| \le \delta_s \qquad (7.18b)$$

$$1 - \left(\delta_p/K\right) \le \left| H_N\left(0 \to f_p\right) \right| \le 1 + \left(\delta_p/K\right) \qquad (7.18c)$$

where K is the number of stages in the multistage design. If the response does not meet the specifications, then either the filter requirements have to be relaxed or the decimation factor lowered until the criteria defined by Eqs. 7.18 are satisfied. These expressions are easily programmed, which provides an automated approach to evaluating the application of the Comb filter to multirate systems.

The Comb filters are most often used as either the first stage of a multistage decimator or the last stage of a multistage interpolator. Again we stress the need to evaluate the benefits of using the Comb filter versus the hardware and software requirements. The Comb filter requires the accumulation of N inputs normalized by N. If N is a power of two, the normalization can be implemented simply via shifts, and no multiplications are required. The worst-case implementation would require N multiplies by the impulse response coefficients $1/N$. Generally, the filter is implemented without multiplies and the normalization is accounted for via scaling the input to the nearest power of two that avoids overflow expressed by,

$$r = \left\langle \log_2(N) \right\rangle$$

where $\langle X \rangle$ denotes the next higher integer of X. The resultant value can then be multiplied by the factor $2^r/N$ to provide the exact normalization required. In many cases, this multiplication can be neglected or compensated for in the subsequent operations. The computation and storage requirements, assuming an implementation requiring one multiply, are expressed by

$$\text{MPS}(k) = F(k-1)/N(k) = F(k)$$
$$\text{APS}(k) = F(k-1) * (N(k) - 1)/N(k)$$
$$\text{Storage}(k) = N(k) + 1$$

We assumed that one register was required to hold the partial sum as the operation is performed. The use of a Comb filter in the design of a multirate filter is left as a problem.

7.2 MULTIRATE DSP INTERPOLATION PROCESS

Interpolation is the process of increasing the sampling rate. We will discuss sampling rate increases of integer values represented by a factor I. The operation consists of adding $I - 1$ zeros between the pairs of input sample values $[x(n), x(n + 1)]$ and filtering the resultant signal to band-limit the spectrum such that the images produced by the sampling rate increase operation are attenuated. The filter is typically called an anti-imaging filter. The spectral interpretation of the interpolation process is shown in Figure 7.13 for an interpolation factor of four. The input signal must be band-limited to $F/2$ prior to the interpolation process as shown in Figure 7.13a. The output following the addition of the zeros shows the images produced in Figure 7.13b. The anti-imaging filter requires a gain of I in the passband to compensate for the gain reduction due to the process. The filter passband cutoff frequency will be based on the input signal passband representing the highest frequency of interest in the input signal. As shown in Figure 7.13c, the passband cutoff is confined to a value given by

$$f_p < F/2 \tag{7.19}$$

and the lowest stopband frequency is given by

$$f_s = F/2 \tag{7.20a}$$

The stopband extends to $F * I/2$ in order to attenuate the entire imaging band. The interpolator output is shown in Figure 7.13d. Note that if we relax the stopband frequency then a portion of the images would remain. If the input signal spectrum is confined below $f_c < F/2$, then the stopband can be relaxed to

$$f_s = F - f_c \tag{7.20b}$$

As discussed for the decimator, passband and stopband deviation requirements depend on the filter application and the required performance. The passband deviation will generally be in tenths of decibels or smaller, and the stopband deviation can vary over a wide range of decibels (-10s to -100s).

If the input signal spectrum is confined to a small portion of the input baseband ($f_c \ll F/2$), then the anti-imaging filter can be designed as a multiband filter to account for the don't care bands. This is shown in Figure 7.14. In this case, the stopband regions are given by

$$SB_n = \begin{cases} nF \pm f_c & n = 1, 2, \ldots, \text{Int}\left((I + 1)/2\right) - 1 \\ nF - f_c \text{ to } nF & n = \text{Int}\left((I + 1)/2\right) \end{cases} \tag{7.21}$$

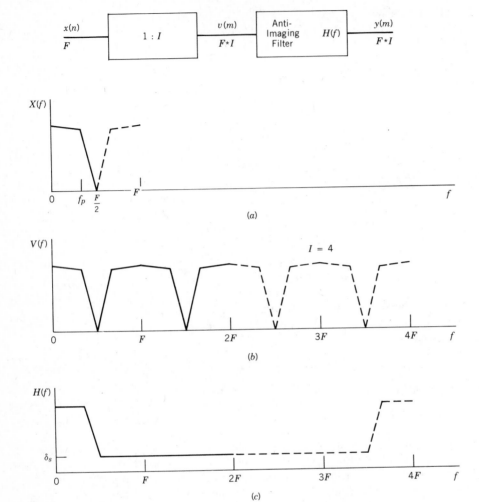

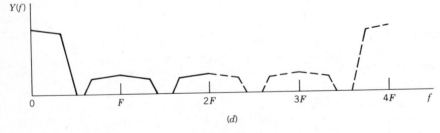

Figure 7.13 Multirate interpolation process.

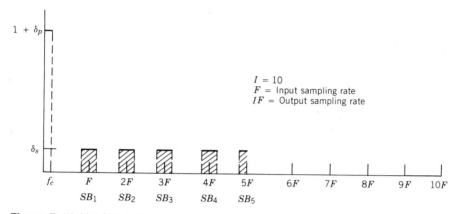

Figure 7.14 Multiband interpolator.

As we found for the decimator multiband filters, significant savings can be obtained for cases where the stopbands represent a small percentage of the frequency above the passband.

7.2.1 Filter Flow Graphs for Integer Factor Interpolators

We presented filter flow graphs for decimators in Section 7.1.1. For interpolators, the sampling rate increase operation occurs before the filter operation. Using the standard direct form FIR filter implementation, and commuting the sampling rate increase operation into the filter, results in a nonrealizable flow graph as shown in Appendix 7.A (i.e., the delays result in fractional values). Therefore, an alternative form for the interpolator FIR filter structure is required in order to develop an efficient implementation.

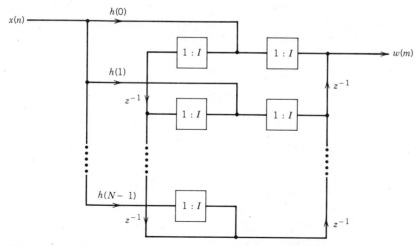

Figure 7.15 Multirate interpolator FIR filter implementation.

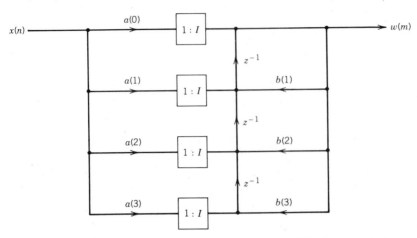

Figure 7.16 Multirate interpolator IIR filter numerator coefficients commuted.

Using the dual property (see Appendix 7.A) between interpolators and decimators, the interpolator flow graph can be found by performing the transpose of the filter flow graph developed in Figure 7.4b. The transpose of a filter flow graph is described in Appendix 7.A. Note that the decimator transpose results in the interpolator filter shown in Figure 7.15, which has the sampling rate increase operation commuted within the filter after the multiplications. Also, since we used the symmetric form of the FIR decimator flow graph, the interpolator form also takes advantage of the symmetry of the coefficients.

For IIR filters the sampling rate increase operation is commuted with the feedback portion of the canonic implementation. Again, a direct form implementation is required, which requires additional coefficient accuracy. Figure 7.16 presents a direct form IIR implementation of an interpolator developed as a transpose of the IIR decimator direct form implementation given in Figure 7.5a.

7.2.2 Multistage Interpolators

The multistage interpolator form is presented in Figure 7.17. The input sampling rate is defined by F. For stage k of a K-stage design, the sampling

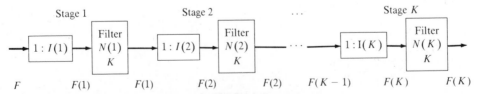

Figure 7.17 Multistage interpolation process (K stages).

rate and filter requirements are given by

$$I = \prod_{k=1}^{K} I(k) \tag{7.22a}$$

$$F(k) = F(k-1) * I(k) \qquad \text{where } F(0) = F \tag{7.22b}$$

$$f_p(k) = f_p \tag{7.22c}$$

$$\delta_p(k) = \frac{\delta_p}{K} \tag{7.22d}$$

$$f_s(k) = F(k-1) - \frac{F}{2} \tag{7.22e}$$

$$\delta_s(k) = \delta_s \tag{7.22f}$$

The computational and storage requirements assuming that each stage is implemented using a symmetric FIR interpolator implementation is given by

$$\text{MPS}(k) = F(k-1) * [N(k) + 1]/2 \tag{7.23a}$$

$$\text{APS}(k) = F(k-1) * [N(k) - I] \tag{7.23b}$$

$$\text{Storage}(k) = 2 * [N(k) + 1] \tag{7.23c}$$

Note that the multiplications are performed at the sampling rate prior to the increase by $I(k)$. The number of additions is dependent on the stage interpolation factor I and the implementation of the flow graph. The additions are performed after the interpolation factor, but since $I - 1$ zeros have been added, several trivial operations result. The increase in control complexity to avoid the trivial operations must be traded off versus the additional multiplies required. Equation 7.23b assumes that only the nontrivial additions are performed.

7.3 MULTIRATE DSP BY RATIO OF INTEGERS

The previous discussions for decimators and interpolators was limited to either decreasing the sampling rate by an integer value D or increasing the sampling rate by integer value I. For a process that requires either an increase or a decrease by a noninteger value we must cascade the operations as shown in Fig. 7.18a. Note that the interpolation process is shown first followed by the decimation process. This ordering allows the anti-imaging and the antialiasing filters to be combined as shown in Figure 7.18b. The filter must satisfy the composite requirements of the two filters.

$$f_p < \text{MIN} \left(F * I/2D; f_c \right) \tag{7.24a}$$

$$f_s = \text{MIN} \left(\frac{F * I}{D} - f_p; F - f_c \right) \tag{7.24b}$$

$$f_c \leq \frac{F}{2} \tag{7.24c}$$

The filter must be designed at the interpolated rate, $F * I$.

(*a*) Cascaded interpolation and decimation filters.

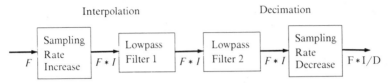

(*b*) Composite filter for anti-imaging and antialiasing.

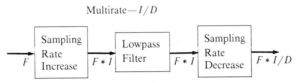

Figure 7.18 Sampling rate change by ratio of integers I / D.

Figure 7.19 presents a matrix representation of the filtering process for a sampling rate change of 1.5 resulting from the sampling rate increase factor $I = 3$ and the sampling rate decrease factor $D = 2$. To implement the sampling rate increase operation, two zeros were added between each input sample (i.e., $I - 1$). To implement the decimation factor, the interpolated input samples are shifted by two prior to each calculation. The zeros do not have to be multiplied by the filter impulse response coefficients. We have assumed that a FIR filter of order 12 provides the desired frequency response characteristics. It is desirable to establish a procedure for efficiently performing this operation. The process can be implemented as a time-varying filter at the filter output rate provided the input samples and the filter coefficients are properly controlled.

Matrix illustration for $I = 3$, $D = 2$, and $N = 12$; therefore $Q = 4$

$$y(m) = \sum_{n=0}^{3} g(m \text{ MOD } I, n) * x\{[mD/I] - n\}$$

$$
\begin{bmatrix} y_0 \\ y_1 \\ y_2 \\ y_3 \\ y_4 \\ y_5 \\ y_6 \\ y_7 \\ y_8 \\ y_9 \\ y_{10} \\ y_{11} \end{bmatrix}
=
\begin{bmatrix}
x_0 & 0 & 0 & x_{-1} & 0 & 0 & x_{-2} & 0 & 0 & x_{-3} & 0 & 0 \\
0 & 0 & x_0 & 0 & 0 & x_{-1} & 0 & 0 & x_{-2} & 0 & 0 & x_{-3} \\
0 & x_1 & 0 & 0 & x_0 & 0 & 0 & x_{-1} & 0 & 0 & x_{-2} & 0 \\
x_2 & 0 & 0 & x_1 & 0 & 0 & x_0 & 0 & 0 & x_{-1} & 0 & 0 \\
0 & 0 & x_2 & 0 & 0 & x_1 & 0 & 0 & x_0 & 0 & 0 & x_{-1} \\
0 & x_3 & 0 & 0 & x_2 & 0 & 0 & x_1 & 0 & 0 & x_0 & 0 \\
x_4 & 0 & 0 & x_3 & 0 & 0 & x_2 & 0 & 0 & x_1 & 0 & 0 \\
0 & 0 & x_4 & 0 & 0 & x_3 & 0 & 0 & x_2 & 0 & 0 & x_1 \\
0 & x_5 & 0 & 0 & x_4 & 0 & 0 & x_3 & 0 & 0 & x_2 & 0 \\
x_6 & 0 & 0 & x_5 & 0 & 0 & x_4 & 0 & 0 & x_3 & 0 & 0 \\
0 & 0 & x_6 & 0 & 0 & x_5 & 0 & 0 & x_4 & 0 & 0 & x_3 \\
0 & x_7 & 0 & 0 & x_6 & 0 & 0 & x_5 & 0 & 0 & x_4 & 0
\end{bmatrix}
\begin{bmatrix} h_0 \\ h_1 \\ h_2 \\ h_3 \\ h_4 \\ h_5 \\ h_6 \\ h_7 \\ h_8 \\ h_9 \\ h_{10} \\ h_{11} \end{bmatrix}
$$

Figure 7.19 Multirate processing: FIR structure for sampling rate conversion by 3 / 2.

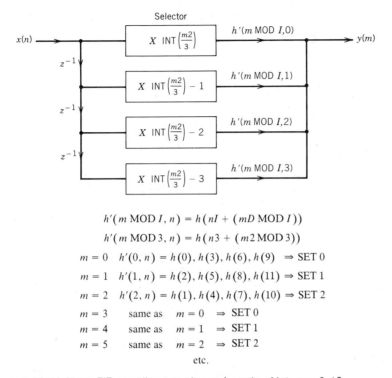

$$h'(m \, \text{MOD} \, I, n) = h(nI + (mD \, \text{MOD} \, I))$$

$$h'(m \, \text{MOD} \, 3, n) = h(n3 + (m2 \, \text{MOD} \, 3))$$

$$m = 0 \quad h'(0, n) = h(0), h(3), h(6), h(9) \Rightarrow \text{SET } 0$$

$$m = 1 \quad h'(1, n) = h(2), h(5), h(8), h(11) \Rightarrow \text{SET } 1$$

$$m = 2 \quad h'(2, n) = h(1), h(4), h(7), h(10) \Rightarrow \text{SET } 2$$

$$m = 3 \quad \text{same as} \quad m = 0 \Rightarrow \text{SET } 0$$

$$m = 4 \quad \text{same as} \quad m = 1 \Rightarrow \text{SET } 1$$

$$m = 5 \quad \text{same as} \quad m = 2 \Rightarrow \text{SET } 2$$

etc.

Figure 7.20 Multirate FIR sampling rate change by ratio of integers $3/2$.

Illustration for $I = 3$, $D = 2$, and $N = 12$; therefore $Q = N/I = 4$

m	$[2m/3]$	m MOD 3	Input buffer	$y(m) = \sum\limits_{n=0}^{3} g(m \, \text{MOD} \, I, n) * x\{[mI/D] - n\}$
0	$x(0)$	g_0	x_0 0 0 0	$y_0 = x_0 h_0$
1	$x(0)$	g_1	x_0 0 0 0	$y_1 = x_0 h_2$
2	$x(1)$	g_2	$x_1 x_0$ 0 0	$y_2 = x_1 h_1 + x_0 h_4$
3	$x(2)$	g_0	$x_2 x_1 x_0$ 0	$y_3 = x_2 h_0 + x_1 h_3 + x_0 h_6$
4	$x(2)$	g_1	$x_2 x_1 x_0$ 0	$y_4 = x_2 h_2 + x_1 h_5 + x_0 h_8$
5	$x(3)$	g_2	$x_3 x_2 x_1 x_0$	$y_5 = x_3 h_1 + x_2 h_4 + x_1 h_7 + x_0 h_{10}$
6	$x(4)$	g_0	$x_4 x_3 x_2 x_1$	$y_6 = x_4 h_0 + x_3 h_3 + x_2 h_6 + x_1 h_{11}$
7	$x(4)$	g_1	$x_4 x_3 x_2 x_1$	$y_7 = x_4 h_2 + x_3 h_5 + x_2 h_8 + x_1 h_{11}$
8	$x(5)$	g_2	$x_5 x_4 x_3 x_2$	$y_8 = x_5 h_1 + x_4 h_4 + x_3 h_7 + x_2 h_{10}$

Samples enter buffer based on $(2m/3)$ value and coefficient set is selected based on the value of $\{m \, \text{MOD} \, 3\}$. Subscripts and parenthesis used interchangeably.
$I = 3$ therefore three unique coefficient sets:

$$\left. \begin{array}{l} g_0 = h_0 \quad h_3 \quad h_6 \quad h_9 \\ g_1 = h_2 \quad h_5 \quad h_8 \quad h_{11} \\ g_2 = h_1 \quad h_4 \quad h_7 \quad h_{10} \end{array} \right\} \begin{array}{l} \text{Note order of coefficients} \\ \text{each set is every } I = 3 \text{ value of } h(n) \\ \text{starting at } (nI + 2m \, \text{MOD} \, 3) \end{array}$$

Figure 7.21 FIR structure for sampling rate conversion by $3/2$.

The time-varying filter output is given by

$$y(m) = \sum_{n=0}^{N-1} h'(m \text{ MOD } I, n) x \left[\text{INT} \left(\frac{m * D}{I} \right) - n \right] \qquad (7.25)$$

where the h' represent I sets of coefficients that are repeated with period I. The h' are given by

$$h'(m \text{ MOD } I, n) = h(nI + mD \text{ MOD } I) \qquad \begin{array}{l} n = 0, 1, 2, \ldots N - 1 \\ m = 0, 1, 2, \ldots \end{array} \qquad (7.26)$$

Implementation of this filter is illustrated in Figure 7.20 for the sampling rate conversion by 1.5 illustrated in Figure 7.19. Again, we have assumed that a FIR filter of order 12 provides the desired frequency response characteristics. Therefore, three sets of four filter coefficients are obtained as shown in the figure. Note that a selector is required prior to each multiplier to implement the sampling rate change process. The selector updates the value for input to the multiplier based on the integer portion of $2m/3$ represented by

$$\text{INT} \left(\frac{m * 2}{3} \right) \qquad (7.27)$$

as shown in Eq. 7.25. Now if we evaluate Eq. 7.25 for the first several outputs $y(m)$, we obtain the results shown in Figure 7.21.

Crochiere and Rabiner[7] developed a diagram of a program implementation for the time-varying sampling rate conversion by a ratio of integers. Figure 7.22 provides a computer program flowchart for this process. A program has been developed to perform this operation.[12]

7.4 SUMMARY

Multirate DSP techniques provide a means of modifying the sampling rate. The use of multirate processing can provide significant computational savings. We have shown that via commutation of the decimation (interpolation) operation within the filter structure the multiplications can be implemented at the lower rate. Several filter designs that are useful in multirate processing were discussed. An approach for the optimization of a multirate system into cascaded filter sections was defined. General rules for determining the decimation factors for the cascaded filter sections were presented. Crochiere and Rabiner[7] provide a comprehensive treatment of multirate processing techniques.

The multirate DSP field is still evolving. Many decisions during the DSP design result from available technologies which influence the processor architecture. We discussed some basic conclusions using multiplications, additions and storage as processor resource parameters for comparison of design alternatives. The final design of the signal processing must consider the actual performance of the hardware and software implementation. The advances in

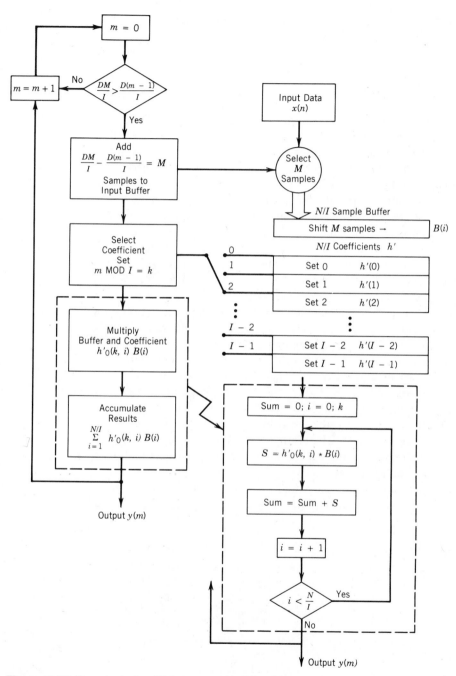

Figure 7.22 Flowchart of multirate process by ratio of integers.

technologies play a significant role in the architectures. These issues will be addressed in Chapter 10.

APPENDIX 7.A
FLOW GRAPH PROPERTIES AND
OPERATIONS FOR MULTIRATE SYSTEMS

For decimators and interpolators, it is important to understand the relationship between the order of the sampling rate change operations with the typical flow graph operations. **Commutation** is the process of interchanging the order of two cascade operations without affecting the input-to-output response of the system, as illustrated in Figure 7.A.1. Typical operations that are equivalent representations for multirate signal flow graphs are shown in Figure 7.A.2. Figure 7.A.2a illustrates the reduction obtained by interchanging the order of operations between scalars and the sampling rate decrease operation. The interchanging of the delay operator and modulator with the sampling rate change processes is shown in Figures 7.A.2b and c, respectively. Crochiere and Rabiner present a detailed discussion of filter structures and network theory for multirate digital systems.[7]

A fifth-order FIR filter direct-form symmetric implementation flow graph is shown in Figure 7.A.3a, followed by a sampling rate decrease operation. Using Figure 7.A.2a, the sampling rate decrease operation can be commuted with the scalar operations of addition and multiplication within the flow graph, as shown in Figure 7.A.3b. Therefore, these operations can be performed at the reduced rate, providing significant computational savings. Using the same fifth-order filter for an interpolator results in the flow graph presented in Figure 7.A.3c. Note that this flow graph is not implementable since it requires fractional delays. Therefore, a different flow graph is required for the interpolator implementation in order to commute the sampling rate increase operation within the FIR filter.

Decimators and interpolators are dual systems. The **dual** of a system is one that performs an operation complimentary to the original system. Given

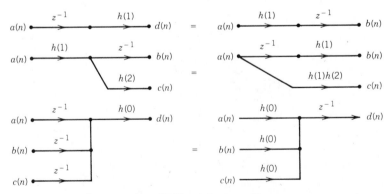

Figure 7.A.1 Commutation of DSP$_1$ scalar operations.

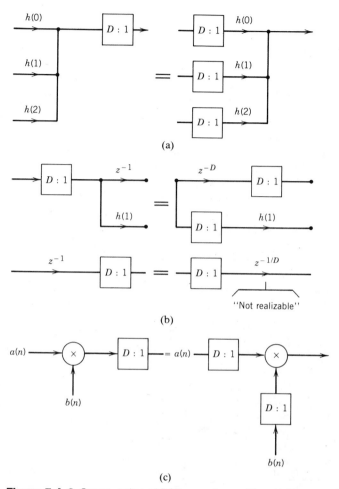

Figure 7.A.2 Commutation of DSP operations with sampling rate decrease.

any linear time-invariant system or linear time-varying system, its dual can be constructed by transposing the original system. The **transpose** of a flow graph is obtained by performing the following operations.

- Reverse the direction of all branches in the signal flow graph.
- Interchange the inputs and outputs.
- Reverse the roles of all nodes in the flow graph.
  - •• Summing points become branching points.
  - •• Branching points become summing points.
- Linear time-invariant branch operations remain unchanged. (Note that the direction was reversed.)
- Replace sampling rate decrease (increase) operations with sampling rate increase (decrease) operations.

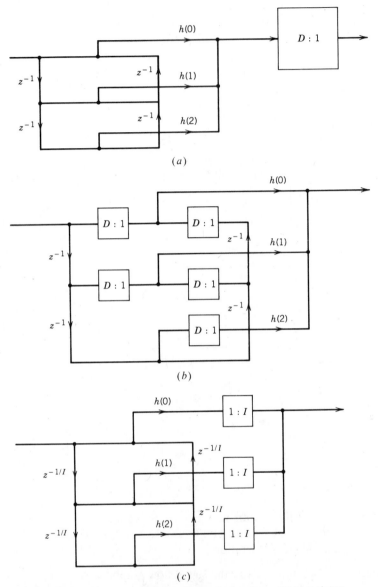

Figure 7.A.3 Commutation of multirate decimator and interpolator.

An interpolation process can be obtained from the transpose of a decimation process. The interpolator given in Figure 7.A.3*c* was not implementable. The interpolation shown in Figure 7.A.4*a* is the transpose of Figure 7.A.3*b* obtained using the rules just defined. Figure 7.A.4*b* is identical to Figure 7.A.4*a* redrawn in the standard form showing the input on the left and the output on the right. This flow graph implements an interpolation process with

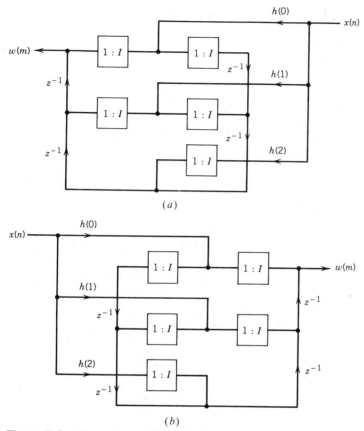

Figure 7.A.4 Transpose of FIR decimator to develop dual FIR interpolator.

the sampling rate increase operation commuted to the other side of the coefficient multiplications. This provides significant savings over the noncommuted implementation.

A key property of the transpose operation is that the number of multipliers and the rate of multiplier operation remains unchanged. Therefore, the transpose of a flow graph that is efficient with respect to the multiply operations will also be efficient with respect to the multiply operations. Hence, it is often more convenient to perform the transpose of a decimator, (interpolator) filter that has been optimized to find the optimized interpolator (decimator) filter.

References

1. R. E. Crochiere and L. R. Rabiner, "Optimum FIR Digital Filter Implementations for Decimation, Interpolation, and Narrow-Band Filtering," *IEEE ASSP*, 23(5):444–456, 1975.

2. R. E. Crochiere and L. R. Rabiner, "A Novel Implementation for Narrow-Band FIR Digital Filters," *IEEE ASSP*, 23(5):457–464, 1975.

3. R. E. Crochiere and L. R. Rabiner, "Further Considerations in the Design of Decimators and Interpolators," *IEEE ASSP*, 24(4):296–311, 1976.

4. H. G. Martinez and T. W. Parks, "A Class of Infinite-Duration Impulse Response Digital Filters for Sampling Rate Reduction," *IEEE ASSP*, 27(2):154–162, 1979.

5. M. G. Bellanger and G. Bonnerot, "Premultiplication Scheme for Digital FIR Filters with Application to Multirate Filtering," *IEEE ASSP*, 26(1):50–55, 1978.

6. J. F. Kaiser, "Nonrecursive Digital Filter Design Using the I_0-Sinh Window Function," IEEE Proc. Int. Symp. Circuits and Systems, April 22–25, 1974, pp. 20–25.

7. R. E. Crochiere and L. R. Rabiner, *Multirate Digital Signal Processing*, Prentice–Hall, Englewood Cliffs, N.J., 1983.

8. L. R. Rabiner, J. F. Kaiser, and R. W. Schafer, "Some Considerations in the Design of Multiband Finite-Impulse-Response Digital Filters," *IEEE ASSP-22(6):*, 1974.

9. J. H. McClellan, T. W. Parks, and L. R. Rabiner, "A Computer Program for Designing Optimum FIR Linear Phase Digital Filters," *IEEE Trans. Audio Electroacoustics*, vol. AU-21:506–526, 1973.

10. A. Peled and B. Liu, *Digital Signal Processing: Theory, Design, and Implementation*, Wiley, New York, 1976.

11. Digital Signal Processing Committee, *Programs for Digital Signal Processing*. IEEE ASSP Society, IEEE Press, New York, 1979.

12. R. E. Crochiere, "A General Program to Perform Sampling Rate Conversion of Data by Rational Ratios." In *Programs for Digital Signal Processing*, New York, IEEE Press, 1979.

13. F. Mintzer and B. Liu, "An Estimate of the Order of an Optimal FIR Bandpass Digital Filter," Proc. IEEE Int. Conf. ASSP, April 1978, pp. 483–486.

Additional Readings

BELLANGER, M. G. *Digital Processing of Signals, Theory and Practice*. Wiley, New York, 1984.

BELLANGER, M. G. "Computation Rate and Storage Estimation in Multirate Digital Filtering with Half-Band Filters," *IEEE ASSP*, ASSP-25(4):August 344–346, 1977.

MARTINEZ, H. G. and PARKS, T. W. "Design of Recursive Digital Filters with Optimum Magnitude and Attenuation Poles on the Unit Circle," *IEEE ASSP*, ASSP-26: 150–157, 1978.

MINTZER, F., and LIU, B. "Aliasing Error in the Design of Multirate Filters," *IEEE ASSP*, ASSP-26: 76–88, 1978.

MINTZER, F., and LIU, B. "The Design of Optimal Multirate Bandpass and Bandstop Filters," *IEEE ASSP*, ASSP-26: 534–543, 1978.

CHU, S., and BURRUS, S. "Optimum FIR and IIR Multistage Multirate Filter Design," *Circuits, Systems and Signal Processing*, 3:3, 1983.

PROBLEMS

7.1.1 Determine the stopband critical frequency for the following filters using the criteria that permits aliasing within the transition band.

	f_p	F	D	δ_p	δ_s
a.	10	5000	100	0.1	0.01
b.	75	1000	5	0.1	0.01
c.	25	7500	50	0.1	0.01
d.	200	4000	8	0.1	0.01
e.	5	800	20	0.1	0.01
f.	15	500	10	0.1	0.01
g.	40	800	8	0.1	0.01
h.	30	700	7	0.1	0.01

7.1.2 Repeat Problem 7.1.1 using the criteria that does not permit aliasing to occur within the transition region.

7.1.3 Determine the PM FIR filter order required to implement the filters given by Problem 7.1.1. Use the simpler approximation form (Eq. 5.92) discussed.

7.1.4 Repeat Problem 7.1.3 using the more accurate PM FIR approximation form (Eq. 5.91).

7.1.5 Calculate the computational requirements (MPS, APS, and Storage) for the filters defined by Problem 7.1.1 with the order determined using the simpler approximation form.

7.1.6 For the following PM FIR filters determine the decimation factor for minimizing each of the computation requirements (MPS, APS, Storage).

	f_p	F	δ_p	δ_s	aliasing criteria
a.	10	1000	0.1	0.01	aliasing allowed
b.	50	10,000	0.1	0.01	aliasing allowed
c.	20	4000	0.1	0.01	aliasing allowed
d.	20	1000	0.1	0.01	aliasing allowed
e.	10	1000	0.05	0.001	aliasing allowed
f.	50	10,000	0.05	0.001	aliasing allowed

7.1.7 For the following filters determine a two-stage multistage decimator design resulting in the decimation given. Assume a PM FIR implementation that allows aliasing in the transition band.

	f_p	F	D	δ_p	δ_s
a.	10	5000	100	0.1	0.01
b.	25	7500	50	0.2	0.005
c.	200	4000	8	0.1	0.01
d.	5	800	20	0.05	0.001
e.	15	500	10	0.1	0.01
f.	40	800	8	0.4	0.0005

7.1.8 Determine the computational requirements for the designs developed in Problem 7.1.7.

7.1.9 Determine the multistage implementation to minimize the number of mutiplications required for the filters given in Problem 7.1.7. Assume a PM FIR implementation with no aliasing permitted in the transition band.

7.1.10 Repeat Problem 7.1.9 for a Butterworth IIR filter design.

7.1.11 Repeat Problem 7.1.9 for a Chebyshev IIR filter design.

7.1.12 Repeat Problem 7.1.9 for an elliptic IIR filter design.

7.1.13 Determine the multiband stopband frequencies for the filters given by Problem 7.1.1. The first 10 are sufficient for $D > 20$.

7.1.14 Perform a PM FIR multiband design for Problem 7.1.7d. What computational savings result compared to Problem 7.1.7?

7.1.15 Show that a halfband filter is always odd and that the even coefficients are zero. Use the frequency-response constraint.

7.1.16 For the following filter specifications, design a multistage halfband filter implementation. Assume aliasing allowed in the final transition band.

	f_p	F	D	δ_p	δ_s
a.	30	400	4	0.01	0.01
b.	15	500	8	0.01	0.01
c.	40	800	8	0.01	0.01
d.	200	4000	8	0.01	0.01
e.	5	800	24	0.01	0.01
f.	25	7500	64	0.01	0.01
g.	10	5000	96	0.01	0.01

7.1.17 Derive the frequency response function for a Comb filter of order N.

7.1.18 For the following filter specifications, design a multirate filter using a Comb filter.

	f_p	F	D	δ_p	δ_s
a.	4	400	40	0.1	0.01
b.	15	5000	100	0.1	0.01
c.	40	8000	80	0.1	0.01
d.	40	4000	40	0.1	0.01
e.	5	2000	100	0.1	0.01
f.	25	7500	64	0.1	0.01
g.	10	5000	96	0.1	0.01

Comment on the passband and stopband variation compared to the desired values.

7.1.19 Design a multirate filter to implement the following bandpass filters using decimation followed by interpolation to accomplish the filter implementation. Note that the sampling rate at the output equals the input rate. Multirate techniques are used only for computational savings.

	$f_p(1)$	$f_p(2)$	$f_s(1)$	$f_s(2)$	F	δ_p	δ_s
a.	5	10	2	15	400	0.1	0.01
b.	15	100	5	200	5000	0.1	0.01
c.	40	80	25	120	8000	0.1	0.01
d.	40	80	25	120	4000	0.1	0.01
e.	10	25	5	40	2000	0.1	0.01
f.	25	50	15	75	7500	0.1	0.01
g.	10	15	5	20	5000	0.1	0.01

Use the bandpass order equation given in Chapter 5 (Eq. 5.361) and developed by Mintzer and Liu.[13] What are the computational savings over a conventional design.

7.2.1 Design an interpolation filter to meet the following filter specifications.

	f_p	F	I	δ_p'	δ_s
a.	100	300	4	0.1	0.01
b.	15	50	10	0.1	0.01
c.	40	100	20	0.1	0.01
d.	40	100	40	0.1	0.01
e.	5	20	10	0.1	0.01
f.	25	75	6	0.1	0.01
g.	10	80	8	0.1	0.01

7.2.2 Develop a flow graph for a sixth order FIR interpolation filter showing the interpolation commuted within the filter structure.

7.2.3 Design a multistage interpolator for the filter specifications given in Problem 7.2.1.

7.2.4 Linear interpolation between two consecutive sample points of sequence $x(n)$ is to be performed. Therefore, the values interpolated between the two consecutive sample points lie on a straight line. Develop an equation relating the interpolated output to the input as a function of I interpolated points. Determine the FIR filter coefficients as a function of I. Derive the filter transfer function and frequency response. Show the results of $I = 2$ and $I = 5$. If the input spectrum is given by Figure 7.P.1, then plot the filter frequency response and the signal spectrum for $L = 2$ and $L = 5$. Discuss the image spectrum that results.

7.3.1 Design a multirate filter to implement the following filter specifications. Design two filters to implement the design via interpolation and decimation. Then design a common filter to perform the operation. Discuss the differences in the computational requirements. How many sets of coefficients are required, and list their indexes.

	f_p	F	D	I	δ_p	δ_s
a.	100	300	5	4	0.1	0.01
b.	15	50	7	10	0.1	0.01
c.	40	100	6	20	0.1	0.01
d.	40	100	17	40	0.1	0.01
e.	5	20	12	10	0.1	0.01
f.	25	75	8	6	0.1	0.01
g.	10	80	12	8	0.1	0.01

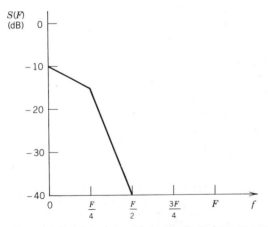

Figure 7.P.1 Input signal log magnitude frequency spectrums.

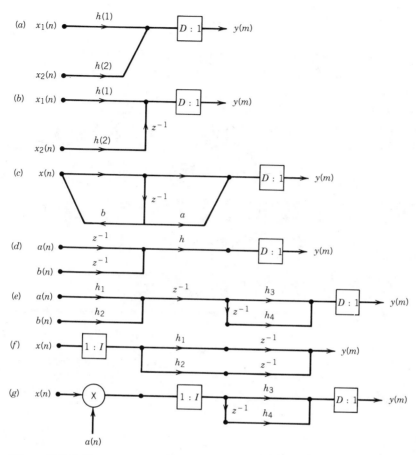

Figure 7.P.2 Flow graphs for commutation.

7.A.1 Develop alternative flow graphs for the figures given in Figure 7.P.2 using the commutation rules. Discuss the relative merits of the alternative implementations.

7.A.2 Develop the transpose of the figures given in Figure 7.P.2 and the commuted figure. Discuss the computational advantages of the alternative implementations.

7.A.3 Draw a general symmetric FIR filter flow graph for an even-order filter. If the filter is followed by a decimation, D, show the form with the decimation operation commuted.

8

RESPONSE OF LINEAR SYSTEMS TO DISCRETE-TIME RANDOM PROCESSES, POWER SPECTRUM ESTIMATION, AND DETECTION OF SIGNALS IN NOISE

In previous chapters we focused on the analysis of discrete-time deterministic signals in the time and frequency domains. We now extend the concepts of digital signal processing to consider the more realistic problem of processing random signals.

8.0 INTRODUCTION

The analytical methods considered in the previous chapters focused on the processing of deterministic discrete-time signals. However, the signals that are frequently encountered for communication, radar, and sonar applications are derived from random processes. For example, when estimating system performance for a sonar system the analyst must consider that the sonar signals are corrupted by ambient ocean noise, which may be stationary or nonstationary, depending on the degree of statistical regularity. At the receiver, additional processing noise is added to the signal causing further distortion. Typical examples of processing noise are quantization noise caused by analog-to-digital conversion and finite arithmetic operations. All these random noise sources provide potential interference to the detection and estimation of signal characteristics. Therefore, when processing random signals the engineer is concerned with the concepts of detection and estimation theory. Detection theory involves the application of probability and statistics to design digital processors that discriminate signals from noise-corrupted signals and noise-only signals. Once the signal is detected, we are usually interested in estimating the signal parameters, such as amplitude, phase, bearing, time of arrival, and frequency. Since entire textbooks have been devoted to the subjects of detection and estimation theory,[7,8] we will only consider those concepts of the subjects that relate to the topics covered in this book.

In this chapter we consider discrete linear systems with random inputs and power spectrum estimation of the input and output processes. The concepts of signal detection will be covered in Section 8.3 where we consider the detection problem of random signals in noise.

It will be assumed that you have an elementary knowledge of probability and random processes; however, an effort has been made in the development of the concepts to make the material self contained.

8.1 RESPONSE OF LINEAR SYSTEMS TO DISCRETE-TIME RANDOM PROCESSES

If a discrete-time random process $x(n)$ is applied as the input to a discrete linear system, the output $y(n)$ is also a discrete-time random process. In this chapter the random processes considered are assumed to have *wide-sense stationary* and *ergodic* properties over some specified time interval. A wide-sense stationary process only requires that the first and second moments are not functions of time, and the autocorrelation function depends only on the time difference. An *ergodic random process* requires that any statistic calculated by averaging over all members of an ergodic ensemble at a fixed time can be calculated by averaging over all time on a single representative member of the ensemble; that is, time averages equal ensemble averages.

Let the autocorrelation functions (acf) of the input and output processes be given by $\phi_{xx}(m)$ and $\phi_{yy}(m)$, that is,

$$\phi_{xx}(m) = E[x(n)x(n+m)] = \lim_{N \to \infty} \frac{1}{2N+1} \sum_{n=-N}^{N} x(n)x(n+m)$$

$$\phi_{yy}(m) = E[y(n)y(n+m)] = \lim_{N \to \infty} \frac{1}{2N+1} \sum_{n=-N}^{N} y(n)y(n+m) \quad (8.1)$$

where the acf $\phi_{xx}(m)$, $m = 0, \pm 1, \pm 2, \ldots$, of a discrete process $x(n)$, $n = 0, \pm 1, \pm 2, \ldots$, is defined as a statistical measure of the interdependence between the instantaneous value of the process x at one time, m to that of another time, $n + m$. The time separation m is also called the time lag. In Section 8.2 the acf given by Eq. 8.1 will be estimated by performing the summation over finite limits.

Let $x(n)$ be a real wide-sense stationary random process. Since for linear systems the response is related to the input by the convolution summation, $\phi_{yy}(m)$ can be expressed in terms of $\phi_{xx}(m)$ and the unit impulse response of the system, that is,

$$y(n) = \sum_{i=-\infty}^{\infty} h(i)x(n-i)$$

Then the acf of the output process can be expressed by

$$\phi_{yy}(m) = E[y(n)y(n+m)]$$

$$= E\left[\sum_{i=-\infty}^{\infty} h(i)x(n-i) \sum_{j=-\infty}^{\infty} h(j)x(n+m-j)\right]$$

Letting $j = i + j$ in the second summation, and taking the expected value of the random process, we obtain

$$\phi_{yy}(m) = \sum_{i=-\infty}^{\infty} \sum_{j=-\infty}^{\infty} h(i)h(i+j)E[x(n-i)x(n-i+m-j)]$$

Finally, since the process is assumed to be wide-sense stationary, the autocorrelation function depends only on the time difference; therefore, the acf of the system output process can be expressed by

$$\phi_{yy}(m) = \sum_{i=-\infty}^{\infty} \sum_{j=-\infty}^{\infty} h(i)h(i+j)\phi_{xx}(m-j) \tag{8.2}$$

On examining Eq. 8.2 we see that the autocorrelation of the output of a linear system is the convolution of the input autocorrelation with the autocorrelation of the system impulse response.

It will be assumed throughout the remainder of this chapter that the input process is stationary and that $E[x(n)]$ and $E[x^2(n)]$ exist and are finite.

The cross-correlation function (ccf) of two discrete-time random processes is a measure of the relationship of $x(n)$ to $y(n)$ as a function of time. If $x(n)$ and $y(n)$ are jointly wide-sense stationary random processes, then their cross-correlation function is given by

$$\phi_{xy}(m) = E[x(n)y(n+m)] \tag{8.3}$$

The ccf of the input $x(n)$ and response $y(n)$ of a discrete-time linear system is given by Eq. 8.3. Using the convolution summation, and taking the expected value of the random process, Eq. 8.3 can be expressed as

$$\phi_{xy}(m) = E\left[x(n) \sum_{j=-\infty}^{\infty} h(j)x(n+m-j)\right]$$

$$= \sum_{j=-\infty}^{\infty} h(j)E[x(n)x(n+m-j)]$$

$$= \sum_{j=-\infty}^{\infty} h(j)\phi_{xx}(m-j) \tag{8.4}$$

Example 8.1
Determine the acf of the sum of two random processes $w(n) = x(n) + y(n)$.

Solution:
From Eqs. 8.1 and 8.3, $\phi_{ww}(m)$ is given by

$$
\begin{aligned}
\phi_{ww}(m) &= E\left[w(n)w(n+m)\right] \\
&= E\left\{\left[x(n) + y(n)\right]\left[x(n+m) + y(n+m)\right]\right\} \\
&= E\left[x(n)x(n+m)\right] + E\left[y(n)y(n+m)\right] + E\left[x(n)y(n+m)\right] \\
&\quad + E\left[y(n)x(n+m)\right] \\
&= \phi_{xx}(m) + \phi_{yy}(m) + \phi_{xy}(m) + \phi_{yx}(m) \quad \blacksquare
\end{aligned}
$$

8.1.1 Autocovariance and Cross-Covariance Sequences

In the previous section the concept of the correlation function was described; that is, it was shown that correlation processing can be used to measure waveform similarities. In autocorrelation, a signal is compared with a time-shifted version of itself; whereas, in cross-correlation the similarities between two signals is determined as a function of the time shift between them.

The autocovariance (acvf) and cross-covariance (ccvf) functions of two stationary random processes $x(n)$ and $y(n)$ are given by,

$$
\gamma_{xx}(m) = E\left\{\left[x(n) - m_x\right]\left[(x(n+m) + m_x)\right]\right\} = \phi_{xx}(m) - m_x^2
$$

$$
\gamma_{xy}(m) = E\left\{\left[x(n) - m_x\right]\left[(y(n+m) - m_y)\right]\right\} = \phi_{xy}(m) - m_x m_y
$$

The Z-transform of $\gamma_{xx}(m)$ and $\gamma_{xy}(m)$ are given by $\Gamma_{xx}(z)$ and $\Gamma_{xy}(z)$, respectively. From Eq. 8.1 we can see that the correlation functions equal the covariance functions only when $m_x = 0$. For random processes, as the time lag m becomes large the random variables will effectively become independent and the acf and ccf will equal m_x^2 and $m_x m_y$, respectively, whereas the acvf and ccvf approach zero in the limit. As a result, since these sequences tend to converge for large m, the Z-transforms exist only when $m_x = 0$.

8.1.2 Power Spectral Density and Cross-Power Spectral Density for Discrete-Time Random Processes

The power spectral density (PSD) provides a measure for the distribution of the average power of random processes. The Z-transform and inverse Z-transform of the autocorrelation function of a zero-mean wide-sense stationary discrete-time random process $x(n)$ form a transform pair, as shown by

$$\Phi_{xx}(z) = \sum_{m=-\infty}^{\infty} \phi_{xx}(m) z^{-m}$$

$$\phi_{xx}(m) = \frac{1}{2\pi j} \oint_c \Phi_{xx}(z) z^{m-1} dz \tag{8.5}$$

where the PSD is defined as the Z-transform of the acf with $z = e^{j2\pi fT}$, that is

$$P_{xx}(f) = \Phi_{xx}(e^{j2\pi fT}) = \sum_{m=-\infty}^{\infty} \phi_{xx}(m) e^{-j2\pi fmT} \tag{8.6}$$

The PSD is purely real and positive, and phase information is not preserved. It should be noted, however, that the acf and the PSD of a random process present different aspects of the correlation information about the process.

As described in Section 8.1.1, the Z-transform of the acf only exists when $m_x = 0$. Therefore, if m_x is not equal to zero the Fourier transform will result in an impulse at $\omega = 0$. For a further discussion, refer to Chapter 8 of Oppenheim and Schafer.[1]

The Z-transform of the ccf is given by

$$\Phi_{xy}(z) = \sum_{m=-\infty}^{\infty} \phi_{xy}(m) z^{-m} \tag{8.7}$$

where the cross-power spectral density (CPSD) of two discrete processes is the Z-transform of their ccf with $z = e^{j2\pi fT}$, that is

$$P_{xy}(f) = \sum_{m=-\infty}^{\infty} \phi_{xy}(m) e^{-j2\pi fmT} \tag{8.8}$$

The CPSD, in general, assumes both positive and negative values, and the relative phase between the signals is preserved. A zero value of CPSD indicates that one or both of the individual spectra are zero at that frequency. A relatively large value for the CPSD likewise indicates that both the individual spectra have large values at that frequency. Therefore, the CPSD indicates the relationship between two signals.

If $\Phi_{xx}(z)$, $\Phi_{yy}(z)$, and $H(z)$ are the transforms of $\phi_{xx}(m)$, $\phi_{yy}(m)$, and $h(n)$, respectively, it can be shown that

$$\Phi_{yy}(z) = H(z) H(z^{-1}) \Phi_{xx}(z) \tag{8.9}$$

or

$$P_{yy}(f) = \Phi_{yy}(e^{j2\pi fT}) = |H(e^{j2\pi fT})|^2 \Phi_{xx}(e^{j2\pi fT}) \tag{8.10}$$

that is, the PSD of the output process of a discrete-time linear system is equal to the PSD of the input process times the squared magnitude response of the system.

The CPSD of the output process of a discrete-time linear system is equal to PSD of the input process times the transfer function of the system, that is

$$\Phi_{xy}(z) = H(z)\Phi_{xx}(z) \tag{8.11}$$

or

$$P_{xy}(f) = H(e^{j2\pi fT})\Phi_{xx}(e^{j2\pi fT}) = H(e^{j2\pi fT})P_{xx}(f) \tag{8.12}$$

Example 8.2
Using Eq. 8.2 verify Eq. 8.9.

Solution:
Taking the Z-transform of Eq. 8.2 we obtain

$$\sum_{m=-\infty}^{\infty} \phi_{yy}(m)z^{-m} = \sum_{m=-\infty}^{\infty} \sum_{i=-\infty}^{\infty} \sum_{j=-\infty}^{\infty} h(i)h(i+j)\phi_{xx}(m-j)z^{-m}$$

Since $z^{-m} = z^{i}z^{-i-j}z^{-m+j}$, and defining $m = i + j$ and $n = m - j$ and using the results of Eq. 8.5 we obtain

$$\Phi_{yy}(z) = \sum_{i=-\infty}^{\infty} h(i)z^{i} \sum_{m=-\infty}^{\infty} h(m)z^{-m} \sum_{n=-\infty}^{\infty} \phi_{xx}(n)z^{-n}$$

$$= H(z^{-1})H(z)\Phi_{xx}(z)$$

which agrees with Eq. 8.9. ∎

Consider the linear time-invariant systems shown in Figure 8.1, where $x(nT)$ and $y(nT)$ are jointly wide-sense stationary random processes. It is of interest to determine the cross-correlation and CPSD function of the output processes of the two systems. From the convolution summation the filter responses are given by

$$v(n) = \sum_{k=-\infty}^{\infty} h_{1}(k)x(n-k)$$

$$w(n) = \sum_{k=-\infty}^{\infty} h_{2}(k)y(n-k)$$

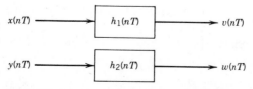

Figure 8.1 Filtering discrete-time random processes.

The acf is obtained from

$$\phi_{vw}(m) = E[v(m+n)w(n)]$$

$$= E\left[\sum_{k=-\infty}^{\infty} h_1(k)x(m+n-k) \sum_{p=-\infty}^{\infty} h_2(p)y(n-p)\right]$$

Taking the expected value of the random processes results in

$$\phi_{vw}(m) = \sum_{k=-\infty}^{\infty} h_1(k) \sum_{p=-\infty}^{\infty} h_2(p)E[x(m+n-k)y(n-p)]$$

$$= \sum_{k=-\infty}^{\infty} h_1(k) \sum_{p=-\infty}^{\infty} h_2(p)\phi_{xy}(m-k+p)$$

Taking the Z-transform of both sides of $\phi_{vw}(m)$ we obtain

$$\sum_{m=-\infty}^{\infty} \phi_{vw}(m)z^{-m} = \sum_{m=-\infty}^{\infty} \sum_{k=-\infty}^{\infty} h_1(k) \sum_{p=-\infty}^{\infty} h_2(p)\phi_{xy}(m-k+p)z^{-m}$$

Since $z^{-m} = z^{-k}z^p z^{-(m-k+p)}$, we can write

$$\sum_{m=-\infty}^{\infty} \phi_{vw}(m)z^{-m} = \sum_{k=-\infty}^{\infty} h_1(k)z^{-k} \sum_{p=-\infty}^{\infty} h_2(p)z^p \sum_{m=-\infty}^{\infty} \phi_{xy}(m)z^{-m}$$

Finally, the CPSD of the output processes is given by

$$\Phi_{vw}(z) = H_1(z)H_2(z^{-1})\Phi_{xy}(z) \tag{8.13}$$

8.1.3 White Noise

White noise is a random process whose power spectrum is constant over the entire frequency range. Thus, ideal white-noise contains an infinite amount of power and is accordingly not physically realizable. In practice, however, the noise may have a constant PSD over a wide but still finite range; that is, the noise spectrum must be constant over a sufficiently broad band relative to the frequency band of interest. From the foregoing definition, the autocorrelation function and PSD of white noise can be expressed by

$$\phi_{xx}(m) = \sigma_x^2 \delta(m)$$

$$P_{xx}(f) = \Phi_{xx}(e^{j2\pi fT}) = \sigma_x^2 \tag{8.14}$$

It should be noted that white-noise can be described as either Gaussian or uniform depending on the amplitude distribution.

As an illustration, we determine the response of a digital filter to a white-noise input. Since $\Phi_{xx}(e^{j\omega}) = \sigma_x^2$, we obtain from Eq. 8.10

$$P_{yy}(\omega) = \sigma_x^2 |H(e^{j2\pi fT})|^2 \tag{8.15}$$

which is real and nonnegative. Therefore, as shown in Figure 8.2*a*, the PSD of the digital filter can be determined by passing white noise through the filter and evaluating the output spectrum using the FFT algorithm.

The results of averaging 1, 2, and 20 independent-detected FFT records are shown in Figures 8.2*b*, *c*, *d*. We can see that the digital filter magnitude squared frequency response improves by averaging the independent FFT records. This results because the noise terms are independent and the variance is reduced as the reciprocal of the number of samples averaged. We will consider this process in detail in Section 8.2 when the concept of power spectrum estimation is covered.

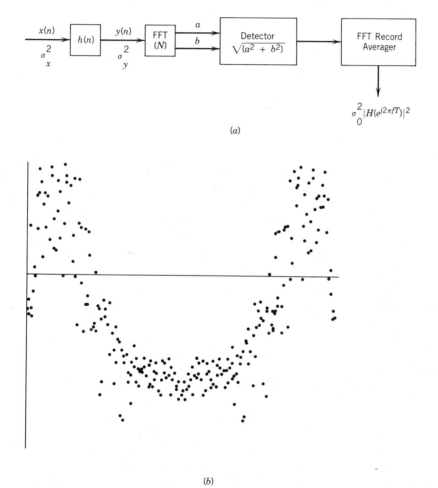

(a)

(b)

Figure 8.2 Determination of a digital filter power spectrum density function: (*a*) Signal flow diagram of power spectrum estimation; (*b*) power spectrum after processing one FFT record; (*c*) power spectrum after averaging 2 FFT records; (*d*) power spectrum after averaging 20 FFT records.

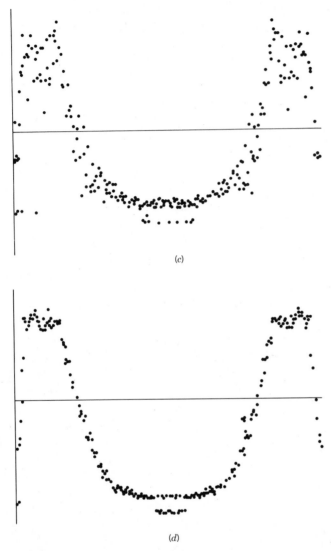

(c)

(d)

Figure 8.2 Continued

Example 8.3

The input to a discrete-time linear system is a white-noise sequence with zero mean and variance (second moment) σ_x^2. Show that the response of the system with impulse response $h(n)$ is given by

$$\sigma_y^2 = \sigma_x^2 \sum_{n=0}^{\infty} h^2(n) \qquad (8.16)$$

Solution:

$$\sigma_y^2 = E\left[y^2(n)\right] = E\left[\sum_{m=-\infty}^{\infty} h(m)x(n-m) \sum_{k=-\infty}^{\infty} h(k)x(n-k)\right]$$

taking the expected value of the random process, and using Eq. 8.14, we obtain

$$\sigma_y^2 = \sum_{m=-\infty}^{\infty} \sum_{k=-\infty}^{\infty} h(m)h(k)E[x(n-m)x(n-k)]$$

$$= \sum_{m=-\infty}^{\infty} \sum_{k=-\infty}^{\infty} h(m)h(k)\sigma_x^2\delta(m-k)$$

Finally, for $m = k$, and then letting $m = n$, we obtain

$$\sigma_y^2 = \sigma_x^2 \sum_{n=-\infty}^{\infty} h^2(n)$$

which corresponds to the integral taken over the limits 0 to F of Eq. 8.10. It should be noted that this equation is only true if the filter input is white noise.

∎

Example 8.4

The transfer function of a digital filter is given by

$$H(z) = \frac{1}{1 - 0.9z^{-1}}$$

Determine the average output power of the filter to a white-noise random process with zero mean and variance σ_x^2.

Solution:

Since successive samples of a white-noise random process are statistically independent with zero mean, for m an integer not equal to zero the autocorrelation function is given by

$$\phi_{xx}(m) = E[x(n)x(n+m)] = E[x(n)]E[x(n+m)] = 0$$

For $m = 0$ we obtain the average power in the input, that is,

$$\phi_{xx}(0) = E[x^2(n)] = \frac{1}{2\pi j}\oint_c \Phi_{xx}(z)z^{-1}\,dz = \sigma_x^2$$

which is consistent with Eq. 8.14. Then from Eq. 8.9 the average output power is given by

$$\phi_{yy}(0) = \frac{1}{2\pi j}\oint_c \sigma_x^2 H(z^{-1})H(z)z^{-1}\,dz$$

$$= \frac{\sigma_x^2}{2\pi j}\oint_c \frac{1}{(z-0.9)(1-0.9z)}\,dz = \frac{\sigma_x^2}{0.19}$$

where the contour of integration encloses only the pole $z = 0.9$. ■

Example 8.5

Consider two linear time-invariant filters connected in cascade as shown in Figure 8.3.

Let $x(n)$ be a wide-sense stationary process with unit variance and autocorrelation function $\phi_{xx}(n)$. The impulse response of each section is given by $h_1(n) = h_2(n) = a^n$ and $0 < a < 1$. Determine the following:

1. The impulse response of the overall system.
2. The PSD function of the system output process.
3. The acf of the system output process.
4. The average power in the passband.

Solution:

1. The system transfer function is obtained by taking the Z-transform of the impulse response and forming the product of the transfer functions of each section of Figure 8.3. Since $H_1(z) = H_2(z)$ we obtain

$$H(z) = H_1(z)H_2(z) = \frac{(1-a)^2 z^2}{(z-a)^2}$$

then from Eq. 3.23 the impulse response of the overall system is given by

$$h(n) = b^2(na^n + a^n) \qquad n \geq 0$$

where the normalization constant $b = 1 - a$

2. The PSD obtained from Eqs. 8.6 and 8.9 is given by

$$P_{yy}(f) = H(z)H(z^{-1})\big|_{z=e^{j2\pi fT}} = \frac{b^4}{(1 + a^2 - 2a\cos\omega T)^2}$$

Figure 8.3 Cascade of first-order digital filters.

3. The acf can be obtained by noting that

$$\Phi_{yy}(z) = H(z)H(z^{-1}) = \frac{b^4 z^2}{(z-a)^2(1-az)^2}$$

then from Eq. 3.23 we obtain

$$\phi_{yy}(n) = b^4 \left[\frac{(1-a^2)(n+1)a^n + 2a^{n+2}}{(1-a^2)^3} \right]$$

4. The average output power in the passband is obtained from

$$\phi_{yy}(0) = E[y^2(n)] = b^4 \left[\frac{1+a^2}{(1-a^2)^3} \right]$$

8.2 POWER SPECTRUM ESTIMATION

Estimation of the PSD of a discrete-time random process is currently based on techniques using the fast Fourier transform (FFT). The two fundamental problems associated with processing finite data records in estimating the PSD are spectral leakage and smearing. Spectral leakage is a result of processing finite duration data records of NT seconds. As a result, frequencies that are not integer multiples of the FFT bin width F/N will "leak out" throughout the entire spectral record (see Section 6.6.2). Spectral smearing results from "windowing" the data record to reduce leakage, that is, spectral smearing represents the trade-off between desired frequency resolution, which we define as the FFT analysis bandwidth given by

$$BW_a = k_w \frac{F}{N} \tag{8.17}$$

and statistical accuracy of the spectral estimate. As described in Chapter 6, optimal window selection is a trade-off between sidelobe level and the equivalent noise bandwidth coefficient k_w. Windows with low sidelobe levels have large values of k_w, which result in wide analysis bandwidths. As a consequence, frequency smearing (aliasing) across FFT bins results. If windows are selected on the basis of narrow analysis bandwidths, then the statistical accuracy of the frequency estimate is degraded due to leakage resulting from sidelobe level increase. It should be noted that for a rectangular window $k_w = 1$; therefore, for this case the analysis bandwidth is equal to the FFT bin width.

In Section 8.2.1 we investigate the periodogram method of estimating the PSD. This PSD estimate is obtained by computing the squared magnitude of the DFT of a discrete-time sequence. In Section 8.2.3 we consider a method developed by Welch[2] of sectioning an indefinite duration data sequence into finite length segments, which may be contiguous or overlapped, and then taking the arithmetic average of modified periodograms at each frequency.

The PSD can also be estimated based on a modeling approach. The assumed model of the process is usually an autoregressive (all pole) or an autoregressive-moving average model. It has been shown that the modeling approach is computationally efficient and results in high spectral resolution. This approach to power spectrum estimation is covered in Chapter 11.

8.2.1 Periodogram Method of Power Spectrum Estimation

Consider a zero-mean wide-sense stationary random process with autocorrelation $\phi_{xx}(m)$ and PSD $P_{xx}(f)$. An estimate of the true acf $\phi_{xx}(m)$ using an observation of N samples can be obtained from

$$c_{xx}(m) = \frac{1}{N} \sum_{n=0}^{N-|m|-1} x(n)x(n+|m|) \qquad m = 0, \pm 1, \ldots, \pm(N-1) \quad (8.18)$$

which is referred to as the **sampled acf function**. For $m \geq N$, $c_{xx}(m) = 0$; that is, there are no products available with this lag. We also note that $c_{xx}(-m) = c_{xx}(m)$.

For the special case of a lag of $m = 0$ we obtain

$$c_{xx}(0) = \sigma_x^2 = E[x^2(n)] = \frac{1}{N} \sum_{n=0}^{N-1} x^2(n) \qquad (8.19)$$

which is the mean-squared value (or power) in the signal. We now examine the mean and variance of this estimate.

For large N compared to the lag m, the estimate given by Eq. 8.18 has a mean equal to

$$E[c_{xx}(m)] = \frac{1}{N} \sum_{n=0}^{N-|m|-1} E[x(n)x(n+|m|)]$$

$$= \frac{N-|m|}{N}\phi_{xx}(m) = \phi_{xx}(m) - \frac{|m|}{N}\phi_{xx}(m) \qquad (8.20)$$

From Eq. 8.20 we can see that the sample acf is biased since it does not equal the true acf $\phi_{xx}(m)$ at lag m. For an estimator to be acceptable, the difference between the true value of the quantity and the expected value of its estimator should be small. This difference is usually called the bias of the estimator. However, we can see that $c_{xx}(m)$ is asymptotically unbiased since the term $|m|/N$ approaches zero as N approaches infinity.

It can be shown[1] that for large N the variance of the sampled acf is approximately

$$\text{var}[c_{xx}(m)] = E[c_{xx}^2(m)] - E^2[c_{xx}(m)]$$

$$= \frac{1}{N} \sum_{i=-(N-|m|-1)}^{N-|m|-1} [\phi_{xx}^2(i) + \phi_{xx}(i+m)\phi_{xx}(i-m)] \quad (8.21)$$

and is therefore a *consistent estimator* since the variance decreases as the number of observations becomes large. An estimator is said to be *consistent* if both the bias and variance tend to zero as the number of observations becomes large (see Problem 8.2.10).

THE PERIODOGRAM

An estimate of the PSD can be obtained by taking the Z-transform of the biased acf estimate, that is, defining the Z-transform of $c_{xx}(m)$ by $I_N(z)$ we obtain

$$I_N(z) = \sum_{m=-\infty}^{\infty} c_{xx}(m) z^{-m} \tag{8.22}$$

where the true PSD and acf are related by Eq. 8.5. Let us now modify the observed sequence $x(n)$ by multiplying it by a rectangular window function given by

$$a(n) = \begin{cases} 1 & n = 0, 1, \ldots, N-1 \\ 0 & \text{otherwise} \end{cases} \tag{8.23}$$

then we obtain the modified (finite) sequence

$$x_N(n) = x(n) a(n) \tag{8.24}$$

where the sequence x_N is of finite duration. We can now write

$$c_{xx}(m) = \frac{1}{N} \sum_{n=-\infty}^{\infty} x_N(n) x_N(n+m)$$

where $x_N(n)$ is defined by Eq. 8.24. Equation 8.22 can then be expressed by

$$I_N(z) = \frac{1}{N} \sum_{m=-\infty}^{\infty} \sum_{n=-\infty}^{\infty} x_N(n) x_N(n+m) z^{-m}$$

$$= \frac{1}{N} \sum_{n=-\infty}^{\infty} x_N(n) \sum_{m=-\infty}^{\infty} x_N(n+m) z^{-m}$$

$$= \frac{1}{N} X_N(z^{-1}) \sum_{k=-\infty}^{\infty} x_N(k) z^{-k}$$

$$= \frac{1}{N} X_N(z) X_N(z^{-1})$$

Finally, letting $z = e^{j2\pi fT}$ we obtain

$$I_N(f) = \frac{1}{N} \left| X_N(e^{j2\pi fT}) \right|^2 = \frac{1}{N} \left| \sum_{n=0}^{N-1} x_N(n) e^{-j2\pi fnT} \right|^2 \tag{8.25}$$

The function $I_N(f)$ is known as the **periodogram** of the finite sequence $x_N(n)$;

that is, the periodogram is the magnitude squared function of the DFT of the data sequence. Therefore, the spectral estimate $I_N(f)$ can be determined from the DFT of the sampled acf function or computed directly from the data using Eq. 8.25.

Equation 8.25 represents an estimate of the PSD. To demonstrate how well it approximates the true PSD we must determine the statistical properties of the periodogram. The expected value of the periodogram is obtained from Eq. 8.22

$$E[I_N(f)] = \sum_{m=-\infty}^{\infty} E[c_{xx}(m)] e^{-j2\pi fmT}$$

Substituting Eq. 8.20 where we consider both positive and negative lag estimates, we obtain

$$E[I_N(f)] = \sum_{m=-(N-1)}^{N-1} \left(1 - \frac{|m|}{N}\right) \phi_{xx}(m) e^{-j2\pi fmT} \qquad (8.26)$$

Let us now define the function in parenthesis of Eq. 8.26 by

$$a_B(m) = 1 - \frac{|m|}{N} \qquad |m| \le N - 1 \qquad (8.27)$$

which is a triangular function (Bartlett window function). The expected value of the periodogram can now be expressed as the infinite sum

$$E[I_N(f)] = \sum_{m=-\infty}^{\infty} a_B(m) \phi_{xx}(m) e^{-j2\pi fmT} \qquad (8.28)$$

where from Eq. 8.27 the Bartlett window function is defined over finite limits. We see that $I_N(f)$ given by Eq. 8.22 or Eq. 8.25 provides a biased estimate of the true PSD $P_{xx}(\omega)$ (see Eq. 8.6). Specifically, $E[I_N(f)]$ differs from $P_{xx}(f)$ by the window function $a_B(m)$. Thus, the mean of the periodogram is the Fourier transform of the product of the true acf and the Barlett window function. Since multiplication in the time domain is equivalent to convolution in the frequency domain, an equivalent relationship for Eq. 8.28 is

$$E[I_N(f)] = \frac{1}{F} \int_{-F/2}^{F/2} P_{xx}(\lambda) A_B(f - \lambda) \, d\lambda \qquad (8.29)$$

which is equal to the true PSD observed through the Bartlett spectral window where $A_B(f)$ is the Fourier transform of the Bartlett window, that is

$$A_B(f) = \frac{1}{N} \left(\frac{\sin N\pi f}{\sin \pi f}\right)^2 \qquad (8.30)$$

For large N, the spectral window $A_B(f)$ approaches an impulse function so that

$$E\left[I_N(f)\right] \approx P_{xx}(f) \qquad (8.31)$$

Therefore, the periodogram is asymptotically unbiased since the Bartlett window function vanishes for large N.

An expression for the variance of the periodogram[3] for large N is given by

$$\text{var}\left[I_N(f)\right] \approx RP_{xx}^2(f) \qquad (8.32)$$

where R is a constant that depends on the window function used; for example, for this case we considered the Bartlett window. Equation 8.32 shows that as N increases, the variance becomes proportional to the square of the spectrum and does not converge to the true spectrum. In fact, as the DFT record length N increases the amplitude and rate of the fluctuations in the periodogram increase. As a result, this estimate can produce vastly different spectral estimates for different sets of sequences obtained from the same random process even for large N. Therefore, the periodogram is not a consistent estimate.

We will show in Section 8.2.3 that this problem can be circumvented by averaging short sequences of identically distributed periodograms. The resultant averaged periodogram will have the same mean value as any of the individual spectral estimates; however, the variance of the averaged of K identically distributed random variables will be $1/K$ of the individual spectral variances.

8.2.2 Data and Spectral Windows

Processing indefinitely long sequences of data for the purpose of spectral analysis is conventionally performed by dividing the data into short sections. The samples of each section are usually multiplied by an appropriate data window function in the time domain or, equivalently, the DFT of the data convolved with the corresponding spectral window in the frequency domain. A window function is applied to the data section to control leakage, which causes a bias in the estimate; that is, frequencies that are not integer multiples of (ΔF) will leak out throughout the entire DFT frequency axis. However, good performing windows, that is, windows with sidelobe levels more than 70 dB down, exhibit mainlobes with single-sided bandwidths of approximately four FFT bins. Therefore, for this case to keep the required FFT bin width alias free it is necessary to increase the sampling frequency by a factor of four. It will be shown in Section 8.2.3 that this is equivalent to performing the FFT processing with 75% overlap of data records. This is also referred to as four-to-one redundancy processing.

From Eq. 8.29 we see that the expected value of the periodogram is given by the frequency domain convolution of the true PSD and the Bartlett spectral window. Similarly, the expected value of an arbitrary spectral window function with the PSD is obtained by replacing $A_B(f)$ with $A(f)$ in Eq. 8.29. The result of windowing is therefore a trade-off between spectral smoothing (reduced leakage) and reduced frequency resolution (due to the increased width of the mainlobe of the spectral window). The bias can therefore be interpreted in terms of the effect the window function has on spectral resolution.

Let us now consider an example of windowing in the frequency domain: the Fourier coefficients at the output of an FFT processor are given by $a_k + ib_k$, where k is the frequency index, that is, $k = 0, 1, \ldots, N - 1$. If we consider the Hamming spectral window, then the Hamming Fourier coefficients for DFT processing are given by

$$A_k = -0.23a_{k-1} + 0.54a_k - 0.23a_{k+1}$$
$$B_k = -0.23b_{k-1} + 0.54b_k - 0.23b_{k+1}$$

Notice that we have performed a three-point convolution of the Fourier coefficients $a_k + jb_k$ with the Hamming coefficients: -0.23, 0.54, -0.23. Finally, the detected output of the FFT processor is $A_k^2 + B_k^2$. Alternatively, we could have obtained the same result in the time domain by multiplying the input data sequence $x(n)$ by the window function given by Eq. 6.19.

An excellent discussion of window functions that are widely used in spectral analysis have been published by Harris,[6] and were discussed in Chapter 6. Additional properties of spectral windows will be considered in Section 8.2.3 with respect to averaging modified periodograms.

8.2.3 Estimation of the PSD by the Method of Averaging Modified Periodograms

The method of PSD estimation by averaging modified periodograms was originally proposed by Welch.[2] Since this method is based on the application of the FFT, it is a computationally efficient technique for power spectrum estimation. Essentially the method involves sectioning the data record into either overlapping or nonoverlapping sections. Each data section is then modified by multiplying the data sequence by an appropriate window function, before computing the periodogram. Windowing can be performed in the frequency domain by convolving the spectral estimate with the appropriate window coefficients (Chapter 6). Finally, the modified periodograms are averaged and the resulting spectral estimate is asymptotically unbiased and consistent.

Let $x(n)$, $n = 0, 1, \ldots, N - 1$ be a sample from a wide-sense stationary process. Assume that $E[x(n)] = 0$, and let $x(n)$ have PSD $P_{xx}(\omega)$. For overlap processing we take sections of length M with starting points D units apart. Let $x_1(n) = x(n)$, $n = 0, 1, \ldots, M - 1$ be the first such section. Then

$$x_{i+1}(n) = x(n + iD) \qquad \begin{aligned} i &= 0, 1, \ldots, K - 1 \\ n &= 0, 1, \ldots, M - 1 \end{aligned} \tag{8.33}$$

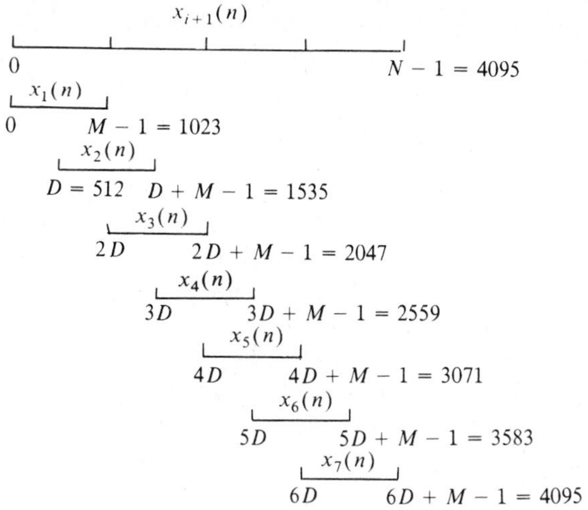

Figure 8.4 Overlap processing of data sections.

Therefore,

$$x_1(n) = x(n) \qquad\qquad n = 0, 1, \ldots, M - 1$$
$$x_2(n) = x(n + D) \qquad\quad n = 0, 1, \ldots, M - 1$$
$$x_3(n) = x(n + 2D) \qquad\; n = 0, 1, \ldots, M - 1$$
$$\vdots \qquad\qquad\qquad\qquad \vdots$$
$$x_K(n) = x[n + (K - 1)D] \quad n = 0, 1, \ldots, M - 1$$

Finally, we have K sections, $x_1(n), \ldots, x_K(n)$ covering the entire data record of length N. Notice that for $D < M$ the sections overlap, and for $D \geq M$ the sections do not overlap. This process is illustrated in Figure 8.4 for the case of $N = 4096$ and $D = 512$, that is, 50% overlap processing of data sections.

From Figure 8.4 it can be seen that the total record length is given by

$$N = (K - 1)D + M \qquad\qquad (8.34)$$

where

$$K = \text{number of overlapped sections averaged}$$
$$M = \text{section length (FFT length)}$$
$$D = (1 - r)M = \text{section overlap starting point}$$
$$r = (M - D)/M = \text{fractional overlap}$$

Solving Eq. 8.34 for K we get

$$K = \frac{(N/M) - r}{1 - r} = \frac{\Delta FT_i - r}{1 - r} \qquad\qquad (8.35)$$

where $N = FT_i$ and T_i is the integration (averaging) time in seconds; then the number of independent FFTs averaged can be obtained from

$$K_{IND} = \frac{N}{M} = \frac{FT_i}{M} = \Delta FT_i \qquad (8.36)$$

where from Eq. 8.36 the FFT frequency resolution is given by $\Delta F = F/M$.

As an example, let the FFT length $M = 1024$, the total record length $N = 131,072$ samples, and the fractional overlap $r = 0.5$. From Eq. 8.35 we obtain a total of $K = 255$ overlapped sections averaged, and from Eq. 8.36 $K_{IND} = 128$ independent sections averaged. For a sampling frequency $F = 256$ samples/second the integration time is given by $T_i = N/F = 512$ seconds. Therefore, we obtain 128 independent samples (FFT sections) averaged in 512 seconds.

We now develop the *Welch method* for estimating the PSD. For each section of length M we calculate a modified periodogram; that is, we first multiply the data sequence $x(n)$ by an appropriate data window function $a(n)$, $n = 0, 1, \ldots, M - 1$, forming the sequences $x_1(n)a(n), \ldots, x_k(n)a(n)$ as described previously (see Fig. 8.4). The DFT of the K-modified sequences are then obtained using the FFT algorithm as follows.

$$Y_i(f_k) = \frac{1}{M} \sum_{n=0}^{M-1} x_i(n)a(n)e^{-j2\pi nk/M} \qquad \begin{matrix} i = 1, 2, \ldots, K \\ k = 0, 1, \ldots, M - 1 \end{matrix} \qquad (8.37)$$

where i is the section (time) index and k is the frequency index. The data window $a(n)$ is selected to achieve desired specification requirements,[4] and $Y_i(f_k)$ is the DFT of the modified input sequences. Finally, we obtain the K-modified periodograms.

$$J_i(f_k) = \frac{M}{U}|Y_i(f_k)|^2 \qquad \begin{matrix} i = 1, 2, \ldots, K \\ k = 0, 1, \ldots, M - 1 \end{matrix} \qquad (8.38)$$

where

$$U = \frac{1}{M} \sum_{n=0}^{M-1} a^2(n) \qquad (8.39)$$

and the discrete FFT bin frequencies are

$$f_k = k\frac{F}{M} = k\Delta F \qquad k = 0, 1, \ldots, M - 1 \qquad (8.40)$$

The spectral estimate is finally obtained by averaging these modified periodograms, that is

$$\hat{P}_{xx}(k) = \frac{1}{K} \sum_{i=1}^{K} J_i(k) = \frac{1}{K} \sum_{i=1}^{K} |Y_i(k)|^2 \qquad k = 0, 1, \ldots, M - 1 \quad (8.41)$$

Time Index (i)   Magnitude-Squared Spectral Estimates: $J_i(k)$

1		$J_1(0)$	$J_1(1)$	$J_1(2)$	$\ldots$	$J_1(M-1)$
	S					
2	U	$J_2(0)$	$J_2(1)$	$J_2(2)$	$\ldots$	$J_2(M-1)$
	M					
$\vdots$		$\vdots$	$\vdots$	$\vdots$	$\vdots$	$\vdots$
K		$J_K(0)$	$J_K(1)$	$J_K(2)$	$\ldots$	$J_K(M-1)$
		$\hat{P}_{xx}(0)$	$\hat{P}_{xx}(1)$	$\hat{P}_{xx}(2)$	$\ldots$	$\hat{P}_{xx}(M-1)$

Figure 8.5 Computation of spectral estimates using the Welsh method.

where in Eq. 8.41 we replace $f_k = k\Delta F$ by k, noting that all frequency samples are separated by ΔF Hz. We can see that the spectral estimate given by Eq. 8.41 is the time average of the modified periodograms $J_i(f_k)$; that is, we form the time average of M magnitude squared estimates, as illustrated in Figure 8.5. It should be noted that the caret (^) is used to indicate that $\hat{P}_{xx}(k)$ is an estimate of $P_{xx}(k)$, where $\hat{P}_{xx}(k)$ is consistent if as the number of observations becomes large it approaches the true value of $P_{xx}(k)$.

It can be shown that the expected value of the modified periodogram is

$$E\left[\hat{P}_{xx}(f_k)\right] = \frac{1}{F} \int_{-F/2}^{F/2} A(f) P_{xx}(f_k - f) \, df \qquad (8.42)$$

where

$$A(f) = \frac{1}{MU} \left| \sum_{n=0}^{M-1} a(n) e^{-j2\pi f n T} \right|^2 \qquad (8.43)$$

Therefore, the expected value of the estimate is the true PSD convolved with the magnitude-squared window transform. The normalizing factor U is required for the estimate $P_{xx}(f_k)$ to be asymptotically unbiased.

THE VARIANCE OF THE AVERAGED PERIODOGRAM ESTIMATOR

For overlapped sections, the transforms are correlated and the reduction in variance obtained by averaging correlated data is not proportional to the number of averages. It is shown in Appendix 8.A that the reduction in variance to be obtained from averaging overlapped spectral estimates is given by the ratio of the variance of the average of the modified periodograms to the variance of an independent measurement, that is

$$\frac{\text{var}\left[\hat{P}_{xx}(f_k)\right]}{\text{var}\left[J_i(f_k)\right]} = \frac{1}{K}\left\{1 + 2\sum_{l=1}^{L-1}\left(1 - \frac{l}{K}\right)c^2[(L-l)s]\right\} \qquad (8.44)$$

where

$$L = 1/s$$
$$s = 1 - r$$
$$r = \text{fractional overlap}$$
$$c[(L - l)s] = \text{correlation coefficient}$$
$$\text{(function of data window)}$$
$$T_i = \text{integration time in seconds}$$

From Eq. 8.44 we can see that for K independent (nonoverlapped) sections the correlation between sections is zero. Therefore, when we average K identically distributed independent measurements, the variance of the average is related to the individual variance of the measurements by

$$\frac{\text{var}\left[\hat{P}_{xx}(f_k)\right]}{\text{var}\left[J_i(f_k)\right]} = \frac{1}{K} \tag{8.45}$$

We can see from Eqs. 8.44 and 8.45 that the Welsh estimate is a consistent estimate since the variance approaches zero as K gets large.

We can now define the effective number of independent samples as the reciprocal of Eq. 8.44, that is

$$K_{\text{EFF}} = \frac{\text{var}\left[J_i(f_k)\right]}{\text{var}\left[\hat{P}_{xx}(f_k)\right]}$$

$$= \frac{K}{1 + 2\sum_{l=1}^{L-1}c^2(1 - ls) - \dfrac{2}{K}\sum_{l=1}^{L-1}lc^2(1 - ls)} \tag{8.46}$$

where for $\Delta F T_i > 10$ we can approximate Eq. 8.46 by

$$K_{\text{EFF}} \approx \frac{K}{1 + 2\sum_{l=1}^{L-1}c^2(1 - ls)} \tag{8.47}$$

CORRELATION COEFFICIENT FUNCTION
The correlation coefficient function[6] is given by

$$c(r) = \frac{\sum_{n=0}^{rN-1}a(n)a[n + (1 - r)N]}{\sum_{n=0}^{N-1}a^2(n)} \tag{8.48}$$

As an example, calculate $c(0.5)$ for the Hanning window function where $r = 0.5$ and $N = 8$, where the causal Hanning window function is given by

$$a_H(n) = 0.5(1 - \cos 2n\pi/N) \qquad n = 0, 1, \ldots, N - 1 \tag{8.49}$$

TABLE 8.1
OVERLAP CORRELATION COEFFICIENTS $c(r)$

r	Rectangular	Hanning	Hamming	Bartlett	Riesz	Blackman-Harris
0.0625	0.0625	0.000008	0.001311	0.000481	0.001128	0.000000
0.1250	0.1250	0.000257	0.004530	0.003891	0.008542	0.000003
0.1875	0.1875	0.001878	0.012018	0.013159	0.026964	0.000034
0.2500	0.2500	0.007512	0.026856	0.031216	0.059513	0.000241
0.3125	0.3125	0.021430	0.052798	0.060990	0.107820	0.001202
0.3750	0.3750	0.049072	0.093846	0.105410	0.172140	0.004607
0.4375	0.4375	0.096028	0.153490	0.167410	0.251470	0.014350
0.5000	0.5000	0.166667	0.233770	0.249910	0.343670	0.037602
0.5625	0.5625	0.262680	0.334310	0.353920	0.445570	0.084864
0.6250	0.6250	0.381890	0.451670	0.472580	0.553060	0.167720
0.6875	0.6875	0.517680	0.579090	0.597100	0.661260	0.293750
0.7500	0.7500	0.659150	0.706910	0.718690	0.764590	0.459990
0.8125	0.8125	0.792350	0.823630	0.828570	0.856900	0.648130
0.8750	0.8750	0.902110	0.917490	0.917930	0.931580	0.825500
0.9375	0.9375	0.974620	0.978420	0.978010	0.981670	0.953330

We then obtain from Eq. 8.48

$$c(0.5) = \frac{0.25 \sum_{n=0}^{3}(1 - \cos n\pi/4)[1 - \cos(n + 4)\pi/4]}{0.25 \sum_{n=0}^{7}(1 - \cos n\pi/4)^2}$$

$$= \frac{(0 + 0.5 + 1.0 + 0.5)}{(0 + 0.08578 + 1.0 + 2.9142 + 4 + 2.9142 + 1.0 + 0.08578)} = \frac{1}{6}$$

The correlation coefficients for the rectangular, Hanning, Hamming, Bartlett (triangular), Riesz, and Blackman-Harris (four-term, -92 dB) windows are given in Table 8.1.

Consider the case where a record length $N = FT_i$ is required to perform spectral analysis with postdetection averaging of overlapped sections. We now define the overlap processing gain (OPG) by

$$\text{OPG} = \frac{K_{\text{EFF}}}{K_{\text{IND}}} \tag{8.50}$$

It can be seen that the OPG is independent of system parameters, that is, ΔFT_i, and can be used for estimating system SNR performance. Figure 8.6 presents plots of OPG versus fractional overlap for various window functions. Also, Table 8.2 gives the optimum fractional overlap requirements as a function of window function. Therefore, for good performing windows (see Section 8.2.2) we can realize variance reductions for overlapped processing of up to 0.75 overlap; however, for $r > 0.75$ we are merely consuming processor time. For a further discussion on optimal window functions refer to the IEEE paper published by Harris.[6]

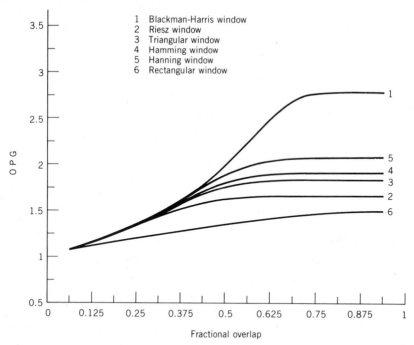

Figure 8.6 Overlap processing gain versus fractional overlap.

TABLE 8.2
OPG VERSUS OPTIMUM FRACTIONAL OVERLAP

Window	OPG	r
Bartlett	1.85	2/3
Hann	2.08	2/3
Hamming	1.88	2/3
Riesz	1.66	3/4
Blackman-Harris (four-term, -92 dB)	2.81	3/4

Example 8.6

Determine the effective number of independent samples and overlap processing gain for a narrowband signal processor given the following system parameters.

$$M = 4096 \text{ (FFT length)}$$

$$F = 1024 \text{ Hz (sampling frequency)}$$

$$T_i = 300 \text{ sec (integration time)}$$

$$r = 0.75 \text{ (fractional overlap)}$$

Assume a Hamming window function will be used to reduce spectral leakage.

Solution:

From Eq. 8.46 and Table 8.1 we obtain

$$K_{EFF} = \frac{297}{1 + 2\left[c^2(0.75) + c^2(0.5) + c^2(0.25)\right] - \dfrac{2}{297}\left[c^2(0.75) + 2c^2(0.5) + 3c^2(0.25)\right]}$$

where from Eq. 8.35, $K = 297$, and from Table 8.1 the correlation coefficients are

$$c(0.75) = 0.706910$$
$$c(0.5) = 0.233770$$
$$c(0.25) = 0.026856$$

Solving we obtain

$$K_{EFF} = \frac{297}{1 + 2(0.5551) - \dfrac{2}{297}(0.6111)} = 141.02 \approx 141$$

Since $\Delta FT_i = 75 > 10$, we could have neglected the negative term, that is

$$K_{EFF} \approx 140.75 \approx 141$$

We can now determine OPG from Eq. 8.50

$$OPG = \frac{141}{75} = 1.88$$

which could have been obtained from Figure 8.6 or Table 8.2. ∎

8.3 DETECTION PERFORMANCE ANALYSIS

This section presents a brief discussion of some of the basic concepts of detection theory and shows how this theory relates to estimating the performance requirements of digital systems. For a comprehensive discussion of this subject, refer to references 4, 7, and 8.

Consider the detection problem where the received signal has been observed for some period of time. The signal at the receiver is assumed to contain sinusoids buried in white Gaussian noise. The detection process consists of choosing between the signal-plus-noise hypothesis (H_1) or the noise-only hypothesis (H_0). At the receiver the analog signal is processed by a presampling filter, analog-to-digital converter, and spectral analyzer using the FFT algorithm. Following the FFT processing, frequency weighting, magnitude detection, and integration of the sections are performed (see Fig. 8.7). At

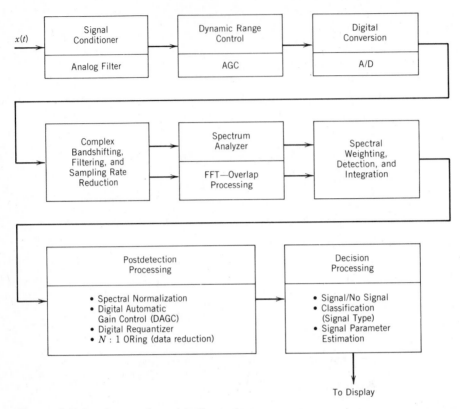

Figure 8.7 Spectrum analyzer detection system.

the output of the integrator, the probability densities on the two hypotheses are as shown in Figure 8.8.

The *probability of false alarm* (P_{fa}) equals the area under the noise-only density function to the right of the detection threshold setting, α. Thus,

$$P_{fa} = \int_{\alpha}^{\infty} P_0(x)\, dx \qquad (8.51)$$

The *probability of detection* (P_d) equals the area under the signal-plus-noise density function to the right of the threshold setting, that is

$$P_d = \int_{\alpha}^{\infty} P_1(x)\, dx \qquad (8.52)$$

The purpose of the detection process is to decide whether a signal is present or absent subject to some criterion. It is also of interest to consider the signal to be detected with no predetermined or a priori probability of occurrence. For this case the criterion that we use is known as the *Neyman-Pearson criterion* for optimum detection.[8,11] This criterion is based on specifying a fixed

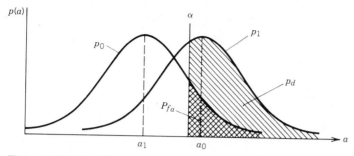

Figure 8.8 Probability density function distributions of noise and signal plus noise.

threshold, α, in accordance with Eq. 8.51. The probability of detection is then maximized with the given probability of false alarm constraint. The solution to this problem is determined by the likelihood-ratio test, which is the ratio of the signal-plus-noise and noise-only probability density functions, that is

$$\frac{P_1(x)}{P_0(x)} \geq \beta \tag{8.53}$$

where β is a function of the threshold setting α, and in the Neyman-Pearson sense the selection of the threshold will depend on the probability of false alarm. This ratio may be used to decide whether or not the signal is present. The greater the likelihood ratio, the more probable it is that the receiver input is due to signal-plus-noise rather than to noise alone.

The values of P_{fa} and P_d in the form of receiver operation characteristic (ROC) curves completely specify the receiver performance. Typical specifications for P_d and P_{fa} are 0.9 and 0.000001, respectively. These values are specified after integrating for T_i seconds, which normally ranges from 300 to 600 seconds. Performance of a receiver that uses a linear or square-law detector can be determined by using (ROC) curves. These curves express the probability of detection versus the probability of false alarm with signal-to-noise ratio as a parameter.

8.3.1 Linear Envelope Detector

The signal-to-noise ratio (SNR) required at the detector input to satisfy the P_d, P_{fa}, and T_i specifications has been calculated for a linear envelope detector by Robertson.[7,9] The linear-detection ROC curves relate P_d versus P_{fa} for a family of SNRs in 1-dB increments and for integration times from 1 independent sample to 8192 independent samples. The relationship between the effective number of independent samples and T_i is given by Eqs. 8.36 and 8.50, that is

$$K_{\text{EFF}} = \text{OPG}\, K_{\text{IND}} \tag{8.54}$$

Robertson's linear-detection characteristic curves have been expressed in a closed-form function by Albersheim,[10] whose accuracy is within 0.2 dB over a wide range of the parameters. Expressed in decibels, the solution for the signal-to-noise ratio at the input to the detector, whereby a sinusoid is detectable in the presence of random noise, is

$$\text{SNR}_{DI} = -5 \log_{10} K_{\text{EFF}}$$
$$+ \left(6.2 + \frac{4.54}{\sqrt{(K_{\text{EFF}} + 0.44)}} \right) \log_{10} (A + 0.12 AB + 1.7B)$$

where

$$A = \ln \left(\frac{0.62}{P_{fa}} \right)$$

$$B = \ln \left(\frac{P_d}{1 - P_d} \right) \tag{8.55}$$

Using Eq. 8.55, the SNR at the input to the detection process can be easily computed for the desired P_d, P_{fa}, and K_{EFF}. The equation can be programmed on a personal computer and combined with other performance calculations to determine system performance. Figure 8.9 is a plot of SNR versus the number of independent samples for a $P_d = 0.5$ and a $P_{fa} = 0.0001$. Note that as the number of samples is increased by a factor of two, the SNR required is reduced by approximately 1.5 dB. Given the ROC performance

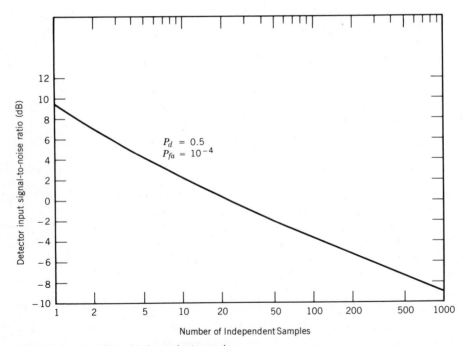

Figure 8.9 Number of independent samples

data, the system performance can be quantified to assure that requirements are met.

8.3.2 System Performance: Narrowband Spectral Analysis Detection System

Narrowband detection processing performance can be estimated by determining the *minimum detectable SNR (MDSNR)* in a 1-Hz band at the input to the system. Thus, MDSNR expressed in decibels is given by

$$\text{MDSNR} = \text{SNR}_{DI} + 10\log_{10} BW_a + SL + PL \qquad (8.56)$$

where

$$BW_a = k_w \Delta F \qquad (8.57)$$

is referred to as the FFT analysis bandwidth. The factor k_w is referred to as the equivalent noise bandwidth factor (see Section 6.6.4 and Table 6.2), and $\Delta F = F/N$ is the FFT bin width. The parameter SL is the FFT scallop loss (see Table 6.2), and PL includes typical signal-processing losses such as

RC Integrator versus linear averager

A/D noise

Filtering noise

FFT noise

Analog noise

ORing

Requantization

As an example of a typical narrowband performance calculation, consider the following system parameters.

FFT window function	= Hanning
FFT section fractional overlap (r)	= 0.5
Sampling frequency (F)	= 80 Hz
FFT length (M)	= 512
Integration time (T_i)	= 300 sec
Probability of false alarm (P_{fa})	= 0.0001
Probability of detection (P_d)	= 0.5
Processing losses (PL)	= 0 dB

The solution is

k_w	= 1.5
SL	= 1.42 dB
OPG	= 1.9
K_{EFF}	= 87.9
SNR_{DI}	= -3.4 dB
MDSNR (per Hz)	= -8.3 dB

8.3.3 Narrowband Spectral Processing

A simple example of the gain in discrete frequency detection performance resulting from averaging (that is, integrating) FFT output bins over time was presented in Section 1.1.3. For this example, a sinusoidal signal buried in noise was analyzed via an FFT spectrum analyzer. The complex spectral outputs were detected (linear envelope) and linearly integrated for M samples. This process is illustrated in Figure 1.11, where a sinusoidal signal is placed exactly at the center of FFT bin 8 in order to avoid the effects of spectral leakage across bins. As a result, spectral weighting was not necessary to reduce sidelobe leakage effects across FFT bins. The simulation was performed for $M = 1$, 4, and 16 integrated samples for each bin. As illustrated in Figure 1.11b, for $M = 1$ the detector output represents the system output for no averaging. Note the large variations across the detected frequency outputs. The effect of averaging is shown in Figure 1.11d and e. After integrating four FFT outputs per bin the variation has decreased significantly and some potential peaks are discernable. Following the 16-sample integrated output, a single peak is easily detected in Figure 1.11e. Also note the symmetry in the

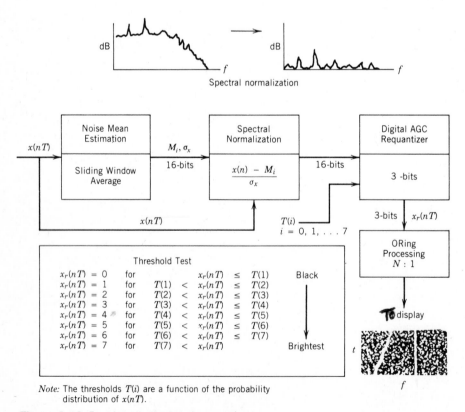

Figure 8.10 Postdetection spectral processing.

upper and lower half of the outputs. This is because a real input signal was used; therefore the output folds around half the sampling rate.

This example was used to illustrate the key concepts of spectral processing to detect discrete frequency components imbedded in Gaussian noise. You should experiment with similar simulations to gain insight into the effects of varying signal amplitudes, signal frequencies on and off FFT bins, FFT length variations, variance reduction associated with section overlap processing, and weighting of the FFT output using data or spectral windows.

To aid the decision process, the detected frequency bin amplitudes are further processed, as shown in Figure 8.10. Spectral noise normalization determines a time-varying estimate of the background noise. This processing is employed to estimate spectral shapes and slopes, which can be used to reduce system dynamic range requirements. Digital automatic gain control and requantization is employed for display formatting; for example, as shown in Figure 8.10 the three-bit output of the automatic gain control (AGC) requantizer is formatted to eight shades of gray. ORing (peak-picking) processing is included to reduce the amount of data to be displayed, and finally, display marking to 8 shades of gray is achieved by performing automatic detection threshold tests.

Figure 8.10 shows the displayed spectral data, where the FFT bin outputs are shown amplitude-modulated by threshold processing. The spectrogram illustrates the detection of two time-varying tonals, which are indicated by the bright frequency-time trace[13] in a noise background.

8.4 SUMMARY

In this chapter we discussed the response of linear systems to random signals. In particular, we showed that if a discrete-time random process is applied as an input to a discrete-time linear system the output is also a random process. The input/output relationship was developed in terms of autocorrelation functions and power spectral densities.

Estimation of the sampled PSD was considered, where the method of averaging overlapped modified periodograms was considered to provide a substantial reduction in variance over averaging contiguous data sections. As shown, although the overlapped periodograms are more statistically dependent, the increased number of sections averaged results in an improved estimate.

Finally, the detection of signals in noise was considered. A narrowband detection system was used to illustrate the concepts of the signal detection process.

APPENDIX 8.A
VARIANCE OF THE SAMPLE MEAN

It is frequently of interest to determine the sample mean and the variance of the sample mean for statistically independent and dependent random vari-

Figure 8.A.1 Computation of the sample mean of a random sample of length N.

ables. Since the problem of estimating the sample mean for a random variable is essentially identical to that of estimating the sample mean of a wide-sense stationary random processes, we consider both cases together.

Independent Discrete Samples. Suppose N measurements are obtained of a sample function from a random process. Let $\{x_n\}$, $n = 1, 2, \ldots, N$ be the value of the nth random sample sequence. We further assume that the random sample sequences are statistically independent. The mean value of the process can be estimated as shown in Figure 8.A.1

The sample mean of size N of a random process is, therefore, an estimate of the true mean m_x of the random process. As the number of measurements becomes large, then the expected value of the sample mean approaches the true mean, that is

$$E[\hat{m}_x] \approx E\left[\frac{1}{N}\sum_{n=1}^{N} x_n\right] = \frac{1}{N}\sum_{n=1}^{N} E[x_n] = m_x \qquad (8.A.1)$$

Therefore, if the expected value of the sample mean is equal to the expected value of the random process (true mean), the bias is equal to zero. The sample mean is therefore an unbiased estimator for the mean.

The expected value of the variance of the sample mean is given by

$$\sigma_y^2 = \text{var}[m_x] = E[\hat{m}_x^2] - (E[\hat{m}_x])^2 \qquad (8.A.2)$$

where

$$E[\hat{m}_x^2] = E\left[\left(\frac{1}{N}\sum_{n=1}^{N} x_n\right)^2\right] = \frac{1}{N^2} E\left[\sum_{n=1}^{N}\sum_{m=1}^{N} x_n x_m\right] = \frac{1}{N^2}\sum_{n=1}^{N}\sum_{m=1}^{N} E[x_n x_m] \qquad (8.A.3)$$

since x_n and x_m are independent random variables we obtain

$$E[x_n x_m] = E[x_n]E[x_m] = m_x^2 \qquad \text{for } n \neq m$$
$$= E[x_n^2] = \sigma_x^2 + m_x^2 \qquad \text{for } n = m \qquad (8.A.4)$$

Finally, the expected value of the second moment of the sample mean is

$$E[\hat{m}_x^2] = \frac{1}{N^2}\sum_{\substack{n=1 \\ n \neq m}}^{N}\sum_{m=1}^{N} m^2 + \frac{1}{N^2}\sum_{n=1}^{N} \sigma_x^2 = m_x^2 + \frac{\sigma_x^2}{N} \qquad (8.A.5)$$

Thus, the variance of the sample mean is given by

$$\text{var}\left[\hat{m}_x\right] = m_x^2 + \frac{\sigma_x^2}{N} - m_x^2 = \frac{\sigma_x^2}{N} \qquad (8.\text{A}.6)$$

We can see that as the number of measurements increases, the variance of the sample mean decreases, and since the bias is zero, the sample mean is a consistent estimator.

Dependent Discrete Samples. In digital signal processing it is usually necessary to consider the case where the analysis will include sample means of dependent samples. Separating the double summation of Eq. 8.A.3 into two terms we obtain

$$E\left[\hat{m}_x^2\right] = \frac{1}{N}\left(\sum_{n=1}^{N} E\left[x_n^2\right] + \sum_{\substack{n=1 \\ n \neq m}}^{N}\sum_{m=1}^{N} E\left[x_n x_m\right]\right) \qquad (8.\text{A}.7)$$

For a stationary process the expected value expression in the second term on the right side of Eq. 8.A.7 may be expressed as the correlation between the nth and mth samples of the sequence. This correlation is simply the acf of the process, that is,

$$E\left[x_n x_m\right] = \phi\left(\tau_{nm}\right) = \phi\left(lT\right) \qquad (8.\text{A}.8)$$

where we will define

$$\tau_{nm} = (n - m)T = lT \qquad (8.\text{A}.9)$$

Let us now examine the double summation in Eq. 8.A.7., for $N = 4$ we can see that

$$\sum_{\substack{n=1 \\ n \neq m}}^{4}\sum_{m=1}^{4} \phi(|n - m|) = \begin{array}{l} \phi(1) + \phi(2) + \phi(3) \\ +\phi(1) + \phi(1) + \phi(2) \\ +\phi(2) + \phi(1) + \phi(1) \\ +\phi(3) + \phi(2) + \phi(1) \end{array}$$
$$= 6\phi(1) + 4\phi(2) + 2\phi(3) \qquad (8.\text{A}.10)$$

or

$$2\sum_{l=1}^{3}(4 - l)\phi(l) = 2[3\phi(1) + 2\phi(2) + \phi(3)] \qquad (8.\text{A}.11)$$

Therefore, we can deduce

$$\sum_{\substack{n=1 \\ n \neq m}}^{N}\sum_{m=1}^{N} E\left[x_n x_m\right] = \sum_{l=1}^{N-1} 2(N - l)\phi(lT) \qquad (8.\text{A}.12)$$

In general it can be seen there are $N(N-1)$ terms; that is, for this case two terms of $\phi[(N-1)T]$, four terms of $\phi[(N-2)T]$, and $2(N-1)$ or six terms of $\phi(T)$.

It is helpful to consider a 4-by-4 array of correlation elements making up the double sum of Eq. 8.A.3, that is

$$\begin{bmatrix} E[x_1 x_1] & E[x_1 x_2] & E[x_1 x_3] & E[x_1 x_4] \\ E[x_2 x_1] & E[x_2 x_2] & E[x_2 x_3] & E[x_2 x_4] \\ E[x_3 x_1] & E[x_3 x_2] & E[x_3 x_3] & E[x_3 x_4] \\ E[x_4 x_1] & E[x_4 x_2] & E[x_4 x_3] & E[x_4 x_4] \end{bmatrix} \qquad (8.A.13)$$

From Eq. 8.A.13 the following can be observed: First, we note that

$$E[x_n x_m] = E[x_m x_n] \qquad (8.A.14)$$

That is, every term above the principle diagonal has an equal counterpart below that diagonal. Further, since the process is stationary, each element $E[x_n x_m]$ is equal to an element $E[x_{n+j} x_{m+j}]$, and hence all the $(N-k)$ elements in the kth diagonal above the principle diagonal are equal. Finally, each of the principle diagonal terms $E[x_n x_n] = \sigma_x^2$.

Proceeding with the derivation, we note that the variance of the sample mean is given by

$$\sigma_x^2 = E[x_n^2] - m_x^2 \qquad (8.A.15)$$

Substituting Eq. 8.A.12 into Eq. 8.A.7, we obtain

$$E[\hat{m}_x^2] = \frac{\sigma_x^2 + m_x^2}{N} + \sum_{l=1}^{N-1} \frac{2(N-l)}{N^2} \phi(lT) \qquad (8.A.16)$$

From Eq. 8.A.15 the variance of the sample mean can be expressed by

$$\sigma_x^2 = \frac{\sigma_x^2}{N} + \frac{(1-N)m_x^2}{N} + \frac{2}{N^2} \sum_{l=1}^{N-1} (N-l)\phi(lT) \qquad (8.A.17)$$

Moving the m_x^2 inside the summation we obtain

$$\sigma_y^2 = \frac{\sigma_x^2}{N} + \frac{2}{N^2} \sum_{l=1}^{N-1} (N-l)[\phi(lT) - m_x^2] \qquad (8.A.18)$$

where we used the identity

$$\frac{(1-N)}{N} m_x^2 = -\frac{2}{N^2} \sum_{l=1}^{N-1} (N-l)m_x^2 \qquad (8.A.19)$$

For independent input samples, Eq. 8.A.18 reduces to

$$\sigma_y^2 = \frac{\sigma_x^2}{N} \qquad (8.A.20)$$

since $\phi(lT) = E[x_n x_m] = E[x_n]E[x_m] = m_x^2$, Eq. 8.A.20 is equivalent to Eq. 8.A.6.

In Eq. 8.A.18 m_x, σ_x^2 are the mean and variance in the input random process, and the correlation function of the input at time displacement (of data sequences) ΔT is given by $\phi(l\Delta T)$. Let us now define the normalized autocorrelation function for a stationary process by

$$c^2(lT) = \frac{\phi(lT) - m_x^2}{\sigma_x^2} \qquad (8.A.21)$$

Expressing Eq. 8.A.21 as

$$\sigma_x^2 c^2(lT) = \phi(lT) - m_x^2 \qquad (8.A.22)$$

and substituting into Eq. 8.A.18 we obtain the final result

$$\sigma_y^2 = \frac{\sigma_x^2}{N}\left[1 + \frac{2}{N}\sum_{l=1}^{N-1}(N-l)c^2(lT)\right] \qquad (8.A.23)$$

which is equivalent to Eq. 8.44 for the case where $\Delta F T_i = 1$.

The function $c^2(lT)$ represents the correlation coefficients of the window function. We can see that for $c^2(lT) = 0$, Eq. 8.A.23 reduces to Eq. 8.A.20.

References

1. A. V. Oppenheim and R. W. Schafer, *Digital Signal Processing*, Prentice–Hall, New York, 1975.

2. P. D. Welch, "The Use of Fast Fourier Transform for the Estimation of Power Spectra," *IEEE Trans. Audio Electroacoustics*, AU-15: 70–73, June 1967.

3. G. M. Jenkins and D. G. Watts, *Spectral Analysis and Its Applications*, Holden–Day, San Francisco, 1969.

4. M. Schwartz and L. Shaw, *Signal Processing: Discrete Spectral Analysis Detection and Estimation*, McGraw–Hill, New York, 1975.

5. S. A. Tretter, *Discrete-Time Signal Processing*, Wiley, New York, 1976.

6. F. J. Harris, "On the Use of Windows for Harmonic Analysis with the Discrete Fourier Transform," *Proc. IEEE*, 66: 51–83, January 1978.

7. A. D. Whalen, *Detection of Signals in Noise*, Academic Press, New York, 1971.

8. H. L. Van Trees, *Detection, Estimation, and Linear Modulation Theory*, Wiley, New York, 1968.

9. G. H. Robertson, "Operating Characteristics for a Linear Detector of CW Signals in Narrowband Gaussian Noise," *Bell Syst. Tech. J.*, No. 4: 755–774, 1967.

10. W. J. Alberhseim, "A Closed-form Approximation to Robertson's Detection Characteristics," *Proc. IEEE*, 69: 839, July 1981.

11. M. I. Skolnik, *Introduction to Radar Systems*, McGraw–Hill, New York, 1962.

12. M. Abramowitz and I. A. Stegum, *Handbook of Mathematical Functions*, Dover, New York, 1964.

13. W. C. Knight, R. G. Pridham, S. M. Kay, "Digital Signal Processing for Sonar, *Proc. IEEE*, 69: 1451–1506, November 1981.

PROBLEMS

8.1.1 Show that the following equalities are true.

a. $\gamma_{xx}(m) = \phi_{xx}(m) - m_x^2$

b. $\gamma_{xy}(m) = \phi_{xy}(m) - m_x m_y$

c. $\phi_{xx}(0) = E[x(n)^2]$

d. $\gamma_{xx}(0) = \sigma_x^2$

e. $\phi_{xx}(m) = \phi_{xx}(-m)$

f. $\gamma_{xx}(m) = \gamma_{xx}(-m)$

g. $\phi_{xy}(m) = \phi_{yx}(-m)$

h. $\gamma_{xy}(m) = \gamma_{yx}(-m)$

8.1.2 A random process is given by the sum of two independent random processes, that is,

$$w(n) = x(n) + y(n)$$

If the acf of $x(n)$ and $y(n)$ are given by

$$\phi_{xx} = \alpha^{|n|} \text{ and } \phi_{yy} = \beta^{|n|}$$

determine the acf $\phi_{ww}(m)$ of $w(m)$. Assume $m_x = m_y = 0$.

8.1.3 Determine the acf of the output of a discrete linear system if the impulse response of the system is given by

$$h(n) = \begin{cases} \alpha^n & \text{for } n \geq 0, \quad \text{and } 0 < \alpha < 1 \\ 0 & \text{for } n < 0 \end{cases}$$

and the input to the system is a white-noise sequence with zero mean and variance σ_x^2.

8.1.4 Determine the ccf between the input and output of Problem 8.1.3.

8.1.5 The ccf of the input and output of a discrete-time linear system is given by Eq. 8.4. Show that the Z-transform of Eq. 8.4 is given by Eq. 8.11.

8.1.6 Determine the PSD of a random process whose acf is given by

$$\phi_{xx}(m) = \alpha^{|m|} \quad 0 < \alpha < 1, \quad \text{and} \quad -\infty \leq m \leq \infty$$

8.1.7 A white-noise process $x(n)$ is passed through a filter whose transfer function is given by

$$H(z) = \frac{1}{1 - az^{-1}} \quad 0 < a < 1$$

From Eq. 8.5 find the acf of the output, that is, $\phi_{yy}(m)$, and determine the power in the passband. Also find the ccf, $\phi_{xy}(m)$, between the input and output processes.

8.1.8 The Z-transform of $\phi_{xx}(m)$ is given by

$$\Phi_{xx}(z) = \frac{1}{(1 - az^{-1})(1 - az)(1 - bz^{-1})(1 - bz)}$$

where $0 < a < 1$ and $0 < b < 1$. Find $\phi_{xx}(m)$.

8.1.9 Show that the CPSD given by Eq. 8.13 can also be expressed by

$$\Phi_{vw}(z) = H_2(z^{-1})\Phi_{vy}(z)$$

8.1.10 Verify all steps of Example 8.5.

8.1.11 The PSD of a random process given by

$$\Phi_{xx} = (1 - z^{-1})(1 - z)$$

is passed through a digital filter whose transfer function is

$$H(z) = \frac{1}{1 - 0.9z^{-1}}$$

If the output random signal is $y(n)$, find the output PSD $\Phi_{yy}(z)$ and the acf $\phi_{yy}(n)$.

8.1.12 A signal $x(nT)$ is corrupted by additive noise $v(nT)$. To enhance the signal-to-noise ratio (SNR) the observed signal

$$y(nT) = x(nT) + v(nT)$$

is processed by a lowpass digital filter with transfer function

$$H(z) = \frac{1}{1 - 0.8z^{-1}}$$

Let the sampled PSD of the signal and noise be given by

$$\Phi_{xx}(z) = \frac{1}{(1 - 0.9z^{-1})(1 - 0.9z)}$$

$$\Phi_{vv}(z) = \sigma_v^2$$

Since the noise and signal are independent we have

$$\Phi_{xv}(z) = 0$$

a. Find the signal-to-noise ratio at the filter input, that is,

$$\text{SNR}_i = \frac{E[x^2(nT)]}{E[v^2(nT)]}$$

b. Let the component of the filter output resulting from the signal $x(nT)$ be $w_x(nT)$, and the component resulting from the noise $v(nT)$ be $w_v(nT)$. Find the output SNR

$$SNR_0 = \frac{E[w_x^2(nT)]}{E[w_v^2(nT)]}$$

c. Determine the SNR improvement

$$E = \frac{SNR_0}{SNR_i}$$

8.1.13 Derive Eq. 8.11.

8.2.1 A measurement of N samples of a stationary zero mean random process are to be used to estimate its acf. Show that $c_{xx}(0) = \sigma_x^2$.

8.2.2 Using the results of Problem 8.2.1, deduce Eq. 8.18.

8.2.3 Derive Eq. 8.42. *Hint*: Take the expected value of both sides of Eq. 8.38 and make use of the fact that if $x(n)$ is a zero mean process, then

$$\phi_{xx}(m) = \frac{1}{F} \int_{-F/2}^{F/2} P_{xx}(f) e^{j2\pi fmT} df$$

8.2.4 The Fourier coefficients of two signals are given by

$$S_x(f) = A_x(f) + jB_x(f)$$
$$S_y(f) = A_y(f) + jB_y(f)$$

show the following:

a. The autopower spectrum $P_{xx}(f)$ is purely real and positive, and phase information is not preserved.

b. The cross-power spectrum $P_{yx}(f)$ can assume both positive and negative values, and the relative phase between the signals is preserved.

c. Show that

$$H(f) = \frac{P_{yx}(f)}{P_{xx}(f)}$$

8.2.5 *Ensemble Averaging:* In the real world, signals are usually corrupted by noise. Noise, however, is random and independent of the signals of interest. If repeated time or frequency records of noisy signals are averaged in the limit, the noise will average to zero, thereby enhancing the desired signal component. In this problem we investigate the effect averaging has on computational noise in calculating the transfer function magnitude.

a. Auto-Spectra Transfer Function: From Figure P8.2.5 for $S_N(f) = 0$ we have

$$S_y(f) = S_x(f)H(f)$$

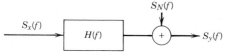

Figure P8.2.5 Coherence function measurement.

or in terms of the power spectrum

$$P_{yy}(f) = P_{xx}(f)|H(f)|^2$$

Otherwise with noise

$$S_y(f) = S_x(f)H(f) + S_N(f)$$

Determine $P_{yy}(f)$ and $|H(f)|^2$ in the presence of internal system noise.

b. Cross Spectra Transfer Function: For $S_N(f) = 0$, we have

$$P_{yx}(f) = S_y(f)S_x(f)$$

and from Eq. 8.12

$$H(f) = \frac{P_{yx}(f)}{P_{xx}(f)}$$

Determine $P_{yx}(f)$ and $|H(f)|^2$ in the presence of internal system noise, and comment on the result obtained in part a.

HINT: Note that the term $S_N(f)S_x^*(f)$ (* denotes complex conjugate) approaches zero as the number of averages becomes large.

8.2.6 *Coherence Function:* For the determination of the cross-power spectrum of system outputs and inputs, the relationship between input and output is not sufficient information to determine an accurate transfer function. In addition, we must know whether or not the system output was totally caused by the system input. That is, noise and/or nonlinear distortion can cause large outputs, and therefore large values of the cross-spectrum, at various frequencies, and thus introduce errors in calculating the transfer function. The coherence function used to determine system causality is defined as

$$C_{yx}^2(f) = \frac{|P_{yx}(f)|^2}{P_{xx}(f)P_{yy}(f)} \qquad 0 \le C_{yx}^2(f) \le 1$$

If the coherence is 1, the system has perfect causality. Low coherence at given frequencies indicates that the transfer function has inaccuracies at those frequencies.

a. Show that for the zero-system noise and/or nonlinear distortion the coherence function is equal to unity.

b. Show that the coherence function for the system given in Figure P8.2.5 is given by

$$C_{yx}^2(f) = \frac{|H(f)|^2 \bar{P}_{xx}(f)}{|H(f)|^2 \bar{P}_{xx}(f) + \bar{P}_{NN}(f)}$$

where $\bar{P}_{xx}(f)$ and $\bar{P}_{NN}(f)$ denotes the average PSD of the input signal and noise source, respectively.

HINT: For ensemble averaging of frequency records, terms of the form $S_N S_x^*$ become negligible compared to $S_N S_N^*$; that is, if the sources that make up the noise source S_N are uncorrelated with the noise sources at the input to the system, these terms will average to zero.

8.2.7   Write a computer program to compute the correlation coefficients for the Hanning, Hamming, and Blackman-Harris (four-term, -92 dB) data windows given by Eqs. 6.17, 6.19, and Table 6.3, respectively. Determine the correlation coefficients for fractional overlaps of 0.125, 0.25, 0.375, 0.5, 0.675, 0.625, 0.75, and 0.875. It should be noted that the accuracy obtained is a function of the number of terms N used in evaluating Eq. 8.48. Therefore, it is recommended that N be increased (e.g., in powers of two) to obtain desired accuracy. Compare your results with Table 8.1.

8.2.8   The reduction in variance of the PSD estimate for processing overlap data sections is given by Eq. 8.44. Determine the reduction in variance for $r = 0.5$, 0.75, and 0.875. Assume a Hann window function, the data record length $N = 16,384$, and the FFT length $M = 1024$.

8.2.9   Repeat Problem 8.2.8 using the Hamming window function. Let $N = 16,384$, $M = 1024$, and $r = 0.5$, 0.75, and 0.875.

8.2.10   Derive Eq. 8.21

8.3.1 a. For Gaussian noise statistics the probability density function for signal-plus-noise under hypothesis H_1 is given by

$$p_1(x) = \frac{1}{\sqrt{2\pi}\,\sigma} \exp\left[-\frac{(x_i - \mu)^2}{2\sigma_x^2}\right]$$

and for the noise-only case under hypothesis H_0

$$p_0(x) = \frac{1}{\sqrt{2\pi}\,\sigma_x} \exp\left(-\frac{x_i^2}{2\sigma_x^2}\right)$$

Determine the log-likelihood ratio, and solve for the threshold by equating the test statistic x to the threshold, that is, let $x = \alpha$.

b. Repeat part a if multiple-independent samples of the received signal are available. For this case, the joint density functions of the Gaussian process are the products of the individual densities, that is

$$p_1(x) = \left(\frac{1}{2\pi\sigma_x^2}\right)^{m/2} \prod_{i=1}^{m} \exp\left[-\frac{(x_i - \mu)^2}{2\sigma_x^2}\right]$$

$$p_0(x) = \left(\frac{1}{2\pi\sigma_x^2}\right)^{m/2} \exp\left[-\sum_{i=1}^{m} \frac{x_i^2}{2\sigma_x^2}\right]$$

Using the log-likelihood ratio, determine the threshold α.

c. For Gaussian noise the probability of false alarm can be expressed as

$$P_{fa} = \int_{\alpha/\sqrt{2}\,\sigma_n}^{\infty} \frac{e^{-x^2}}{\sqrt{\pi}}\, dx = 0.5\left(1 - \text{erf}\,\frac{\alpha}{\sqrt{2}\sigma_n}\right)$$

From the handbook of mathematical functions given in reference 12 determine the threshold α/σ_n for false-alarm probabilities of 10^{-1}, 10^{-2}, and 10^{-3}.

8.3.2 If Gaussian noise were passed through a narrowband filter and the filter output was processed through an envelope detector, then the envelope of the processed noise would have a Rayleigh probability density function given by

$$P_0(x) = \begin{cases} \dfrac{x}{\sigma^2} \exp\left(-\dfrac{x^2}{2\sigma^2}\right), & x \geq 0 \\ 0 & x < 0 \end{cases}$$

a. Show that the probability that the noise envelope will exceed the threshold α is given by

$$\text{Probability}\,(\alpha < x < \infty) = P_{fa} = \exp\left(-\frac{\alpha^2}{2\sigma^2}\right)$$

or

$$\frac{\alpha^2}{2\sigma^2} = \ln\left(1/P_{fa}\right)$$

b. It can be shown[11] that if a sine-wave signal with amplitude A plus Gaussian noise were passed through a narrowband filter then the probability density function at the output of the envelope detector would be given by

$$P_s(x) = \frac{x}{\sigma^2} \exp\left(-\frac{x^2 + A^2}{2\sigma^2}\right) I_0\left(\frac{xA}{\sigma^2}\right)$$

This density is referred to as a generalized Rayleigh or Rician. The function $I_0(z)$ is the modified Bessel function of zero order and argument z, which can be expressed in an asymptotic expansion.[11] Then probability of detection

$$P_d = \int_T^\infty P_s(x)\, dx$$

can be expressed as a function of the signal-to-rms noise ratio and $\ln(1/P_{fa})$ (see part a), that is,

$$P_d = 0.5[1 - \mathrm{erf}(A - B)] + \frac{\exp[-(A^2 - 2AB + B^2)]}{4\sqrt{\pi}\,B}$$

$$\times \left(1.25 - \frac{A}{4B} + \frac{1 + 2A^2 - 4AB + 2B^2}{16B^2}\right)$$

where

$$A = \left[\ln(1/P_{fa})\right]^{0.5}$$

and

$$B = \left(\frac{S}{N}\right)^{0.5}$$

$$\mathrm{erf}(z) = \frac{2}{\sqrt{\pi}} \int_0^z \exp[-u^2]\, du$$

$$z = A - B$$

Using this relationship, write a computer program that expresses the probability of detection as a function of the signal-to-noise ratio with the probability of false alarm as a parameter. The results of this program can be plotted in the form of ROC curves showing the probability of detection for a sine wave in noise. The results of the program can be compared with the ROC curves given in reference 11. An approximation formula for $\mathrm{erf}(z)$ is given in reference 12, page 299 (Eq. 7.1.26).

8.3.3 a. Write a computer program to determine the SNR at the input to the detector using Eq. 8.55. Check your results using the ROC curves of reference 7.

  b. Using the computer program developed in part a, determine the SNR at the input to the detector for a narrowband spectral analysis detection system consisting of five processing channels. The system requirements are as follows.

General System Requirements

Spectral window	Blackman-Harris (four-term, -92 dB)
Fractional overlap (r)	0.75
Integration time (T_i)	300 sec
P_{fa}	0.0001
P_d	0.5

FFT Channel Requirements

	Sampling Frequency (F) (Hz)	FFT Size
Channel 1:	819.2	512
Channel 2:	409.6	512
Channel 3:	204.8	512
Channel 4:	102.4	512
Channel 5:	51.2	512

c. Using the values of SNR_{DI} determined in part b, calculate the MDSNR for each of the five narrowband channels, where $PL = 2$ dB.

8.3.4 Repeat Problem 8.3.3 for $r = 0.5$, and compare results with solution obtained from Problem 8.3.3.

8.3.5 The output of an ORing device used for data reduction is mathematically defined as $Y = \text{MAX}(X_1, X_2, \ldots, X_N)$. Consider N statistically independent identically distributed random variables as $X_1, X_2, \ldots, X_N$, and their density and distribution functions by $f(x)$ and $F(x)$, respectively. The pdf of

$$Y = \text{MAX}(X_1, X_2, \ldots, X_N)$$

is given by

$$q_0(y_N) = Nf(y_N)[F(y_N)]^{N-1}$$

where $f(y)$ and $F(y)$ define the original distributions (see H. Lass and P. Gottlieb, *Probability and Statistics*, Addison–Wesley, Reading, Mass., pp. 301–309.) Consider the case for a Rayleigh distribution given by

$$f(y) = ye^{-y^2/2}, \quad 0 \le y \le \infty$$

determine $q_0(y_N)$ for $N = 1$ and 2. Also determine the mean and variance for the pdf's $q_0(y_1)$ and $q_0(y_2)$.

9

Finite Register Length Effects in Digital Signal Processing

DSP system design requires a thorough analysis of finite arithmetic effects to ensure that system performance is not degraded.

9.0 INTRODUCTION

Input signals encountered by sonar or radar signal processing systems are continuous and must first be sampled in time and digitized in amplitude prior to processing by digital systems. Since the hardware complexity of digital signal processing systems is directly related to the digital wordlength, it is important to limit the number of bits in the various processing elements of the system. This requires an error analysis at each point in the system to assure that the system implementation does not degrade system performance, that is, degrade dynamic range by increasing the system noise level.

It is assumed in the following sections that you are familiar with binary number representation. The following are some of the issues related to the implementation of algorithms in special-purpose hardware.

1. Analog-to-digital conversion.
2. IIR digital filter finite wordlength effects.
   a. Product-quantization errors.
   b. Coefficient-quantization errors.

c. Dynamic range considerations.

d. Zero input limit cycle behavior.

3. FIR digital filter wordlength effects.

4. FFT finite wordlength effects.

The error analysis presented herein will concentrate primarily on fixed-point arithmetic operations. The approach, however, can be extended to analyze the effects of floating-point operations.

9.1 QUANTIZATION NOISE INTRODUCED BY ANALOG-TO-DIGITAL CONVERSION

For most engineering applications, the input signals to be processed are continuous in time. Therefore, prior to discrete-time processing the continuous signals (voltage or current) must be quantized in both time and magnitude. The uniform sampling theorem has already been covered in Chapter 1, where it was shown that the continuous (analog) signal can be recovered from the digital signal if the sampling frequency of the analog-to-digital (A/D) converter is greater than twice the highest frequency content of the continuous signal, that is, $F > 2f_h$. This frequency is usually set by prefiltering the continuous signal by an analog lowpass filter with cutoff frequency equal to the specified highest frequency of interest, f_h. At the output of the analog lowpass filter the analog signal is usually processed by an automatic gain control (AGC) device. The purpose of this device is to adjust the signal level so that the digital conversion process will not result in a degradation of the signal-to-noise ratio (SNR); that is, the AGC device will automatically control the dynamic range of the input signal. The process of converting an analog signal to a digital signal is shown in Figure 9.1. For additional information on the design and analysis of A/D converters refer to references 1 and 2.

In this section we cover magnitude quantization and saturation error resulting from the A/D converter process. The A/D conversion process results in two main sources of error:

1. Quantization error: This error results from the representation of the continuous signal amplitudes by a fixed number of digital levels within a predefined maximum and minimum analog range.

2. Saturation error: This error results from the analog signal levels that exceed the predefined dynamic range of the A/D converter.

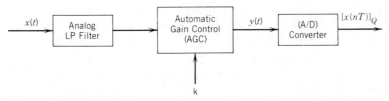

Figure 9.1 Block diagram of the analog-to-digital converter process.

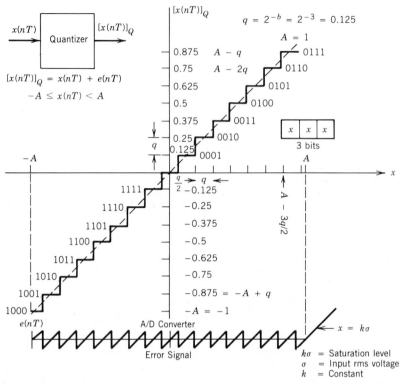

Figure 9.2 Number quantization: Two's-complement rounding for $b = 3$.

Figure 9.2 represents a three-bit A/D converter and the associated error signal. Notice that within the range of the minimum and maximum saturation levels the error is equally divided across the quantization interval, q. If the input is above the maximum or minimum saturation levels, the error is unbounded. Hence it is desirable to avoid saturation effects. Also, if the input signal voltage is within the dynamic range of the A/D converter, then only the quantization noise will represent the error source.

In digital processing of analog signals, the quantization error is commonly viewed as an additive noise signal, that is

$$[x(nT)]_Q = x(nT) + e(nT) \tag{9.1}$$

Therefore, the A/D converter output is the sum of the input signal $x(nT)$ and the roundoff error signal $e(nT)$; that is, the signal at the output of the sampler is encoded to $(b + 1)$ bits (one sign bit to the left of the binary point, and b fractional bits to the right of the binary point) to yield the quantized output sequence $[x(nT)]_Q$. Numerous simulations have shown that the error sequence $e(nT)$ can be closely approximated by a white-noise sequence uncorrelated with $x(nT)$, which is uniformly distributed over $[-q/2, q/2]$. Therefore, for a uniform random process with zero mean the variance is given by

$$\sigma_e^2 = \frac{q^2}{12} = \frac{2^{-2b}}{12} \tag{9.2}$$

where $q = 2^{-b}$ is referred to as the width of quantization. We can now define the signal-to-quantization noise ratio as

$$SNR_Q = 10 \log_{10} \left(\frac{\sigma_x^2}{\sigma_e^2} \right) \tag{9.3}$$

where σ_x^2 is the signal power, and σ_e^2 is the quantization noise power. As described earlier, the AGC device maximizes the SNR_Q by controlling the dynamic range of the input signal. The signal level at the input to the A/D converter should be made as large as possible without exceeding the maximum and minimum limits of the A/D converter. Automatic gain control can be obtained by scaling the input by a gain that is inversely proportional to the standard deviation of the input, that is

$$y(t) = \frac{x(t)}{k\sigma_x} \tag{9.4}$$

where k is the AGC constant. It is shown in Section 9.1.2 that $k\sigma_x$ is the saturation level of the A/D converter for Gaussian signals. From Eq. 9.4 the variance of $y(t)$ is given by

$$\sigma_y^2 = E[y^2(t)] = \frac{E[x^2(t)]}{k^2\sigma_x^2} = \frac{1}{k^2} \tag{9.5}$$

where we assume the mean of $x(t)$ is zero. Substituting σ_y^2 for σ_x^2 in Eq. 9.3 we obtain

$$SNR_Q = 10 \log_{10} \frac{1}{k^2\sigma_e^2} \approx 6b + 10.8 - 20 \log_{10} k \tag{9.6}$$

As an example, for $k = 4$ we obtain

$$SNR_Q \approx 6b - 1.24 \text{ dB} \tag{9.7}$$

For this case an 8-bit A/D converter (7 magnitude bits plus sign) would result in an $SNR_Q = 40.76$ dB. This indicates that the quantization noise power is approximately 41 dB below the signal and background noise power. Therefore, the noise added by the A/D converter can be considered negligible.

It should be noted that the SNR_Q is also referred to as the dynamic range (DR) of the quantizer. The DR is a measure of the ability of the quantizer to pass amplitude information, that is, the smallest level sinusoid that can be detected without being buried in the quantization noise. In Section

9.1.2 we show that the AGC constant k is a function of the number of A/D converter bits.

9.1.1 Signal Quantization Noise Due to A / D Conversion

We now demonstrate that the dynamic range of an A/D converter can be determined by the smallest input sinusoidal component discernable in the noise spectrum of the quantizer. In Section 8.1.3 we showed that the power spectrum of a white-noise process is given by

$$P_q(f) = \frac{q^2}{12} \tag{9.8}$$

where $q = 2^{-b}$ is defined as the quantization step size. Therefore, the error signal $e(nT)$ can be represented as a broadband process with PSD given by Eq. 9.8. Then the dynamic range of the quantizer for a sinusoidal signal can be expressed by

$$DR = -10 \log_{10} \frac{q^2}{12} \approx 6b + 10.8 \tag{9.9}$$

From Eq. 9.9 one can determine the number of binary bits that must be used to represent the data when the dynamic range is specified. This relationship can be verified through a simulation using an A/D with an input signal consisting of a sum of 20 sinusoids. The input signal can be expressed by

$$x(n) = \sum_{i=0}^{20} A_i \cos(2\pi f_i nT) \tag{9.10}$$

where $f_i = 60 + 50i$ Hz and $A_i = 10^{-5i/20}$. That is, the sinusoidal tones were spaced at 50 Hz intervals with the amplitude of each succeeding tone being decreased by 5 dB. The dynamic range could then be roughly measured by determining how many tones could be distinguished from the quantization noise in the spectral output. The results of this test are plotted in Figures 9.3a through d as a function of 6-, 8-, 10- and 12-bit A/D converters. By inspection, it can be seen that these results approximate Eq. 9.9.

9.1.2 Quantization and Saturation Noise Due to A / D Conversion

A detailed noise analysis of A/D converters with uniform step size, which identifies both quantization and saturation noise, has been performed by Gray and Zeoli.[3] As described in Section 9.1, quantization error results from the A/D converter process of sampling and quantizing a continuous input signal, such that all signal amplitudes lying within a given range are arbitrarily assigned the same amplitude. Saturation occurs when the input signal exceeds

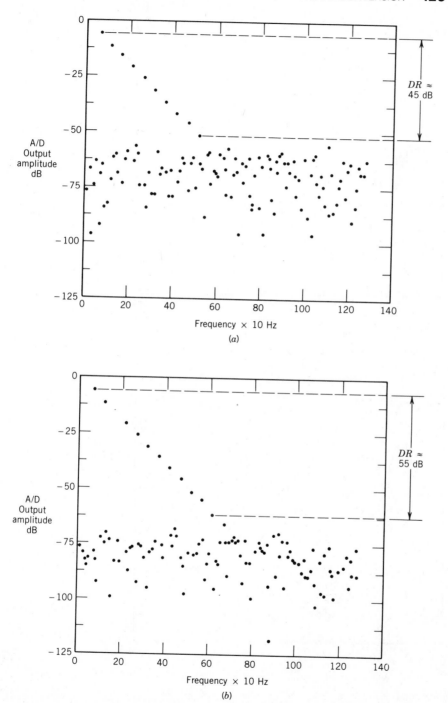

Figure 9.3a Analog-to-digital converter output spectrum for dynamic-range determination: (*a*) 6-bit spectrum; (*b*) 8-bit spectrum; (*c*) 10-bit spectrum; (*d*) 12-bit spectrum.

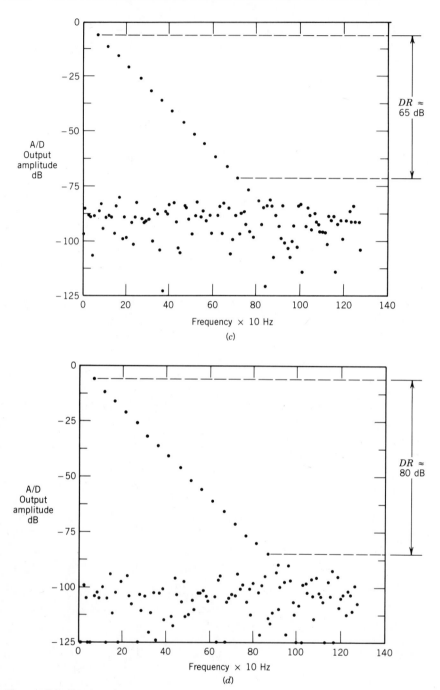

Figure 9.3 Continued

the upper bound of the voltage range of the A/D converter. The following analysis assumes that the input to the A/D converter is a Gaussian-distributed random process.

The error signal resulting from A/D converter quantization and saturation noise depend on both the number of bits and the rms input level relative to the saturation level of the converter. For a given number of bits, the signal input level can be adjusted to minimize the sum of these two noise powers. The analysis determines the optimum AGC level setting for a specified number of bits, as well as the sensitivity of the total noise power to input level.

Figure 9.2 shows the transfer function of a 3-bit A/D converter, where the saturation level is set at k times the input signal rms voltage σ_x. Gray and Zeoli have shown that the total noise power PN due to the error signal $e(nT)$ can be computed from

$$PN = 2 \int_0^\infty e^2(x) f(x) \, dx \qquad (9.11)$$

where $f(x)$ is a Gaussian probability density function. The total noise power can be separated into two components, one due to quantization (QN), and one due to saturation (SN). These terms are given by

$$QN = 2 \int_0^{k\sigma} e^2(x) f(x) \, dx \qquad (9.12)$$

and

$$SN = 2 \int_{k\sigma}^\infty e^2(x) f(x) \, dx \qquad (9.13)$$

where the error signal is uniformly distributed over the interval $-q/2$ to $q/2$, that is, $e^2(x) = q^2/12$. Using the Gaussian pdf,

$$f(x) = \frac{1}{\sigma\sqrt{2\pi}} e^{-x^2/2\sigma^2} \qquad (9.14)$$

and the fact that the quantization step size is given by

$$q = \frac{k\sigma_x}{2^b - 1} \qquad (9.15)$$

where $k\sigma_x$ is the A/D converter saturation setting and b is the number of bits exclusive of the sign bit. Equations 9.12 and 9.13 can be expressed in terms of k, $N = 2^b$, and the tabulated complement of the error function[4]

$$F(k) = \frac{1}{\sqrt{2\pi}} \int_k^\infty e^{-u^2/2} \, du \qquad (9.16)$$

where the function $F(k)$ can be approximated from[4]

$$F(x) = Z(x)[A1 * T + A2 * T^2 + A3 * T^3 + A4 * T^4 + A5 * T^5] \qquad (9.17)$$

where

$$Z(x) = e^{-x^2/2}$$

$A1 = 0.127414796$
$A2 = -0.142248368$
$A3 = 0.710706871$
$A4 = -0.726576013$
$A5 = 0.530702714$
$P = 0.2316419$ $\qquad$ (9.18)

$$T = \frac{1}{1 + P * x} \qquad (9.19)$$

Finally, the quantization and saturation noise components are given by

$$\frac{QN}{\sigma^2} = 2\left[\frac{k^2}{12(N-1)^2}[0.5 - F(k)]\right] \qquad (9.20)$$

and

$$\frac{SN}{\sigma^2} = 2\left[(k^2 + 1)F(k) - \frac{k}{\sqrt{2\pi}}e^{-k^2/2}\right] \qquad (9.21)$$

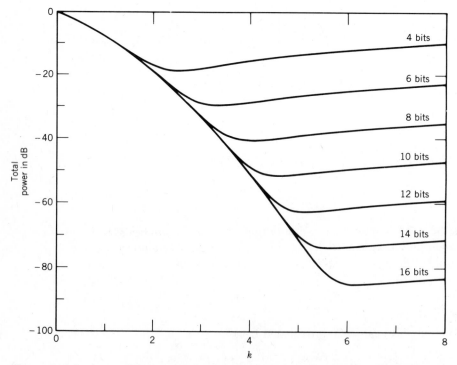

Figure 9.4 Analog-to-digital converter noise power versus AGC constant for $M = 4$ to 16 bits.

TABLE 9.1
OPTIMUM AGC CONSTANT AND NOISE POWERS AS A FUNCTION OF
A / D CONVERTER BIT SIZE M.

M	k (OPT)	Quantization Noise Power	Saturation Noise Power	Total Noise Power	Total Noise Power (db)
2	1.2	$0.9238E - 01$	$0.9549E - 01$	$0.1879E + 00$	-7.26
3	1.9	$0.3151E - 01$	$0.1543E - 01$	$0.4693E - 01$	-13.29
4	2.5	$0.1050E - 01$	$0.2399E - 02$	$0.1290E - 01$	-18.90
5	2.9	$0.3103E - 02$	$0.5912E - 03$	$0.3694E - 02$	-24.33
6	3.3	$0.9434E - 03$	$0.1283E - 03$	$0.1072E - 02$	-29.70
7	3.6	$0.2720E - 03$	$0.3766E - 04$	$0.3097E - 03$	-35.09
8	3.9	$0.7858E - 04$	$0.1039E - 04$	$0.8879E - 04$	-40.51
9	4.2	$0.2261E - 04$	$0.2720E - 05$	$0.2533E - 04$	-45.96
10	4.5	$0.6462E - 05$	$0.6805E - 06$	$0.7143E - 05$	-51.46
11	4.8	$0.1835E - 05$	$0.1641E - 06$	$0.1999E - 05$	-56.99
12	5.1	$0.5173E - 06$	$0.3812E - 07$	$0.5554E - 06$	-62.55
13	5.4	$0.1449E - 06$	$0.8510E - 08$	$0.1534E - 06$	-68.14
14	5.6	$0.3895E - 07$	$0.3042E - 08$	$0.4199E - 07$	-73.77
15	5.9	$0.1081E - 07$	$0.6188E - 09$	$0.1143E - 07$	-79.42
16	6.2	$0.2888E - 08$	$0.2059E - 09$	$0.3094E - 08$	-85.09

For a given value of N, Eqs. 9.20 and 9.21 have been evaluated and the total noise power $PN/\sigma^2 = QN/\sigma^2 + SN/\sigma^2$ has been plotted versus k. From these curves, the values of k corresponding to the minimum of PN/σ^2 can be determined for different values of M (number of magnitude bits plus sign) Figure 9.4 presents curves of A/D converter dynamic range (DR) versus the AGC constant k for 4 through 16 bits. Table 9.1 presents the optimum values of k, as well as the values of QN/σ^2, SN/σ^2, and PN/σ^2 at k_{opt} for various values of M. Figure 9.4 shows the sensitivity of the total noise power to changes in k.

Example 9.1
Consider an 8-bit (includes sign bit) A/D converter. From the noise power curves of Figure 9.4 (or Table 9.1), (a) determine the minimum total noise power and the corresponding AGC constant k; then compute the quantization noise given by

$$QN = \frac{q^2}{12} \tag{9.22}$$

where

$$q = \frac{k\sigma_x}{2^b - 1} \tag{9.23}$$

and compare the results with Table 9.1 for $M = 8$. (b) Repeat the process; however, this time set the AGC constant $k = 2$ and compare the total noise power with the calculated value of QN/σ^2.

Solution:

a. For an 8-bit A/D converter $k_{opt} = 3.9$. Then the quantization step size is given by

$$q = \frac{3.9\sigma_x}{2^7 - 1} = 0.0307087\sigma$$

Then $q^2/12 = 0.0000786\sigma^2$, and from Table 9.1 $PN/\sigma^2 = 0.00008879$. Therefore,

$$\frac{PN}{\sigma^2} \approx \frac{q^2}{12}$$

b. From Figure 9.4, for $k = 2$ the total noise power is approximately -19.5 dB, which converts to $PN/\sigma^2 = 0.01122$. We then obtain

$$q = \frac{2\sigma_x}{2^7 - 1} = 0.015748\sigma$$

$$\frac{q^2}{12} = 0.0000021\sigma^2$$

It is evident that saturation noise can have a significant effect on the total noise power.

9.1.3 Analog-to-Digital Converter Output Noise Power

Consider the output of an A/D converter as the input to a digital filter. From Eq. 9.1, the output of an A/D converter consists of two components: the unquantized input signal $x(nT)$ and the quantization error signal $e_i(nT)$. The output of a digital filter when the input signal is corrupted by A/D converter noise is shown in Figure 9.5 where

$$z(nT) = y(nT) + e_0(nT) \tag{9.24}$$

From Eq. 9.24, $y(nT)$ is the filter response to the A/D converter output signal component $x(nT)$ and the random process $e_0(nT)$ is the response of

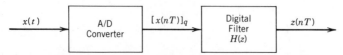

Figure 9.5 Quantization noise at the output of a digital filter.

the filter to the input error signal $e_i(nT)$. Using Parseval's theorem (see Eq. 3.32 and Problem 3.3.6), the steady-state output noise variance due to the quantization error signal is given by

$$\sigma_{e0}^2 = \sigma_e^2 \sum_{n=0}^{\infty} h^2(nT) = \frac{\sigma_e^2}{2\pi j} \oint_c H(z)H(z^{-1})z^{-1}\, dz \qquad (9.25)$$

where the closed contour of integration is around the unit circle $|z| = 1$, in which case only the poles that lie inside the unit circle are evaluated using the residue theorem.

Therefore, the steady output noise variance due to quantization noise can be obtained in the time or frequency domain. Also, for $z = e^{j2\pi fT}$, that is, the contour of integration is the unit circle, σ_{0e}^2 can be obtained by integrating the magnitude-squared function of the digital filter (see Eq. 3.32). Thus

$$\sigma_{e0}^2 = \sigma_e^2 \frac{1}{2\pi} \int_0^{2\pi} \left| H(e^{j2\pi fT}) \right|^2\, df \qquad (9.26)$$

Example 9.2

The output signal of an A/D converter is passed through a first-order lowpass filter with transfer function given by

$$H(z) = \frac{(1 - \beta)z}{z - \beta} \qquad 0 < \beta < 1$$

Find the steady-state output noise variance (power), due to quantization, at the output of the digital filter.

Solution:

From Eq. 9.25 we obtain

$$\sigma_{e0}^2 = \sigma_e^2 \frac{1}{2\pi j} \oint_c \frac{(1 - \beta)^2 z^{-1}\, dz}{(z - \beta)(z^{-1} - \beta)} = \sigma_e^2 \left[\frac{(1 - \beta)^2}{1 - \beta^2} \right]$$

where $\sigma_e^2 = 2^{-2b}/12$. Therefore, in general the output noise process can be measured in terms of the output variance given by Eq. 9.25. ∎

9.2 FINITE REGISTER LENGTH EFFECTS IN CASCADED REALIZATIONS OF IIR SECOND-ORDER SECTIONS

For fixed-point arithmetic the product of two M-bit numbers results in numbers $2M$-bits long. In digital signal processing applications, using a fixed-point processor, it is necessary to round this product to an M-bit result.

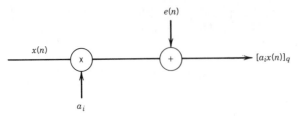

Figure 9.6 Statistical model of product quantization.

It will be shown that the analysis of product roundoff noise is similar to A/D conversion noise. However, it will be necessary to define the proper transfer function from the noise source to the filter output.[5-7]

The output of a finite wordlength multiplier can be expressed as

$$[a_i x(n)]_q = a_i x(n) + e(n) \tag{9.27}$$

where $a_i x(n)$ is the exact product and $e(n)$ is the quantization error signal. The multiplier of a fixed-point arithmetic processor can be modeled as shown in Figure 9.6.

Using the product quantization model given in Figure 9.6, consider the second-order filter sections shown in Figure 9.7. As illustrated in Figure 9.7, if product quantization is carried out using rounding arithmetic, an error signal $e_i(n)$ is added at the sum node for each multiplier. The error signal can be regarded as a random process with uniform probability density function; that is, the mean (dc value), mean square (variance or average power), and autocorrelation function are given by

$$E[e_i(n)] = 0$$

$$\sigma_e^2 = E[e_i^2(n)] = \frac{q^2}{12} = \frac{2^{-2b}}{12}$$

$$\phi_{ei}(m) = E[e_i(n)e_i(n + m)] \tag{9.28}$$

Since $e(n)$ and $e(n + m)$ are statistically independent, we obtain

$$\phi_{ei}(m) = \frac{q^2}{12}\delta(m) \tag{9.29}$$

Therefore, from Eq. 8.14 the power spectral density of the error signal can be expressed as

$$P_{ei}(f) = \frac{q^2}{12} \tag{9.30}$$

As described in Section 8.1.3, we can see that $e_i(n)$ is a white-noise process where $\sigma_e = q^2/12$. For a white-noise process the PSD is a constant at all

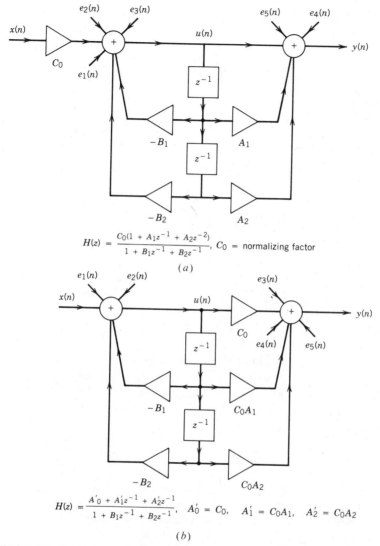

$$H(z) = \frac{C_0(1 + A_1 z^{-1} + A_2 z^{-2})}{1 + B_1 z^{-1} + B_2 z^{-1}}, \quad C_0 = \text{normalizing factor}$$

(a)

$$H(z) = \frac{A'_0 + A'_1 z^{-1} + A'_2 z^{-1}}{1 + B_1 z^{-1} + B_2 z^{-1}}, \quad A'_0 = C_0, \quad A'_1 = C_0 A_1, \quad A'_2 = C_0 A_2$$

(b)

Figure 9.7 Product quantization noise models.

frequencies, and from Eq. 9.29 the autocorrelation function is an impulse at the origin.

9.2.1 Effect of Product Quantization in Cascaded IIR Sections Using Fixed-Point Arithmetic

The effects of rounding due to multiplication in cascaded IIR second-order sections can be determined by computing the response of the filter due to the noise signal $e_k(n)$. Thus, if $h_k(m)$ is the filter's impulse response from the noise source to the filter output, the response to $e_k(n)$ is given by the convolution summation

$$e_{ok}(n) = \sum_{m=0}^{n} h_k(m) e_k(n-m) \tag{9.31}$$

assuming that samples from the same noise source are uncorrelated, and each noise source is uncorrelated with the input sequence, the variance of $e_{ok}(n)$ is given by

$$\sigma_{ok}^2(n) = \sigma_e^2 \sum_{m=0}^{n} h_k^2(m) \tag{9.32}$$

For the IIR digital filter case, if the poles are inside the unit circle in the z-plane, the impulse response approaches zero as n approaches infinity. It then follows that the steady-state noise variance can be obtained from (see Example 8.3)

$$\sigma_{ok}^2 = \sigma_e^2 \sum_{m=0}^{\infty} h_k^2(m) \tag{9.33}$$

where $\sigma_e^2 = 2^{-2b}/12$.

Now consider the filter structure of Figure 9.7. It can be seen that the output due to any steady-state noise source will be given by Eq. 9.33, where $h_k(m)$ is the impulse response from the noise source to the filter output, and σ_e^2 is the variance of the injected noise due to product quantization. The total steady-state noise variance at the output of the filter sections due to each multiplication is then given by

$$\sigma_0^2 = \sum_{i=1}^{M} \sigma_{oi}^2 \tag{9.34}$$

where M is the number of cascaded digital filter sections to be analyzed. It should be noted that any two noise sources, associated with different multipliers, are uncorrelated.

The evaluation of the steady-state output noise variance involves the infinite summation of h_k^2. As shown by Eq. 9.25, this summation can be evaluated by

$$\sum_{m=0}^{\infty} h_k^2(m) = \frac{1}{2\pi j} \oint_c H_{k,M}(z) H_{k,M}(z^{-1}) z^{-1} dz \tag{9.35}$$

where $H_{k,M}(z)$ is defined as the noise transfer function (NTF), that is, the transfer function from the noise source to the filter output. Clearly, when analyzing a system the noise transfer function depends on the structure of the digital network.

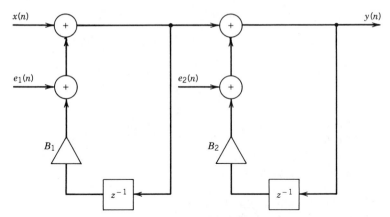

Figure 9.8 Product quantization noise generated in a cascade of two first-order sections.

The method of analysis will now be demonstrated by examining the cascade realization of the two first-order digital lowpass filters shown in Figure 9.8.

The system transfer function is given by

$$H(z) = H_1(z)H_2(z) \tag{9.36}$$

where

$$H_1(z) = \frac{1}{1 - B_1 z^{-1}} \tag{9.37}$$

and

$$H_2(z) = \frac{1}{1 - B_2 z^{-1}} \tag{9.38}$$

where $0 < B_1 < 1$ and $0 < B_2 < 1$. From the realization we note that the NTF seen by noise source $e_2(n)$ is $H_2(z)$, whereas the NTF seen by $e_1(n)$ is $H(z)$. The total steady-state noise variance can now be determined from Eqs. 9.34 and 9.35. Thus, from Eq. 9.34 we obtain

$$\sigma_{eo}^2 = \sigma_{o1}^2 + \sigma_{o2}^2 \tag{9.39}$$

where the steady-state noise power attributed to $e_{o2}(n)$ is given by

$$\sigma_{o2}^2 = \frac{\sigma_e^2}{2\pi i} \oint_c \frac{1}{(1 - B_2 z^{-1})(1 - B_2 z)} z^{-1} \, dz \tag{9.40}$$

and the NTF is $H_2(z)$. Since the noise source $e_{01}(n)$ sees the NTF $H(z)$, the steady-state noise power attributed to $e_{01}(n)$ is given by

$$\sigma_{o1}^2 = \frac{\sigma_e^2}{2\pi i} \oint_c \frac{1}{(1 - B_1 z^{-1})(1 - B_1 z)(1 - B_2 z^{-1})(1 - B_2 z)} z^{-1} \, dz \tag{9.41}$$

Since the closed contour of integration is taken to be the unit circle, only the poles within the unit circle are considered when solving the contour integral using the residue theorem. Solving Eqs. 9.40 and 9.41, we obtain from Eq. 9.39

$$\sigma_{eo}^2 = \frac{2^{-2b}}{12}\left[\frac{1}{1 - B_1^2} + \frac{(1 + B_1 B_2)}{(1 - B_1 B_2)(1 - B_1^2)(1 - B_2^2)}\right] \quad (9.42)$$

From the foregoing example it can be seen that for more complicated digital networks this approach of analyzing roundoff noise would become quite tedious. In the following sections we develop a more general approach by evaluating the NTF in the frequency domain.

9.2.2 Computation of the Steady-State Output Noise Variance for a Second-Order IIR Digital Filter Section

The Z-transform of the response function for a second-order IIR section is given by

$$Y(z) = C_0\left[X(z) + A_1 X(z)z^{-1} + A_2 X(z)z^{-2}\right]$$
$$- \left[B_1 Y(z)z^{-1} + B_2 Y(z)z^{-2}\right] \quad (9.43)$$

where C_0 is the second-order section normalization constant (see Section 4.6). Using the second-order section noise model shown in Figure 9.7a, the output power spectral density of the error signal can be derived. The first sum node response is given by

$$U(z) = E_{o1}(z) + C_0 X(z) - B_1 U(z)z^{-1} - B_2 U(z)z^{-2} \quad (9.44)$$

where $E_{o1}(z)$ is the Z-transform of the sum of the noise sources at the output of the first sum node, that is,

$$E_{o1}(z) = \sum_{i=1}^{r} E_i(z) \quad (9.45)$$

and r is the number of noise sources added to the output of the first sum node. For example, by examining Figure 9.7a three noise sources are added to the first sum node output, whereas, the noise model of Figure 9.7b shows two noise sources added to the output of the first sum node.

The second sum node response is given by

$$Y(z) = U(z) + E_{o2}(z) + A_1 U(z)z^{-1} + A_2 U(z)z^{-2} \quad (9.46)$$

where $E_{o2}(z)$ is the sum of the noise sources at the output of the second sum node, that is

$$E_{o2}(z) = \sum_{i=1}^{s} E_i(z) \quad (9.47)$$

where s is the sum of the noise sources at the output of the second sum node. Since the transfer function $U(z)$ is given by

$$U(z) = \frac{E_{o1}(z) + C_0 X(z)}{1 + B_1 z^{-1} + B_2 z^{-2}} \tag{9.48}$$

We obtain for the response at the output of a second-order section, which includes noise terms, that is

$$Y(z) = X(z)H(z) + \frac{1}{C_0} E_{o1}(z)H(z) + E_{o2}(z) \tag{9.49}$$

Thus the response of a second-order section due the product quantization is given by the last two terms of Eq. 9.49, that is,

$$Y_{on}(z) = \frac{1}{C_0} E_{o1}(z)H(z) + E_{o2}(z) \tag{9.50}$$

Consider two statistically independent discrete noise signals $e_i(n)$ and $e_j(n)$ combined to form the sum

$$f(n) = e_i(n) + e_j(n) \tag{9.51}$$

Then the autocorrelation function of $f(n)$ is given by

$$
\begin{aligned}
\phi_f(m) &= E\left\{\left[e_i(n) + e_j(n)\right]\left[e_i(n+m) + e_j(n+m)\right]\right\} \\
&= E\left[e_i(n)e_i(n+m)\right] + E\left[e_i(n)e_j(n+m)\right] \\
&\quad + E\left[e_j(n)e_i(n+m)\right] + E\left[e_j(n)e_j(n+m)\right] \\
&= \phi_{ei}(m) + \phi_{ej}(m)
\end{aligned} \tag{9.52}
$$

Since $e_i(n)$ and $e_j(n)$ are statistically independent and have zero means, we obtain

$$
\begin{aligned}
E\left[e_i(n)e_j(n+m)\right] &= E\left[e_i(n)\right]E\left[e_j(n+m)\right] = 0 \\
E\left[e_j(n)e_i(n+m)\right] &= E\left[e_j(n)\right]E\left[e_i(n+m)\right] = 0
\end{aligned} \tag{9.53}
$$

Now from Eq. 8.5 the Z-transform of Eq. 9.52 is given by

$$\Phi_f(z) = \Phi_{ei}(z) + \Phi_{ej}(z) \tag{9.54}$$

Referring to Fig. 9.7 and using Eq. 8.9, we obtain the following result for the output power spectral density for the noise terms of Eq. 9.50

$$\Phi_{ny}(z) = \frac{1}{C_0^2} H(z)H(z^{-1}) \sum_{i=1}^{r} \Phi_{ei}(z) + \sum_{i=1}^{s} \Phi_{ei}(z) \tag{9.55}$$

Finally, the output power spectral density of the noise terms is given by

$$\sigma_{eo}^2 = \Phi_{ny}(z) = \frac{q^2}{4C_0^2} H(z)H(z^{-1}) + \frac{q^2}{6} \qquad (9.56)$$

For $z = e^{j2\pi fT}$, Eq. 9.56 can be expressed as

$$\sigma_{eo}^2 = \frac{q^2}{4C_0^2} |H(e^{j2\pi fT})|^2 + \frac{q^2}{6} \qquad (9.57)$$

Since the error source attributed to E_{o2} occurs directly at the filter output, the second term of Eq. 9.57 is not affected by the filter transfer function.

9.2.3 Steady-State Output Noise Variance for Cascaded Second-Order Sections

We now consider a cascade of M second-order sections as illustrated in Figure 9.9 for the case $M = 4$. The transfer function of the cascaded sections is given by

$$H(z) = H_1(z)H_2(z)\ldots H_k(z)\ldots H_M(z) \qquad (9.58)$$

where the kth second-order section is expressed by

$$H_k(z) = C_k \frac{1 + A_{1k}z^{-1} + A_{2k}z^{-2}}{1 + B_{1k}z^{-1} + B_{2k}z^{-2}} \qquad (9.59)$$

For cascaded sections, the noise transfer function for the noise source to the filter output is needed; that is, in general the average noise power at the kth section passes through all the poles and zeros of the $(k + 1)$th and higher sections of the system. Figure 9.9 presents a detailed illustration of the roundoff errors generated by a cascade implementation of four second-order sections. Notice that the calculation of the total computational noise generated at the output of the fourth section requires the summation of all the individual errors caused by each internal error source. As a result of the preceding discussions, we can formulate a general expression of the output steady-state noise variance, that is,

$$\sigma_{eo}^2 = P_n \left\{ \left[r \sum_{k=1}^{M} \frac{1}{C_k^2} \sum_{n=0}^{(N/2)-1} |H_{k,M}(e^{j2\pi n/N})|^2 \right] \right.$$
$$\left. + \left[s \sum_{k=2}^{M} \sum_{n=0}^{(N/2)-1} |H_{k,M}(e^{j2\pi n/N})|^2 \right] + s \right\} \qquad (9.60)$$

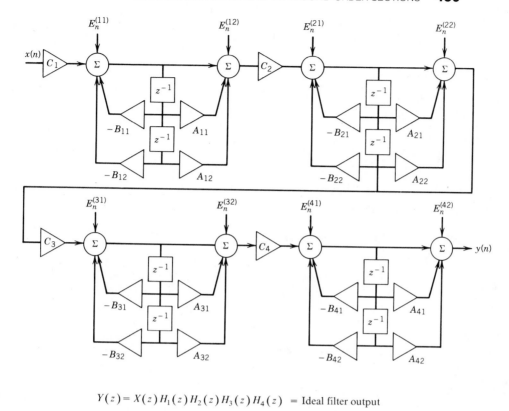

$$Y(z) = X(z)\,H_1(z)\,H_2(z)\,H_3(z)\,H_4(z) \quad = \text{Ideal filter output}$$

$$+ \frac{E_n^{11}(z)}{C_1}\,H_{14}(z) + E_n^{12}(z)\,H_{24}(z) = \text{Output error for section 1}$$

$$+ \frac{E_n^{21}(z)}{C_2}\,H_{24}(z) + E_n^{22}(z)\,H_{34}(z) = \text{Output error for section 2}$$

$$+ \frac{E_n^{31}(z)}{C_3}\,H_{34}(z) + E_n^{32}(z)\,H_{44}(z) = \text{Output error for section 3}$$

$$+ \frac{E_n^{41}(z)}{C_4}\,H_{44}(z) + E_n^{42}(z) \qquad\quad = \text{Output error for section 4}$$

Figure 9.9 Roundoff error response of four second-order sections in cascade.

where

$$P_n = 2^{-2b}/6N$$

$r =$ number of noise sources at the first sum node

$s =$ number of noise sources at the second sum node

$N =$ number of spectral points taken to compute the magnitude of the noise transfer function (NTF)

$n =$ frequency index

$M =$ number of second-order sections

$k =$ second-order section index

The computation of the magnitude response of the noise transfer function can be obtained from Eq. 3.51, that is,

$$\left|H\left(e^{j2\pi fT}\right)\right|^2 = \frac{C_k^2(A + 2B\cos 2\pi fT + 2C\cos 2\pi fT)}{D + 2E\cos 2\pi fT + 2F\cos 2\pi fT} \tag{9.61}$$

where

$$C_k = \text{normalization coefficient}$$
$$A = A_{0k}^2 + A_{1k}^2 + A_{2k}^2$$
$$B = A_{0k}A_{1k} + A_{1k}A_{2k}$$
$$C = A_{0k}A_{2k}$$
$$D = B_{0k}^2 + B_{1k}^2 + B_{2k}^2$$
$$E = B_{0k}B_{1k} + B_{1k}B_{2k}$$
$$F = B_{0k}B_{2k} \tag{9.62}$$

The NTF is then given by the product of the section magnitude-squared response function, that is,

$$\left|H_{kM}\left(e^{j2\pi n/N}\right)\right|^2 = \left|H_k\left(e^{j2\pi n/N}\right)\right|^2\left|H_{k+1}\left(e^{j2\pi n/N}\right)\right|^2 \cdots \left|H_M\left(e^{j2\pi n/N}\right)\right|^2 \tag{9.63}$$

If the magnitude-squared response function is given in decibels, the product in Eq. 9.63 would result in a summation operation.

9.2.4 Sequential Pairing and Ordering of Second-Order Cascade Sections

Equation 9.60 implies that optimal ordering of the numerator and denominator factors in the cascade structure can minimize the output roundoff noise. It can be shown that there are $(M!)^2$ possible orderings of second-order sections.[8]

Evaluating Eq. 9.60 for all possible orderings can be time consuming; for example, 36 evaluations are needed for $M = 3$, 576 evaluations for $M = 4$. It should be noted, however, that for Butterworth and Chebyshev LP, HP, and BP filters the numerator factors are respectively equal, and for these cases only $M!$ orderings would be necessary.

It has been shown[7] that optimal pole-zero pairing can be accomplished by pairing poles with zeros that are closest to them in the z-plane. A description of a heuristic approach for the optimization of the cascade realization of fixed-point digital filters has been developed by Liu and Peled.[8] In this chapter, the optimum solution is obtained by first employing the optimal pole-zero pairing, which will optimally pair the numerator and denominator second-order sections and then evaluate Eq. 9.60 for $M!$ arrangements of the paired second-order sections.

Consider the generalized second-order section transfer function given by

$$H(z) = \frac{\displaystyle\prod_{k=1}^{M} \left(1 + A_{1k}z^{-1} + A_{2k}z^{-2}\right)}{\displaystyle\prod_{k=1}^{M} \left(1 + B_{1k}z^{-1} + B_{2k}z^{-2}\right)} = \frac{\displaystyle\prod_{k=1}^{M} N_k(z)}{\displaystyle\prod_{k=1}^{M} D_k(z)}$$

A cascade arrangement of second-order sections can be expressed by

$$H(z) = \frac{N_i(z)}{D_m(z)} \times \frac{N_j(z)}{D_n(z)} \times \frac{N_k(z)}{D_0(z)} \times \frac{N_l(z)}{D_p(z)}$$

where i, j, k, l, m, n, o, p are integer combinations of $1, 2, 3, 4$. We will return to this example later after considering dynamic range, and scaling requirements for optimal filter implementation.

DYNAMIC RANGE

The dynamic range (DR) of a digital filter is defined as the number of available bits not corrupted by computational noise. Once the output noise variance has been determined, the DR of the filter can be expressed in decibels by

$$DR = (BIT_N - 1)20\log_{10}2 \qquad (9.64)$$

where BIT_N is the least significant bit position of the computational noise buildup given by

$$BIT_N = \text{INT}\left(\frac{-\log_{10}\sigma_{eo}}{\log_{10}2}\right) \qquad (9.65)$$

In Eq. 9.65, INT denotes the integer part of the number. We will return to DR considerations in Section 9.3.1, where the effects of potential summation node overflow are addressed.

Example 9.3

a. Find the output noise variance of an IIR digital filter implemented with four cascaded second-order sections. Assume $N = 2000$, $b = 15$, $r = 3$, and $s = 2$.

b. Determine the bit position of the computational noise buildup and the DR of the filter. Assume the steady-state noise standard deviation $\sigma_{eo} = 0.0000749$.

Solution:

a. The steady-state noise variance is obtained by expanding Eq. 9.60 as follows.

$$\sigma_{eo}^2 = \frac{2^{-2b}}{6N} \left\{ \left[\frac{3}{C_1^2} \sum_{n=0}^{999} \left| H_{14}(e^{j2\pi n/N}) \right|^2 \right] + \left[2 \sum_{n=0}^{999} \left| H_{24}(e^{j2\pi n/N}) \right|^2 \right] \right.$$

$$+ \left[\frac{3}{C_2^2} \sum_{n=0}^{999} \left| H_{24}(e^{j2\pi n/N}) \right|^2 \right] + \left[2 \sum_{n=0}^{999} \left| H_{34}(e^{j2\pi n/N}) \right|^2 \right]$$

$$+ \left[\frac{3}{C_3^2} \sum_{n=0}^{999} \left| H_{34}(e^{j2\pi n/N}) \right|^2 \right] + \left[2 \sum_{n=0}^{999} \left| H_{44}(e^{j2\pi n/N}) \right|^2 \right]$$

$$\left. + \left[\frac{3}{C_4^2} \sum_{n=0}^{999} \left| H_{44}(e^{j2\pi n/N}) \right|^2 \right] + 2 \right\} \tag{9.66}$$

b. The DR of the filter is given by

$$DR = (BIT_N - 1)20 \log_{10} 2 = (BIT_N - 1)6.0206 \tag{9.67}$$

where

$$BIT_N = INT\left(-\frac{\log_{10} 0.0000749}{0.30103} \right) = 13 \text{ (LSB) bit position} \tag{9.68}$$

Finally, the DR is obtained by

$$DR = (13 - 1)6.0206 = 72.25 \text{ dB} \tag{9.69}$$

From Figure 9.10 we can see that 72.25 dB of dynamic range is equivalent to 12-bits. ∎

In the foregoing example internal summation node overflows were not considered. In Section 9.3 we show that potential overflows can occur at the output of the summation nodes of the cascaded second-order sections causing large errors in the filter output signal.

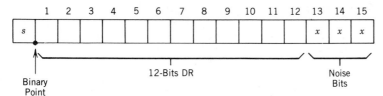

Figure 9.10 Finite arithmetic fractional wordlength for $b = 15$.

9.3 SCALING FORMULATION FOR CASCADED SECOND-ORDER SECTIONS

The evaluation of second-order section normalization coefficients, C_k, was covered in Section 4.6. Implementation of the normalization coefficients assures us that the output of the sections will not overflow; however, they do not guard against overflow at internal summation nodes of the digital network. We now extend the analysis to investigate the potential signal overflow at the output of the digital network summation nodes. The method of analysis that was used to determine the second-order section normalization (scaling) coefficients and steady-state product roundoff error is illustrated in Figure 9.11.

Figure 9.12 shows the section model to be considered in determining potential overflows at internal summation nodes of cascaded second-order sections. It should be noted, however, that the approach is general and can be used to analyze the scaling requirements of any digital network. For the case to be considered, the transfer functions at the indicated summation nodes of Figure 9.12 were determined to be

$$G_{1i}(z) = \frac{F_{1i}(z)}{F_i(z)} = \frac{\left(1 + B_{1i}z^{-1}\right)}{1 + B_{1i}z^{-1} + B_{2i}z^{-2}} \tag{9.70}$$

$$G_{2i}(z) = \frac{F_{2i}(z)}{F_i(z)} = \frac{1}{1 + B_{1i}z^{-1} + B_{2i}z^{-2}} \tag{9.71}$$

$$G_{3i}(z) = \frac{F_{3i}(z)}{F_i(z)} = \frac{1 + A_{1i}z^{-1}}{1 + B_{1i}z^{-1} + B_{2i}z^{-2}} \tag{9.72}$$

$$G_i(z) = \frac{F_{i+1}(z)}{F_i(z)} = \frac{1 + A_{1i}z^{-1} + A_{2i}z^{-2}}{1 + B_{1i}z^{-1} + B_{2i}z^{-2}} \tag{9.73}$$

Signal scaling and roundoff noise analysis will now be illustrated using four cascaded second-order sections. In order to clarify the method of analysis, the magnitude responses of Eqs. 9.70 thru 9.73 were programmed on a digital computer, and signal scaling was determined at the first, second, and third summation nodes of Figure 9.12, using L_∞-norm scaling. As shown in Table 9.2, the summation node overflow equations are a measure of the peak filter response from the ith sum node of the first section to the ith sum node

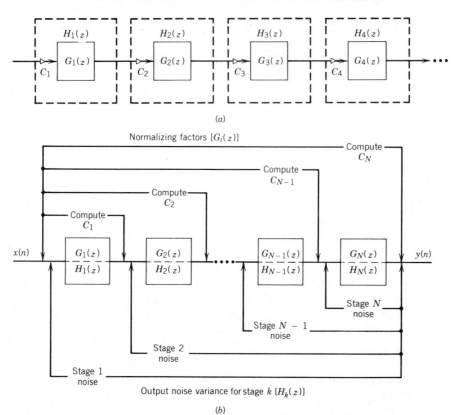

(a)

(b)

Figure 9.11 Cascaded second-order-sections normalization and roundoff noise model: (a) cascade realization showing relationship between transfer functions $H_i(z)$ and $G_k(z)$; (b) cascade realization showing the relationship between the calculation of the section normalization coefficients and noise transfer function.

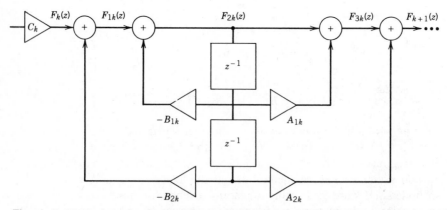

Figure 9.12 Flow graph representation for internal sum-node analysis.

TABLE 9.2
SUMMATION NODE OVERFLOW ANALYSIS FOR FOUR SECOND-ORDER
CASCADED SECTIONS

First Summer

Section 1: $M_{11} = C_1 \max |G_{11}|$
Section 2: $M_{12} = C_1 C_2 \max |G_1 G_{12}|$
Section 3: $M_{13} = C_1 C_2 C_3 \max |G_1 G_2 G_{13}|$
Section 4: $M_{14} = C_1 C_2 C_3 C_4 \max |G_1 G_2 G_3 G_{14}|$

Second Summer

Section 1: $M_{21} = C_1 \max |G_{21}|$
Section 2: $M_{22} = C_1 C_2 \max |G_1 G_{22}|$
Section 3: $M_{23} = C_1 C_2 C_3 \max |G_1 G_2 G_{23}|$
Section 4: $M_{24} = C_1 C_2 C_3 C_4 \max |G_1 G_2 G_3 G_{24}|$

Third Summer

Section 1: $M_{31} = C_1 \max |G_{31}|$
Section 2: $M_{32} = C_1 C_2 \max |G_1 G_{32}|$
Section 3: $M_{33} = C_1 C_2 C_3 \max |G_1 G_2 G_{33}|$
Section 4: $M_{34} = C_1 C_2 C_3 C_4 \max |G_1 G_2 G_3 G_{34}|$

---

1. For simplicity of notation, $G_{ik} = G_{ik}(e^{j2\pi fT})$.
2. G_{ik} denotes the transfer function of the ith summer of the kth section defined by Eqs. 9.70, 9.71, and 9.72.
3. G_k is the section transfer function defined by Eq. 9.73.
4. For compactness of notation

$$\max |G_{ik}| = \max_{0 \le f \le F/2} \left| G_{ik}(e^{j2\pi fT}) \right|$$

of the kth section; for example, the magnitude response of the first sum node of section 1 is given by $M_{ik} = M_{11}$; likewise, the magnitude response of the first sum node of section 2 is given by M_{12}. The results of a computer simulation of the equations given in Table 9.2 are presented in Figure 9.13e.

To protect against internal summation node overflows, it is necessary to modify the normalization coefficients. The modified normalization coefficients, denoted by S_k, can readily be computed using the information given in Figure 9.13e. In Figure 9.13e the peak response $M_{i,k}$ for the first, second, and third summation node is given as a function of frequency. From these results the modified normalization coefficients, which will be referred to as **scale coefficients**, are determined as a function of the **normalization coefficients** C_k. It can be shown that the scale coefficients can be calculated as a function of the normalization coefficients C_k as follows:

$$S_k = \frac{C_k M_{i,k-1}}{M_{i,k}} \tag{9.74}$$

where $k = 1, 2, 3, 4$ is the section scale factor index, $i = 1, 2, 3$ is the summation node index, and $M_{1,0} = 1$. A complete illustration of signal scaling and quantization noise analysis is presented in Example 9.4.

Example 9.4

Consider the following specifications of an elliptic bandpass filter.

$$A_p = 1 \text{ dB} \qquad f_{p1} = 900 \text{ Hz} \qquad f_{s1} = 800 \text{ Hz} \qquad F = 6000 \text{ Hz}$$
$$A_s = 45 \text{ dB} \qquad f_{p2} = 1100 \text{ Hz} \qquad f_{s2} = 1200 \text{ Hz}$$

It is desired to determine the standard deviation of the steady-state computational noise buildup and the overflow potential at the summation nodes shown in Fig. 9.12 for a cascade realization.

Solution:

The results of a computer simulation of Eqs. 9.60 and 9.70–9.73 are presented in Figures 9.13a through f. The results given in Figure 9.13e show that the

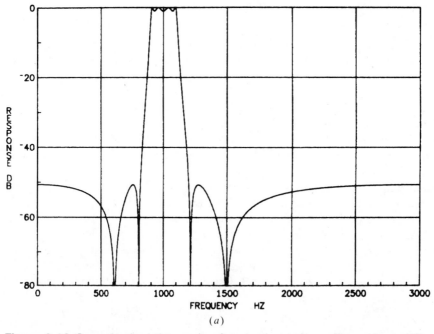

```
IIR BP ELLIPTIC FILTER
REQUIRED SPECIFICATIONS:   PASSBAND ATTENUATION (DB): 1
                           STOPBAND ATTENUATION (DB): 45
                           PASSBAND CUTOFF FREQ (HZ): 900 1100
                           STOPBAND CUTOFF FREQ (HZ): 800 1200
                           SAMPLING FREQ (HZ): 6000
                           ORDER (SPECIFIED): 4    REQUIRED :   3.67
```

(a)

Figure 9.13 Computer-simulation results of summation node overflow and roundoff noise analysis for four cascaded second-order sections: (a) bandpass filter magnitude response; (b) bandpass filter digital and analog poles, zeros, and digital filter coefficients; (c) passband response; (d) passband ripple analysis; (e) sum node overflow and finite wordlength analysis; (f) sum node analysis using modified normalization (scale) coefficient multipliers.

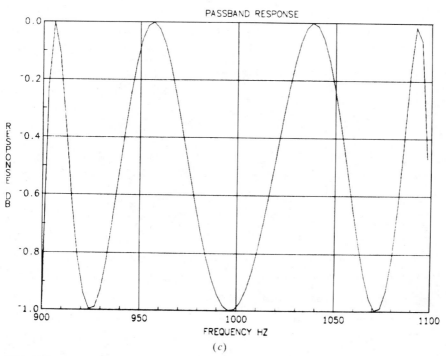

H[Z]=N[Z]/D[Z]
N[Z]=A0+A1Z+⁻1+A2Z+⁻2
D[Z]=B0+B1Z+⁻1+B2Z+⁻2

DIGITAL COEFFICIENTS		(32 BITS)	
A0:	1.000000000	B0:	1.000000000
A1:	⁻.007283537	B1:	⁻.888660370
A2:	1.000000000	B2:	.926866771

A0:	1.000000000	B0:	1.000000000
A1:	⁻1.602667672	B1:	⁻1.046605824
A2:	1.000000000	B2:	.930605671

A0:	1.000000000	B0:	1.000000000
A1:	⁻.588557671	B1:	⁻.804891912
A2:	1.000000000	B2:	.973853912

A0:	1.000000000	B0:	1.000000000
A1:	⁻1.331115818	B1:	⁻1.160308948
A2:	1.000000000	B2:	.976782432

POLES AND ZEROES

POLE	ANALOG	⁻.02597497+J	.6066865
	DIGITAL	.44433019+J	.8540711
DOUBLE ZERO	ANALOG	.00000000+J	.9963648
	DIGITAL	.00364177+J	.9999934

POLE	ANALOG	⁻.02330850+J	⁻.5444068
	DIGITAL	.52330291+J	⁻.8104071
DOUBLE ZERO	ANALOG	.00000000+J	⁻.3320969
	DIGITAL	.80133384+J	.5982174

POLE	ANALOG	⁻.00940931+J	.6485300
	DIGITAL	.40244596+J	.9010500
DOUBLE ZERO	ANALOG	.00000000+J	.7384186
	DIGITAL	.29427884+J	.9557196

POLE	ANALOG	⁻.00740099+J	⁻.5101076
	DIGITAL	.58015447+J	⁻.8001270
DOUBLE ZERO	ANALOG	.00000000+J	⁻.4481058
	DIGITAL	.66555791+J	⁻.7463462

(*b*)

PASSBAND RESPONSE

(*c*)

Figure 9.13 Continued

standard deviation of the steady-state output roundoff noise is given by $\sigma_{oe} = 0.000074909$, which corresponds to a computational noise buildup of 3-bits; that is, bit positions 13, 14, and 15 are corrupted by computational noise buildup. The DR of the filter is determined from Eq. 9.64, that is, $DR = 72.3$ dB. Therefore, as a result of computational noise buildup, the DR of the filter has degraded by approximately 18 dB.

```
PASSBAND RIPPLE ANALYSIS
MAXIMUM DB:      .00000 AT 1038.000000 HZ
MINIMUM DB:     ‾.99969 AT  900.000000 HZ
MAXIMUM PASSBAND RIPPLE              ‾.99969 DB
••••••••••••••••••••••••••••••••••••••••••••••••
STOPBAND ATTENUATION ANALYSIS:
MAXIMUM DB: ‾50.63534 AT      .000000 HZ
MINIMUM DB: ‾80.00000 AT  609.000000 HZ
MINIMUM STOPBAND ATTENUATION  ‾50.63534 DB
••••••••••••••••••••••••••••••••••••••••••••••••
NORMALIZING FACTORS
 .0706239
 .2290807
 .3133776
 .6793442
```

(*d*)

```
••• OVERFLOW ANALYSIS •••
          RESP(DB)        FREQ(HZ)
FIRST  SUMMER
STAGE  1      .598        1044.00000
  (OVERFLOW IN STAGE  1)
STAGE  2     5.227         966.00000
  (OVERFLOW IN STAGE  2)
STAGE  3    12.530        1095.00000
  (OVERFLOW IN STAGE  3)
STAGE  4    15.603         903.00000
  (OVERFLOW IN STAGE  4)
SECOND SUMMER
STAGE  1      .736        1041.00000
  (OVERFLOW IN STAGE  1)
STAGE  2     5.314         963.00000
  (OVERFLOW IN STAGE  2)
STAGE  3    12.594        1095.00000
  (OVERFLOW IN STAGE  3)
STAGE  4    15.655         903.00000
  (OVERFLOW IN STAGE  4)
THIRD  SUMMER
STAGE  1      .707        1041.00000
  (OVERFLOW IN STAGE  1)
STAGE  2     8.008         963.00000
  (OVERFLOW IN STAGE  2)
STAGE  3    11.949        1095.00000
  (OVERFLOW IN STAGE  3)
STAGE  4    16.496         903.00000
  (OVERFLOW IN STAGE  4)
```

```
••••• FINITE WORDLENGTH ANALYSIS •••••
STD. DEV. OF NOISE IS     .000074909
NOISE BUILDS UP THROUGH 13 th BIT
DYNAMIC RANGE OF FILTER IS 54.1854 DB

STRUCTURE OF REGISTER FOR NO OVERFLOW:
1 BIT        SIGN BIT
3 BIT(S)     OVERFLOW BITS
9 BIT(S)     OF DYNAMIC RANGE
3 BIT(S)     OF COMPUTATIONAL NOISE
TOTAL: 16

NORMALIZING FACTORS, BY STAGE
    .07062389
    .22908069
    .31337765
    .67934421

NUMBER OF NOISE SOURCES AT FIRST SUMMER 3
NUMBER OF NOISE SOURCES AT SECOND SUMMER 2
```

(*e*)

```
••• OVERFLOW ANALYSIS •••
          RESP(DB)        FREQ(HZ)
FIRST  SUMMER
STAGE  1     ‾.109        1044.00000
STAGE  2    ‾2.781         966.00000
STAGE  3      .581        1095.00000
  (OVERFLOW IN STAGE  3)
STAGE  4     ‾.893         903.00000
SECOND SUMMER
STAGE  1      .029        1041.00000
  (OVERFLOW IN STAGE  1)
STAGE  2    ‾2.694         963.00000
STAGE  3      .645        1095.00000
  (OVERFLOW IN STAGE  3)
STAGE  4     ‾.841         903.00000
THIRD  SUMMER
STAGE  1      .000        1041.00000
STAGE  2      .000         963.00000
  (OVERFLOW IN STAGE  2)
STAGE  3      .000        1095.00000
  (OVERFLOW IN STAGE  3)
STAGE  4      .000         903.00000
  (OVERFLOW IN STAGE  4)
```

(*f*)

Figure 9.13 Continued

The summation node overflow analysis is obtained by determining the magnitude of the peak response at the sum nodes as shown in Figure 9.13e. The results of this analysis were determined from a computer solution of the equations given in Table 9.2. We can see that the peak response (in decibels) measured from the input of the first section to the indicated summation node is essentially the same for all summation nodes, where positive decibels are considered overflows. That is, numbers ranging from zero to 6 dB require one overflow bit (one bit to the left of the binary point), numbers ranging from 6 dB to 12 dB require two overflow bits to the left of the binary point, numbers ranging from 12 dB to 18 dB require three overflow bits to the left of the binary point, and so on. To determine the actual scale factors necessary to prevent summation node overflow, the sum of the peak magnitude responses for the three summer nodes of Fig. 9.13e were determined. The third summation node was found to be the greatest, 37.2 dB. As a result, the second-order section scale coefficients were calculated using Eq. 9.74. From Figure 9.13e the third summation node gains for sections 1, 2, 3, and 4 are given by 0.707 dB, 8.008 dB, 11.949 dB, and 16.496 dB, respectively. Converting to real numbers we obtain

$$M_{31} = 10^{0.707/20} = 1.0848$$

$$M_{32} = 10^{8.008/20} = 2.5142$$

$$M_{33} = 10^{11.949/20} = 3.95776$$

$$M_{34} = 10^{16.496/20} = 6.68036 \tag{9.75}$$

The scale coefficients can now be determined from Eq. 9.74, that is,

$$S_1 = \frac{C_1}{M_{31}} = \frac{0.070623}{1.0848} = 0.065103$$

$$S_2 = \frac{C_2 M_{31}}{M_{32}} = \frac{(0.229081)(1.0848)}{2.5142} = 0.098841$$

$$S_3 = \frac{C_3 M_{32}}{M_{33}} = \frac{(0.313378)(2.5142)}{3.95776} = 0.199076$$

$$S_4 = \frac{C_4 M_{33}}{M_{34}} = \frac{(0.679344)(3.95776)}{6.68036} = 0.402475 \tag{9.76}$$

To prevent internal summation node overflows, the modified scale coefficients S_k were substituted for the normalization coefficients C_k, and the error and overflow analysis was repeated. We can see from the results shown in Figure 9.13f that the magnitude of the overflow in each section was reduced to less than 1 dB; that is the magnitude of the largest overflow, which occurred at the second summation node of section 3, is 0.645 dB, which is a fraction of one bit. Since the infinite norm assumption represents the possibility of an overflow resulting from a unit sinusoid at the peak response frequency of 903 Hz, an overflow is unlikely to occur for the usual assumption of signals buried in environmental noise. ■

The overflow analysis shows that to prevent against potential overflow the input data is attenuated by the modified scale coefficients by 2 to 3 bits, resulting in degradation of DR; that is, from Figure 9.13e the largest gain of 16.496 dB obtained at the third summer of section 4 requires the input data sequence to be right-shifted. The analysis shows a worst-case shift of 3 bits, resulting in the indicated dynamic range of 54 dB. It should further be noted that the difference in the products of the scale coefficients S_k and the normalization coefficients C_k is given by 49.26 dB − 65.75 dB = −16.5 dB, that is, the total frequency response of the filter is attenuated by 16.5 dB.

The insertion of the scale coefficients, S_k, between cascaded second-order sections to protect against probability of overflows was based on using L_∞ scaling, that is, a unit sine wave at a frequency equal to the maximum response of the internal sum nodes will result in a unit sine wave at the filter output. This type of scaling is referred to as upper-bound scaling and can result in unnecessary loss in system DR, especially when anticipated signal levels are very low. For this case a preferred scaling method may be the L_2-norm or energy scaling (see Section 2.1.2). It should be noted that in the actual implementation of the cascaded filter in hardware the scale factors can be approximated by a power of 2, thereby requiring a right shift rather than a multiply.

9.3.1 Dynamic Range Considerations

Equation 9.67 defined the DR of a cascaded second-order filter. If we are to include any required register overflow bits in the implementation, or equivalently scale the data to eliminate potential overflows, then a degradation in DR will result. The degradation in DR results from the fact that the modified scale coefficients reduce the amplitude of the signal at the input to the filter sections. The DR degradation can be expressed by

$$DR = (BIT_N - 1 - BIT_{OF})6.0206 \qquad (9.77)$$

where BIT_{OF} is the number of overflow bits required to prevent internal summation node overflow. The number of overflow bits required to prevent overflow can be obtained from Figure 9.13e. A typical 16-bit register that can provide for overflow at the summation nodes is shown in Figure 9.14, where the assumed binary point was shifted right by three bits, thus allowing for potential internal sum node overflows. An equivalent approach to eliminate summation node overflows is to insert the scale coefficients preceding the cascaded second-order sections; that is, the input data is optimally scaled

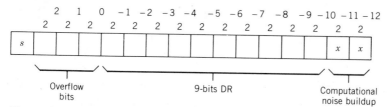

Figure 9.14 Finite arithmetic wordlength allowing for summation node overflows.

(reduced) in the fractional register of Figure 9.14 according to the selected L_p-norm scaling method.

We now return to the analysis required for optimal implementation of the four second-order sections of Example 9.4. The finite wordlength and overflow analysis shown in Figure 9.13e was iterated for numerous numerator and denominator pairing and ordering combinations, and a summary of results are given in Table 9.3. The pole-zero pairing considered as part of the analysis is shown in Figure 9.15, where we can see that the poles were matched to the zero nearest its location.

The numerator and denominator second-order sections corresponding to this pairing are given by

$$
\begin{aligned}
N1: &\ 1 - 0.007283537z^{-1} + z^{-2} \\
D1: &\ 1 - 0.888660370z^{-1} + 0.926866771z^{-2} \\
N2: &\ 1 - 1.602667672z^{-1} + z^{-2} \\
D2: &\ 1 - 1.046605824z^{-1} + 0.930605671z^{-2} \\
N3: &\ 1 - 0.588557671z^{-1} + z^{-2} \\
D3: &\ 1 - 0.804891912z^{-1} + 0.973853912z^{-2} \\
N4: &\ 1 - 1.331115818z^{-1} + z^{-2} \\
D4: &\ 1 - 1.160308948z^{-1} + 0.976782432z^{-2}
\end{aligned}
\tag{9.78}
$$

which corresponds to the section arrangement of Figure 9.13b. This arrangement was considered to be optimal[11] with respect to the pole-zero pairing shown in Figure 9.15.

To determine the validity of this optimal pairing arrangement, numerous other arrangements were analyzed to determine the minimum roundoff noise buildup. Table 9.3 represents a summary of the analysis. Case 1 is the ordering presented in the computer output listing shown in Figure 9.13b and illustrated in Figure 9.15. Case 2 is the ordering that gave the minimum roundoff noise buildup, and finally case 5 was evaluated to be the worst-case arrangement of numerator and denominator pairs used in this analysis. Therefore, as a result of this analysis it was determined that the pairing shown in Figure 9.15 was optimal, and system performance can degrade by as much as 12 dB by improper pairing and ordering of the numerator and denominator second-order sections.

TABLE 9.3
RESULTS OF EXAMPLE 9.4 FOR POLE-ZERO PAIRING AND ORDERING OF
SECOND-ORDER SECTIONS

Case	Section Ordering	Output Noise Standard Deviation	Overflow Bits	Roundoff Noise LSB Buildup	DR (dB)
1	N1 N2 N3 N4	0.0000749	3	3	54.2
	D1 D2 D3 D4	(-82.51 dB)			(9 bits)
2	N2 N3 N4 N1	0.0000644	3	3	54.2
	D2 D3 D4 D1	(-83.82 dB)			(9 bits)
3	N3 N1 N4 N2	0.0000796	3	3	54.2
	D3 D1 D4 D2	(-81.98 dB)			(9 bits)
4	N1 N2 N3 N4	0.0001490	3	4	48.16
	D2 D1 D4 D3	(-76.54 dB)			(8 bits)
5	N1 N2 N3 N4	0.0002676	3	5	42.1
	D4 D3 D2 D1	(-71.45 dB)			(7 bits)
6	N4 N1 N2 N3	0.0001247	2	4	54.2
	D4 D3 D2 D1	(-78.08 dB)			(9 bits)

Notes:
  1. The second-order factors $N1$, $N2$, $N3$, $N4$, $D1$, $D2$, $D3$, and $D4$ are given by Eq. 9.78.
  2. For the roundoff noise least significant bit (LSB) calculation see Eqs. 9.65.
  3. The overflow bit entry was determined from the worst-case result shown in Figure 9.13e; that is, since 6 dB corresponds to 1 bit, 16.496 dB translates to 3 bits.
  4. Dynamic range entries were determined from Eq. 9.77.

9.3.2 Floating-Point and Fixed-Point Arithmetic Considerations

A comparison of roundoff noise in floating-point and fixed-point arithmetic was performed by Weinstein and Oppenheim.[13] The results showed that floating-point arithmetic leads to lower noise-to-signal ratios than does fixed-point if the floating-point mantissa is equal in length to the fixed-point word. If the total floating-point wordlength including the exponent is constrained to have the same number of bits as the fixed-point word, then the comparison depends on the signal power spectral density and filter frequency response. Therefore, a comparison must take into account signal statistics and hardware complexity in implementing floating-point arithmetic and adding additional bits to a fixed-point arithmetic processor.

9.4 IIR COEFFICIENT QUANTIZATION ERRORS

The final consideration in dealing with finite wordlength effects in IIR digital filter design is coefficient quantization. In the previous sections we assumed that the filter coefficients were implemented with infinite accuracy. However, since digital networks operate with a finite number of bits, the effects of coefficient quantization must be considered. As we have shown in Chapter 4, the frequency response of a digital filter is characterized by the coefficients of the transfer function. Therefore, as a result of the finite wordlength each coefficient is implemented with $(b + 1)$ bits, which in turn alters the frequency response of the filter by perturbing the poles and zeros of the transfer

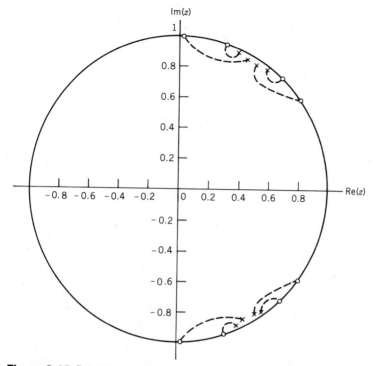

Figure 9.15 Pole-zero pairing.

function. It should be emphasized that coefficient quantization and product quantization are independent. Coefficient wordlength is chosen to satisfy frequency response specifications, whereas processor wordlengths are chosen to satisfy dynamic range and signal-to-noise ratio specifications.

In general, rounding the coefficients of the difference equation to $(b + 1)$ bits will result in errors in the positions of the poles and zeros of the transfer function. It has been shown[9] that pole-zero errors are far more pronounced when high-order systems are realized in direct form than realizations of parallel or cascade second-order sections. The effect of coefficient rounding

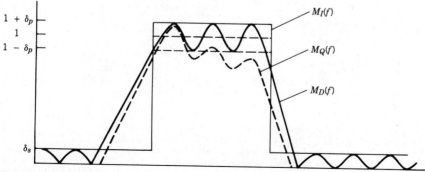

Figure 9.16 Effects of coefficient quantization on magnitude response.

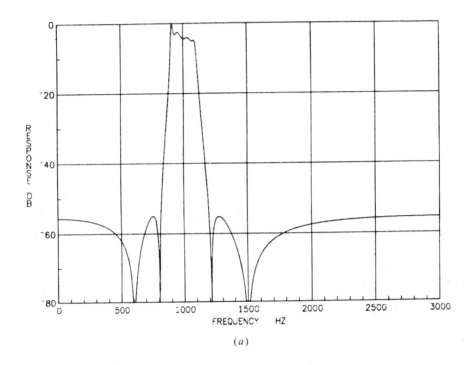

```
IIR BP ELLIPTIC FILTER
REQUIRED SPECIFICATIONS:    PASSBAND ATTENUATION (DB): 1
                            STOPBAND ATTENUATION (DB): 45
                            PASSBAND CUTOFF FREQ (HZ): 900 1100
                            STOPBAND CUTOFF FREQ (HZ): 800 1200
                            SAMPLING FREQ (HZ): 6000
                            ORDER (SPECIFIED): 4    REQUIRED :   3.67
```

(a)

```
H[Z]=N[Z]/D[Z]
N[Z]=A0+A1Z+¯1+A2Z+¯2
D[Z]=B0+B1Z+¯1+B2Z+¯2
DIGITAL COEFFICIENTS      ( 8 BITS)        POLES AND ZEROES
A0:   1.000000000   B0:    1.000000000     POLE         ANALOG  ¯.02597497+J   .6066865
A1:    .000000000   B1:   ¯.890625000                   DIGITAL  .44433019+J   .8540711
A2:   1.000000000   B2:    .921875000      DOUBLE ZERO ANALOG   .00000000+J   .9963648
                                                       DIGITAL  .00364177+J   .9999934

A0:   1.000000000   B0:    1.000000000     POLE         ANALOG  ¯.02330850+J  ¯.5444068
A1:  ¯1.609375000   B1:   ¯1.046875000                  DIGITAL  .52330291+J  ¯.8104071
A2:   1.000000000   B2:    .937500000      DOUBLE ZERO ANALOG   .00000000+J  ¯.3320969
                                                       DIGITAL  .80133384+J  ¯.5982174

A0:   1.000000000   B0:    1.000000000     POLE         ANALOG  ¯.00940931+J   .6485300
A1:   ¯.593750000   B1:   ¯.812500000                   DIGITAL  .40244596+J   .9010500
A2:   1.000000000   B2:    .968750000      DOUBLE ZERO ANALOG   .00000000+J   .7384186
                                                       DIGITAL  .29427884+J   .9557196

A0:   1.000000000   B0:    1.000000000     POLE         ANALOG  ¯.00740099+J  ¯.5101076
A1:  ¯1.328125000   B1:   ¯1.156250000                  DIGITAL  .58015447+J  ¯.8001270
A2:   1.000000000   B2:    .984375000      DOUBLE ZERO ANALOG   .00000000+J  ¯.4481058
                                                       DIGITAL  .66555791+J  ¯.7463462
```

(b)

Figure 9.17 Effects of coefficient quantization on magnitude response: (a) filter coefficients of Example 9.4 quantized to 8 bits; (b) elliptic bandpass filter poles, zeros, and
second-order section coefficients; (c) passband response; (d) passband response analysis; (e) bandpass filter sum-node overflow and finite wordlength analysis.

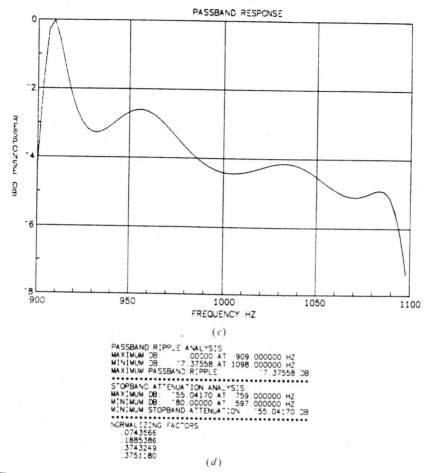

PASSBAND RESPONSE

(c)

```
PASSBAND RIPPLE ANALYSIS
MAXIMUM DB:        .00000 AT   909.000000 HZ
MINIMUM DB:      -7.37558 AT  1098.000000 HZ
MAXIMUM PASSBAND RIPPLE          -7.37558 DB
.............................................
STOPBAND ATTENUATION ANALYSIS:
MAXIMUM DB:     -55.04170 AT   759.000000 HZ
MINIMUM DB:     -80.00000 AT   597.000000 HZ
MINIMUM STOPBAND ATTENUATION    -55.04170 DB
.............................................
NORMALIZING FACTORS
  .0743566
  .1885386
  .3743249
  .3751180
```

(d)

Figure 9.17 Continued

can be determined by comparing the frequency response of the ideal filter implemented with 32-bit coefficients with that of the actual filter implemented with quantized coefficients of $(b + 1)$ bits.

The error resulting from coefficient quantization can be described in terms of the deviation of the desired frequency response to the actual frequency response using finite wordlengths. Figure 9.16 shows the ideal, desired, and quantized tolerance specifications of a bandpass digital filter, that is

$$M_I(f) = \text{ideal magnitude response}$$

$$M_Q(f) = \text{quantized magnitude response (less than 32 bits)}$$

$$M_D(f) = \text{desired magnitude response (32 bits)}$$

$$\delta_p = \text{passband tolerance}$$

$$\delta_s = \text{stopband tolerance}$$

```
••• OVERFLOW ANALYSIS ••••
           RESP(DB)          FREQ(HZ)
FIRST  SUMMER
STAGE  1       .478          1041.00000
     (OVERFLOW IN STAGE  1)
STAGE  2      5.158           963.00000
     (OVERFLOW IN STAGE  2)
STAGE  3     12.158          1089.00000
     (OVERFLOW IN STAGE  3)
STAGE  4     15.481           909.00000
     (OVERFLOW IN STAGE  4)
SECOND SUMMER
STAGE  1       .619          1038.00000
     (OVERFLOW IN STAGE  1)
STAGE  2      5.248           963.00000
     (OVERFLOW IN STAGE  2)
STAGE  3     12.236          1089.00000
     (OVERFLOW IN STAGE  3)
STAGE  4     15.501           909.00000
     (OVERFLOW IN STAGE  4)
THIRD  SUMMER
STAGE  1       .619          1038.00000
     (OVERFLOW IN STAGE  1)
STAGE  2      7.975           963.00000
     (OVERFLOW IN STAGE  2)
STAGE  3     11.566          1089.00000
     (OVERFLOW IN STAGE  3)
STAGE  4     16.375           909.00000
     (OVERFLOW IN STAGE  4)
```

```
••••• FINITE WORDLENGTH ANALYSIS •••••
STD. DEV. OF NOISE IS      .000061427
NOISE BUILDS UP THROUGH 13 th BIT
DYNAMIC RANGE OF FILTER IS 54.1854 DB

STRUCTURE OF REGISTER FOR NO OVERFLOW:
 1 BIT        SIGN BIT
 3 BIT(S)     OVERFLOW BITS
 9 BIT(S)     OF DYNAMIC RANGE
 3 BIT(S)     OF COMPUTATIONAL NOISE
TOTAL: 16

NORMALIZING FACTORS, BY STAGE
     .07435662
     .18853859
     .37432491
     .37511802

NUMBER OF NOISE SOURCES AT FIRST SUMMER 3
NUMBER OF NOISE SOURCES AT SECOND SUMMER 2
```

(e)

Figure 9.17 Continued

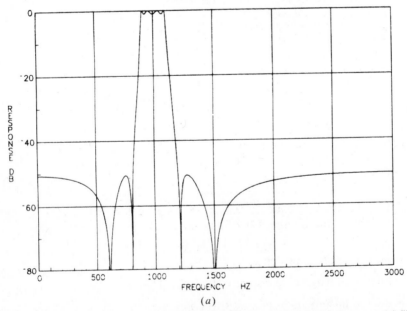

```
IIR BP ELLIPTIC FILTER
REQUIRED SPECIFICATIONS:    PASSBAND ATTENUATION (DB): 1
                           STOPBAND ATTENUATION (DB): 45
                           PASSBAND CUTOFF FREQ (HZ): 900 1100
                           STOPBAND CUTOFF FREQ (HZ): 800 1200
                           SAMPLING FREQ (HZ): 6000
                           ORDER (SPECIFIED): 4     REQUIRED :   3.67
```

(a)

Figure 9.18 Effects of IIR filter coefficient quantization on magnitude response; (a) filter coefficients of Example 9.4-quantized to 16 bit. (b) elliptic bandpass filter poles, zeros, and second-order section coefficients; (c) passband response; (d) passband response analysis; (e) bandpass filter sum-node overflow and finite wordlength analysis.

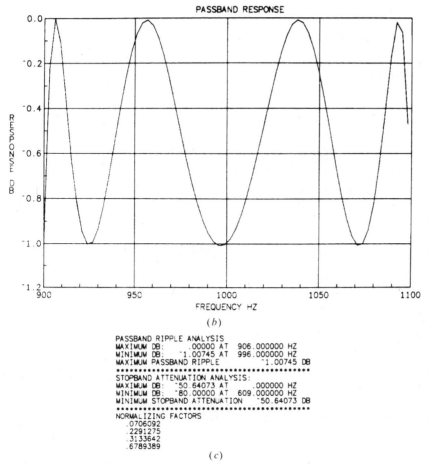

PASSBAND RESPONSE

(b)

(c)

Figure 9.18 Continued

It should be noted that, in general, filters implemented with 16 bits including sign bit, integer bit, and 14 fractional bits will show negligible degradation from the use of 32-bit wordlengths. The effects of coefficient quantization will be demonstrated with the following example: The frequency response of the bandpass filter of Example 9.4 is shown for 8-, and 16-bit coefficients in Figures 9.17 and 9.18, respectively. For 8 bits, the pass-band ripple degrades to -7.3 dB at 1098 Hz and fails the 1-dB passband specification. However, we can see that for 16-bit coefficients the passband ripple is -1.00745 at 996 Hz. The degradation of 0.00745 dB can be acceptable, depending on the required specification tolerence If this passband degradation is outside the specified tolerence, the coefficients must be implemented with additional bits. For example, results were obtained for 18- and 20-bit coefficients, where for 18-bit coefficients the maximum passband response was -1.00188 dB at 900 Hz and for 20 bits the maximum passband response was -0.99979 dB.

```
H[Z]=N[Z]/D[Z]
N[Z]=A0+A1Z+¯1+A2Z+¯2
D[Z]=B0+B1Z+¯1+B2Z+¯2
DIGITAL COEFFICIENTS    (16 BITS)
AO:    1.000000000      BO:    1.000000000
A1:    ¯.007263184      B1:   ¯.888671875
A2:    1.000000000      B2:    .926879883

•••••••••••••••••••••••••••••••••••••••••••
AO:    1.000000000      BO:    1.000000000
A1:   ¯1.602661133      B1:   ¯1.046630859
A2:    1.000000000      B2:    .930603027

•••••••••••••••••••••••••••••••••••••••••••
AO:    1.000000000      BO:    1.000000000
A1:    ¯.588562012      B1:   ¯.804870605
A2:    1.000000000      B2:    .973876953

•••••••••••••••••••••••••••••••••••••••••••
AO:    1.000000000      BO:    1.000000000
A1:   ¯1.331115723      B1:   ¯1.160339355
A2:    1.000000000      B2:    .976806641

•••••••••••••••••••••••••••••••••••••••••••
```

```
POLES AND ZEROES
POLE          ANALOG   ¯.02597497+J   .6066865
              DIGITAL   .44433019+J   .8540711
DOUBLE ZERO ANALOG     .00000000+J   .9963648
              DIGITAL   .00364177+J   .9999934
•••••••••••••••••••••••••••••••••••••••••••
POLE          ANALOG   ¯.02330850+J  ¯.5444068
              DIGITAL   .52330291+J  ¯.8104071
DOUBLE ZERO ANALOG     .00000000+J  ¯.3320969
              DIGITAL   .80133384+J   .5982174
•••••••••••••••••••••••••••••••••••••••••••
POLE          ANALOG   ¯.00940931+J   .6485300
              DIGITAL   .40244596+J   .9010500
DOUBLE ZERO ANALOG     .00000000+J   .7384186
              DIGITAL   .29427884+J   .9557196
•••••••••••••••••••••••••••••••••••••••••••
POLE          ANALOG   ¯.00740099+J  ¯.5101076
              DIGITAL   .58015447+J  ¯.8001270
DOUBLE ZERO ANALOG     .00000000+J  ¯.4481058
              DIGITAL   .66555791+J   .7463462
•••••••••••••••••••••••••••••••••••••••••••
```

(d)

```
••• OVERFLOW ANALYSIS •••
            RESP(DB)      FREQ(HZ)
FIRST  SUMMER
STAGE  1      .598        1044.00000
   (OVERFLOW IN STAGE  1)
STAGE  2     5.226         966.00000
   (OVERFLOW IN STAGE  2)
STAGE  3    12.533        1095.00000
   (OVERFLOW IN STAGE  3)
STAGE  4    15.606         903.00000
   (OVERFLOW IN STAGE  4)
SECOND SUMMER
STAGE  1      .736        1041.00000
   (OVERFLOW IN STAGE  1)
STAGE  2     5.313         963.00000
   (OVERFLOW IN STAGE  2)
STAGE  3    12.597        1095.00000
   (OVERFLOW IN STAGE  3)
STAGE  4    15.658         903.00000
   (OVERFLOW IN STAGE  4)
THIRD  SUMMER
STAGE  1      .707        1041.00000
   (OVERFLOW IN STAGE  1)
STAGE  2     8.007         963.00000
   (OVERFLOW IN STAGE  2)
STAGE  3    11.952        1095.00000
   (OVERFLOW IN STAGE  3)
STAGE  4    16.499         903.00000
   (OVERFLOW IN STAGE  4)
```

```
••••• FINITE WORDLENGTH ANALYSIS •••••
STD. DEV. OF NOISE IS    .000074896
NOISE BUILDS UP THROUGH 13 th BIT
DYNAMIC RANGE OF FILTER IS 54.1854 DB

STRUCTURE OF REGISTER FOR NO OVERFLOW:
  1 BIT       SIGN BIT
  3 BIT(S)    OVERFLOW BITS
  9 BIT(S)    OF DYNAMIC RANGE
  3 BIT(S)    OF COMPUTATIONAL NOISE
TOTAL: 16

NORMALIZING FACTORS, BY STAGE
      .07060916
      .22912750
      .31336422
      .67893891

NUMBER OF NOISE SOURCES AT FIRST SUMMER 3
NUMBER OF NOISE SOURCES AT SECOND SUMMER 2
```

(e)

Figure 9.18 Continued

This example shows that the finite wordlength coefficients may significantly alter the frequency response of the filter from the original design specifications. It should also be noted that the coefficient error is independent of the roundoff error. That is, coefficient error only affects the frequency response of the filter, whereas roundoff error affects the filter dynamic range.

9.5 FIR DIGITAL FILTER FINITE WORDLENGTH EFFECTS

In Chapter 5 we described the direct-form realization of FIR digital filters. The direct-form realization, and difference equation for the case N-odd are given by Figure 5.11 and Eq. 5.81, respectively. It is seen that owing to coefficient symmetry only $[(N + 1)/2]$ multiplies are required to be implemented. It has been shown[10,11] that the roundoff noise at the output of a direct-form FIR filter depends on the location points in the filter where rounding is performed. For this analysis, two cases will be considered.

Referring to Fig. 5.11, if all the multiplication products of two $(b + 1)$-bit numbers are represented by a $2(b + 1)$-bit number, and rounding to $(b + 1)$-bits is performed only after all the products are summed, then the roundoff noise source is modeled as a discrete stationary white noise random process

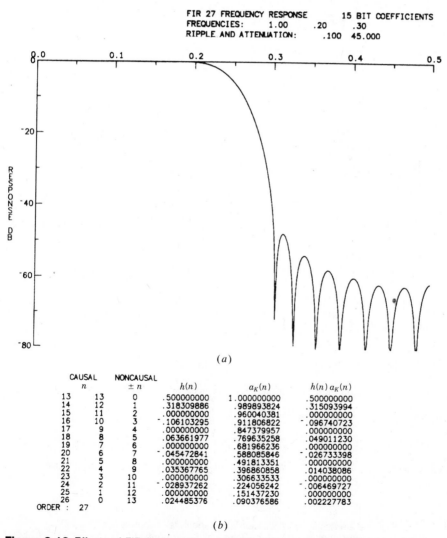

FIR 27 FREQUENCY RESPONSE   15 BIT COEFFICIENTS
FREQUENCIES: 1.00   .20   .30
RIPPLE AND ATTENUATION: .100   45.000

(a)

CAUSAL n	NONCAUSAL ± n	h(n)	$a_K(n)$	$h(n) a_K(n)$
13	13	0	.500000000	1.000000000	.500000000
14	12	1	.318309886	.989893824	.315093994
15	11	2	.000000000	.960040381	.000000000
16	10	3	-.106103295	.911806822	-.096740723
17	9	4	.000000000	.847379957	.000000000
18	8	5	.063661977	.769635258	.049011230
19	7	6	.000000000	.681966236	.000000000
20	6	7	-.045472841	.588085846	-.026733398
21	5	8	.000000000	.491813351	.000000000
22	4	9	.035367765	.396860858	.014038086
23	3	10	.000000000	.306633533	.000000000
24	2	11	-.028937262	.224056242	-.006469727
25	1	12	.000000000	.151437230	.000000000
26	0	13	.024485376	.090376586	.002227783

ORDER : 27

(b)

Figure 9.19 Effects of FIR filter coefficient quantization on magnitude response, (a) magnitude response for filter coefficients quantized to 16-bits, (b) 16-bit quantized coefficients.

uniformly distributed over $\pm q/2$. For this case, the roundoff noise has zero mean and variance given by

$$\sigma_{eo}^2 = \frac{q^2}{12} = \frac{2^{-2b}}{12} \tag{9.79}$$

That is, the only noise source is present at the filter output.

For the second case, if all multiplication products are rounded to b bits before they are summed, then the computational noise will be the sum of $(N + 1)/2$ uncorrelated noise sources. Since each noise source adds directly to

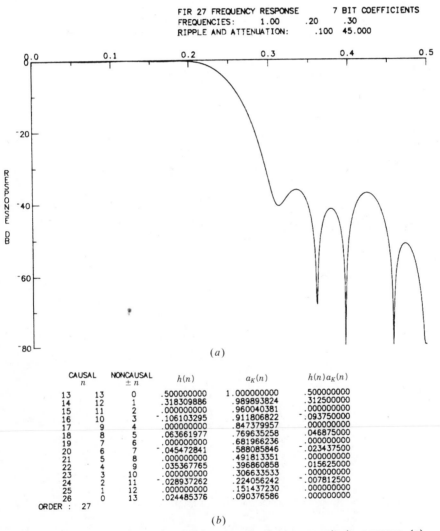

FIR 27 FREQUENCY RESPONSE   7 BIT COEFFICIENTS
FREQUENCIES: 1.00   .20   .30
RIPPLE AND ATTENUATION: .100   45.000

(a)

CAUSAL n	NONCAUSAL $\pm n$	$h(n)$	$a_K(n)$	$h(n)a_K(n)$
13	13	0	.500000000	1.000000000	.500000000
14	12	1	.318309886	.989893824	.312500000
15	11	2	.000000000	.960040381	.000000000
16	10	3	-.106103295	.911806822	-.093750000
17	9	4	.000000000	.847379957	.000000000
18	8	5	.063661977	.769635258	.046875000
19	7	6	.000000000	.681966236	.000000000
20	6	7	-.045472841	.588085846	-.023437500
21	5	8	.000000000	.491813351	.000000000
22	4	9	.035367765	.396860858	.015625000
23	3	10	.000000000	.306633533	.000000000
24	2	11	-.028937262	.224056242	-.007812500
25	1	12	.000000000	.151437230	.000000000
26	0	13	.024485376	.090376586	.000000000

ORDER : 27

(b)

Figure 9.20 Effects of FIR filter coefficient quantization on magnitude response, (a) magnitude response for filter coefficients quantized to 7-bits, (b) 7-bit quantized coefficients.

the output, we can write

$$\sigma_{eo}^2 = \left(\frac{N+1}{2}\right)\frac{2^{-2b}}{12} \tag{9.80}$$

Notice that the noise signal is independent of the filter coefficients since the noise is not recursively processed by the digital network.

FIR SCALING CONSIDERATIONS

To prevent against overflow for the direct-form implementation, the input signal should be scaled by a L_1 norm constant given by

$$C = \frac{1}{\displaystyle\sum_{n=0}^{N-1} |h(n)|} \tag{9.81}$$

This scaling is appropriate for wideband signals,[13] however, it may be too conservative for narrowband signals. For this case the reciprocal of the infinite norm of the magnitude response can be used.

9.5.1 FIR Coefficient Quantization Errors

The expected effect of finite wordlength on the frequency response can be qualitatively determined by computing the frequency response for different coefficient bit lengths. The frequency response for 15 and 7 magnitude bits are shown in Figures 9.19 and 9.20. As expected for 16 bits (sign bit plus 15 magnitude bits), the filter design specifications are preserved. For the 7-magnitude bit case we can see sidelobe distortion where the highest sidelobe is well above the 45-dB specification.

9.6 IIR DIGITAL FILTER LIMIT CYCLE OSCILLATIONS

When the input to an IIR digital filter, implemented with infinite precision arithmetic, is excited by an input sequence that is a constant at sample time n_0T and zero for $nT > n_0T$, the output will ideally decay to zero. However, with an implementation using finite register lengths outputs may oscillate to a nonzero amplitude range. This effect is referred to as *zero-input limit cycle oscillations* and is due to the nonlinear nature of the arithmetic quantization.

Consider a first-order IIR filter for which the difference equation is

$$y(n) = x(n) - 0.5y(n-1) \tag{9.82}$$

Assume the data register length is 3 bits plus a sign bit. If the input is an impulse specified by

$$x(n) = \begin{cases} 0.875 & n = 0 \\ 0 & \text{otherwise} \end{cases}$$

and rounding is applied after the arithmetic operation, then on successive iterations of the filter, the output will be

$$y(0) = 0.875 \qquad\qquad 0.1\ 1\ 1|0 \ \text{roundbit}$$

$$y(1) = -0.5\ (0.875) \qquad 1.0\ 1\ 1|1$$

$$\qquad = -0.5\ \text{(rounded)} \qquad 1.1\ 0\ 0|0$$

$$y(2) = -0.5\ (-0.5) \qquad 0.0\ 1\ 0|0$$

$$\qquad = 0.25\ \text{(rounded)} \qquad 0.0\ 1\ 0|0$$

$$y(3) = -0.5\ (0.25) \qquad 1.0\ 0\ 1|0$$

$$\qquad = -0.125\ \text{(rounded)} \qquad 1.0\ 0\ 1|0$$

$$y(4) = -0.5\ (-0.125) \qquad 0.0\ 0\ 0|1$$

$$\qquad = 0.125\ \text{(rounded)} \qquad 0.0\ 0\ 1|0$$

$$y(5) = -0.5\ (0.125) \qquad 1.0\ 0\ 0|1$$

$$\qquad = -0.125\ \text{(rounded)} \qquad 1.0\ 0\ 1|0$$

$$\vdots$$

Therefore, the steady-state output will oscillate between 0.125 and -0.125.

From this example it can be seen that for values of n in the limit cycle

$$|y(n-1)| = |\alpha y(n-1)|_Q \qquad\qquad (9.83)$$

the magnitude of the effective value of the filter coefficient α is unity, which corresponds to the pole being on the unit circle. These steady-state periodic outputs are bounded by $-2^{-b} \le y(n) \le 2^{-b}$, which is referred to as the *deadband*. For example, for $b = 3$ the output would oscillate between ± 0.125. Thus, the system will remain in a zero input limit cycle until an input is applied to carry the output out of the deadband. Clearly, limit cycle oscillations can be reduced by increasing the data register length.

A more severe type of limit cycle can occur due to summation node overflow in IIR filters implemented with two's-complement arithmetic. This limit cycle results in large periodic oscillations, which can be avoided by properly scaling the input data (see Section 9.3).

For the case of cascade realizations of second-order sections, the limit cycle behavior is difficult to analyze; however, it has been shown that the limit cycle noise measured under zero input conditions never exceeded the roundoff noise predicted by computer simulation under nonzero input conditions.[17]

9.7 FAST FOURIER TRANSFORM ERROR ANALYSIS

In this section we consider the FFT algorithm implemented by special-purpose digital hardware using fixed-point arithmetic. The FFT processor performance will be measured in terms of the specified dynamic range (DR), that is, the number of bits required for the stored sine/cosine table, and the arithmetic unit. Therefore, the specified system DR influences the decision about the word size of the FFT implementation. The effect of fixed-point arithmetic on the roundoff noise in FFT computations has been studied by Welch,[12] Weinstein and Oppenheim,[13] and Thong and Liu.[14] The following results will be based on the study performed by Weinstein and Oppenheim for the radix-2 decimation-in-time (DIT) version of the algorithm using rounding arithmetic.

Also presented are the results of research performed by other investigators. This includes the results of an FFT fixed-point error analysis, which takes into account the correlation between the roundoff errors for the radix-2 decimation-in-frequency (DIF) version of the algorithm. Also presented are expressions for the radix-4 DIT and DIF versions of the algorithm. Finally, the effect of arithmetic roundoff errors when the FFT is implemented using floating-point arithmetic will be considered.

Since the input signal increases in magnitude as it propagates through the butterfly stages of the FFT, it is necessary to scale the intermediate arrays to protect against array overflows in the arithmetic unit (AU). The error resulting from array scaling for an FFT implemented using fixed-point arithmetic will be considered in the analysis.

9.7.1 Fixed-Point FFT Error Analysis

For the DIT FFT radix-2 algorithm the basic computation is described by the signal flow graph of Figure 9.21. Consider the FFT butterfly equations given by

$$X_{m+1}(p) = X_m(p) + W^r X_m(q)$$
$$X_{m+1}(q) = X_m(p) - W^r X_m(q) \tag{9.84}$$

where the indices p and q represent the memory locations of the complex numbers in each array m. It should be noted that $m = 0$ is the initial input (time domain) array to the FFT, and $m = \mu$ refers to the final (frequency domain) output array. From Eq. 9.84 we can see that each butterfly of the

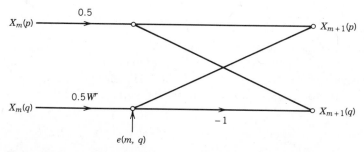

Figure 9.21 Radix-2 butterfly signal flow graph showing scaling con - stant of 0.5, and associated fixed point roundoff noise source.

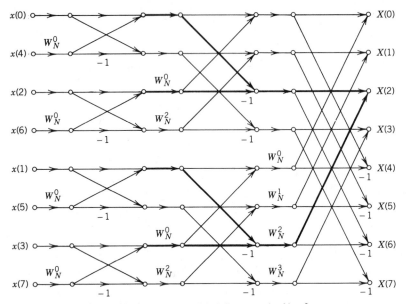

Figure 9.22 FFT decimation-in-time signal flow graph, $N = 8$.

FFT computation requires a multiplication involving W^r, that is, a complex multiply involving cosine and sine terms and subsequent additions. The scaling factor of $1/2$ preceding the butterfly computation is inserted to protect against overflow. Also shown in Figure 9.21 are the noise sources generated resulting from the complex multiply and the scaling operations.

Let us now consider the FFT signal flow graph for $N = 8$ shown in Figure 9.22 where at each stage $N/2$ butterfly computations are performed to produce the next stage array. Similar to the digital filter analysis described in the previous sections, we assume that each complex multiplication performed in the FFT algorithm introduces an error that is statistically independent. The numbers are represented by $(b + 1)$ bits, the first bit being the sign bit, and the binary point immediately following the sign bit. Assuming uniform distribution over the range $2^{-2b}/2 < e(n) < 2^{-2b}/2$, we obtain the variance of the roundoff error as given by

$$\sigma_e^2 = \frac{1}{12} 2^{-2b} \qquad (9.85)$$

where for rounding arithmetic the noise has zero mean.

Since the input data and the twiddle factors are complex, four real multiplications are associated with each complex multiply. Therefore, since there are four real multiplies for every complex multiply the variance of the roundoff error is given by

$$\sigma_{ce}^2 = 4\sigma_e^2 = \frac{2^{-2b}}{3} \qquad (9.86)$$

From Figure 9.22, $(N - 1) = 7$ noise sources propagate to each output node, which results in an output noise variance given by

$$\sigma_{oe}^2 = (N - 1)\sigma_{ce}^2 \tag{9.87}$$

This result shows that the variance of the output noise is independent of the node position in the output array and is proportional to N. It should be noted that multiplications involving twiddle factors given by 1 and j are performed noiselessly. However, it can be shown that for large N the analysis will still give desirable results. For this analysis it was assumed that the noise sources were uncorrelated. Since the uncorrelated assumption gives desirable results when compared to simulations, this aspect of the analysis will not be considered at this point. We should note, however, that expressions for output noise-to-signal ratio have been derived for DIF step-by-step scaling[15] and will be presented in Section 9.7.4.

9.7.2 FFT Scaling

To avoid overflows in the radix-2 version of the algorithm it has been proposed[12, 13] that, assuming fractional arithmetic and $|x(n)| < 1$, the inter-mediate arrays be shifted 1 bit to the right before the next array is computed. We refer to this scaling method as *automatic array scaling* (AAS). This method is considered worst-case block-floating point. For a block-floating point FFT, a shift is done only when overflow is detected. If an overflow is detected, the entire array is shifted right 1 bit and the computation is continued. The number of shifts are counted to determine the scale factor for the entire array. We refer to this method as *conditional array scaling*. In this section we consider AAS, which is considered worst-case or upper bound analysis.

With stage-by-stage scaling it can be shown that the output noise variance due to scaling is given by

$$\sigma_s^2 = \frac{2^{-2b}}{8} \tag{9.88}$$

where we note that if scaling is to be accomplished by rounded right shifts, and the bit to be shifted off is a 1, the rounding error is 0 with probability $1/2$, $-(1/2)2^{-b}$ with probability $1/4$, and $(1/2)2^{-b}$ with probability $1/4$. Since shifts of both real and imaginary parts are needed we have

$$\sigma_{es}^2 = \sigma_{ce}^2 + 4\sigma_s^2 = \frac{5}{6}2^{-2b} \tag{9.89}$$

which is slightly greater than σ_{ce}^2. Equation 9.89 is the noise variance due to multiplication by the twiddle factors and scaling. As a result of the attenuation of $1/2$ at each stage of the FFT, the noise variance σ_{es}^2 will be attenuated.

Finally, the output noise variance is given by

$$\sigma_{eo}^2 = \sigma_{es}^2 \sum_{k=0}^{\mu-1} \left(\frac{1}{2}\right)^k \approx \frac{5}{3} 2^{-2b} \tag{9.90}$$

where $N = 2^\mu$.

To determine the noise-to-signal variance at the output of the FFT, we assume that the input signal $x(n)$ is white, with real and imaginary parts uniformly distributed in $(-1/N\sqrt{2}, 1/N\sqrt{2})$. We then obtain for the output signal variance

$$\sigma_X^2 = \frac{1}{3N} \tag{9.91}$$

Finally, the noise-to-signal ratio for AAS to prevent overflows is obtained from Eqs. 9.90 and Eq. 9.91, that is,

$$\frac{\sigma_{eo}^2}{\sigma_X^2} = (5N)2^{-2b} \tag{9.92}$$

From Eq. 9.92 we note that the rms output noise-to-signal ratio increases as $\sqrt{N}$, or a half a bit per stage. Also, the assumption of a white-noise input signal is not essential in the analysis. Furthermore, it has been shown that a 3-bit improvement can be obtained for a block floating point implementation.[13] To reduce the noise-to-signal ratio, we would like to make σ_x^2 as large as possible; however, we are limited by the constraint $|x(n)| < 1$. In general, with scaling at each stage, the output signal variance is related to the variance at the input array by

$$\sigma_X^2 = \frac{1}{N}\sigma_x^2 \tag{9.93}$$

Combining Eqs. 9.90 and 9.93 we obtain

$$\frac{\sigma_{eo}^2}{\sigma_X^2} = \frac{(5/3)N2^{-2b}}{\sigma_x^2} \tag{9.94}$$

or the rms output noise-to-signal ratio is given by

$$\frac{\sigma_{eo}}{\sigma_X} = \frac{1.29\sqrt{N}\,2^{-b}}{\sigma_x} \tag{9.95}$$

9.7.3 FFT Coefficient Quantization

The computation of the FFT involves multiplication by complex coefficients (twiddle factors) of the form

$$W_N^r = e^{-j2\pi r/N} = \cos\frac{2\pi r}{N} - j\sin\frac{2\pi r}{N} \tag{9.96}$$

In many FFT processors, these coefficients are quantized to $(b + 1)$ bits and stored in memory for table look up. Quantization of the stored sine/cosine table introduces noise into the output of the FFT system. Weinstein[13] developed a simple expression for the ratio for the mean square output error to the mean square output signal of the form

$$\frac{\sigma_{eo}^2}{\sigma_X^2} = \mu \frac{2^{-2b}}{6} \tag{9.97}$$

We can see from Eq. 9.97 that the error variance increases very slowly with N.

9.7.4 Results of Other FFT Error Analyses

The rms output noise-to-signal ratio is determined by Welch[12] for the case of DIT and AAS to prevent overflow. For the cases of a sign-magnitude machine and a two's-complement machine the ratio of the rms error to the rms of the initial array is given by

$$\frac{\sigma_{eo}}{\sigma_X} \leq \frac{C\sqrt{N}\,(2^{-b})}{\sigma_x} \tag{9.98}$$

where $C = 1.13$. Welch also performed the analysis for a sign-magnitude machine and a two's-complement machine with truncation. The results showed that for these cases the constant C in Eq. 9.98 was higher. That is, for sign-magnitude truncation $C = 1.7$ and for two's-complement truncation $C = 2.55$.

Welch also noted that if transforms are generated to estimate spectra and subsequently averaged over time in a sequence of periodograms, then this averaging will decrease the computational error as well as the input noise error associated with the signal. Finally, Oppenheim and Weinstein have shown that if we are taking a transform and then its inverse, the errors in the two transforms are not independent.[13]

The minimum noise-to-signal ratio determined by Thong and Liu[14] for radix-2 DIT two's-complement rounding arithmetic with AAS is given by

$$\frac{\text{Noise}}{\text{Signal}} = 2^{-2b}N^2\left[\frac{8}{3}N - \left(2 + \frac{7\mu}{3}\right)\right] \tag{9.99}$$

The analysis was also performed using radix-2 DIF.

The mean square value of the total error for AAS and DIT radix-4 algorithm has been proposed by Prakash and Rao,[16] where the total error is given by

$$E_T = 2^{-2b}\left[\frac{4}{3}(N - 4\mu) + \frac{5}{6}(N - 1) + \frac{1}{2}(5N - 6\mu - 5)\right] \tag{9.100}$$

The total error is defined as the sum of the total multiplication error variance, total shifting error variance, and total error component due to the mean of shifting errors. The results of the analysis is included in a table that gives the minimum noise-to-signal ratio for values of N up to 1024. Included in the table is the radix-4 simulated and predicted results. For comparison, the predicted result of Thong and Liu of the radix-2 case is included. The results of the analysis showed that the radix-4 error performance is marginally better than the corresponding radix-2 FFT.

Reddy and Sundaramurthy have derived an expression for the output noise-to-signal ratio, taking into account the correlation between errors, when the FFT is computed with fixed-point arithmetic. The analysis is given for the radix-2 DIF form of the FFT algorithm using AAS to prevent overflow.[15]

The expression for the output noise-to-signal ratio taking into account the correlation among the truncation errors is given by

$$\frac{\sigma_{eo}^2}{\sigma_X^2} = 2^{-2b}\left(\frac{37}{6}N - 3\mu - 7 + \frac{4}{3N}\right) \tag{9.101}$$

When truncation errors are assumed uncorrelated the output noise-to-signal ratio becomes

$$\frac{\sigma_{eo}^2}{\sigma_X^2} = 2^{-2b}\left(\frac{19}{6}N - 4 + \frac{4}{3N}\right) \tag{9.102}$$

The analysis shows that the predicted results accounting for correlation between errors fall closer to the experimental results than the expression that is derived with the assumption that the errors are uncorrelated.

9.7.5 FFT System Dynamic Range Considerations

A typical FFT system is shown in Figure 9.23, where the essential elements are an automatic gain control (AGC), analog-to-digital converter (ADC), weighting function, and an FFT processor consisting of a stored $(b + 1)$ bit sine/cosine table and $(b + 1)$ bit arithmetic unit (AU). Since each of these system elements must be implemented by finite-length digital hardware, each contributes to error at the output. The effect of this error is to limit the DR of the FFT processor.

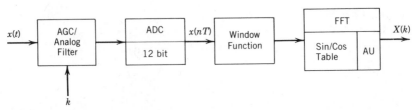

Figure 9.23 Spectral analysis system.

The fundamental source of error in an FFT system is its ADC, which samples and quantizes the continuous input signal.[19] The sources of ADC error are the quantization, saturation, and sampling error. The quantization error arises from the sampling and quantization of a continuous input signal such that all signal amplitudes lying within a given range are arbitrarily assigned the same amplitude. Saturation error arises when the input signal exceeds the voltage range of the ADC. Sampling error arises from the fact that the continuous signal is sampled at $x(nT - \mu_n)$ rather than at $x(nT)$, where μ_n are deviations of the sampling times nT. In this development, we consider these deviations to be negligible. The theoretical relationship for the DR of the ADC has already been developed in Section 9.1.2 for the quantization and saturation error.

The theoretical maximum DR that can be attained by an FFT processor with stored sine/cosine table is given by

$$DR_c = -10\log_{10}\left[0.17\mu 2^{-2b}\right] \tag{9.103}$$

where b is the number of bits (not including the sign bit) to which the data values of the sine/cosine table is stored. The other error source in FFT processors that limits the DR of a system is the AU. The theoretical maximum DR for rounded arithmetic using AAS to prevent overflows is given by Eq. 9.104.

$$DR_r = -10\log_{10}\left(5N2^{-2b}\right) \tag{9.104}$$

Example 9.5

The theoretical maximum DR that can be obtained by an FFT processor with stored coefficient table and AAS scaling is given by Eqs. 9.103 and 9.104 for the coefficients and arithmetic unit (AU) wordlength.

a. Express the number of bits required for the stored coefficients and AU as a function of the number of radix-2 FFT stages (μ) and the required DR, respectively.

b. Assume that for a given application the FFT processor specifications require a DR of at least 60 dB and an FFT size of 1024. From the results of part a, determine the number of bits required for the stored sine/cosine table and the AU in order to meet the specified DR.

c. Show that the DR of a 12-bit A/D converter can be determined using Eq. 9.6.

Solution:

a. From Eq. 9.103 the number of bits (not including the sign bit) to which the FFT coefficient sine/cosine table is rounded is expressed by

$$b_c = 1.66\log_{10}\mu - 1.29 + 0.166DR \tag{9.105}$$

Similarly, from Eq. 9.104 the required AU wordlength as a function of the FFT size and DR is given by

$$b_r = 1.16 + 0.5\mu + 0.166DR \qquad (9.106)$$

b. To achieve the specified requirements for an FFT processor the DR of the A/D converter must be at least 60 dB. From Table 9.1 we can see that the DR (forced to be positive) of a 12-bit A/D converter is 62.55 dB. Therefore, we must design the remainder of the system such that we maintain this DR. The coefficient wordlength that meets the required DR is given by

$$b_c = 1.66\log_{10}10 - 1.29 + 0.166(62.55 \text{ dB}) = 10.75 \text{ bits} \qquad (9.107)$$

If we include the sign bit, the FFT coefficient table should be at least 12 bits, where we note that the A/D wordlength is equal to $b + 1$. The AU wordlength is given by

$$b_r = 1.16 + 0.5(10) + 0.166(62.55 \text{ dB}) = 16.54 \text{ bits} \qquad (9.108)$$

If we include the sign bit, a 18-bit AU will meet the required DR, where AAS was assumed to prevent array overflows.

c. From Eq. 9.6, the DR of the A/D converter is given by

$$DR = SNR_Q = 6b + 10.8 - 20\log_{10}k = 62.65 \text{ dB} \qquad (9.109)$$

where from the Table 9.1 the AGC constant for a 12-bit A/D converter is $k = 5.1$. We can see that the result obtained from Eq. 9.6 is approximately equal to the value given in Table 9.1 (62.55 dB). ∎

9.7.6 Fixed-Point FFT Simulation

A computer program was written to determine the performance of a 16-bit FFT processor. The simulation is shown in Figure 9.24, where the 16-bit fixed-point processor is compared to a 64-bit floating-point processor (which was considered ideal).

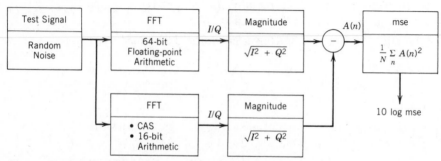

Figure 9.24 Computation of FFT processor mean square error (mse).

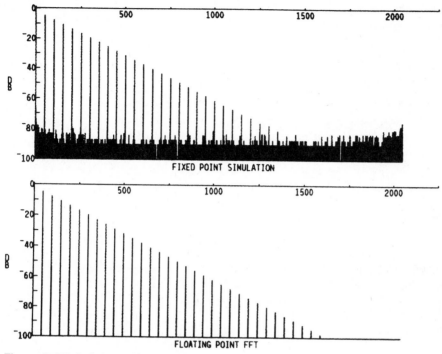

Figure 9.25 Output of FFT fixed-point and floating-point simulation.

The accuracy of the FFT processor can be expressed in terms of the mean-squared error defined as

$$\text{mse} = \frac{1}{N} \sum_{n=1}^{N} \left[|I(n) + jQ(n)| - |I'(n) + jQ'(n)| \right]^2 \qquad (9.110)$$

where $I(n)$ and $Q(n)$ are the real (in-phase) and imaginary (quadrature) components of the fixed-point simulator. Likewise, $I'(n)$ and $Q'(n)$ are the real and imaginary components of the floating-point simulator. The mse was calculated for independent records of random noise and was determined to be less than -75 dB for a 16-point FFT processor.

The effects of 16-bit arithmetic on the output of the fixed-point processor is illustrated by processing a ramp of tonals as input to the fixed-point and floating-point FFT processors. The results of the simulation are presented in Figure 9.25, where the dynamic range (DR) of the fixed-point FFT processor is shown as the range for which the tonals are above the computational noise floor (less than -75 dB).

It should be emphasized that the DR of an FFT system is limited by the A/D converter, that is, the A/D converter saturation and quantization error limits the DR of the processor, and we need only to maintain this DR throughout the processing elements of the system.[20]

9.7.7 FFT Floating-Point Considerations

FFT floating-point error analysis has been studied by Weinstein,[18] who presents both theoretical and simulation results. The theoretical results are derived for rounded arithmetic and a white-noise input signal (flat spectrum). The result, derived to predict output noise-to-signal ratio for a floating-point FFT processor is given by

$$\frac{\sigma_{eo}^2}{\sigma_X^2} = (0.42\mu)2^{-2b} \tag{9.111}$$

where b is the number of bits in the mantissa of the floating-point processor and μ is the number of radix-2 FFT stages.

In this section we compare the DR of a fixed-point FFT processor and the DR of a floating-point FFT processor for a total wordlength of 24 bits. From Eqs. 9.104 and 9.111 we obtain: Fixed point (AAS)

$$DR_{FIP} = 6.02b_r - 3.01\mu - 6.99 \tag{9.112}$$

Floating point

$$DR_{FLP} = 6.02b_r - 10\log_{10}\mu + 3.77 \tag{9.113}$$

where for the fixed-point processor $b_r = 23$ fractional bits, and for the floating-point processor $b_r = 18$ mantissa bits. For the floating-point processor, we allowed 1 bit for the sign of the mantissa, 1 bit for the sign of the exponent, and 4 bits were made available for the exponent. A comparison of the DR of a fixed-point FFT processor versus a floating-point FFT processor as a function

TABLE 9.4
FIXED-POINT VERSUS FLOATING POINT FFT PROCESSOR PERFORMANCE

μ	DR_{FIP} (dB)	DR_{FLP} (dB)	Difference (dB)
8	107.39	103.10	−4.29
9	104.38	102.59	−1.79
10	101.37	102.13	0.76
11	98.36	101.72	3.36
12	95.35	101.34	5.99
13	92.34	100.99	8.65

of the FFT size was performed. The results of this analysis are given in Table 9.4. From Table 9.4 we can see that an FFT length of $N \geq 1024$ is necessary before the performance of the floating-point FFT processor exceeds the fixed-point processor.

9.8 SUMMARY

The effects of finite arithmetic on DSP system performance can result in significant losses. We have presented a detailed approach to assessing the level of noise resulting from filtering and FFT operations. For IIR filters we showed that significant noise buildup can result from the recursive structure. For FIR filters the noise is limited by the number of coefficients (i.e., the impulse response length).

The quantization of the filter coefficients results in an error in the frequency response of the filter. Errors in the IIR coefficients can result in instability if the magnitude of the pole values exceeds unity. For FIR filters, very long impulse response lengths can require long register lengths. FFT twiddle factor quantization was addressed, and we found the coefficient wordlength requirement increases slowly with increasing FFT lengths.

The procedures defined can be easily automated and included as part of the DSP design process. For fixed-point arithmetic processors scaling is a key consideration of the overall DSP design. Advances in hardware technologies are making floating-point processing a viable alternative for real-time processors. This simplifies the design process because it eliminates the complexity of the signal scaling design requirement.

References

1. D. F. Hoeschele, *Analog-to-Digital and Digital-to-Analog Conversion Techniques*, Wiley, New York, 1968.

2. H. Schmid, *Electronic Analog/Digital Conversions*. New York, Van Nostrand, 1970.

3. G. A. Gray and G. W. Zeoli, "Quantization and Saturation Noise Due to Analog-to-Digital Conversion," *IEEE Trans. AES*, January 1971, pp. 222–223.

4. M. Abramowitz and I. A. Stegun, *Handbook of Mathematical Functions*, Dover, New York, 1965.

5. L. B. Jackson, "Roundoff Noise Analysis for Fixed-Point Digital Filters Realized in Cascade or Parallel Form," *IEEE Trans. AGS*, June 1970, pp. 107–122.

6. S. A. Tretter, *Introduction to Discrete-Time Signal Processing*, Wiley, New York, 1976.

7. B. Gold and C. M. Rader, *Digital Processing of Signals*, McGraw–Hill, New York, 1969.

8. A. Peled and B. Liu, *Digital Signal Processing: Theory, Design and Implementation*, Wiley, New York, 1976.

9. J. F. Kaiser, *Digital Filters in Systems Analysis by Digital Computer*, Wiley, New York, 1966.

10. D. S. K. Chan and L. R. Rabiner, "Analysis of Quantization Errors in the Direct Form for FIR Digital Filters," IEEE Trans. AES, August 1973, pp. 354–366.

11. L. R. Rabiner and B. Gold, *Theory and Application of Digital Signal Processing*, Prentice–Hall, Englewood Cliffs, N.J., 1975.

12. P. Welch, "A Fixed-Point FFT Error Analysis," *IEEE Trans. AGS*, June 1969, pp. 153–157.

13. A. V. Oppenheim and C. J. Weinstein, "Effects of Finite Register Length in Digital Filtering and the FFT," *Proc. IEEE*, August 1972, pp. 957–976.

14. T. Thong and B. Liu, Fixed-Point FFT Error Analysis, *IEEE Trans. ASSP*, December 1976, pp. 563–573.

15. V. U. Reddy and M. Sundaramurthy, "New Results in Fixed-Point FFT Error Analysis," *IEEE Trans. Comput.*, March 1977, pp. 120–125.

16. S. Prakash and V. V. Rao, "Fixed-Point Error Analysis of Radix-4 FFT," *Signal Processing*, vol. 3, April 1981, North-Holland, Amsterdam, pp. 123–133.

17. R. B. Kieburtz, "An Experimental Study of Roundoff Effects in a Tenth-Order Recursive Digital Filter," *IEEE Trans. Comm.*, June 1973, pp. 757–763.

18. C. J. Weinstein, "Roundoff Noise in Floating Point FFT Computation, *IEEE Trans. AES*, September 1969, pp. 209–215.

19. J. H. Glisson, C. J. Black and A. P. Sage, "The Digital Computation of Discrete Spectra Using the Fast Fourier Transform," *IEEE Trans. AES*, September 1970.

20. E. O. Brigham and L. R. Cecchini, "A Nomogram for Determining FFT System Dynamic Range," IEEE International Conference on ASSP, May 1977.

21. J. P. Dugre, A. A. L. Beex, L. L. Schraf, "Generating Covariance Sequences and the Calculation of Quantization and Rounding Error Variances in Digital Filters," *IEEE Trans. ASSP*, 28:102–104, 1980.

22. J. P. Dugre, E. I. Jury, "A Note on the Evaluation of Complex Integrals Using Filtering Interpretations," *IEEE Trans.*, *ASSP*, 30:804–807, 1982.

23. E. I. Jury, *Theory and Application of the Z-Transform Method*, Wiley, New York, 1964.

Additional Readings

PARKS T. W., AND BURRUS C. S., Digital Filter Design. Wiley, New York, 1987.

This book provides an extended discussion of finite word-length effects. A design example of a fourth-order elliptic filter is presented and analyzed in detail. Finally, an assembly language program is provided for implementing the filter on a signal-processing chip (TMS32010 from Texas Instruments).

PROBLEMS

9.1.1 The output signal of an 8-bit A/D converter is passed through a first-order digital filter with transfer function

$$H(z) = \frac{z}{z - 0.9}$$

Find the steady-state output noise power (variance) due to A/D converter quantization.

9.1.2 Show that the total noise power (PN/σ^2) listed in Table 9.1 can be approximated by Eq. 9.6, that is

$$SNR_Q \approx PN/\sigma^2$$

for $M = 2, 4, \ldots, 16$.

9.1.3 It is desired to write a computer program to generate tables of the A/D converter noise power versus AGC constant shown in Figure 9.4. The tables should be generated for $M = 2, 3, 4, \ldots, 16$, and the results listed according to the following format.

M BIT QUANTIZER			
k	QN/σ^2	SN/σ^2	PN/σ^2 (dB)
0.2	xxx	xxx	xxx
$\vdots$	$\vdots$	$\vdots$	$\vdots$

let k range from 0.2 through 8.0 in increments of 0.2.

9.2.1 Derive Eq. 9.42.

9.2.2 Determine the total steady-state output noise variance σ_{eo}^2 for the two first-order sections shown in Figure 9.8 if the sections are interchanged.

9.2.3 Find the section ordering that minimizes the output noise power caused by fixed-point finite wordlength arithmetic for the cascaded sections of Problem 9.2.2. Let $B_1 = 0.5$ and $B_2 = 0.9$. This problem shows that the order in which the sections are cascaded influences the output noise power due to roundoff. This problem is further developed in Section 9.3.1 for cascaded second-order sections.

9.2.4 Verify the output noise variance σ_{eo}^2 given in Example 9.2.

9.2.5 The transfer function of a second-order section is given by

$$H(z) = \frac{1}{1 + B_1 z^{-1} + B_2 z^{-2}}$$

whose complex-conjugate poles are given by

$$z_{1,2} = re^{\pm j\phi} = a \pm jb$$

Therefore,

$$B_1 = -2a \qquad B_2 = a^2 + b^2 = r^2$$

and $|r| < 1$ for stability. Determine the output noise power given by Eq. 9.25.

9.2.6 *A closed-form solution to the contour integral*: The computation of the steady-state output noise variance due to input signal quantization requires the evaluation of the noise power gain (NPG) defined by Eq. 9.25. The method developed by Dugre and Jury[21-23] is summarized here. Let $H(z)$ denote the transfer function of an IIR digital filter:

$$H(z) = \frac{\displaystyle\sum_{m=0}^{M} A_m z^{-m}}{1 + \displaystyle\sum_{n=1}^{N} B_n z^{-n}} \qquad M \le N$$

Then the NPG can be obtained from the determinants of the matrices describing numerator and denominator coefficients as follows:

$$NPG = \frac{|\Omega_1|}{B_0 |\Omega|}$$

where Ω is represented by the following matrix of denominator coefficients:

$$\Omega = \begin{vmatrix} B_0 & B_1 & B_2 & B_3 & \cdots & & & & B_N \\ B_1 & B_2 + B_0 & B_3 & B_4 & \cdots & & & B_N & 0 \\ B_2 & B_3 + B_1 & B_4 + B_0 & B_5 & \cdots & & B_N & 0 & 0 \\ B_3 & B_4 + B_2 & B_5 + B_1 & B_6 + B_0 & \cdots & B_N & 0 & 0 & 0 \\ \vdots & \vdots & \vdots & \vdots & & & & & \\ B_N & B_{N-1} & B_{N-2} & B_{N-3} & \cdots & B_3 & B_2 & B_1 & B_0 \end{vmatrix}$$

where $B_0 = 1$ and Ω_1 is the numerator matrix formed from Ω by replacing the first row of Ω by the following vector.

$$\mathbf{c} = \begin{vmatrix} \displaystyle\sum_{i=0}^{N} A_i^2 \\ 2\displaystyle\sum_{i=0}^{N-1} A_i A_{i+1} \\ 2\displaystyle\sum_{i=0}^{N-2} A_i A_{i+2} \\ \vdots \\ 2\displaystyle\sum_{i=0}^{N-(N-1)} A_i A_{i+N-1} \\ 2A_0 A_N \end{vmatrix}$$

The element in the ith and jth column of matrix Ω is given by

$$B_{i,j} = \frac{B_{i-1} \qquad\qquad\quad \text{for } j = 1}{B_{i+j-2} + B_{i-j} \quad \text{for } j > 1}$$

$$i = j = 1, 2, \ldots, N + 1$$

where $B_k = 0$ for $k < 0$, and $k > N$. The matrix Ω is nonsingular for stable filters. For example, a (5×5) Ω matrix is found to be

$$
\Omega = \begin{vmatrix}
B_0 & B_1 & B_2 & B_3 & B_4 \\
B_1 & B_2 + B_0 & B_3 & B_4 & 0 \\
B_2 & B_3 + B_1 & B_4 + B_0 & 0 & 0 \\
B_3 & B_4 + B_2 & 0 & B_0 & 0 \\
B_4 & B_3 & B_2 & B_1 & B_0
\end{vmatrix}
$$

a. Given the first-order section transfer function

$$
H(z) = \frac{A_0 z + A_1}{B_0 z + B_1}
$$

Show that the NPG can be expressed as

$$
NPG_1 = \frac{(A_0^2 + A_1^2) B_0 - 2 A_0 A_1 B_1}{B_0 (B_0^2 - B_1^2)}
$$

b. Given the second-order section transfer function

$$
H(z) = \frac{A_0 z^2 + A_1 z + A_2}{B_0 z^2 + B_1 z + B_2}, \qquad B_0 = 1
$$

show that the NPG can be expressed by

$$
NPG_2 = \frac{F_0 B_0 E - F_1 B_0 B_1 + F_2 (B_1^2 - B_2 E)}{B_0 \left[(B_0^2 - B_2^2) E - (B_0 B_1 - B_1 B_2) B_1 \right]}
$$

where

$$
\begin{aligned}
F_0 &= A_0^2 + A_1^2 + A_2^2 \\
F_1 &= 2(A_0 A_1 + A_1 A_2) \\
F_2 &= 2 A_0 A_2 \\
E &= B_0 + B_2
\end{aligned}
$$

9.2.7

a. Given the following second-order section filter coefficients determine the NPG_2 given in Problem 9.2.6b.

$A0 = 0.07062389$ $B0 = 1$

$A1 = -0.00051439171159$ $B1 = -0.88866037$

$A2 = 0.07062389$ $B2 = 0.926866771$

b. The steady-state output noise variance of a second-order section is given by

$$
\sigma_{eo}^2 = \frac{2^{-2b}}{12} \left[\frac{rNPG}{C^2} + s \right]
$$

$$
\begin{aligned}
r &= 3 \\
s &= 2 \\
C &= 0.07062389 \\
b &= 15 \text{ bits}
\end{aligned}
$$

Using the NPG obtained in part a, determine σ_{eo}^2, where r is the number of noise source at the input sum node of the filter, s is the number of noise sources at the output sum node of the filter, and C is the filter normalization constant.

9.2.8 Given the two cascaded second-order sections

$$H(z) = H_1(z)H_2(z)$$

$$= \frac{A_{01}z^2 + A_{11}z + A_{21}}{B_{01}z^2 + B_{11}z + B_{21}} \times \frac{A_{02}z^2 + A_{12}z + A_{22}}{B_{02}z^2 + B_{12}z + B_{22}}$$

a. Show that the convolution of the coefficients A_{i1} with A_{i2}, and B_{i1} with B_{i2} results in the numerator and denominator coefficients for the fourth-order section transfer function

$$H(z) = \frac{C_0z^4 + C_1z^3 + C_2z^2 + C_3z + C_4}{D_0z^4 + D_1z^3 + D_2z^2 + D_3z + D_4}$$

b. Find the noise power gain of the fourth-order section transfer function[23] using the method described in Problem 9.2.6.

HINT: See Ref. 23, page 299.

9.2.9 The lowpass elliptic digital filter coefficients of two second-order sections are given by

Section 1

$A01$: 1	$B01$: 1
$A11$: 1.765403788	$B11$: -1.113385321
$A21$: 1	$B21$: 0.402397151

Section 2

$A02$: 1	$B02$: 1
$A12$: 0.969623665	$B12$: -0.903967676
$A22$: 1	$B22$: 0.760028928
$C_1 = 0.0767503$	$C_2 = 0.2721642$

a. Determine the NPG of a fourth-order section using the closed-form solution obtained in Problem 9.2.8b.

b. Given the following steady-state output noise variance

$$\sigma_{eo}^2 = r\sigma_e^2 \left[\frac{NPG12}{C_1^2} + \frac{NPG22}{C_2^2} \right] + s\sigma_e^2 [NPG22 + 1]$$

$$r = 3 \qquad \sigma_e^2 = 2^{-2b}/12$$
$$s = 2 \qquad b = 15$$

where NPG12 is the noise power gain obtained for the fourth-order section obtained in part a, and NPG22 is the noise power gain of second-order section 2 (see Problem 9.2.6) determine σ_{eo}^2.

9.2.10 Given the first-order realization

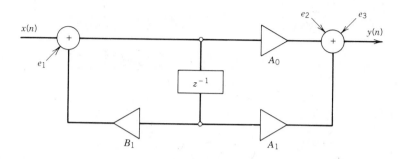

Find the output noise variance.

9.3.1 Derive Eqs. 9.70 through 9.73.

9.3.2 Equation 9.74 defines the modified normalization (scale) coefficients. Using Table 9.2 show that

$$S_1 = \frac{1}{\max|G_{31}|}$$

$$S_k = \frac{C_k M_{i,k-1}}{M_{i,k}} = \frac{1}{\prod_{n=1}^{k-1}S_n \max|G_{i,k} \prod_{n=1}^{k-1}G_n|}$$

$$k = 2,3,4 \quad \text{(Scale factor index)}$$
$$i = 1,2,3 \quad \text{(Sum node index)}$$

9.3.3 Show that second-order section scale coefficients given in Problem 9.3.2 can also be expressed as

$$S_1 = \frac{C_1}{M_{31}}$$

$$S_k = \frac{\prod_{n=1}^{k}C_n}{M_{i,k}\prod_{n=1}^{k-1}S_n}$$

$$k = 2,3,4 \ (\text{Scale coefficient index})$$
$$i = 1,2,3 \ (\text{Sum node index})$$

Using the foregoing equation, determine the scale coefficients S_k for the third sum node given in Figure 9.13e. Note that the results are identical with Eq. 9.76.

10

Signal Processing
System Design

Our signal processing system design methodology provides a systematic approach to specifying requirements, analyzing signals, and developing signal processing designs. The methodology results in efficient signal processing hardware and software implementations.

10.0 INTRODUCTION

The digital signal processing (DSP) techniques presented in the preceding chapters have applications to may diverse fields such as acoustics, radar, sonar, seismology, speech, data communication, and biomedical engineering. This chapter presents a DSP system design methodology and illustrates the use of the methodology in solving DSP system application problems. The methodology uses design and analysis approaches discussed in the preceding chapters. Implementation of the methodologies using computer-aided design and analysis tools provides an efficient approach to handling complex application designs.

Signal processing system design consists of understanding the system application problem, defining signal processing algorithms, implementing the algorithms in hardware and software, and performing a thorough design analysis resulting in a system hardware and software configuration satisfying all requirements. In Chapter 1 we introduced the DSP design methodology consisting of five steps: application requirements analysis, signal analysis, signal processing design, resource analysis, and configuration analysis. In this chapter we expand on the introduction of DSP system design. We discuss the interrelationship between the DSP operations, the corresponding arithmetic requirements, and the signal processor architecture design.[1-4]

Many DSP applications require real-time processing over long periods of time. For real-time processing, successive input samples must be processed to produce an output data stream within a relatively short time delay. This places maximum requirements on the design since the data cannot be stored and subsequently processed over a time period much longer then the data period (non-real-time). The processor must execute the algorithm fast enough to avoid any loss of input data. The requirements may dictate very extreme environmental conditions under which the processor must operate (e.g., military applications). In addition, the space, weight, and power requirements must be minimized. A key design constraint is the system cost.

Often, the DSP algorithms and/or applications offer inherent parallelism in the computation requirements that can be exploited by the processor design. The signal processor architecture and the corresponding signal processing implementation must be complementary. For some applications, the signal processor is given and the job is to develop efficient signal processing designs that satisfy the application requirements. The goal is to develop a signal processing design such that a minimum amount of the overall processor resources are used. This leaves unused resources for growth to meet future processing requirements. Processor resources refer to all measurable parameters such as multiplication rate, memory capacities, control instruction capacities, and I/O data rates. In the case of an existing processor, we must live with the arithmetic precision/wordlengths of the processor. If the processor has not been specified, then we must select an existing processor or design a new processor to meet the requirements. A new design gives the most flexibility, but development costs must be assessed along with the overall cost for the total planned production volumes. In either case, the methodology presented provides a procedure for arriving at a design that satisfies all requirements.

First we present a generic signal processor architecture and discuss the key resource parameters associated with the signal processing system design. Next we present the DSP design methodology. Then a brief description of a narrowband spectral analysis detection system is given, and the performance parameters influencing the design are presented. Finally, a detailed implementation is derived for the detection system in order to illustrate the use of the methodology described. Implementation considerations in hardware and software are discussed. Personal-computer-based techniques for performing the designs are shown.

10.1 SIGNAL PROCESSOR APPLICATION CONSIDERATIONS

The task of developing an efficient signal processing system design is dependent on the signal processor hardware and software architecture, including the data flow and the resource capacities of the processor. The architecture of the signal processor used to implement the application has a significant impact on the resultant signal processing design. For applications where neither the signal processor nor the signal processing are specified, the designer has the flexibility to perform trade-offs between the architecture and the processing operations to determine the best system design. We define a digital signal processing flow graph format, define a generic signal processor architecture, and discuss the resource parameters that must be analyzed during the design process.

10.1.1 Signal Processor Application Environment

The following items provide a brief introduction to a total processing application environment. The interaction between the system elements shown in Figure 10.1 must be understood to clearly identify the signal processor requirements.

Sensor System: The *sensor system* provides the input digital data streams to the signal processors. We will discuss signal conditioning considerations during the design process, but will assume that the sensor system performs these operations (frequency selectivity, gain control, analog-to-digital conversion, etc.).

System Manager: The *system manager* performs the overall management of the application. This includes processing option selections, mode selections, mode parameter changes, input/output channel selections,

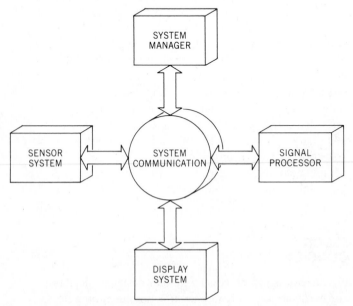

Figure 10.1 Signal processing system.

sensor system control, display system control, and fault monitoring. The application requirements must be translated into system manager requirements and subsequently used to define the requirements of the other system components.

Signal Processor: The *signal processors* perform the primary DSP functions required by the application. Of course, the signal conditioning performed in the sensor system and the data manipulation/processing performed by the display system are also signal processing operations. As part of the design process, trade-offs are required to establish the partitioning of the processing functions between these systems. Operations that are unique to specific sensors and are very special purpose high speed operations are often partitioned to the sensor system. Low data rate general purpose operations associated with the display of the data or special purpose display operations are often partitioned to the display system.

Display System: The *display system* provides the user/machine interfaces to the application. The display processors must provide the required data processing and formatting for presentation. Many forms of postprocessing algorithms have been defined, including automatic detection and enhancing techniques. These processes must be partitioned between the system manager, the signal processor, and the display system.

System Communication: The *system communication* approach is dependent on a broad set of system level requirements. Multiple operator stations that each require the flexibility to view any of the signal processors outputs may be required. Also, multiple sensor inputs that must be routed under system control to any of the signal processors may be required. Input data types, formats, and rates are key resource parameters.

All the foregoing system level items play an important role in the overall system design. We focus on the digital signal processor design herein. The following sections provide a brief introduction of signal processor architecture definitions and design considerations.

10.1.2 Signal Processor Architecture Definitions

In order to introduce signal processor architecture design, we present a basic set of definitions.

Processor Architecture: We define the *processor architecture* as the signal processor data flow, arithmetic capabilities, memory capabilities, I/O capabilities, programmability, and instruction set.

Processor Resource: A *processor resource* is a parameter that can be quantified in terms of maximum allowable utilization by the processing design. Examples include data memory capacity in words and in word-length, data transfer bandwidths, arithmetic computation capacity, processing capacity in cycles per second, and instructions per second.

Operating System: The *operating system* software provides the overall signal processor control and processing mechanisms under which the application programs run.

Data Stream: The *data stream* defines the input signal to be processed. A single data (SD) stream is supported by a signal processor with one

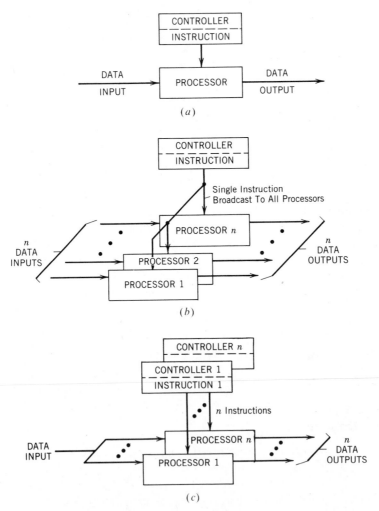

Figure 10.2 Instruction / data processor configurations.

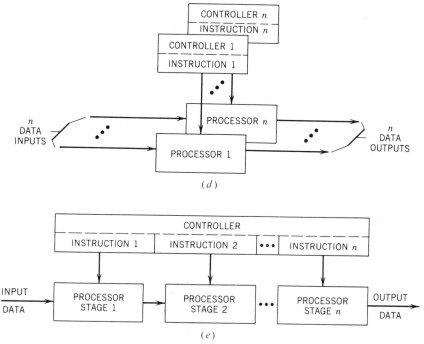

Figure 2 Continued

input. Multiple data (MD) streams require a signal processor that supports multiple input sources.

Instruction: The *instruction* defines the operation to be executed by the signal processor at each machine cycle. A single instruction (SI) architecture supports a processor architecture that executes only one instruction per machine cycle and a multiple instruction (MI) architecture supports execution of multiple instructions per machine cycle.

Instruction Set Architecture (ISA): The *ISA* defines the total set of basic operations that the processor is capable of performing (e.g., a computer is defined by its assembly level set of instructions). The ISA includes all instructions for input/output, memory addressing, arithmetic processing, data manipulation, and logical operations in a manner that optimizes use of memory and time of execution.

Single Instruction Single Data Stream (SISD) Architecture: The *SISD architecture* is the traditional computer architecture (e.g., Von Neumann). This does not generally support the requirements of real-time signal processing applications. It may satisfy the control and general purpose (GP) processing requirements (Fig. 10.2a).

Single Instruction Multiple Data Stream (SIMD) Architecture: A processor architecture that executes the same instruction across multiple input channels is called a *SIMD* processor. This implies parallel hardware using a shared instruction controller (Fig. 10.2*b*).

Multiple Instruction Single Data Stream (MISD) Architecture: The data input is processed by multiple instructions in the *MISD* architecture. This could be in the form of parallel hardware operating on the same input or serial stages of hardware where each stage executes a different instruction on the data as it streams through the stages. The latter form represents the case for a pipeline design (Fig. 10.2*c*).

Multiple Instruction Multiple Data Stream (MIMD) Architecture: The *MIMD* architecture represents the maximum flexibility by providing a complete multiprocessing environment. Different instructions can be executed on different input data streams. This could represent different data channel inputs from the sensor system and different processing modes. Very often, the design could share a memory source or data interconnections. This architecture clearly represents a parallel array-processing structure (Fig. 10.2*d*).

Pipeline Signal Processor: A *pipeline signal processor* is configured to take advantage of the inherent decomposition of DSP algorithms into multiple serial operations. The pipeline provides multiple stages through which the data progress at the basic cycle rate of the hardware. At each successive stage in the pipeline an operation (multiply, add, shift, delay, etc.) is performed. Designs can be achieved for specific DSP operations such as the FFT that approaches the maximum capability of the processor (Fig. 10.3*a*)

Parallel Signal Processor: A *parallel signal processor* is configured to take advantage of the inherent parallelism of DSP algorithms and/or applications. All processors in the parallel configuration can receive different data streams and execute different instructions at each cycle, but the control mechanisms and data-handling complexities must be accounted for in the design. For this reason, total flexibility may be sacrificed to simplify the architecture provided the application requirements can be satisfied (Fig. 10.3*b*).

Array Signal Processor: An *array signal processor* is configured to process array operations efficiently. Multiple processing elements are configured to provide the processing. Generally, the architecture is driven by a type of application that requires a certain high-speed processing such as handling large matrices. The continued rapid increase in the density of technologies has resulted in considerable work on

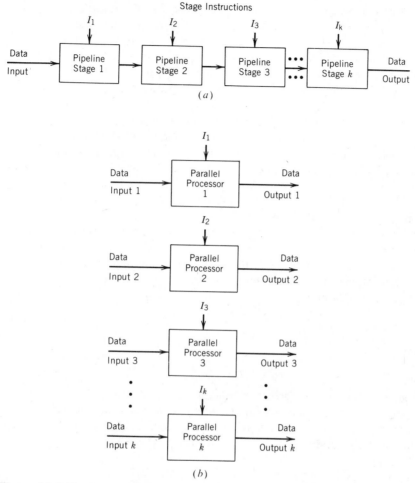

Figure 10.3 Pipeline and parallel array processor configurations.

mapping of algorithms onto array structures. This includes basic operations such as multiplications and higher level operations such as FFT butterflies. Both pipeline and parallel processing techniques provide array processors with high concurrencies and therefore high efficiency. Another definition is a processor that operates on arrays as a normal mode of operation regardless of its architecture.

Vector Signal Processor: A *vector signal processor* provides a complete instruction set to operate on data vectors. Therefore, the conventional conversion of data vectors into scalar form for operation on by *DO loops* is avoided.

Systolic Array Processor: A *systolic array processor* is an array of processors of which each executes a simple operation providing a rhythmic flow of data through the array. The entire array must be synchronized, which presents a problem for large arrays. The key to mapping an operation to a systolic array is the ability to derive an array structure with high concurrency using a pipeline approach.[4]

Throughput (Thruput) Capacity: *Throughput capacity* is a measure of the raw capability of the processor to perform number crunching. If a processor was configured with a multiply and accumulate capability at a basic rate of 10 MHz, then the throughput of the processor would be 10 million multiplications per second (MMPS) and 10 million additions per second (MAPS). Often the individual capacities are combined to state an overall throughput capacity in operations per second. The most accurate measure is the machine cycles it takes to perform each specific algorithm.

Efficiency: *Efficiency* is the ratio of the ideal processing requirement to the actual processing utilization. During the design process we strive to attain an efficiency of one.

$$E = \frac{\text{Resource requirement}}{\text{Resource utilization}} < 1$$

Overhead: *Overhead* is the percent of time that the processor cannot be used owing to processes such as input/output data loads, inability to provide coefficients at a fast enough rate, or poor mapping of the algorithm to the processor architecture. Overhead is given by

$$OH = \frac{\text{Resource utilization} - \text{Resource requirement}}{\text{Resource requirement}} \times 100$$

$$= \frac{1 - E}{E} \times 100$$

Latency: The processor *Latency* is the inherent delay caused by the processor architecture between input samples and corresponding outputs.

10.1.3 Generic Signal Processor Overview

Figure 10.4 illustrates a generic signal processor configuration showing a processing engine, a mass memory, a data transfer network, an input/output processor, and a control processor. Distributed processing engines and mass

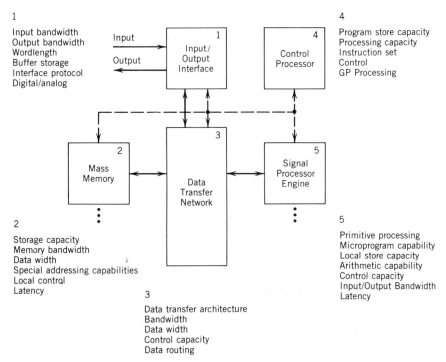

Figure 10.4 Generic signal processor.

memories are easily envisioned and will influence the data interconnections required by the design. Data transfer and memory conflicts, along with control of the processing, provide potential bottlenecks in the design. The modularity of the design will evolve to minimize the bottlenecks based on the application requirements. Localizing specific control/data flow requirements can provide significant performance gains and must be assessed versus centralized approaches. A key part of the signal processing design process is the partitioning and allocation of the DSP functions to the processor. For a new processor design this includes hardware/software design alternatives. Key resource parameters that must be considered for each element type are shown on the figure. The application design requires a thorough evaluation of the utilization of each processor resource parameter. This ensures that all processing can be accomplished within the allocated time.

A brief description of each of the generic signal processor element functions shown in Figure 10.4 is presented. Corresponding resource parameters are discussed for each element.

CONTROL PROCESSOR

The control processor handles overall control of the signal processor. Functions performed include interpreting commands received from external devices (e.g., system manager, operator entry devices), controlling data transfers, control of the signal processor's elements, and control of data inputs and outputs. Some controller functions can be distributed to the other elements of the processor. For example, the processing engine generally has a controller to handle the control of the data flow and arithmetic computations within the element.

The control processor executes the application program that tailors the signal processor to perform the algorithms required. General-purpose processing requirements of relatively low data rates, such as tracking algorithms, are sometimes performed in the control processor. The architecture of the control processor is generally like a general-purpose computer. Specific instruction-set architectures exist and are used as standards by various organizations. If one of these architectures is specified, then your design alternatives will be restricted. A programming language is usually specified. Ada* has emerged as a standard for the Department of Defense (DoD). The language specified can limit performance if features are not supported that would benefit the processor architecture. Support software such as compilers must be available or developed for your processor design.

The resources of the control processor include the storage capacity for programs and data, the processing capacity in terms of instructions per second, the instruction set, and the arithmetic capacity. Although the instructions per second gives an initial estimate of the capability of the control processor, the analysis must consider the actual mix of instructions required by the application. Standard benchmarks have been established for assessing the performance of general purpose computers. For each application, the designer must assure that the benchmarks adequately test the usage. If not, then a set of benchmark tests should be established based on the expected requirements.

PROCESSING ENGINE

The signal processor engine executes the real-time signal processing algorithms necessary to meet the application requirements. The architecture of the engine determines the processing capacity for various DSP operations. The resources of the signal processor engine include the local storage capacity, the input/output data bandwidth, the engine's instruction set and execution times, the arithmetic capabilities, and the inherent delay required to process data through the processor.

Figure 10.5 shows a pipelined processor engine architecture that provides a cache store (local store) feeding an adder that is followed by a multiplier and accumulator. **Cache store** is used to define a storage that is specifically designed to operate at rates supporting the processing engine. A sine/cosine generator is included to provide trigonometric functions for bandshift operations and twiddle factor FFT coefficients. This architecture matches the direct-form symmetric FIR filter algorithm requirements. No attempt has been

---

*Ada is a trademark of the U.S. Department of Defense.

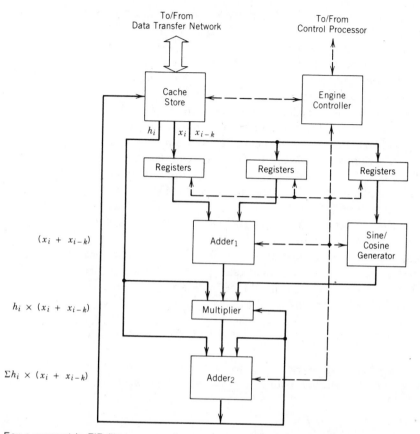

For a symmetric FIR filter, symmetric components are summed in adder 1 multiplied by filter coefficient and summed in adder 2 with partial results.

Figure 10.5 Pipeline processing engine.

made to provide an exact representation of all architectural elements, such as registers, data flow paths, arithmetic, and control.

The processor engine local store provides autonomous operation once the data are transferred from mass memory into the engine. The local store is configured such that input or output transfers can occur on one half the store concurrently with internal operations occurring on the other half of the store. For a FIR filter implemented to take advantage of symmetric coefficients, adder 1 provides the summation of the symmetric data samples fetched from local store, the multiplier performs the multiplication of the sum of the symmetric samples from adder 1 by the corresponding filter coefficient, and adder 2 provides the accumulation over the filter impulse response length. Each stage in the pipe is executed at the clock cycle rate of the engine hardware.

The engine can take on many forms, depending on the algorithm set processing requirements. For example, if the engine's primary function will be large-size FFTs, then large local stores with addressing capabilities compatible with the FFT implementation should be provided. Also, the computational flow should provide for efficient implementation of the FFT computational unit for the radix chosen. Data scaling and special requirements such as the FFT twiddle factors must also be considered. If a more general algorithm set is required, then the individual requirements must be analyzed as part of the signal analysis in step 2 and processing elements developed as necessary in step 3 as part of the signal processing design to meet the requirements.

An estimate of a processor's capability is often given in terms of one or more of the following: million multiplications per second (MMPS); million operations per second (MOPS); million complex operations per second (MCOPS); execution time to perform an N point FFT. These numbers could be misleading because they may not adequately represent the capability of the processor to implement the specific algorithms required by the application under design. Therefore, these estimates should only be used as rough order approximations, and a thorough design should be performed assessing all resource parameters as described herein.

Example 10.1

Characterize the processing engine given in Figure 10.5 versus the processing requirements of a symmetric FIR filter algorithm given by (see Chapter 5, Eq. 5.81)

$$y(n) = h\left(\frac{N-1}{2}\right)x\left[n - \left(\frac{N-1}{2}\right)\right]$$
$$+ \sum_{k=0}^{(N-3)/2} h(k)[x(n-k) + x(n-N+1+k)] \quad (10.1)$$

Two input data streams represent channels that are to be processed by the filter independently. Assume the input sampling rate is F. What is the minimum engine processing rate to handle the processing load for the two channels.

Solution:

The multiplication and addition requirements for a symmetric FIR filter were given in Chapter 2 as a function of the sampling rate.

$$\text{MPS} = F(N + 1)/2$$
$$\text{APS} = FN$$

This assumes that the additions for the symmetric components were performed prior to the multiplications as defined in Eq. 10.1. Now suppose the basic cycle rate of the engine is K. **Cycle** is used to refer to the basic sampling rate of the engine. Then the engine can perform K multiplications per second and $L = 2K$ additions per second. The two channels of data require a multiplication rate of 2MPS and an addition rate of 2APS. If $K > 2$MPS and $L > 2$APS, then the basic engine capability appears to satisfy the requirements and our initial observation is that the processing requirements are satisfied. This is a common error made when trying to assess processing engine capabilities by using operations per second as the basis for analysis.

Let us examine this result a little closer. The limiting case would be when $K = 2$MPS and $L = 2$APS. Under these conditions, we should be concerned that we probably do not have enough resources to process the data since some amount of overhead is always required during the process. **Overhead** represents the time when the arithmetic portion of the pipeline is idle owing to other required operations that cannot be performed in parallel with the arithmetic operations. Overhead items often include items such as loading/unloading data, setting up the engine to process the algorithm, and providing special coefficients required by the algorithm. For this example, loading the initial data block for the first channel must be completed before the engine can begin the arithmetic operations. The cycles used for this process will use up some of the cycles required to complete the processing. This illustrates the need for the processor capability to be above the actual processing needs.

Without detailed analysis of the pipeline and specification of the size of the data blocks to be processed, a good rule of thumb is to provide 20 to 30% for overhead. That is, let $K = (1.2 \text{ to } 1.3) * 2$MPS and $L = (1.2 \text{ to } 1.3) * 2$APS. Generally, the local store design provides capability for loading/unloading in parallel once the operations have started. After several designs have been performed, the processor characterization will be established for the processing algorithms frequently used. This will provide a basis for accurate estimation of the overhead. Prior to actual implementation of the processing, the best approach for estimating the resources is via the development and use of an accurate simulator. This can provide a valuable aid during the architecture design process, resulting in an optimum design.

Equation 10.1 requires that the processor be able to implement the algorithm, taking advantage of the reduced multiplication rate, or that the processor be able to process twice the multiplication rate. The engine shown in Figure 10.5 supports this algorithm configuration. Now we take a case where the configuration in Figure 10.5 is modified to delete the adder 1 portion of the pipeline. In this case we can no longer perform the additions of the symmetric components prior to the multiplication without cycling through the pipeline to adder 2. Therefore we require that $K = (2.2 \text{ to } 2.3) * 2$MPS and $L = (1.2 \text{ to } 1.3) * 2$APS. In this case we would implement a direct-form FIR without taking advantage of the symmetry. Therefore, it would take twice as many cycles of the engine to implement the same filter owing to the architecture of the engine. For this processor, if we had used Eq. 10.1 to determine the multiplication requirements, our analysis would have been off by a factor of two. For this reason, operations per second can cause significant errors in the design if used independently of the architecture for processor resource analysis. ■

From Example 10.1 we can easily see that comparing processing engines is not a simple process. The architecture of the engine must be considered. The same is true for the entire signal processor. A better measure is obtained by estimating the time required to process a representative set of benchmark operations. The term representative applies to the specific application use intended for the processor. Ultimately, a thorough resource analysis based on the exact operations must be performed to ensure that the processor meets the requirements.

In Reference 1, Chapter 9 covers special purpose hardware for digital filtering and signal generation and Chapter 10 covers special purpose hardware for the FFT. These chapters provide an introductory description of the considerations that are required in developing architectures to handle specific signal processing algorithms. Technology will continue to play a key role in the evolution of the engine architecture and the ability to process advanced algorithms.

DATA TRANSFER NETWORK

As shown in Figure 10.4, all high-speed data transfers between elements (input, output, and internal) are handled by the data transfer network. The data transfer network can be a bottleneck (i.e., limiting resource) to achieving higher processing throughput. Therefore, the bandwidth of the element must support the entire application requirement. In a design with several distributed elements the rates can become very high and will depend on how the signal processing design partitions the signal processing tasks to the distributed elements. Both the hardware and software architecture for this portion of the signal processor are key to achieving an efficient implementation. Note that this could become a single point of failure that would result in the loss of the entire signal processor capability unless multiple data paths or redundancies are provided.

The resources of the network include the data routing paths, the data bandwidths, and the data wordlength and block sizes supported. The control requirements for specifying input and output data addresses and the detailed data transfer protocols are key to the processor operation. Figure 10.6 illustrates a data network for a configuration showing three mass memories and three processing engines. Note that total connectivity is not achieved in this implementation, as shown in Figure 10.6b where the switch path from point 1 to point 2 is not connected. Therefore, in order to transfer data from one engine to another the data would have to be routed into a mass memory first. This shows the importance of understanding the processing requirements, the potential limitations imposed by the design, and the resultant impact on overall system performance.

The network could be configured as a high-speed system bus connecting all elements of the processor. In this case each element would contend for

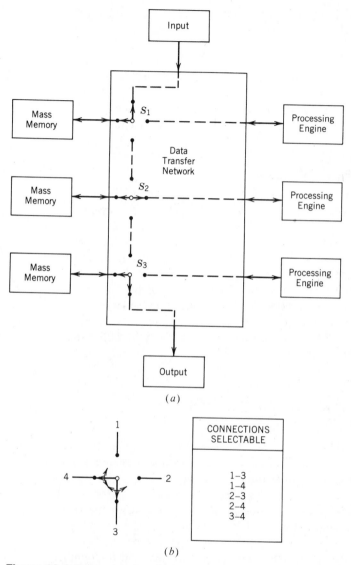

Figure 10.6 Data network architecture.

access to the bus as needed. Therefore, the total bandwidth of all transfers required would have to be supported by the bus. Simultaneous contentions for access to specific elements can occur and may require priority assignments based on the processing requirements.

MASS MEMORY

The mass memory must support the application storage and bandwidth requirements. **Memory bandwidth** is the data transfer rate for accessing and storing of data. Special addressing can be beneficial for specific algorithm requirements. For example, the FFT data reordering required at the input or output is often accomplished by special addressing routines. Also, real-time DSP applications may impose unique architectural design requirements such as dynamic allocation/reallocation of memory during system mode changes. This provides flexibility in the processing and minimizes the memory requirements.

The resources of the mass memory include the storage capacity, the data width, special addresssing modes, and overall memory management control capabilities.

INPUT / OUTPUT DATA INTERFACE

The input/output interface provides the high-speed data inputs and outputs required to perform the signal processing and the subsequent processed data for additional processing and/or display. We assume for this configuration that the input is digital. This allocates all portions of the analog signal conditioning, in the case of an analog input, and analog-to-digital conversion to hardware external to the signal processor (sensor system).

The resources of the I/O data interface include the bandwidth capacity, buffer storage, data wordlength, interface protocols, and number of input and output interfaces supported. Many applications require multiple sensor input ports and multiple output ports for subsequent data analysis and display. Many I/O interface architectures are used, including point-to-point, ring, and star connections.

10.1.4 Signal Processing Design Definitions and Considerations

In order to develop a methodology for the design of DSP systems, a consistent set of definitions is important. The following provide a baseline set of definitions used in defining DSP application requirements.

> **Processing Flexibility:** The *processing flexibility* encompasses the range of application processing options that must be processed, the number of modes of processing that are required per processing option, the processing option and mode concurrency requirements, and the number of channels that must be processed by each assigned mode. For example, for an aircraft control application we could have a communication and radar processing option. The radar processing system could have detection and tracking modes that must be processed concurrently. To achieve the required range and azimuth resolution requirements, processing of several channels for each mode could be required.

> **Processing Concurrency:** *Processing concurrency* refers to the need to process functions during the same time and applies to both hardware and software operations.

Processing Option: The *processing option* defines the major categories of application processing such as communications, navigation, radar, and sonar. A programmer develops an application command program to handle the processing options.

Mode: A *mode* defines the basic level of assignment under the processing option. The application programmer would develop a program to command the mode operation. Mode parameters can be used to provide changes in the processing within the mode.

Channel: A *channel* defines a basic stream of input data that is assigned to be processed by a mode.

Mode Switching: *Mode switching* is the requirement to process different modes under control of the system manager. Critical requirements include the maximum time allowed to accomplish the mode switch and the need to retain past mode information (data or signal processing setup control) for subsequent reassignment.

Mode Flexibility: The set of parameters that must be dynamically controllable during the execution of the mode define *mode flexibility*. For example, for the radar problem, we can have multiple range resolutions that can be switched dynamically by an operator resulting in changes in the mode processing requirements.

Functional Flow Diagram: The *functional flow diagram* describes the input/output requirements along with the intermediate signal processing functions that must be performed for each mode. The functions are

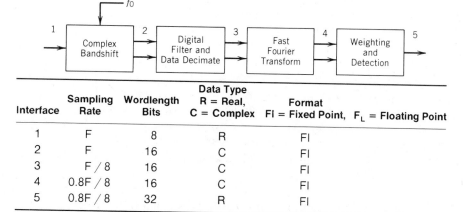

Interface	Sampling Rate	Wordlength Bits	Data Type R = Real, C = Complex	Format FI = Fixed Point, F_L = Floating Point
1	F	8	R	FI
2	F	16	C	FI
3	F / 8	16	C	FI
4	0.8F / 8	16	C	FI
5	0.8F / 8	32	R	FI

Figure 10.7 Spectral analysis signal processing functional diagram.

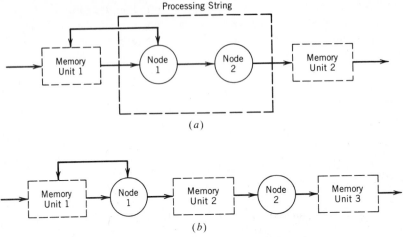

Figure 10.8 Digital signal processing flow graph (DFG) formats.

described at a high level. This is the first step of the signal processing design. For each mode of operation, a signal processing functional diagram is developed. Figure 10.7 shows a typical signal processing functional diagram for a narrowband spectral detection system.

DSP Flow Graph (DFG): The *DSP flow graph* is an unambiguous method of representing the signal processing flow at a level at which the actual primitive operations are executed within the signal processor. The DFG is made up of nodes connected by queues along with memory units (MU). MUs represent points in the DFG where data are transferred to/from memory over the data transfer network. Figure 10.8 illustrates alternative DFG implementations for two nodes. The alternative shown in *a* requires less data network bandwidth because no intermediate transfers are required between nodes. Alternative *b* requires the additional data transfers between nodes, but has the added flexibility of partitioning the nodes to separate processing engines.

Primitive Operation: A *primitive operation* represents the level at which the signal processor has been preprogrammed. Generally, the signal processor architecture provides a microprogramming capability for implementing the primitive operations. Examples of potential primitive operations are multiplication of a scalar with a data vector, a FIR filter of order 9, a quadratic stage canonic IIR filter, and a FFT four-point computational unit.

Node: *Nodes* represent the execution of primitive operations or strings (chains) of primitive operations. Each node must have at least one input and one output queue. A read amount and a consume amount must be specified along with the data type (real, complex) and the data format (e.g., fixed-point 16-bit real and 16-bit imaginary packed into 32-bit words). The *read* amount is the number of data samples to be read from the queue. The *consume* amount is the number of samples that can be erased from the queue. For output queues, a produce amount must be specified. Memory unit storage areas must be specified for all input and output queues that are stored external to the processing element.

Queue: *Queues* define the inputs and outputs of the nodes. Queues represent coefficients, data, and data history required for the current execution of the primitive operation. For input or output queues external to the element executing the operation, a source or sink memory unit storage area must be defined.

String: A *string* is a collection of primitive operations constrained for implementation without requiring intermediate input data transfers from other elements during execution. Other terms used for strings include segments and chains. Strings reduce the data transfer requirements, which can be a limiting factor in achieving the processing requirements.

Memory Units: The *memory units* represent source/destination points in the processing flow where the data are stored in a mass memory storage area. These represent a load/unload to/from the processing engine.

Figure 10.9 illustrates a DFG representation of the Figure 10.7 application. The complex band shift (CBS) primitive operation produces the quadrature signal components that feed the two filter paths. For this application we have assumed that a FIR symmetric filter of order 27 is sufficient to provide the spectral characteristics required to allow the data decimation by a factor of eight. As discussed in Chapter 7, the FIR filter primitive and data decimation primitive can be implemented as one primitive operation, reducing the computations required by the decimation factor. Following the filtering portion of the graph, an FFT of length 256 is performed to produce the spectral signal output, which is then weighted and detected. We implement 75% overlap FFT processing to achieve optimum gain. Therefore, the read amount would be 256 and the consume amount would be 64.

In developing the DFG, each primitive operation must be partitioned to a processing element, and all input and output queue characteristics must be defined. This evolves as part of the DSP design, resource analysis, and configuration analysis steps in our procedure. A key part of the design is to establish partitions at the boundaries where data must be transferred from one element of the processor to another.

Two strings are shown in Figure 10.9. The first string includes the CBS, FIR, and data decimation operations, and the second string includes the

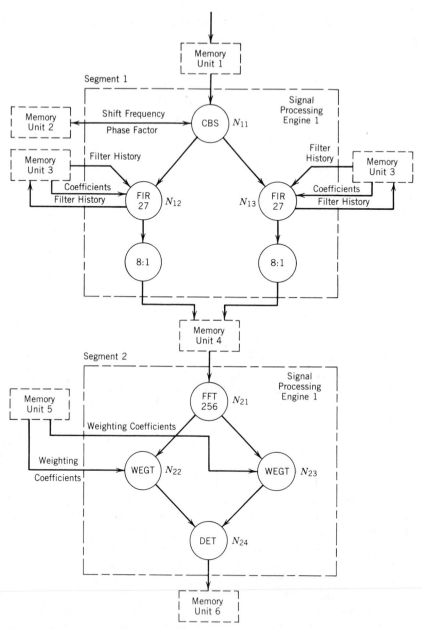

Figure 10.9 Spectral analysis signal processing flow graph.

spectral processing portion of the graph. For this graph we initially partition all memory units to mass memory element 1 and all signal processing operations to signal processor engine 1. To complete the graph design, the queue parameters (read, consume and produce amounts, data types and lengths) must be defined and clearly itemized. If multiple elements are required, the actual partitioning will be developed during the resource and configuration analysis steps of the design process discussed later in this chapter.

10.1.5 Signal Processor Design Considerations

A method for performing the design for real-time DSP applications is detailed in Section 10.2. Developing a new signal processor requires a team of highly skilled designers familiar with the latest hardware and software techniques as well as the latest signal processing techniques. The team must perform a complete signal processing system design. We have briefly discussed the elements of a distributed signal processor and the key design parameters associated with the elements. References have been made to hardware and software architectures. We prefer to address the total processor architecture requirements instead of creating a design limitation by making an early decision on the hardware/software boundaries. In some cases, the processor is specified or already exists and the job requires a thorough understanding of the processor architecture in order to achieve an efficient signal processing design. In this section we present an introduction to the architecture design considerations.

Several basic design alternatives can be defined, including centralized/ distributed, synchronous/asynchronous, pipeline/parallel, data driven/time-line, and high speed bus/multiport switch. These are high-level alternatives and represent multiple choices in some cases (e.g., the control portion of the processor can be centralized, distributed, or a combination of both; the data memory can be centralized or distributed; parallel pipelined processing engines can be used). Lower level alternatives such as the details of address registers and instruction fetch and execute processes are beyond the scope of this overview.

Centralized control was fine for computers that did not have to deal with real-time processing. Each step in the program could be decoded and executed sequentially. For a programmable real-time signal processor, many DFGs can be concurrently assigned. The operations required to schedule and dispatch a node for execution must be multiplied by the number of nodes and the execution rate of the nodes. This number gets very large even for simple graphs, and analysis has shown that the processing requirements exceed available control processor capabilities. In addition, if multiple processing elements need to be controlled, the processing element can be idle while it is waiting for the next control input. Increased capabilities of the processing elements without increasing the control processor capability results in higher idle times (higher overheads). Therefore, some distribution of the control is required.

Synchronous processing throughout the processor is the easiest to implement. This requires all elements to operate at a basic rate or a multiple of the basic rate. Therefore, if any portion of the design does not achieve the performance required, all elements must be proportionately reduced. Multiple clocks can be provided for asynchronous designs at the expense of additional hardware. An asynchronous design allows for different elements to be operating at totally different rates. Obviously, this can complicate the control of the processing, but it provides the most flexibility.

Distributed control is used to off-load the central controller from high rate control tasks that are required by a specific element of the processor. System level controls such as assigning DFGs to be executed by a processing element are still partitioned to the central controller. The boundaries at which the control is partitioned to central and distributed controllers is determined as part of the resource and configuration analysis steps of the design procedure.

Concurrency requirements can significantly influence the architecture design. If concurrent functions are allocated to separate processors then no significant problems exist. However, it is desirable to share common resources to the maximum extent possible in order to reduce hardware/software requirements. This requires architectures that provide the capability to share resources without interference between operations. Methods for switching between operations and maintaining the hardware/software in a consistent state are required. Here concurrency refers to the ability to perform multiple operations at the same logical time frame. Therefore, the hardware/software must operate at a basic cycle rate much higher than the concurrent operations required. Processing of concurrent modes or channels can be provided by parallel hardware at various hardware/software levels. The degree of synchronization necessary will depend upon the specific concurrent processing operations required.

Many applications offer inherent parallelism in the required DSP operations, which has resulted in today's pipelined and parallel signal processor architectures. The signal processor architecture plays a key role in determination of the final signal processing implementation and vice versa. In order to map an algorithm to the processor architecture, a detailed design analysis must be performed.

Pipelined architectures take advantage of overlapping the computation cycles across a set of input samples that are processed in a batch sequentially through the stages of the pipeline. The ability to use all elements of the pipe at each clock cycle provides maximum efficiency. For pipelines with K stages, a delay results equal to the sum of the K delays. For a synchronous design (all stages with equal delay, d), the delay is equal to the product dK. This is the time that it takes to prime the pipe. Outputs occur every clock cycle once the pipe is full, assuming that the algorithm maps to the architecture with full utilization. If only a fraction of the stages, L/K, are occupied, then the outputs occur every dK/L clock cycles.

Parallel processing elements are sometimes used under the same control in applications requiring multiple channels with identical processing. The control architecture for this SIMD processor is referred to as **lock-step control**.

The design of the engine may incorporate multiple identical processing paths including local store. This design, although highly efficient for specific applications, is very inefficient for algorithms that can only use some of the K parallel paths. Parallel paths can be provided with independent control (MISD, MIMD), which results in the greatest flexibility at the expense of the addition control hardware.

Signal processor architectures historically featured time-line scheduling. **Time-line** architectures required that the execution of each operation be performed at a specific point in the processing sequence based on the application requirements. Since many applications require that the processing be changed in real time based on system commands, the control must adapt to the changes. This requires deletion and/or insertion of operations to be scheduled and dispatched for execution. The scheduling algorithms were often complex and required careful implementation to assure proper load balancing.

Data-driven architectures have emerged with the availability of VLSI and the corresponding higher memory densities. The **data-driven** architecture, as the name implies, uses the availability of the input data to schedule and dispatch the operation. For each input queue of a node, a data block size is specified and used as the criteria for subsequent node execution. The sizes are referred to as *threshold values* for the input queues. When the operation is executed, each queue has a *read* amount and a *consume* amount that are part of the description along with the data type (real, complex) and format (number of bits, fixed point, etc.). When the operation is executed, the results are written into output queues. The queues reside in mass memory and represent the boundaries of the signal processing operations that must be handled at the control level above the processing engine.

Implementation of data-driven architectures offers challenges to the designer in assuring that real-time processing is accomplished within the allocated time, without the excessive use of memory and throughput. Several nodes may be ready for execution simultaneously. This requires sufficient buffer storage to store data at each input node for the maximum period of time for which the executions could be delayed. The scheduling algorithm is simply defined by the availability of the specified input data.

Most real-time signal processing applications require very high rates if scheduled and dispatched at the primitive level, and therefore it is usually advantageous to combine groups of primitives together in order to reduce the amount of control required. This permits the processing element to process the group of primitives autonomously without intervention by the controller. The architecture design must provide for task scheduling and dispatching such that the graphs can be executed without loss of data, and without requiring excessive amounts of storage. Combining of several primitive functions into one executable processing element function can significantly reduce the number of data transfers required between mass memory and the processing element. For each executable operation, the buffer sizes are chosen to reduce the amount of data loads and unloads required, which maximizes the amount of data processed through the processing element at each instance of the graph. This provides the highest efficiency.

Latency is the delay from the time a signal processing task is ready for execution to the time that it is executed. Latencies in processing result from dynamic scheduling of the real-time nodes from the DFGs instantiated. These delays occur randomly because the control is dynamic and the assignment of the various processing modes to the processor is done externally by an operator via the system controller over the control interface. A worst-case latency can be calculated by assuming that the tasks with the greatest processing times from each DFG all are ready for dispatching at the same time. Latencies in storage result from the peak processing loads. Data queues must be large enough to hold all data until the task is dispatched for execution and the corresponding data transfer from main memory takes place. The sum of the peak processing times is multiplied by the input data rate for the queue in order to determine the worst-case storage requirements.

Latency presents a problem if the results are required to make a real-time decision in less time then the latency delay. For example, a terrain following/avoidance processing mode must have a result within a short time in order to react to abrupt changes in the terrain. Therefore, the architecture design may require priority scheduling and dispatching of critical tasks to satisfy operational needs. In addition, long delays require larger memory unit storage buffers to ensure data are not lost. The sum of the peak processing times is multiplied by the input data rate for the queue in order to determine the worst-case storage requirements. Providing double buffers at critical points is a method commonly employed to alleviate the potential for lost data due to latency problems.

We discussed the resource analysis for a signal processing flow graph previously. Although a rough-order estimate of throughput can be obtained using multiply and add requirements of the various primitive operations, it is not recommended, as shown by Ex. 10.1. Algorithms with relatively low multiply requirements can require complex data manipulations and operations. These could be significant in terms of utilization of the processing resources and could be significantly underestimated if only multiply and add operations were counted. Also, implementations of the algorithms onto the processor's architecture will result in inefficiencies. Therefore, the only accurate throughput estimate is the actual processor engine cycles required to implement the operation. To this value the overheads required for control, data transfers, and so forth must be evaluated and added. In addition, the use of each resource within the signal processor must be completely quantified.

Once the resource analysis is complete, the results are analyzed to assess the adequacy of the design. For applications with existing hardware, the alternatives are signal processing design modifications and potential new microcoded operations to improve performance. For the case where a new signal processor is being designed along with the processing, the following design alternatives are assessed.

Modify the signal processor configuration:
    Increase the number of elements.
    Add an element type (may require a new design).
    Modify the element architecture.
    Modify the processor architecture.

Modify the signal processing design:
    Develop more efficient primitive operations.
    Increase the data batch sizes processed by primitives.
    Change the processing design (e.g., fast convolution instead of a conventional filtering approach).
    Reduce the application requirements.

The goal is to minimize the overhead, thereby achieving the highest possible utilization of the signal processor. This should minimize the hardware requirements and result in the lowest cost to the customer.

Next we define a signal processor for use during the system design process presented in Section 10.2 and 10.3.

10.1.6 Signal Processor Specification

For this application, a signal processor has been specified with the primitive capabilities and execution times as given in Table 10.1. For each primitive operation, a setup time and a batch time are provided. The **setup** time represents the time it takes to configure the processor to execute the primitive. The **batch** time is the actual time required to perform the operation. Therefore, inherent inefficiencies due to the inability to fully utilize the engine architecture for the different operations are included. Overhead times that result owing to execution of multiple nodes and/or graphs due to lack of sufficient resources as described in the preceding sections must be quantified.

A distributed memory and engine architecture is given as shown in Figure 10.4 for the generic signal processor discussion. The processor engine (Fig. 10.5) has a raw multiply capability of 5 million multiplies per second (MMPS). The product is performed for two 16-bit input values producing a 32-bit result. The processor can generate sine and cosine values at a 5 megahertz rate. It contains a 16-bit preadder before the 16×16-bit multiplier and a 32-bit accumulator following the multiplier. The cache store contains 64 K bytes of memory with a 5 megabytes per second bandwidth. The processing element is microprogrammable, and each stage in the pipeline can be controlled at the basic cycle time of 200 nanoseconds. Eight thousand bytes of storage are provided for primitive programs.

The mass memory element provides 2 million bytes of storage and has a bandwidth of 5 megabytes per second. No special addressing modes will be specified.

The control processor can perform 1 million 16-bit instructions per second (MIPS) and has a program storage capacity of 256K bytes.

TABLE 10.1
PROCESSING ELEMENT PRIMITIVE EXECUTION TIMES

Primitive	Setup Time	Exec Time	Per I/O	Mult / Primitive	Adds / Primitive
CBS	5	0.8	1	2	0
VBS	10	1.6	1	4	2
FIR-11	10	1.8	0	6	11
FIR-27	26	3.6	0	14	27
FIR-31	28	4.2	0	16	31
FIR-39	32	4.8	0	20	39
HFIR-11	7	0.8	0	4	7
HFIR-27	12	1.2	0	8	15
HFIR-31	16	2	0	9	17
HFIR-35	20	2.8	0	10	19
IIR-1	6	1.8	1	2	2
IIR-2	12	2.4	1	4	4
IIR-4	20	4	1	8	8
FFT-256	7	3.8	1	4096	6144
FFT-512	13.8	3.8	1	9216	13824
WEGT	2	3.5	0	3	2
DET	4	1.8	1	2	1
INT	3	1.2	1	1	1

Time = Setup time + Exec time $* M$
where M = No. of input points if per I/O = 1 or outputs if per I/O = 0;
Processor capabilities:

MOPS = 15	MEM-BW = 5 megabytes/sec
MMPS = 5	CONTROL = 1 MIPS
MAPS = 10	DTN = 20 megabytes/sec
MEMORY = 2 megabytes	Sine or cosine at 5 megahertz

The data transfer network is a 16-bit parallel bus capable of transferring data at a 20 megabytes per second rate. The transfers must contend for priority on the bus since no multiple paths are provided for simultaneous transfers between different elements.

The I/O interface has a bandwidth of 10 megabytes per second on each input channel and 5 megabytes per second on each output channel. The input channels are 1-byte parallel words, and the output channels are 2-byte parallel words. Buffering is provided for input data to accommodate processor latency, preventing loss of input data during the processing.

10.2 SIGNAL PROCESSING SYSTEM DESIGN METHODOLOGY

Signal processing system design requires a thorough knowledge of the application problem requirements. Figure 10.10 presents the top-level approach to the signal processing system design. The process is an iterative step-by-step

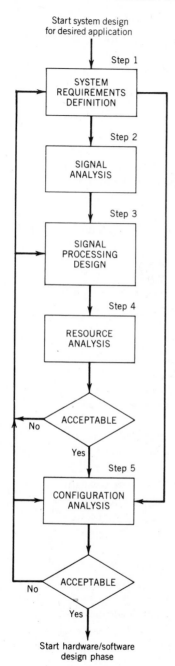

Step 1
User/customer driven
Develop system level requirements
Signal processing
Non-signal processing
System level documentation
Requirements specifications
Interface design specifications

Step 2
Define input signal characteristics
Types
Parameters
Noise sources and distributions
Data rates

Step 3
Develop signal processing graphs
for each processing mode
Specify primitive operations
Initial partitioning of processing
element types
Arithmetic analysis
Iterative process
Results in architecture approach
for new signal processor design

Step 4
Final partitioning of process to
element types
Compute all resource utilizations for
each mode
Memory, control, bandwidth, throughput
Iterative process

Step 5
Define configuration
Based on resource analysis
Based on non-signal-processing
requirements
Partition multimode processing to
configuration
Perform resource analysis
Iterative process

Figure 10.10 Signal processing system design (from Chapter 1 Figure 1.3).

procedure considering all key design parameters. The result of the procedure is a signal processing design and a signal processor configuration that satisfies all application requirements identified.

The first step in the process is to assure that all requirements are defined. Then the signals to be processed are analyzed and the results are used to define and analyze signal processing algorithms as part of the signal processing design. At this point, the signal processor to be used for the application must be identified. If an existing signal processor has not been specified, then one must be selected or a new one developed. The functions performed for the signal processing design are different for an existing signal processor application versus a new design, as itemized in Table 10.2. The final part of the signal processing design is to develop *DSP flow graphs* (DFGs) for each of the processing modes defined in step one.

The initial resource analysis is performed for each individual graph assuming one processing engine and one memory element. All critical resources for the signal processor are estimated for each proposed DFG. Graph parameters can be varied to assess their effect on resource requirements. New primitives and/or strings can be defined and analyzed. Initially the signal processing can be designed without knowledge of the hardware or software design. In this case the resource analysis is based upon the computational requirements of the DFGs and the associated data transfer bandwidths. Bowan and Brown[6] emphasize the use of virtual machine layers during the design process. For new processors, the architecture can be modified to achieve more efficient throughput performance. This provides a basis for selecting the number of elements required and partitioning the graphs to the elements as part of the configuration analysis.

The configuration analysis combines the signal processing and nonsignal processing requirements in arriving at the final system configuration. The nonsignal processing considerations include environment (e.g., aircraft, ship, submarine, laboratory), reliability, maintainability, availability, power, weight, volume, and cost. The designer must understand their importance to the application design and assure that their requirements are satisfied. A resource analysis is performed for the configuration defined, and the results are analyzed. Different partitionings of the DFGs can be tried in order to achieve load balancing and possibly to reduce the number of elements required. If the resource analysis does not meet satisfactory performance, then either the signal processing graphs and/or algorithm definition must be refined or for new signal processors the architecture design may require modification. Also, failure to arrive at an acceptable design results in a review of the requirements for possible reductions.

Exact algorithms for many steps in the design procedure are not available due to the complexity of the design process. Problems such as partitioning signal processing functions to hardware elements require heuristic approaches that limit the number of alternatives to be analyzed to a manageable number. As you gain experience in DSP design you will be able to utilize the knowledge

TABLE 10.2
SIGNAL PROCESSOR DESIGN METHODOLOGY CONSIDERATIONS

Methodology	Processor Methodology Considerations	
	Existing Signal Processor	**New Signal Processor**
Step 1 System Requirements Definition	Define signal processing requirements Define nonsignal processing requirements	Define signal processing requirements Define nonsignal processing requirements
Step 2 Signal Analysis	Define signal characteristics	Define signal characteristics

Step 3 Signal Processing Design	Develop DSP flow graphs
	   Select element types
	   Maximize use of existing primitives
	   Implement new primitives as required
	   Assess arithmetic errors
	   Define strings
	Iterative process with resource analysis
	and configuration analysis
		Develop DSP flow graphs
		   Analyze processing option and
		   processing mode requirements
		   Establish primitive operations
		   Assess arithmetic requirements
		Develop architecture for most efficient
		implementation of requirements
		   Specify element types
		   Map primitives to elements
		   Define strings
		Iterative process with resource analysis
		and configuration analysis
		Perform preliminary hardware/software
		design

Step 4 Resource Analysis	Compute all resource requirements
	   Storage (data, program, parameters)
	   Memory bandwidth
	   I/O bandwidth
	   Control load
	   Throughput load
	   Latency
		Compute all resource requirements
		   Storage (data, program, parameters)
		   Memory bandwidth
		   I/O bandwidth
		   Control load
		   Throughput load
		   Latency

Step 5 Configuration Analysis	Determine configuration
	   Partition options/modes/DFGs
	   to elements
	   Perform resource analysis
	   Assess nonsignal processing
	   performance
	   Iterate process
	Return to signal processing design and
	modify to meet requirements
	Reduce requirements if required
		Determine configuration
		   Partition options/modes/DFGs
		   to elements
		   Perform resource analysis
		   Assess nonsignal processing
		   performance
		   Iterate process
		Return to signal processing design and
		modify to meet requirements
		Reduce requirements if required

obtained to limit the alternatives further. Computers provide the necessary processing capacity to perform the design tradeoffs.

The requirements lend themselves to a hierarchical decomposition from high-level requirements to lower and lower level requirements. An onionskin diagram can be used to represent this decomposition. The outer skin repre-

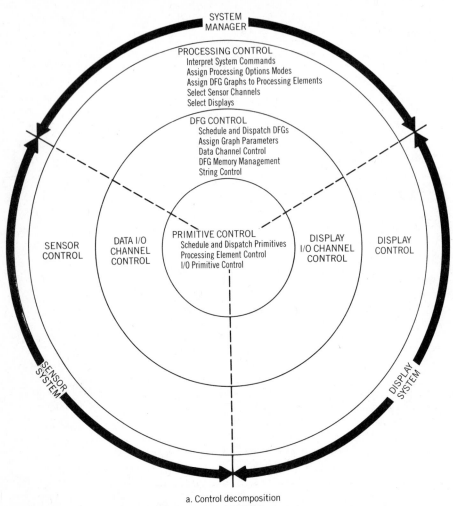

a. Control decomposition

Figure 10.11 Onionskin diagram — hierarchical decomposition of processing requirements: (*a*) control; (*b*) processing.

sents the high level requirement. Figure 10.11 illustrates the concept of these diagrams. Use of the diagram is not essential, but a method of capturing the various levels of requirements is required. Keeping hardware and software boundary decisions out of the design process for as long as possible is desirable in order to keep the design unconstrained. This prevents an early choice which could significantly limit the final design.[6]

Figure 10.11*a* illustrates the decomposition of the control requirements for the processing. At the highest level, we must deal with the external interfaces. Commands from the system manager are received, interpreted, and used to set up the processing control. The next lower level of control decomposition is shown as DFG control, and the lowest level shown is

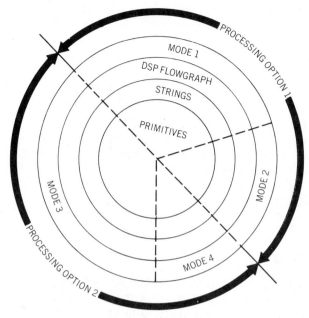

b. Processing decomposition

Figure 10.11 Continued

primitive control. We could have inserted a string control level prior to primitive control or decomposed some of the control levels further. Figure 10.11*b* shows the decomposition of the signal processing at a conceptual level. The more decomposition that is done, the better the chances of arriving at a good hardware/software design.

The methodology covers the application requirements/definition phase of the signal processor development cycle. This usually includes a system requirements review (SRR) and a preliminary design review (PDR) with the customer. Each of the steps shown in Figure 10.10 is described in the following paragraphs.

10.2.1 Step 1— System Requirements Definition

The first step results in a detailed definition of all performance requirements and defines the input/output data requirements. Documentation at this point includes a system requirements specification that defines high-level signal and nonsignal processing requirements. A signal processor may have been specified or the signal processor selection and/or development may be left as part of the design process.

All processing options and modes are specified. The details of the signal characteristics may be given or left as part of the analysis, with only high-level performance requirements given. It is important to specify the processing option and mode concurrencies at this point in the design. All required mode-variable parameters need to be defined. Times available for processing option assignments and mode switching must be specified.

Development of preliminary specifications for interface design specifications (IDS) and a prime item development specification (PIDS) for the signal processor are initiated. IDSs are developed for all interfaces, including signal input sources, system management computers, and system input/output devices. The IDS specifies all interface characteristics, including data rates, data lengths, control, and message protocols. The PIDS specifies all signal and nonsignal processing requirements. These specifications are completed as part of the methodology when step 5 is successfully completed.

Step 1 feeds step 2, signal analysis, with the signal processing requirements, and step 5, configuration analysis, with the nonsignal processing requirements.

10.2.2 Step 2 — Signal Analysis

Signal characteristics were discussed in Chapter 2 Section 2.1, and the signal analysis methodology presented in Figure 10.12 was given in Figure 2.1. Generally, the signal classes have been defined and analyzed in great detail. Therefore, only specific signal properties need to be defined for the application under analysis. This step results in the definition of all input signal characteristics.

For example, a sonar system processes acoustic sounds underwater to extract specific information about the input signal. The goal of the processing may be to detect that the signal is present or, in the case where the signal has been detected, the processing goal may be to localize or track where the signal source is located. Surface ships and submarines emit various acoustic signals into the ocean (e.g., broadband and narrowband). Broadband signals have energy spread across some frequency range, and narrowband signals are discrete frequency sinusoids. The signal classes have been characterized along with the effects of the ocean on the signals.

For the narrowband signals we need to determine the range of signal levels in frequency and amplitude, the noise levels, range of interference levels in frequency and amplitude, and the stability of the signal. This will be based on the specifics of the signal source, the ocean area, the weather, and the season.

The signal source is normally quantified in terms of signal characteristics. Therefore, if a specific surface ship is the signal source, all properties of the narrowband signals emitted into the ocean will be known within predefined limits. The ocean area is characterized by a temperature versus depth profile along with an ocean-bottom parameter. From the ocean area characteristics the effects on the signal transmission can be quantified. These data are discussed in several texts,[7,8] and several programs are available that perform

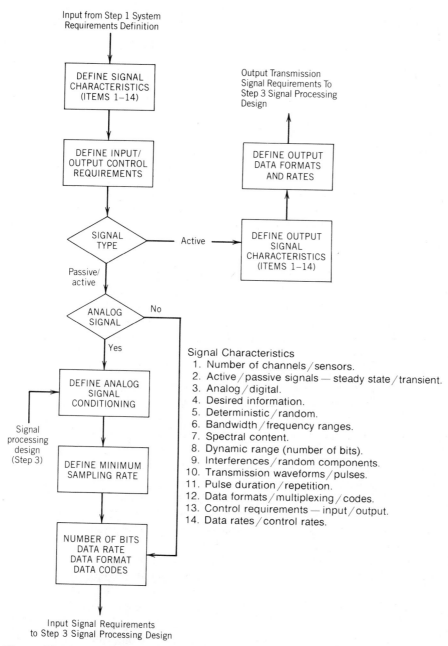

Figure 10.12 Signal analysis methodology.

detailed signal transmission analysis as a function of range, frequency, and depth.

In order to proceed to step 3, signal processing design, the signal type and characteristics must be completely defined.

10.2.3 Step 3 — Signal Processing Design

Step 3 results in a DFG for each processing mode. The DFG is based on the requirements defined in step 1 along with the results of the signal classes analyzed in step 2. The performance specification is mapped into a signal processing flow, ensuring that all processing requirements are met.

The procedure is shown in Figure 10.13. Following definition of the algorithms required, a high-level signal processing flow diagram is produced. Next, the flow is broken into implementable primitive operations. If the input signal is analog, then the design must establish the analog signal conditioning requirements prior to analog-to-digital (A/D) conversion. The sampling rate and A/D bit length are key elements in the design and have significant impact on the dynamic range and processing throughput requirements. The actual analysis process varies based on whether an existing off-the-shelf signal processor is used or a new signal processor is being developed. If the option is left to the designer, then the process may require several iterations selecting and analyzing suitable signal processors or refining the architecture design of a new signal processor.

For existing signal processors, the processing is developed using primitive operations that the processor is programmed to perform. Generally, today's programmable signal processors have a basic set of primitives (sometimes referred to as macros) that they are microprogrammed to perform. Selection of a signal processor is based on its capability to process the set of primitive operations required by the application. The design is developed maximizing the use of existing primitives in order to minimize costs of development. New primitives must be defined as required to implement the algorithm. In this case, the primitive must be mapped to the hardware and preliminary specifications and timing estimates must be developed. For example, if the processor was capable of performing only a limited set of FIR filters, the design should attempt to use those specific capabilities. Subsequent throughput analysis results may determine that the processing throughput performance is not adequate and a new primitive should be tried.

The arithmetic capabilities (i.e., data lengths, multiply/divide operations, additions, scaling, fixed point, floating point, block floating point, transcendental functions) must be completely assessed based on algorithm requirements. This is a key element of this design step.

For designs that require new signal processors to be defined, a much more flexible approach is taken. In this case, a complete set of primitives are defined and used to develop design requirements. Several important architectural features must be considered, including programmability, commandabil-

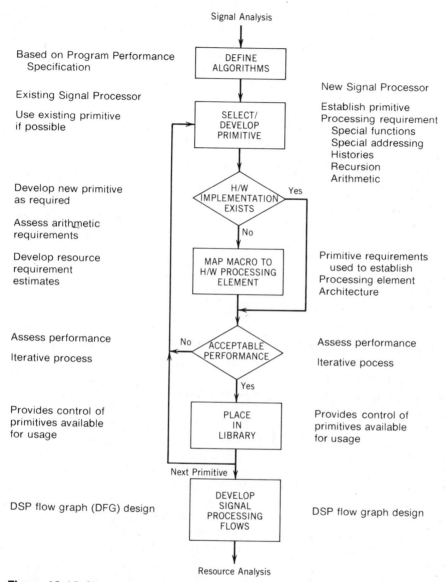

Figure 10.13 Signal processing design.

ity, flexibility, arithmetic capabilities, speed of computations, and all non-processing requirements. The steps in Figure 10.13 are used to arrive iteratively at a baseline signal processor design.

Once a complete set of primitives has been defined and mapped to the hardware, a preliminary signal processing flow graph (DFG) is defined. At this point, all graph parameters must be specified. Alternative DFGs and/or hardware/software allocations are proposed at this point for assessment in

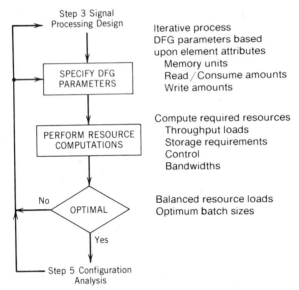

Figure 10.14 Resource analysis methodology.

step 4. During this step, initial partitioning is required between hardware and software boundaries. The DFG is refined during the resource analysis performed in step 4 of the design procedure discussed in the following section.

For a new signal processor a principles of operation (POO) is developed defining the architecture of the processing elements. A computer program performance specification (PPS) is developed to define the operating system requirements for the processor. These documents will evolve to final specifications when the configuration analysis is completed.

10.2.4 Step 4 — Resource Analysis

The resource analysis procedure shown in Figure 10.14 starts with the signal processing flow graphs developed during the signal processing design. This procedure results in all signal processor resources being evaluated for each DFG.

The first step in the procedure is to specify the DFG parameters. The data queue buffer sizes are defined. At this point, an initial throughput assessment is made for the control requirements for managing and dispatching the execution of the nodes (i.e., primitives). The buffer sizes are chosen to reduce the setup time compared to the batch time required and maximizes the amount of data processed through the processing element. Once an acceptable control load and initial throughput analysis result is obtained, a thorough analysis is performed estimating all critical resource parameters. Each of the throughput resource parameters is computed for storage (local and mass), bandwidths (memory, I/O, and internal), control, and computational load (accuracy and throughput). If all parameters are not satisfactory to the designer, then the analysis is iterated. The analysis procedure iterates until the percentage utilization of the processing elements is minimized for the processing defined.

Note that the throughput estimates at this point are for one element of each type. Therefore, if the load for a specific processing element is 150%, then two elements of this type will be required in the final configuration. Also, the additional data transfer network load required if intermediate data transfers are required between the distributed processing elements is assessed as part of the final step of the system design, configuration analysis.

The resource analysis can be performed using a static or dynamic approach. Static analysis refers to estimating the total resources based on summing individual calculations of the resources required to implement portions of the graphs. Contentions and overheads are estimated. Usually an overhead factor is applied based on actual hardware experience or dynamic analysis results. Dynamic analysis models the architecture of the processor and allows analysis of multiple graph implementations and assessment of overheads, latencies, and contentions.

A static analysis provides a quick method of estimating the resource requirements, but requires methods of estimating overheads associated with control of the processing. For the processing engine, a static analysis will give exact times for primitive executions since the number of cycles to implement the operation can be computed. If strings are executed within the engine, then some overhead may be required between execution of successive primitives and must be accounted for.

A dynamic analysis provides the best resource estimates. This analysis requires accurate estimates of all control and execution times. Generally, the simulation is done at the event level (i.e., transfer a block of data, dispatch string 1, etc.) and does not actually process data. The accuracy of the dynamic analysis is dependent on the event level times used in the simulation model. Since the architecture design will be based on the results, it is important that very detailed estimates are made. Tasks that are used often must be optimized. A complete simulation of the architecture at the hardware logic level assures accuracy of the results, but requires large processing capacities and long processing times. Once the hardware is available, measurements are used to provide actual processing times for future resource analyses.

Using the signal processing graph flow shown in Figure 10.9, we develop a resource analysis for the processor defined in Table 10.1.

SIGNAL PROCESSOR RESOURCE ANALYSIS

First we must select an input data block size and ensure that all primitive operations are completely defined. For Figure 10.9, we assume an input sample rate of 8000 samples-per-second (s/s) and a block size of 1024 samples and perform the FFT operation with 50% redundancy in order to meet performance requirements.

Processor Engine Resource Analysis. String 1 and string 2 are both executed in the processor engine. For string 1, the cache store requirements must satisfy the input data length requirements, including coefficients and histories. For the 1024 batch size chosen, the requirements are given by 1024 samples 8-bits each. We will read each 8-bit word into a 16-bit field of the cache store. Therefore, 2048 bytes of storage are required plus 4 bytes for the phase factor and frequency control of the CBS, 28 bytes are required for the FIR filter coefficients (2 bytes per coefficient), and 54 bytes of storage are required for the filter histories from the preceding execution of the string.

The throughput requirements are found by adding up the setup times and batch times for each of the primitive operations in the string. The cache store is configured to allow outputs from the previous operation and inputs for the next operation from one half of its memory during execution of the current string using the other half of the cache memory. Therefore, we must calculate the data transfer times and compare them to the current execution time to assure that the data are available for the next string execution. Otherwise the processing engine will be idle waiting for the data to be available. In the steady state, the time that the engine is not available for processing divided by the ideal processing time is the processing overhead. The overhead will be related to the engine architecture and the class of primitive operations being executed. Computing the time to process each primitive operation for string 1 results in

$$T(\text{CBS}) = 5 + 0.8(1024) \qquad = 824.2 \text{ microseconds}$$

$$T(\text{FIR27}) = 26 + 3.6(1024)/8 = 486.8 \text{ microseconds}$$

$$T(\text{string 1}) = T(\text{CBS}) + 2*T(\text{FIR27})$$

$$T(\text{string 1}) = 1797.8 \text{ microseconds}$$

In order to implement the CBS primitive operation, a sine and cosine value is required every 0.8 microseconds (i.e., at a 1.25 megahertz rate). The engine provides a sine or cosine at 5 megahertz. To compute the throughput percentage required by string 1, the time required to process string 1 is compared to the execution period of string 1.

$$T(\text{string 1 period}) = 1024/8000 = 0.128 \text{ seconds}$$

$$\text{Throughput}(\text{string 1}) = 100*\frac{T(\text{string 1})}{T(\text{string 1 period})} = 1.405\%$$

Next we compute the throughput requirements for string 2 which consists of FFT, WEGT, and DET primitive operations.

$$T(\text{FFT}) = 7 + 3.8(256) = 979.8 \text{ microseconds}$$
$$T(\text{WEGT}) = 2 + 3.5(200) = 702 \text{ microseconds}$$
$$T(\text{DET}) = 4 + 1.8(200) = 362 \text{ microseconds}$$
$$T(\text{string 2}) = T(\text{FFT}) + 2*T(\text{WEGT}) + T(\text{DET})$$
$$T(\text{string 2}) = 2747.8 \text{ microseconds}$$

Here we have assumed that the processor architecture performs the FFT algorithm not-in-place and therefore did not require a bit-reversal operation. The storage required for the not-in-place algorithm is less than or equal to twice the FFT length, 512 complex 16-bit values or 1024 bytes. We need to modify the processor design or take additional engine resources if an in-place FFT implementation is desired. This would be part of the design tradeoffs performed. Note that 200 values were used for the WEGT and DET operations to correspond to the good frequency points within the processing band. The out-of-band points are discarded. Again we can compute the throughput percentage requirements for string 2 and a total throughput percentage for the entire graph.

$$T(\text{string 2 period}) = 128/1000 = 0.128 \text{ seconds}$$
$$\text{Throughput}(\text{string 2}) = 100 * \frac{T(\text{string 2})}{T(\text{string 2 period})} = 2.146\%$$
$$T(\text{total}) = T(\text{string 1}) + T(\text{string 2}) = 4488.6 \text{ microseconds}$$
$$\text{Throughput}(\text{total}) = 100 * (0.0044886/0.128) = 3.551\%$$

For the FFT primitive operation we need 2048 sine and cosine values during the batch time it takes to execute the FFT.

$$T(\text{sine and cosine generation}) = 2048/(3.8*256) = 2.105 \text{ microseconds}$$
$$R(\text{sine and cosine generation}) = 1/2.105 \times 10^{-6} = 0.475 \text{ megahertz}$$

This is within the engine capacity of providing a sine and cosine at 2.5 megahertz.

A general expression for the engine throughput percentage load for a graph with K strings is given by

$$\text{Throughput}(\text{total}) = \sum_{i=1}^{K} \text{Throughput}(\text{string } i) \qquad (10.2a)$$

where the throughput for string i is equal to

$$\text{Throughput(string } i) = \frac{\sum_{j=1}^{J} T(\text{primitive } j)}{\text{String } i \text{ period}} * 100 \qquad (10.2b)$$

and the total time for primitive j is given by

$$T(\text{primitive } j) = \text{Load time} + \text{setup time(primitive } i)$$
$$+ \text{batch time(primitive } j) + \text{unload time} \quad (10.2c)$$

The load time and unload time are zero provided that the data transfers can be hidden by the processing (i.e., they occur while the engine is executing the previous or next string).

Mass Memory Resource Analysis. Two key resource parameters for the mass memory are the storage capacity and bandwidth. The storage capacity is computed from the graph and must take into account the scheduling and dispatching architecture to ensure that enough storage is allocated to the memory units to prevent loss of data due to latency. For time-line scheduling the worst-case time between successive executions of a string is computed and used to specify the data buffer size. For the purpose of this example we assume that a double buffer is sufficient to handle the worst-case latency.

Memory unit 1 = 1024 samples * 2 (double buffer) * 1 byte/sample
= 2048 bytes

Memory unit 2 = 2 bytes (phase factor) + 2 bytes (frequency)
+ 2 bytes (# of samples)

= 6 bytes

Memory unit 3 = 14 coefficients * 2 bytes/coefficient
+ 26 filter history registers * 2 bytes/register

= 80 bytes

Memory unit 4 = 256 samples * 2 (real and complex) * 2 bytes
* 2(double buffer)

= 2048 bytes

Memory unit 5 = 5 coefficients * 2 bytes/coefficient
= 10 bytes

Memory unit 6 = 200 samples * 4 bytes/sample * 2(double buffer)
= 1600 bytes

Then the total storage for M memory units for DFG k is given by

$$\text{DFG } k \text{ MU storage} = \sum_{i=1}^{M} \text{Memory unit}(i) \qquad (10.3)$$

which equals 5792 bytes for the six memory units just discussed.

The memory bandwidth is computed based on the read and write requirements of the graph. For each memory unit the bandwidth is computed.

$$\text{MU1 input} \quad \text{BW} = 8000 \quad \text{samples/second} * 1 \text{ byte/sample}$$

$$= 8000 \quad \text{bytes/second}$$

MU1 output BW = 1024 bytes / 0.128 sec = 8000 bytes/sec

MU2 input BW = 6 bytes / 0.128 sec = 46.875 bytes/sec

MU2 output BW = 2 bytes / 0.128 sec = 15.625 bytes/sec

MU3 input BW = 80 bytes / 0.128 sec = 625 bytes/sec

MU3 output BW = 52 bytes / 0.128 sec = 406.25 bytes/sec

MU4 input BW = 1024 bytes / 0.128 sec = 8000 bytes/sec

MU4 output BW = 1024 bytes / 0.128 sec = 8000 bytes/sec

MU5 input BW = 10 bytes / 0.128 sec = 78.125 bytes/sec

MU6 input BW = 800 bytes / 0.128 sec = 6250 bytes/sec

The total memory bandwidth for M memory units is given by

$$\text{Memory BW} = \sum_{i=1}^{M} \left(\frac{\text{MU}i \text{ input bytes}}{\text{MU}i \text{ input period}} + \frac{\text{MU}i \text{ output bytes}}{\text{MU}i \text{ output period}} \right) \quad (10.4)$$

The total memory bandwidth for this example is 39,422 bytes/sec.

The memory reads and writes occur at the memory hardware clock rate, which was given in Table 10.1 as 5 megabytes-per-second. Therefore, the actual time to read the memory values out of memory unit 1 is given by

$$\text{MU1 instantaneous data transfer time} = 1024 \text{ bytes}/(5 \times 10^6) \text{ bytes/sec}$$

$$= 0.2048 \text{ milliseconds}$$

Five megabytes-per-second is the rate at which the data are available to the data transfer network that provides an aggregate data transfer rate of 20 megabytes-per-second. Since the strings take 1.8 and 2.7 milliseconds to execute, the transfer requirements are a maximum of 12% of the total time.

Data Communications Resource Analysis. The data transfer network operates at 20 megabytes-per-second based on a 16-bit parallel bus operating at 10 megahertz. The network must support the transfer of the data from element to element within the signal processor as defined by the graph. For our example, the transfers shown are between mass memory units 1 through 6 and processor engine 1. The requirements are given by the memory bandwidth computed for this example. The rates should support the transfer of data such that the processor engine does not have to sit idle waiting for data.

The total data transfer network load is given by computing the individual transfer requirements and summing them over the total number of transfers, M, across the DTN shown in the graph.

$$\text{DTN}(\%) \text{ load} = 100 * \left(\sum_{i=1}^{M} \text{DTN}(i) \right) \Big/ \text{DTN capacity} \qquad (10.5)$$

where $\text{DTN}(i)$ equals the bytes-per-second required for the ith data transfer network graph queue.

Control Processor Resource Analysis. For the control processor we need to compute the CP throughput requirements and the program store requirements based on the application. Our example requires a graph control program (GCP) to direct the execution of the graph on the signal processor. The number of program instructions required to implement the GCP must be estimated and the resultant throughput calculated. This will be done by the software organization. For the purpose of this example, we assume that a graph of this complexity can be completely controlled by 500 instructions. The throughput is given by

$$\text{GCP}(1) \text{ load} = 100 * \frac{500/0.128}{5,000,000} = 0.078\%$$

The program store requirements are given by the number of instructions, 500 instructions times 2 bytes per instruction equals 1000 bytes. This equals 0.39% of the total program store available.

In order to compute the total control processor resources the executive program that provides the operating system under which the GCPs run must be included in the resource estimates. Generally, multiple graphs will be operating simultaneously and the total throughput and storage for all the GCPs must be calculated. For M graphs, the percentage loads are given by summing the individual GCP percentage loads with the executive load for both the throughput and storage loading.

$$\text{CP}(\%) \text{ load} = \text{Executive}(\%) + \sum_{i=1}^{M} \text{GCP}(i) \text{ load} \qquad (10.6)$$

Input / Output Processor. For our example DFG the input requirements are minimal. These are based on the input data sample rate and wordlength. The input for the graph requires 8000 samples-per-second at 8-bits per sample. Therefore the interface bandwidth is 8000 bytes-per-second. The I/O processor resource loadings are computed by summing over all input and output channels.

$$\text{I/O}(\%) \text{ load} = 100 * \left(\frac{\sum_{j=1}^{\text{IN}} \text{BW}_{\text{in}}(j)}{\text{BW}_{\text{in}}\text{capacity}} + \frac{\sum_{k=1}^{\text{OUT}} \text{BW}_{\text{out}}(k)}{\text{BW}_{\text{out}}\text{capacity}} \right) \qquad (10.7)$$

RESOURCE ANALYSIS TOOLS

The preceding example illustrates the use of a static level resource analysis of the DFG given in Figure 10.9 for the processor described in Section 10.1.6. For this one DFG the calculations were performed by using an electronic calculator. However, most application design problems will require several DFGs to be analyzed. Therefore, an automated method of computing the resource parameters from the DFG description is desirable. The process described can be implemented using personal computer spread sheet templates. Examples of spread sheet outputs will be shown in Section 10.3.

In order to provide dynamic results, simulation level analysis of the resources is required. The simulation models the processing requirements, especially all frequently used control operations such as the scheduling and dispatching of strings. Timing results are then collected over several iterations of the graphs and used to determine the processor resources. Simulation of the processor architecture requires implementation of complex models, with large processing throughputs, that generally require large mainframe computer resources. Data formatting and analysis tools must be developed in order to simplify the evaluation of the complex data outputs from the simulation. These tools should provide the capability of observing the latencies and contention for bus resources. A plot of each specific task being scheduled, dispatched, and executed as a function of time and processor hardware/software provides an excellent method of seeing idle times, data loads, data unloads, and processor element loading.

10.2.5 Step 5 — Configuration Analysis

The configuration analysis combines the resource analysis results with the nonsignal processing application requirements to assure total system compliance to all requirements. The DFG resource requirements for all processing modes and channels are used to define the signal processor configuration. All DFGs for the required processing modes and channels are partitioned to the signal processor configuration and a thorough resource analysis is performed. The partitioning must consider degraded-mode requirements to assure that the partitioning is done such that critical processing can be accomplished if a hardware failure occurs. If all system requirements are met, then the analysis is complete and the configuration defined. Otherwise, additional trade-off studies must be performed.

If multiple processing elements are required, then the quantity is selected based on the resource analysis results in conjunction with step 3. The signal processing graphs are then partitioned across the elements. This process requires careful analysis based on the DFG throughput requirements. For example, if two adjacent processing operations are partitioned to different processing elements, then the output data from the first operation must be transferred to the second operation before it can be executed. If this is done very often, the data rate on the data transfer network can exceed the processors' capabilities. Therefore, the processing element-to-element data transfer requirements should be minimized.

If multiple channels of the same processing mode are required, the partitioning must be made to the system configuration. Also, if multiple processing modes are required, all modes must be partitioned to the system configuration. The resources are computed based on the partitioning using the equations defined for the resource analysis. Again we have the option of using a static or dynamic resource analysis approach. If confidence in the static model has been achieved it provides the simplest approach. An event level simulation of the processing gives a more accurate estimate of the resources. The simulation should model the dynamic characteristics of the processing. A detailed hardware level simulation provides the actual resources required for the processing.

10.2.6 Signal Processing System Design Summary

The foregoing analysis procedure is intended to describe a representative approach to a signal processing design application problem. The details of each step have not been covered, but it is hoped that enough material has been presented to give you an understanding of the analysis tasks required. Many of the complex analysis and design tasks can be simplified by using personal computers. This will be discussed further as part of the application design. The finite arithmetic analysis and the partitioning of the signal processing graph to the processor are key to achieving efficient application designs. The complexity of the design process continues to increase with the rapid advances in technology. Therefore, more sophisticated analysis and development tools are essential to performing accurate DSP designs.

10.3 SPECTRAL ANALYSIS APPLICATION

The detection of discrete frequency components embedded in broadband spectral noise is encountered in many signal processing applications. The time-domain representation of the composite signal is the summation of the individual noise and discrete frequency components. The spectral energy of the input signal is illustrated in Figure 10.15. Separate noise and target spectral components are shown individually to illustrate the need to resolve the discrete frequency components along the frequency axis. The signal is representative of ocean acoustic signals, which consist of many noise sources and radiated spectral energy from surface ships and submarines, which are generally masked by the higher noise signal component levels. We will perform a DSP system design to determine the processing required for this important application.

To discriminate the target components from the noise components, sufficient signal processing is required to enhance the target-to-noise ratio until the target is detected within a specified level of performance. Spectral estima-

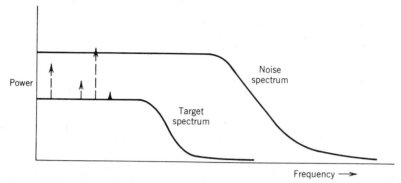

Figure 10.15 Separate input signal spectral components illustrated: broadband noise; broadband target; discrete targets.

tion techniques were presented in Chapter 8 showing the benefits of averaging overlapped FFT outputs to achieve processing gain. This approach will be used to implement the discrete frequency (narrowband) spectral analysis detection system.

The concepts of probability of detection and probability of false alarm are presented in Section 8.3. Several textbooks cover detection theory in detail.[9-11] Normally, the performance requirement for detection processing will specify frequency bandwidths for analysis that result in a specified probability of detection (P_d) after processing for a time period (T_i) and with a probability of false alarm, P_{fa}.

Receiver operating characteristic (ROC) curves are available for a broad class of processing systems that plot P_d versus P_{fa} as a function of signal-to-noise ratio (SNR). The ROC plots for a narrowband spectral analysis detection system are presented in detail in Reference 9, and a description of a closed-form equation[12] for calculating the ROC performance for a linear envelope detector is presented in Section 8.3.1.

From an intuitive perspective, the performance is expected to increase as the analysis resolution is made finer. This is illustrated in Figure 10.16 by looking at the spectral noise energy that passes through an FFT analysis bin. The shaded area shown on the FFT output represents the mainlobe output of one frequency bin. As the bin is made narrower (i.e., FFT length is increased) the output noise power is reduced whereas the discrete frequency component energy remains the same. Therefore, a higher SNR results at the output of the finer analysis bin and a correspondingly higher P_d should be expected, maintaining a constant P_{fa} and processing time. In Chapter 1 we presented a simulated FFT application to a sinusoidal signal embedded within Gaussian random noise. This illustrates the key concept of processing gain realized by averaging of successive FFT outputs.

The following paragraphs present the signal processing system design for the narrowband spectral analysis application. We will consider one signal class

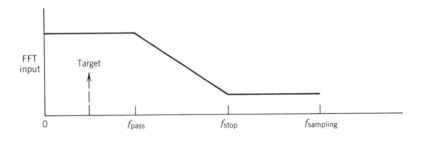

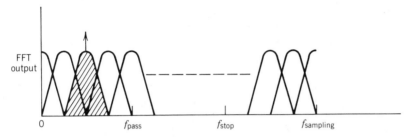

Figure 10.16 FFT analysis illustration.

(ocean acoustics), one processing option (passive), and two processing modes (narrowband and vernier detection) for simplicity. Processing for multiple signal classes, processing options, and processing modes will usually be required in an actual application. The procedures presented in Section 10.2 will be used to perform the system design.

10.3.1 Spectral Processing System Requirements Definition — Step 1

A stationary input signal consisting of broadband Gaussian noise and discrete frequency components from zero to 10,000 Hz is to be processed in order to detect the discrete frequency components that exist between 62.5 Hz and 1000 Hz. The processing requirements are presented in Table 10.3. The requirement is to process four frequency bands starting at 62.5 Hz with the analysis resolutions specified in the table. In addition, selectable vernier frequency bands with a processing bandwidth covering 1/16th of the original frequency band, but with 16 times finer analysis resolution are required for each of the four bands.

The total processing requirement is 40 input channels. The input channels must be individually assignable, and the center frequency of each vernier must be selectable for each channel. All requirements must be defined and documented in this step.

The interface design and system level requirements specifications are started in this step of the design. These specifications are essential to develop-

TABLE 10.3
SPECTRAL ANALYSIS APPLICATION REQUIREMENTS

Parameter	Specification / Requirement
Input Signal $x(t)$	$x(t) = n(t) + s(t)$
$\quad\quad\quad\quad n(t)$	Gaussian $(0, \sigma)$
$\quad\quad\quad\quad s(t)$	Sinusoidal $A\{i\} \sin(\omega\{i\}t + p\{i\})$
$\quad$ Bandwidth	0–10,000 Hz uniform;
	10,000 Hz and above: 6 dB/octave rolloff
$\quad$ Maximum interfering tone level	20 dB/Hz
$\quad$ Dynamic range	50 dB
Performance P_d, P_{fa}, T_i	0.5, 0.0001, 5 minutes
Antialiasing P_d, P_{fa}, T_i	≤ 0.01, 0.0001, 5 minutes
Processing bandwidth	62.5–1000 Hz
Spectral analysis	
Analysis resolutions*	Within $\pm 10\%$ of:
$\quad$ octaves, (Δf)	1.25 $\quad$ Hz from 500 to 1000 Hz
	0.625 $\quad$ Hz from 250 to 500 Hz
	0.3125 $\quad$ Hz from 125 to 250 Hz
	0.15625 Hz from 62.5 to 125 Hz
$\quad$ Verniers	1/16th of octaves
Redundancy	$\geq 50\%$
Noise bandwidth	≤ 1.5 analysis resolution
Scalloping loss	
$\quad$ Maximum	≤ 2 dB
$\quad$ Average	≤ 0.7 dB
Highest sidelobe level	≤ -26 dB
Other losses	≤ 1 dB

*Requirements based on P_d, P_{fa}, T_i for signal levels at the system input.

ing a system that meets all signal and nonsignal processing requirements. Since the signal processor has been specified, then the principles of operation and the unit level specifications form a basis for the design. Otherwise they would be developed as part of the design process. The basis for the signal processing requirements specification is given in Table 10.3. We need to define the system in which the processor is to be configured. This will normally be specified by the customer. For our example we assume that a system manager, sensor system, and display system are specified. The interfaces of each of these systems to the processor must be completely described.

Number of data bits.

Number of control bits.

Control protocol.

Maximum data rate.

Electrical characteristics.

Connector requirements.

These requirements will be captured in the IDS, including all control messages required from the system manager to assign and control all processing options and modes.

The system level signal processing requirements must be defined and documented. This includes the items described in Section 10.1.4.

Processing option 1:	Narrowband detection
Processing mode 1:	Octave processing
Processing mode 2:	Vernier processing band 1
Processing mode 3:	Vernier processing band 2
Processing mode 4:	Vernier processing band 3
Processing mode 5:	Vernier processing band 4
Concurrencies:	All modes concurrent
Number of channels:	Maximum of 40
Mode flexibility:	Per Table 10.3

The nonsignal processing requirements are specified as part of this step. This part of the procedure is not germaine to the DSP treatment herein, but you are reminded of its importance to the overall application design as discussed in Section 10.2. The following list is given to provide some insight into the number of nonsignal processing requirements that must be considered.

Form factor constraints:	Height, width, length
Weight:	Maximum weight
Power:	Type, maximum watts
Cooling:	Forced air, liquid, etc.
Reliability:	Mean time between failure (MTBF)
Availability:	0.999 for 24-hour mission
Maintainability:	Maximum mean time to repair; Mean time to repair (MTTR)
Fault detection:	95% probability of detecting fault
Fault isolation:	90% probability of isolation to 1 module given detection of fault
Fault tolerance:	Must maintain 20 channels of processing
Environmental:	Shock, vibration, thermal, altitude, humidity, etc.

10.3.2 Acoustic Signal Analysis — Step 2

Using the signal analysis procedure (Fig. 10.12), we develop the signal characteristics for the classes of signals defined in step 1. From step 1 the signals to be processed for this application are analog input signals resulting from ocean acoustic sensors. The input signal levels are based on the acoustic energy impinging on the sensor. The energy is a function of weather conditions, the number of sources, the vicinity of the sources to the sensor, the type of sources, and the characteristics of the ocean.

We have assumed that the signal we are dealing with is digital at the input to the signal processor. Therefore, the sensor input must be preconditioned and digitized. The analog signal conditioning is designed to provide a good balance between the analog hardware requirements and the subsequent digital hardware requirements.

First we review each of the items listed in Figure 10.12 and identify all known characteristics/requirements as listed in Table 10.4. The input control requirements are for the system controller to send commands to the signal processor specifying the sensor input and the channel allocation. The output control is generated by the signal processor based on the sensors assigned by the system controller.

TABLE 10.4
SIGNAL DEFINITION

Item	Characteristic / Parameter	Definition
1	Number of channels/sensors	40/40
2	Active/passive	Passive
	Steady state/transient	Steady state
3	Analog/digital	Analog
4	Desired information	Discrete frequency target components
5	Deterministic/random	Random—Ocean (Atlantic and Pacific areas)
6	Bandwidth/frequency ranges	Table 10.3
7	Spectral content	Table 10.3
8	Dynamic range	50 dB
9	Interferences/random components	20 dB discrete components/ shipping noise, sea state 3
10	Waveforms/pulses transmitted	N/A
11	Pulse duration/repetition	N/A
12	Data formats/multiplexing/codes	To be defined
13	Control requirements—Input	Commands from system controller for desired channel/sensor input
	Output	Command to signal source to activate sensor input
14	Data rates/control rates	To be defined

The signal is passive, and therefore no signal transmission is required. Since the signal is analog, we must define the analog signal conditioning, select the minimum sampling rate, and determine the number of bits required to represent the signal adequately. This process is iterated with the digital processing design based on the resource analysis requirements until an acceptable design is achieved. In some cases this may be defined since the sensor system may already exist. Then, this part of the application design would be to analyze the digital input for possible conditioning to reduce the overall throughput requirements.

ANALOG SIGNAL CONDITIONING

The analog signal conditioning is performed to limit the bandwidth and adjust the signal level for subsequent A/D conversion. This function is divided into three operations: analog presampling filter (APF), automatic gain control, and A/D converter.

The filter requirements are based on the input signal characteristics, and the required processing bandwidth. Therefore, since the processing bandwidth is 62.5 Hz to 1000 Hz, the analog input signal can be filtered ideally to 1000 Hz prior to A/D conversion. Based on sampling theory for a real signal (i.e., folding occurs from one-half the sampling frequency and above), the lowest frequency that will fold back into the processing bandwidth is given by

$$f_s = F - f_p \tag{10.8}$$

f_s = lowest stopband frequency

F = Sampling rate

f_p = Highest passband frequency

We refer to this as the critical folding frequency. The APF response characteristics are shown in Figure 10.17, where f_p equals 1000 Hz and the sampling rate, passband tolerance, and stopband tolerance must be developed. The

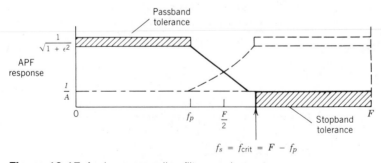

Figure 10.17 Analog presampling filter requirements.

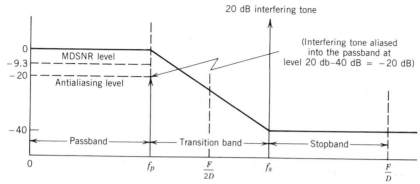

Figure 10.18 Stopband frequency requirements.

actual sampling rate selected must be traded off versus the filter order required (analog hardware) and subsequent digital processing requirements. Selection of higher sampling frequencies will reduce the analog filter requirements, but will result in higher initial digital processing loads. Also, if other processing algorithms are to be applied to the signal, then the bandwidth and sampling rate should be selected to satisfy the maximum processing bandwidth requirement. The passband and stopband tolerances are developed first.

The passband tolerance is selected based on the total allowable variation. From a system performance standpoint, it is necessary to keep the passband deviation as small as possible. This minimizes performance degradations caused by nonuniform outputs. The total variation will be a composite of the individual filter frequency responses through which the signal is passed. Hence a portion of the total allowable variation is allocated to the various processing functions. The composite frequency response of all the filtering must be calculated to determine if the design meets the requirements. For this example, 0.5 dB passband variation has been allocated to the analog processing. As discussed in Chapters 4 and 5, the smaller the allowable passband variation the higher the filter order required to maintain equivalent rolloff and stopband attenuation. Generally a total passband deviation of 1 dB will be acceptable for detection systems.

The stopband (SB) attenuation is calculated to achieve the desired antialiasing performance based on the highest input interfering out-of-band tone. The filter must provide enough attenuation of the out-of-band 20 decibel (dB) tones to reduce their levels below the 0.01 probability of detection levels allowed by the specification. This is determined from the ROC curves by reading the SNR for $P_d = 0.5$ and reading the SNR reduction required to reduce the P_d to 0.01. For a 1% P_d the SNR reduction must be $\geq$ 8 dB for the P_{fa} and T_i values given. The attenuation level calculation computes the difference in dB between the interfering tone (IT) and the MDSNR and adds the dB factor required to meet the antialiasing performance as shown in Figure 10.18. The minimum stopband attenuation required to meet the specifications is given by

$$\text{Min SB atten} = \text{IT} - \text{MDSNR} + (\text{level required below MDSNR})$$

$$= 20 - (-9.1) + 8 = 37.1 \text{ dB} \tag{10.9}$$

Therefore the stopband must at least provide 37.1 dB attenuation from frequency f_s and higher. The MDSNR value used in Eq. 10.9 is calculated using the procedure presented in Chapter 8. This value is the lowest MDSNR for the 62.5 to 125 Hz band. Since f_s actually folds into the upper part of the processing band at 1000 Hz, the MDSNR could have been selected based on the 500–1000 Hz processing performance for frequencies from f_s to $f_s + 500$ Hz. Likewise, the attenuation requirements for all other frequencies can be based on the performance band to which they will fold. For the purpose of this design, 40 dB will be chosen as the attenuation requirement. If the filter requirement were to be slightly above order N, then the designer may choose to reduce this attenuation requirement in order to allow use of a Nth order filter and still meet the minimum 37.1 dB requirement.

The APF order required can now be computed as a function of sampling rate and filter type. For this application a Chebyshev filter will be selected. The designer must trade off all design choices in order to determine the most cost effective design for the application. The Chebyshev filter orders required as a function of the subsequent sampling rate are shown in Table 10.5, where

$$\text{Transition ratio } k = \frac{f_p}{f_s} = \frac{f_p}{F - f_p} = \frac{1}{(F/f_p) - 1}$$

$$D = F/f_p$$

For this design, a sampling rate of 2500 Hz or 2.5 times the highest frequency, 1000 Hz, to be processed has been selected. This results in an APF of order 8.

The designer would normally iterate the sampling rate choice, trading off the resulting analog/digital processing requirements to minimize the overall hardware costs. If there is a significant amount of digital processing reserve, then the choice would be toward a higher sampling rate (i.e., reduced analog hardware). Figure 10.19 presents the spectrum of the input signal and shows the output spectrum after sampling without filtering and with filtering. The necessity for the APF is easily seen from the figure by observing the output levels of the out-of-band tones that have aliased into the processing band. Also note that the in-band SNR has been degraded 3 dB owing to the folding of the noise spectrum of equal level for the unfiltered case. If phase linearity is important to the processing algorithm, then the filter design and other processing operations need to satisfy these requirements.

An important characteristic that must be known is the dynamic range of the input signal as a function of time. This is required in order to determine

TABLE 10.5
CHEBYSHEV FILTER ORDER REQUIREMENT VERSUS THE A / D SAMPLING RATE

Chebyshev Filter Order	$D*$	$k*$	F
∞	2	1	2000
15	2.1	0.901	2100
12	2.2	0.833	2200
10	2.3	0.769	2300
8	2.4	0.714	2400
8	2.5	0.667	2500
7	2.6	0.625	2600
7	2.7	0.588	2700
6	2.8	0.556	2800
6	2.9	0.526	2900
6	3.0	0.5	3000

*For this example, we will use $D = 2.5$; therefore $N = 8$ and $F = 2500$ Hz. The transition ratio, k, is given by

$$k = \frac{f_p}{f_s} = \frac{f_p}{F - f_p} = \frac{1}{(F/f_p) - 1}$$

$$MIN(F) \geq 2f_p$$

Let $F = Df_p$ where $D \geq 2$, then

$$k = \frac{1}{D - 1}$$

the analog-to-digital converter (A/D) design, and usually requires a gain control process preceding the A/D. The gain control is designed to expand the signal level to use the full A/D range when the input energy is small and to reduce the energy level to within the A/D range when the input signal energy is high. Since we are dealing with a stochastic signal, the gain control adjusts the input based on a probabilistic setting (e.g., for a Gaussian input with standard deviation σ, $k\sigma$ will be set at the maximum allowable magnitude of the A/D converter). This process is described in Chapter 9.

Most of today's sonar systems employ an automatic gain control (AGC) algorithm that estimates the standard deviation of the input over a specified time and performs the proper adjustment. The bandwidth of the signal was given in Table 10.3 along with the desired signal components to be detected. Out-of-band interfering tones will be less than or equal to 20 dB/Hz.

The AGC must either contain analog hardware to compute the standard deviation or the standard deviation is calculated as part of the digital processing and analyzed to determine if a new gain value should be applied. The decision is based on the change in the estimate for a specified time period (the AGC time constant) and requires careful selection of constants to assure that it will

(*a*) Without APF — Aliasing

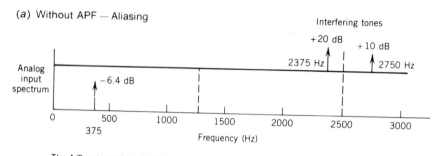

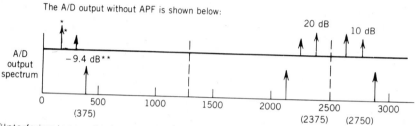

*Interfering tones alias into processing band:
 20 dB tone alias from 2375 Hz to 125 Hz
 10 dB tone alias from 2750 Hz to 250 Hz

**The noise power is also aliased into the processing band and therefore will degrade signal-to-noise ratios of in-band signals and result in performance loss.

(*b*) With APF — Aliasing

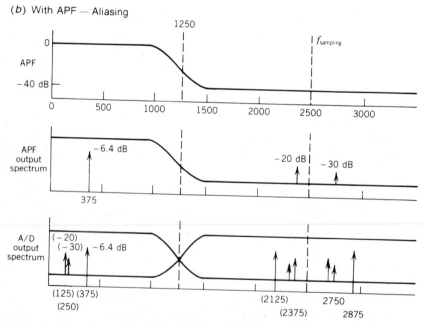

Note that interfering tones have been attenuated below the MDSNR by at least 8 dB, which will provide us with the required P_d of less than 0.05, with $P_{fa} = 10^{-4}$ after 5 minutes of processing.

Figure 10.19 APF Antialiasing illustraton.

not oscillate between gain values. Refer to Chapter 9 for a discussion of AGC settings versus A/D converter bit lengths. The designer must select the desired step size for which the gain adjustment is required across the input dynamic range. Also, if the Gaussian input does not apply, the input signal characteristics must be analyzed to determine the optimum setting based on the signal.

$$V_{max} = \text{Maximum voltage level of the A/D} \qquad (10.10)$$

then,

$$\text{AGC}_{out} = V_{max} * \text{AGC}_{in}/(k * \sigma) \qquad (10.11)$$

represents the ideal output calculation. Instead of this ideal calculation, the analog hardware performs an adjustment by a fixed amount at the desired update rate. Large step sizes greater than 6 dB are not desirable since they put abrupt amplitude changes into the input data. The time constant of the estimate must also be selected to provide a good estimate based on the stability of the input. If the time constant is too short, several changes could occur within one FFT, causing erroneous frequency components. Long time constants could result in clipping or underflows due to large changes in the input signal.

The A/D converter samples the analog signal at the selected sampling rate, 2500 Hz for this application, and quantizes the signal into a binary representation of B bits. The number of bits are selected to handle the expected input signal-to-noise ratios without degrading the SNR due to quantization errors as discussed in Chapter 9. For this problem a fixed-point implementation is assumed. Fixed-point implementations require careful design of the data scaling to ensure that the noise effects do not degrade system performance. Floating-point implementations offer advantages in reduced complexity of the scaling analysis and resultant programming that should be considered versus the larger data lengths and hardware complexities.

Eight bits is sufficient to handle the maximum-to-minimum signal levels to be processed within the noise. The number of bits determination for the A/D is presented in Chapter 9. The output of the A/D is transferred to the main memory via the data transfer network for subsequent digital processing. For each channel to be processed, the required input data bandwidth is given by

$$\text{BW}\{\text{input}\} = 2500 \text{ bytes/second/channel} \qquad (10.12)$$

Therefore, for 40 channels the input data bandwidth equals 100K bytes per second.

10.3.3 Spectral Detection Signal Processing Design — Step 3

The signal class we are dealing with is ocean acoustic signals. The processing option is passive narrowband spectral detection. Five processing modes are required: 1 narrowband processing mode for four frequency bands and four

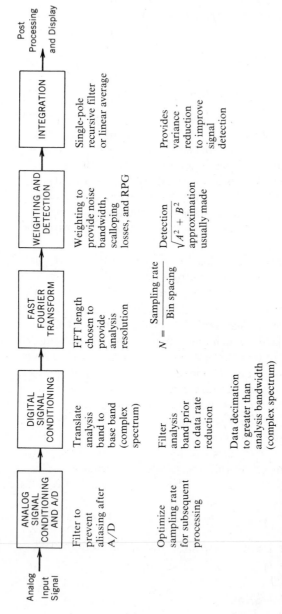

Figure 10.20 Spectral analysis detection system.

vernier processing modes, one for each of the four narrowband frequency bands. The performance requirements were defined in Table 10.3. In some cases the analysis resolutions and bands will be left as part of the design. Then we would determine the required analysis resolutions using Section 8.3.2. We would trade off filtering versus FFT lengths to determine the most efficient design to achieve the required performance.

The analog signal conditioning was defined in step 2, resulting in a digital input signal at a sampling rate of 2500 samples per second per channel. Next we develop a high-level functional flow diagram for the narrowband detection process. Figure 10.20 presents a typical narrowband spectral analysis detection system flow diagram for one channel. The top level processing blocks are based on the algorithm requirements for performing the desired analysis. Important considerations of the analysis/design at each point in the processing are shown in the figure. Once the processing required is understood to this level, the designer can proceed to the detailed development of the signal processing design resulting in a DFG. For each block in Figure 10.20, the detailed processing requirements are developed.

For this application, a signal processor has been specified with the primitive capabilities and execution times as given in Table 10.1. Direct use of the fast Fourier transform (FFT) could be used, but generally requires high computation rates to provide the necessary resolution owing to the uniformly spaced analysis bins across the entire frequency range up to the sampling rate. Chapter 7 discussed multirate processing and the techniques by which the signal of interest can be translated to baseband, filtered to the desired bandwidth, and data-decimated to reduce the sampling rate for subsequent spectral analysis processing. This approach will be developed in the following sections.

Although we have shown one channel of processing in Figure 10.20, 40 channels must be processed. For each channel, processing of the four frequency bands specified along with a vernier selection for each frequency band is required. The digital signal conditioning consists of frequency translation, filtering, and data decimation. Following the digital signal conditioning, FFT processing is performed and the FFT outputs are weighted, detected, and integrated. Weighting in the frequency domain to reduce spectral leakage from outside the mainlobe of the analysis bin into the mainlobe of the analysis bin was discussed in Chapter 6. The final processing operation, Figure 10.20, is integration of the detected bin outputs to achieve the specified processing gain. Following the integration of the detected outputs, additional postdetection processing is usually required to prepare the data for display and for post-processing to compute signal characteristics and thresholding for detection. We will not consider these items for this design example.

Using the signal processing design methodology presented in Figure 10.13, we design the signal processing for the narrowband spectral analysis application.

DIGITAL SIGNAL CONDITIONING

The digital signal conditioning prepares the signal for subsequent analysis processing necessary to implement the algorithm. For this application, the signal must be conditioned for subsequent analysis into the four frequency analysis bands and the four vernier bands given in Table 10.3. The conditioning will be performed using the multirate techniques discussed in Chapter 7. Multirate techniques reduce the processing requirements by allowing data decimation before subsequent processing for applications that require only a portion of the input spectrum. For each band, the signal is translated to baseband, filtered to the desired bandwidth, and data-decimated to twice the real bandwidth within aliasing constraints prior to the spectral analysis.

Figure 10.21 illustrates a multirate signal processing graph to implement the signal conditioning for the 500 to 1000 Hz frequency band. The frequency bands are octave bands. An **octave band** has a lower band frequency that is one half the upper band frequency. The octave band filtering is an ideal relationship for halfband designs. Therefore, halfband filters may provide computational savings if used in this implementation. The processing design must consider each frequency band and vernier.

Frequency Translation. For octave 4, the frequency band translation is performed via multiplication of the real input signal by a complex exponential with frequency equal to the center of the processing band (750 Hz) and is commonly referred to as a complex band shift (CBS) primitive operation. This

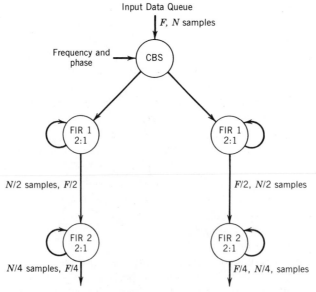

Figure 10.21 Multirate signal processing graph for digital signal conditioning.

translates the input frequencies by -750 Hz (see Chapter 7), centering the 500 to 1000 Hz processing band at zero Hz. The resultant output contains the quadrature components of the complex signal. Implementation of the CBS requires generation of the sine and cosine components of the complex exponential at the desired center frequency of the processing band and subsequent multiplication of the real input by the sine and cosine components given by

$$\text{Real channel (I-channel for in-phase)} = x(nT) * \cos(2\pi f_0 nT)$$

$$\text{Imaginary channel (Q-channel for quadrature phase)} = x(nT) * \sin(2\pi f_0 nT)$$

where f_0 is the center frequency and T is the sampling period equal to the reciprocal of the sampling rate, $1/F$. Several techniques exist for generating the sinusoidal components such as recursive algorithms and addressing a table of stored values. The selected signal processor has the capability to generate sine and cosine components at a 5 megahertz rate. The frequency is required along with a phase increment, which is used to maintain a continuous function from batch to batch.

The CBS for each of the four processing bands and four vernier bands could be implemented directly using the input signal at the 2500 Hz sampling rate. This would require two multiplies per input sample for each of the eight bands, resulting in a multiply per second (MPS) requirement of

$$\text{MPS} = 2 \text{ multiplies/CBS} * 8 \text{ CBS} * 2500 \text{ samples/second}$$

$$= 40{,}000 \text{ mult/sec}$$

Since the processing element provides 5 million multiplies per second (MMPS), the CBS if implemented in this form would require 40,000/5M or 0.8% load per channel plus the time required to set up each band and channel for processing.

The preceding computations on multiply requirements provide a rough-order estimate of the throughput requirement and would be used initially if a processor had not been specified. Since a processor was specified, the operation to be performed is checked against the processor's primitive capabilities and, provided a primitive exists, to perform the operation the processing time is computed using the execution times in Table 10.1. If a primitive operation does not exist for the processor to perform the desired operation, then one is developed and added to the capability chart.

The time to process a batch of data through the primitive is now calculated. The batch is generally chosen large enough to minimize any overhead that may result owing to setting up the control for the operation and data input/output transfer times. Also, the batch size is selected to provide sufficient data for subsequent operations. For this operation a 1024-point batch size will be used. Note that the batch size will also determine the input buffer size required. The buffer must provide storage for enough samples to assure that the next batch of input data will be processed before any samples are written over by new input samples. The time in microseconds to process an N point input sample batch is given by $[5 + (0.8 * N)]$ microseconds, as shown in Table 10.1. For a 1024-sample input batch the time is

$$\text{CBS time}/1024 = 5 + (0.8 * 1024) = 824.2 \ \mu\text{sec}$$

This time is required for each of the octave and vernier bands, resulting in a total time of 6593.6 μsec. At the input sampling rate of 2500 Hz the time per 1024-sample batch is 0.41 seconds, and the load is 1.6% for one channel. Although this represents a relatively small load on the element per channel, it does require 64% of a processing element for the 40 channels. Also, the filtering operations that follow would be required to use the input sampling rate of 2500 Hz.

Figure 10.22 illustrates a more efficient implementation of the process by reducing the bandwidth of the lower three octave input processing bands prior to the CBS operation via a FIR filter followed by a decimation operation. Note that the vernier analysis bands are selected from the octave outputs, further reducing the vernier band shift (VBS) input data rate. The VBS operation must be performed on the complex data from the octave filter outputs, requiring four multiplications and two additions per complex input sample. For this implementation, the CBS multiplication requirements are reduced to

$$\text{MPS} = [2 \ \text{mult}/\text{CBS} * (1 + 0.5 + 0.25 + 0.125) * 2500]$$
$$+ [4 \ \text{mult}/\text{VBS} * (1 + 0.5 + 0.25 + 0.125) * 625]$$
$$= 14{,}062.5 \ \text{mult}/\text{sec}$$

where the quantity in the parentheses represents the reduction in sampling rate before the CBS or VBS operation, and the vernier sampling rate following is one fourth the octave bands. The time to process the verniers using this approach is given by

$$\text{CBS time}/1024 = 5 + 0.8(1024) + 10 + 1.6(256)$$
$$+ 5 + 0.8(512) + 10 + 1.6(128)$$
$$+ 5 + 0.8(256) + 10 + 1.6(64)$$
$$+ 5 + 0.8(128) + 10 + 1.6(32)$$
$$= 20 + 0.8(1920) + 40 + 1.6(480) = 2364 \ \mu\text{sec}$$

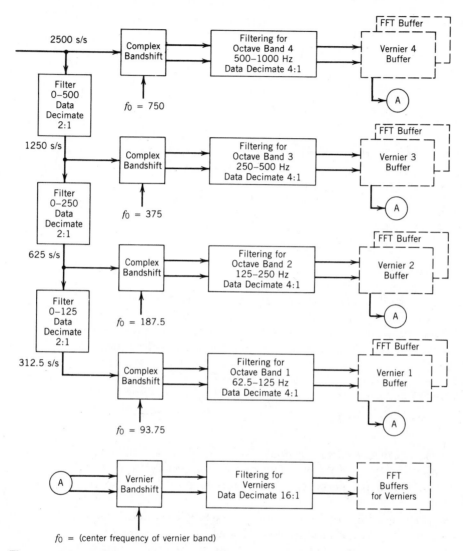

Figure 10.22 Digital signal conditioning for octaves and verniers.

To this the time (multiplications) required for the additional filtering prior to the three lower octave CBS operations must be added. The filter required must reduce the bandwidth such that the data rate can be reduced by two and the aliased spectrum is rejected by at least 40 dB per our previous APF discussion. The filter requirements are developed for the frequency octave band from 500 to 1000 Hz, as shown in Figure 10.23.

The choice of equal ripple factors was made to meet the halfband FIR filter definition. The initial value used is the smaller of the passband and stopband requirements for a single filter stage. Using Eq. 5.43, the passband

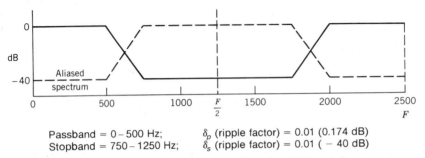

Passband = 0 – 500 Hz; δ_p (ripple factor) = 0.01 (0.174 dB)
Stopband = 750 – 1250 Hz; δ_s (ripple factor) = 0.01 (– 40 dB)

Figure 10.23 Octave band 3 filter requirements prior to CBS operation.

ripple in dB is calculated:

$$Ap = 20 \log\left[(1 + \delta p)/(1 - \delta p)\right] = 0.174 \text{ dB}$$

This will accommodate only two stages of filters to stay within the digital processing passband deviation allocation of 0.5 dB with 0.152 dB of margin. The maximum number of stages are expected for the vernier implementation where the ripple factors should be divided by the number of anticipated stages (refer to Chapter 7). For this design, a factor of 10 will be assumed. Calculating the new ripple factor using Eq. 5.48:

$$\delta p = \left[\{10**(Ap/20) - 1\}/\{10**(Ap/20) + 1\}\right] = 0.002878$$

which as expected equals one tenth the ripple factor for 0.5 dB. Additional trade-offs are made to determine the optimum design implementation from a throughput standpoint in the next step of the procedure as the design is finalized.

The filter order equations for the FIR filter Parks-McClellan designs are given in Chapter 5, Eq. 5.91. For this problem, the simpler form given by Eq. 5.92 will be used in order for you to more easily calculate the results using a hand calculator for other cases. Generally, computer programs are available to perform these calculations, which use the more accurate approximation equation. Using Eq. 5.92, the FIR filter order required to meet the initial requirements is given by

$$N = \frac{-10 \log \delta_p \delta_s - 15}{14(f_s - f_p)/F} + 1 = 26.6 \rightarrow 27$$

Since this filter satisfies the halfband FIR filter definition given in Chapter 5, and therefore the even coefficients, excluding the midpoint, will be zero, and the number of multiplies required is reduced to

$$\text{Filter mult/sec} = [(27 + 5)/4] * (0.5 + 0.25 + 0.125) * 2500$$

$$= 17,500 \text{ mult/sec}$$

This results in a total multiply per second requirement of 31562.5, and the subsequent processing operations have been reduced significantly.

Again, computing the processing time for a 1024-sampling batch through the frequency translation and associated filtering operations results in

$$\text{Filter time/1024} = 3 * 12 + 1.2(512 + 256 + 128) = 1111.2 \ \mu\text{sec.}$$

where from Table 10.1 the time for the halfband filter is $[12 + 1.2M]$ microseconds for each filter operation. M is the number of filter output points after data decimation. The total time for the CBS/VBS and filtering equals 3475.2 microseconds. This represents a 47% reduction in the processing time, and also reduces the subsequent processing requirements. The total load for this approach to frequency translation is 0.85% (0.0034752/0.4096).

Other approaches are possible; for example, the verniers could be implemented before the filtering operation for the octave bands, which would provide better response across the octave boundaries. The designer should consider alternatives that provide more flexibility and may decide to implement a less efficient alternative that provides easier incorporation of future enhancements. This will be addressed in the problems for this chapter.

Octave Band Digital Filtering. The advantage of translating the center of the frequency band to be analyzed to 0 Hz (i.e., baseband) is the ability to use the same approach for multiple processing bands (i.e., quadrature demodulation followed by lowpass filters). Therefore, the control structure for each band is identical with the only changes required for the demodulation frequency and the lowpass filter bandwidth. When the bandwidth and sampling rates are proportionately reduced, the lowpass filters for each band are identical. For the octave bands defined, the filtering strings are all the same.

The analysis proceeds using lowpass digital filtering on each of the quadrature components with a passband cutoff equal to one half the octave bandwidth. The bandwidth is one half the octave band because the quadrature input contains unique information from 0 Hz to the sampling rate. Therefore, the lower half of the octave band appears adjacent to the sampling rate, as shown in Figure 10.24. The shaded areas are equivalent bands of the complex bandshifted signal representing the lower half of the octave band. For the 500 to 1000 Hz octave band this represents the 500 to 750 Hz half of the octave.

The goal of the filtering design is to minimize the processing requirements and to prepare the signal for subsequent spectral analysis processing. Chapters 4 and 5 discussed the design of IIR and FIR digital filters, respec-

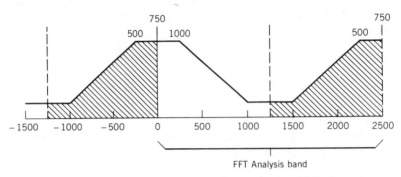

Figure 10.24 Octave band (500–1000 Hz) complex bandshifted spectrum representation.

tively. Chapter 7 addresses the effects of multirate processing on the throughput requirements of the cascaded filtering and decimation operations and forms the basis for our initial choice of FIR filters. Recall that FIR filters can be implemented at the decimated sampling rate following the filter owing to the nonrecursive structure of the FIR filter. Also, FIR filters do not exhibit the noise buildup inherent in the IIR feedback loop (Chapter 9). In Chapter 7 we concluded that the processing requirements were generally reduced via implementing the filtering operation with cascaded stages of filtering and data decimation. In order to determine the optimum design as discussed in Section 10.2, FIR and IIR design alternatives must be developed and traded off. For this example, FIR filters will be selected and used to illustrate the design process. The IIR alternatives will be left as problems for you to perform in order to gain experience and insight into the various filtering alternatives available.

The filter designs must provide sufficient stopband attenuation to meet the antialiasing performance levels and passband ripple levels to satisfy the requirements of Table 10.3. The stopband attenuation requirement is developed from the antialiasing performance requirement and the largest out-of-band signal level that would alias into the processing band. Figure 10.25 illustrates this concept where the data decimation operation has been included. Again, Eq. 10.9 applies and the attenuation required is 37.1 dB assuming worst-case MDSNR (i.e., minimum MDSNR = -9.1 dB). This assumes that the input spectrum is uniform. Otherwise the shape must be accounted for when computing the per hertz levels at the aliasing frequency. Also, the requirements can be relaxed such that the attenuation required is a function of the MDSNR of the octave band that the interference aliases into.

The filter passband ripple is set less-than-or-equal-to the total allowable ripple minus the ripple of the APF. For our design, the total passband ripple is ≤ 1 dB. Since 0.5 dB was allocated for the APF portion, the composite digital filter passband ripple allowable is ≤ 0.5 dB (i.e., 1 dB minus 0.5 dB).

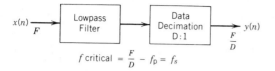

$$f \text{ critical} = \frac{F}{D} - f_p = f_s$$

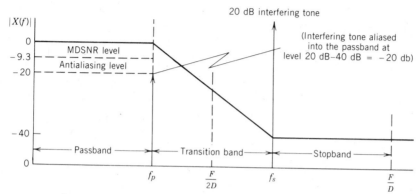

Where $\frac{F}{D}$ represents the sampling rate following the filter and
data decimation operation

Figure 10.25 Filter frequency response requirements for decimation.

The next step is to calculate the required filter order using the stopband attenuation, passband ripple, and sampling rates. The critical folding frequency is given by

$$f_s = (F/D_i) - f_p$$

where F is the sampling rate at the input to the filter and D_i is the data decimation at the output of the ith filter stage. The data decimation factor for each of the bands must be determined. The desired output sampling rate is based on the analysis resolution required and the FFT length since

$$\Delta f = (\text{FFT input sampling rate})/M \qquad (10.14)$$

where M is the FFT length. These values should be traded off versus the A/D sampling rate chosen and the filter/decimation implemented prior to the FFT in order to get the best match to the application requirements.

Using the analysis resolution from Table 10.3 for the 500 to 1000 processing band, the ideal FFT length is obtained. The next higher power of 2 represents the minimum FFT length.

$$M > \text{processing bandwidth}/\Delta f$$
$$> (1000 - 500)/1.25 > 400 \rightarrow 512 \text{ next higher power of 2} \quad (10.15)$$

Next the estimate is compared with the possible FFT input sampling rates

(*a*) Alternative 1

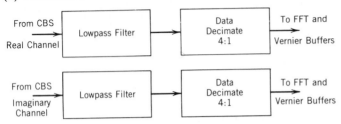

(*b*) Alternative 2

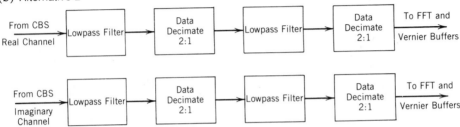

Figure 10.26 Octave band filter graph alternatives.

based on realizable integer data decimation factors given by

$$\Delta f = F/(M * D) \tag{10.16}$$

where D is the total decimation factor from the A/D to the FFT. Solving (Eq. 10.16 to obtain an estimate of the total decimation required results in

$$D \approx F/(M * \Delta f)$$
$$\approx 2500/(512 * 1.25) = 3.9 \tag{10.17}$$

Therefore a value of 4 is chosen, which results in an actual analysis resolution from Eq. 10.14 of 1.22 Hz. The number of analysis bins across the frequency band (500 to 1000) is given by 500 divided by 1.22, which equals 410 bins.

Figure 10.26 presents two filtering alternatives developed for the octave bands. The basis for the FIR filter implementation choice is discussed in Chapter 7. For this application, a signal processor has been specified with the characteristics itemized in Table 10.1. Therefore, the signal processing graph design proceeds trying to optimize the throughput performance using the specified processor's capabilities as defined in Table 10.1. The designer would generally have experience working with the processor and therefore would have apriori information on the best filtering choices for implementation with this processor. In this instance, the hardware implementation for typical filtering primitives such as FIR and IIR filters generally exists. The signal processing design can proceed using the primitives that exist, skipping the primitive definition and sizing steps of the procedure. For each of the alternatives defined, the filtering requirements must be established, the filter orders computed, initial data queue sizes selected, and preliminary processing times computed.

The actual filter designs (i.e., finite arithmetic effects and specifying coefficients) are not necessary until the final graph design has been selected and the implementation is being performed. At this point it is assumed that the processor's arithmetic capabilities are sufficient to meet the application requirements. A thorough analysis of the processor's arithmetic capabilities versus the application requirements is essential. Also, primitive implementations providing more accuracy may be feasible at the expense of throughput. The arithmetic capabilities will be key to the final adequacy of the implementation and therefore must be analyzed in detail to verify the design.

The processing requirements for each of the alternatives shown in Figure 10.26 are developed in the following paragraphs. Alternative 1 uses one FIR filter stage followed by a 4 to 1 data decimation and alternative 2 uses a two-stage FIR decimator.

Alternative 1—Digital Filtering: The processing graph and spectral representation for alternative 1 digital filtering are presented in Figure 10.27. Using Eq. 5.92, the approximate FIR filter order is equal to 39, based on the filter parameters shown in Figure 10.27a. You should also try Eq. 5.91, which provides a more accurate estimate of the order and is easily implemented on a programmable calculator or a personal computer. Note that the filter order equations provide estimates of the order required and do not account for errors in the frequency response due to finite-length coefficients. Therefore, the frequency response using the finite-length coefficients should be plotted to ensure that the filter meets requirements. For the purposes of our design, the equations provide estimates that are sufficient for comparisons. The processing graph for alternative 1 is shown in Figure 10.27b.

The equations can be easily programmed to provide a useful tool for the designer. The IIR filter-order calculations can be performed using equations from Chapter 4, Section 4.14.

Next we compute the time to process a batch of data through the primitive operation. Using the input batch for octave 4 of 1024 samples and the FIR processing time from Table 10.1 as a function of the number of output points, the instantaneous processing time is computed. The FIR39 primitive process time in microseconds is $(32 + 4.8 * M)$, where M is the number of filter output points after data decimation. Computing the time for 1024 input sample points results in

$$\text{FIR time}/1024 = 2 * [32 + 4.8 * (256)] = 2521.6 \ \mu\text{sec}$$

If the remaining three octaves are performed on input batches of 128, 64, and 32, the total time to process all four octaves is computed by summing the four

(*a*) Octave 4 — (500 – 1000 Hz) requirements

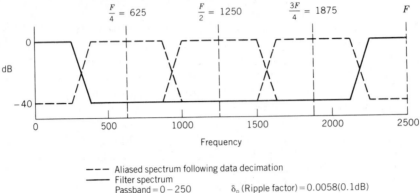

- - - Aliased spectrum following data decimation
——— Filter spectrum
Passband = 0 – 250 δ_p (Ripple factor) = 0.0058(0.1dB)
$f_s = (F/4) – 250 = 375$ (critical frequency)
Stopband = 375 – 1200 δ_s (Ripple factor) = 0.01(– 40dB)

(*b*) Processing graph

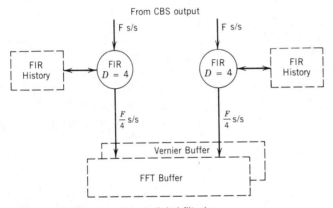

Figure 10.27 Alternative 1 digital filtering.

octave times, which equals 4864 μsec. This represents a 1.19% load for one channel of data. For multiple channels, the time must be multiplied by the number of channels. Remember that at this point we have not considered overhead time, which may be required for setting up control of the primitive, loading and unloading data, and multiple channel operations.

Normally, as the batch size is increased, the throughput is improved because overhead processes such as setups to implement the primitive operation are performed less often. The batch sizes selected will be a function of the cache storage available in the processing engine along with the number of channels processed concurrently. The optimum batch sizes will be determined as part of the resource analysis.

Alternative 2—Digital Filtering: The signal processing graph and spectral representations for alternative 2 processing of Figure 10.26*b* are presented in

From CBS Operation

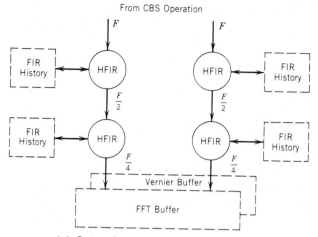

(a) Octave band filtering graph alternative 2

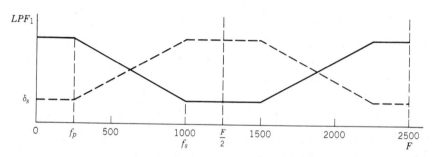

Passband = 0 – 250 δ_p = 0.002878 (0.05 dB)
f_s = 1250 – 250 = 1000
Stopband = 1000 – 1250 δ_s = 0.002878 (–51.1 dB)

(b) Alternative 2 — stage 1 filter requirements

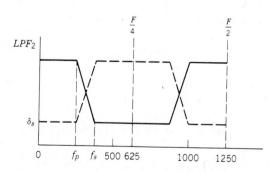

Passband = 0 – 250 δ_p = 0.002878 (0.05 dB)
f_s = 625 – 250 = 375
Stopband = 375 – 625 δ_s = 0.002878 (–51.1 dB)

(c) Alternative 2 — stage 2 filter requirements

Figure 10.28 Octave band filtering alternative 2.

Figure 10.28. This alternative uses a cascade of two filter stages with data decimation of 2 to 1 at the output of each stage to arrive at the desired 4 to 1 factor. Repeating the procedure used for alternative 1 in computing the filter orders results in an order 11 FIR followed by an order 27 FIR. Note that the passband ripple parameter was set equal to the stopband ripple parameter. This provides a FIR filter specification that meets the halfband FIR filter definition discussed in Chapter 5. Therefore, all the even filter coefficients, excluding the center coefficient, will be equal to zero and the processing requirements are reduced.

Since two filter stages are cascaded, the overall passband ripple is maintained by dividing the single stage allocation by 2. The reduction is computed by comparing the processing time for the symmetric FIR primitives and the halfband FIR primitives given in Table 10.1. The calculations for alternative 2 are given by

$$N(i) \approx \frac{-10 \log \delta_p \, \delta_s - 15}{14 \dfrac{(f_s(i) - f_p)}{F(i)/D(i-1)}} + 1$$

$$f_s(i) = \frac{F(i)}{D(i)} \qquad F(1) = F \qquad D(0) = 1 \qquad (10.18)$$

From Eq. 10.18 the first stage order is 8.53, which is implemented as an order 11 halfband FIR. An order 11 was chosen since the estimate is an approximation and for halfbands order 9 has endpoint coefficients equal to zero. If an order 9 design meets the frequency response requirements, then an order 7 implementation would be used. For stage 2 the filter order required is 25.6, which again is rounded up to the next higher implementable order, 27.

You should refer to Chapter 7 for a discussion on the design of cascaded filter/data decimation operations. Remember that the composite ripple must be considered, and therefore as the number of cascaded stages increases, the individual stage ripple factors are divided by the number of stages. For our example we have used 0.05 dB per filter, assuming no more than 10 stages would be used in cascade to implement any of the required bands. If less stages are required, the ripple factor could be increased.

Process time for a 1024-sample batch of input data through the alternative 2 processing graph shown in Figure 10.28a for octave 4 is given by

$$\text{Filter time}/1024 = 7 + 0.8*(512) + 12 + 1.2*(256) = 735.8 \ \mu\text{sec}$$

Multiplying by 2 for the real and imaginary channels equals 1471.6 μsec. Computing the time for the four octaves results in a total time of 2840 μsec, which represents a 0.69% load for alternative 2 compared to a 1.19% load for alternative 1.

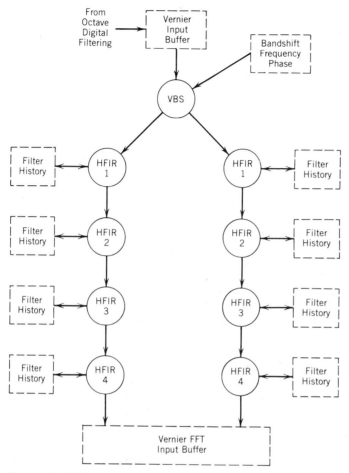

Figure 10.29 Vernier band digital filtering.

Several implementation choices exist. Each primitive can be implemented individually, which requires loading and unloading of the data to/from the processing engine for each primitive operation. The entire frequency translation and digital filter octave processing can be implemented without intermediate data transfers. This eliminates intermediate queue data buffer requirements at the expense of smaller batch sizes for the primitive operations that follow the data decimation operations. These choices must be evaluated during the resource analysis procedure.

Vernier Band Digital Filtering. The vernier filtering is performed via a complex bandshift of the octave filter complex output data followed by a filtering operation and further data decimation. The vernier center frequency is selectable across the octave band, and the vernier bandwidth is one-sixteenth of the octave bandwidth. Figure 10.29 illustrates the proposed processing flow for the

TABLE 10.6
VERNIER FILTER REQUIREMENTS: OCTAVE BAND 4 VERNIER

Vernier Stage	Input Sampling Rate	Data Decimation Factor	Output Sampling Rate	Passband Cutoff f_p	Stopband Frequency f_s	N
1	625	2 : 1	312.5	62.5	250	11
2	312.5	2 : 1	156.25	31.25	125	11
3	156.25	2 : 1	78.125	15.625	62.5	11
4	78.125	2 : 1	39.0625	15.625	23.4375	27

verniers. FIR filters will again be used for the design. You will be asked as part of the problem set to implement the design with IIR filters. Note that the design has used the maximum number of cascaded stages by breaking the 16 to 1 factor into powers of 2, resulting in four cascaded stages. You will also be asked to develop a design using one stage, two stages, and three stages as part of the problem set and assess the resources required.

Given the proposed processing flow of Figure 10.29, the next step is to develop the FIR filter requirements, select the FIR primitive implementation from Table 10.1 provided one exists, select an appropriate batch size for the input data queue, and determine the instantaneous processing time for processing the batch of data through the processing element. Since the bandwidths of the lower octave verniers and the sampling rate are each reduced by one half the filters defined for the next higher octave band vernier, all verniers will have equivalent processing flows.

Table 10.6 presents the filtering requirements and spectral content of each of the cascade vernier stage filters shown in Figure 10.29. The filtering again satisfies the halfband FIR filter definition, which may provide improved throughput. The passband ripple must be added to the APF and octave filtering passband ripples in order to assure that the composite is within $\leq$ 1 dB as specified in Table 10.3. This addition should be performed by adding the actual frequency responses of the filters implemented in order to account for the finite arithmetic effects and the peaks of the ripples occurring at different frequencies. A good design practice is to assume all the peaks line up and that a margin of 15% should be left for implementation considerations.

Since the halfband filters require that the passband ripple factor be equal to the stopband ripple factor, the passband deviation for one stage will usually be much less than the allowable ripple. For cases where several stages are required, such as very narrow (relative to the input band) verniers, care must be taken to maintain reasonable levels. Since we based our filter designs on 10 cascaded stages, and the maximum number of stages is 9 (3 prior to CBS in octave 1, 2 for octave 1 filters, and 4 for the vernier filters), a margin of 10% exists for the worst case. Octaves 2, 3, and 4 verniers require 8, 7, and 6 stages (providing 20%, 30%, and 40% margin), respectively.

The instantaneous processing time for Figure 10.29 is calculated using the processing equations from Table 10.1 for the FIR halfband filter orders required in Table 10.6.

$$\text{Vernier process time}/1024 = 7 + 0.8(512) \rightarrow \text{vernier stage 1}$$
$$+ 7 + 0.8(256) \rightarrow \text{vernier stage 2}$$
$$+ 7 + 0.8(128) \rightarrow \text{vernier stage 3}$$
$$+ 12 + 1.2(64) \rightarrow \text{vernier stage 4}$$
$$= 826.6 \, \mu\text{sec}.$$

Therefore, the time to process four verniers is equal to 3306.4 μsec, which represents a 0.81% instantaneous load on the processing element for one input channel.

The preceding signal conditioning processing has been defined such that each of the octave bands can be processed with the same identical processing and each of the verniers has identical processing flows. The only difference is the rate at which each input queue matures.

SPECTRAL ANALYSIS

Figure 10.30 presents the signal processing graph for spectral analysis processing consisting of an FFT, weighting, detection, and integration. The following discussion describes the primitive operations and computes the instantaneous processing time required to execute the spectral analysis processing operations depicted in Figure 10.30.

FFT Processing. The FFT length for our design approach was computed using Eq. 10.15 to be 512. The redundancy requirement was specified in Table 10.3 to be $\geq 50\%$. Therefore, assuming 50% redundancy, an FFT execution is required when 256 new samples are ready in the FFT input buffer data queue. The rate at which this occurs is calculated by dividing the 256 new samples required by the queue input sample rate.

$$\text{FFT rate} = (\text{Queue input sample rate})/(\text{Number of new samples})$$

Since each octave and vernier has a different output sample rate, the corresponding FFT rates will be different, as shown in Table 10.7. The length of time between FFT executions for an octave or vernier is given by the reciprocal of the corresponding FFT rate. Note that for the vernier channels this time gets very long and may dictate higher FFT rates, depending on display considerations and other postprocessing algorithms. The time to process a 512-point FFT with the execution time from Table 10.1 is given by

$$\text{FFT process time} = 13.8 + 3.8(512) = 1959.4 \, \mu\text{sec}$$

From this an instantaneous percentage load of 0.48% per octave 4 FFT is required.

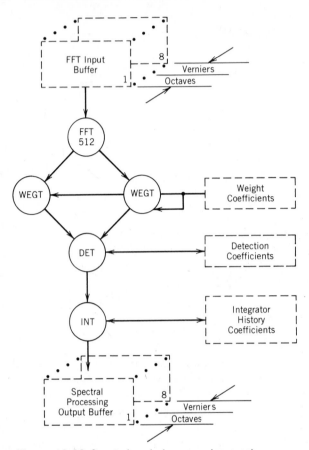

Figure 10.30 Spectral analysis processing graph.

TABLE 10.7
FFT PROCESSING RATES

Octave Band	Filter Output Rate	FFT Rate	FFT Time (Sec)
4	625	2.4414	0.4096
3	312.5	1.207	0.8192
2	156.25	0.6104	1.6384
1	78.125	0.3052	3.2768
Vernier Band			
4	39.0625	0.1526	6.5536
3	19.53125	0.0763	13.1072
2	9.765625	0.0381	26.2144
1	4.8828125	0.0191	52.4288

The FFT primitive operation results in a bit-reversed order for a time-ordered input. Therefore the design must provide for reordering of the data at the input or the output of the process or the algorithm must be implemented not-in-place. This was discussed in Chapter 6. For our sizing purposes, it has been assumed that the hardware provides for the reordering on data transfer, and therefore no time is required within the processing element to reorder the data.

Weighting Processing. The FFT complex output is processed by the WEGT primitive operation, which performs a time weighting of the data implemented via a convolution of the real and imaginary FFT outputs with the frequency coefficients of the time-domain weighting function. The weighting function is designed to provide the desired frequency bin response characteristics, including noise bandwidth, scalloping loss, and sidelobe levels as discussed in Chapter 6.

For our problem Table 10.3 specified a noise bandwidth factor of 1.5, average and maximum scalloping losses of 0.7 dB and 2 dB, respectively, and a highest sidelobe level of -26 dB. To meet these requirements, a Hanning weighting function is chosen. This function has three coefficients $(-0.5, 1, -0.5$; see Chapter 6) to be convolved with the FFT outputs. The WEGT primitive performs a five-point convolution of the FFT outputs with any desired coefficients given by (for Hanning $a_1 = a_5 = 0$)

$$a_i \qquad i = 1, 2, 3, 4, 5$$

where the weighting is performed on the real and imaginary FFT outputs independently using the expression

$$\text{WEGT}(k) = \sum_{i=1}^{5} a_i v(k - 3 + i) \qquad 0 < k < 409$$

Note that two additional bins on each side of the output bins must be carried through this operation in order to compute the endpoint weighted outputs. Otherwise, a unique weighting algorithm must be applied to the endpoints. Also, remember that the bin numbers must be reordered to compensate for the low side of the band appearing at the high end of the FFT output bin numbers owing to the complex band shift, as shown in Figure 10.24. The bin versus frequency outputs are shown in Figure 10.31.

The time to process each FFT output batch with the WEGT primitive is given by $(2 + 3.5 * M)$, where M represents the number of valid FFT outputs required for subsequent processing. The number of valid FFT output bins used equals the bandwidth divided by the resolution (e.g., for octave 4 $500/1.22 = 410$). The time to process the 410 samples with WEGT is equal to 1437 μsec per execution, which represents an octave 4 instantaneous load of 0.35%.

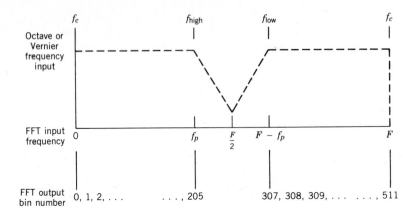

Octave Band 4 $F = 625$, $f_c = 750$, $f_{low} = 500$, $f_{high} = 1000$

Octave Frequency	FFT Frequency	FFT Bin Number
750	0	0
751.2207	1.2207	1
752.4414	2.4414	2
⋮	⋮	⋮
999.023	249.023	204
1000.244	250.244	205
⋮	⋮	⋮
499.755	374.756	307
500.977	375.977	308
⋮	⋮	⋮
748.7793	623.779	511

Figure 10.31 FFT output bins versus frequency band processed.

Detection Processing. The next operation, detection, estimates the magnitude of the complex spectral outputs. The estimate computes the sum of the product of b_1 and the larger of the real and imaginary components with the product of b_2 times the smaller of the real and imaginary components.

$$\text{DET}(k) = b_1 \text{Larger}[\text{RW}(k), \text{IW}(k)] + b_2 \text{ smaller}[\text{RW}(k), \text{IW}(k)]$$
$$k = 0, 1, 2, 3, \ldots, 409$$

$\text{RW}(k) = $ Real weighted FFT output

$\text{IW}(k) = $ Imaginary weighted FFT output

This can be shown to provide a good estimate with the proper selection of b_1 and b_2, while avoiding the requirement to compute the square root of the sum of the squares of the real and imaginary components. The time required to process N samples with the DET primitive is given by $(4 + 1.8 * N)$. For our 410-sample input queue of complex spectral values this results in a process time equal to 742 μsec, which represents an instantaneous load of 0.18% per octave 4 DET execution.

Integration Processing. The last operation defined for this application problem is the integration, INT, of the detected frequency components over time. For this example, the values were integrated using a single pole recursive filter with a difference equation given by

$$y(k, nT) = x(k, nT) + c * y(k, nT - T) \qquad k = 0, 1, 2, \dots, 409$$

where nT represents the current DET output time, x is the DET output, y is the INT output, c is designed to give the desired integration time constant, and k is the detected bin output. This operation is performed for each of the 410 detected spectral outputs.

The time to perform the INT primitive operation using Table 10.1 is given by $(3 + 1.2 * N)$, which for N equal to 410 results in a time 495 μsec. This represents an instantaneous octave 4 processing load of 0.12%.

Many other postdetection processing operations could be specified such as data normalization and requantization for display purposes, auto-detection thresholding, and data oring to reduce quantity of values to be displayed. Since the processing defined demonstrates the use of the system design procedure, these have been omitted for the purpose of simplifying this problem. Refer to Section 8.3.3 for postdetection discussion.

Since all operations have been defined and a signal processing graph developed with instantaneous primitive execution times, and queue requirements defined, the system design continues with the resource analysis procedure shown in Figure 10.14.

10.3.4 Spectral Detection System Resource Analysis — Step 4

The first step of the resource analysis procedure (Fig. 10.14) uses the preliminary results of the signal processing design to finalize the DFGs and to partition the processing to the processor elements.

The processor was specified for this application with one signal processing element type and one control processor type. Therefore, all signal processing operations are allocated to the signal processing element, and, likewise, the graph control operations are allocated to the control processor. The internal graph control is handled by the processing engine local control. For signal processors with multiple element types, allocation of the operations is based on the estimates performed as part of the signal processing definition in the previous step of the design procedure.

(a) STRING (segment) definitions

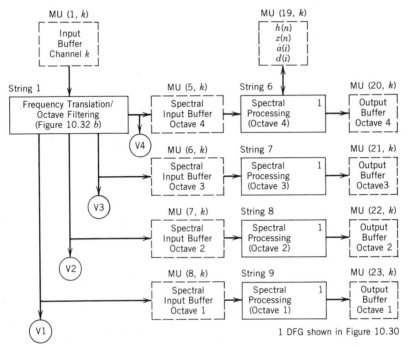

Figure 10.32 Octave band processing DSP flow graph.

Four basic signal processing operations were defined, and signal processing flows developed.

1. Frequency translation (CBS, VBS) (Fig. 10.22).
2. Octave filtering (Fig. 10.28).
3. Vernier filtering (Fig. 10.29).
4. Spectral processing (Fig. 10.30).

Based on the processing flows, segments (i.e. strings) are established to perform the resource analysis.

Figure 10.32a combines the entire frequency translation and prefiltering for the octave bands with the four octave band filtering operations into one executable segment. The segment processing requirements are shown in Figure 10.32b. To complete the octave processing, four spectral processing segments, shown in Figure 10.32a, are defined as per Figure 10.30. Table 10.8 defines the initial graph parameters to be used in the resource analysis. For each of the 40 channels the frequency translation/octave filtering segment is executed followed by execution of four spectral processing segments (one for each of the octave bands). Each memory unit is defined by allocating the total storage required and the input and output queues associated with the MU. All primitive operations must be completely specified, and the graph input batch

(b) STRING 1 DSP flow graph (frequency translation octave filtering)

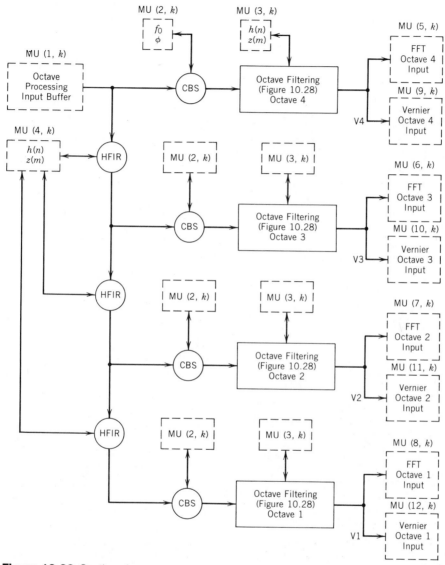

Figure 10.32 Continued

data sizes are specified. Each queue must be named and read, produce and threshold amounts defined.

Figure 10.33 presents the processing segments required to implement the vernier processing portion of the application. Eight segments are required to implement the vernier frequency translation, filtering, and spectral analysis when a vernier is assigned to each of the four octave frequency bands. One

TABLE 10.8
OCTAVE BAND PROCESSING GRAPH PARAMETERS (FIG. 10.32) FOR ONE
INPUT CHANNEL
(STRING 1: Frequency translation/octave filtering channel k—execution rate = once every
1024 input samples)

		Memory Unit Buffers	
	Primitive	Input Buffers	Output Buffers
Primitive	CBS	MU(1), MU(2)	MU(2)
operations	HFIR11	MU(3)	MU(3)
(Figs. 10.28	HFIR27	MU(3)	MU(5), MU(9), MU(3)
and 10.32b)	HFIR11	MU(3)	MU(3)
	HFIR27	MU(3)	MU(5), MU(9), MU(3)
	HFIR27	MU(1), MU(4)	MU(4)
	CBS	MU(2)	MU(2)
	HFIR11	MU(3)	MU(3)
	HFIR27	MU(3)	MU(6), MU(10), MU(3)
	HFIR11	MU(3)	MU(3)
	HFIR27	MU(3)	MU(5), MU(9), MU(3)
	HFIR27	MU(4)	MU(4)
	CBS	MU(2)	MU(2)
	HFIR11	MU(3)	MU(3)
	HFIR27	MU(3)	MU(7), MU(11), MU(3)
	HFIR11	MU(3)	MU(3)
	HFIR27	MU(3)	MU(5), MU(9), MU(3)
	HFIR27	MU(4)	MU(4)
	CBS	MU(2)	MU(2)
	HFIR11	MU(3)	MU(3)
	HFIR27	MU(3)	MU(8), MU(12), MU(3)
	HFIR11	MU(3)	MU(3)
	HFIR27	MU(4)	MU(5), MU(9), MU(3)

Buffers	Memory Unit Specifications
MU(1) Input signal	Digitized real data, 8-bits S.xxxxxxx per sample 2048 bytes allocated (double buffer assuming 1024 sample input batches) Data rate = 2500 samples per second
MU(2) CBS parameters	Frequency code: Real, positive integer, 16 bits Phase code: Real, positive integer, 16 bits 1 each per octave

MU(3)
Octave
filtering
filter
coefficients
and filter
histories
	Filter coefficients:
	HFIR11: Real, signed fraction, 16-bits S.xx . . . x
	  4 coefficients
	HFIR27: Real, signed fraction, 16-bits S.xx . . . x
	  8 coefficients
	Filter histories: Real, signed fraction, 16-bits
	  S.xxxxxxxxxxxxxxx
	  HFIR11: 4 register values per filter
	  HFIR27: 8 register values per filter

MU(4)
Octave band
conditioning
filter
coefficients
and histories
	Filter coefficients:
	  HFIR27: Real, signed fraction, 16-bits S.xx . . . x
	  8 coefficients
	Filter histories: Real, signed fraction, 16-bits
	  S.xxxxxxxxxxxxxxx
	  HFIR27: 8 register values per filter

Buffers	Memory Unit Specifications
MU(5), MU(9) Octave 4 output data	Filter output data: Complex, signed fraction, 16-bits S.xxxxxxxxxxxxxxx per real and imaginary outputs, 256 4-byte outputs per execution of string 1. 1024 4-byte buffer
MU(6), MU(10) Octave 3 output data	Filter output data: Complex, signed fraction, 16-bits S.xxxxxxxxxxxxxxx per real and imaginary outputs, 128 4-byte outputs per execution of string 1. 768 4-byte buffer
MU(7), MU(11) Octave 2 output data	Filter output data: Complex, signed fraction, 16-bits S.xxxxxxxxxxxxxxx per real and imaginary outputs, 64 4-byte outputs per execution of string 1. 768 4-byte buffer
MU(8), MU(12) Octave 1 output data	Filter output data: Complex, signed fraction, 16-bits S.xxxxxxxxxxxxxxx per real and imaginary outputs, 32 4-byte outputs per execution of string 1. 768 4-byte buffer

STRING 6: Spectral processing octave 4 (Fig. 10.30) — execution rate = once every 256 new input samples (Table 10.7).

	Memory Unit Buffers		
	Primitive	Input Buffers	Output Buffers
Primitive operations (Fig. 10.30)	FFT	MU(5), MU(19)	
	WEGT	MU(19)	
	DET		
	INT	MU(19)	MU(19), MU(20)

Buffers	Memory Unit Specifications
MU(5) FFT octave 4 input buffer	As defined above for octave 4 filter output
MU(19) STRING 6 parameters	FFT length: Real integer, 16-Bits. WEGT coefficients: Real, signed fraction, 16-bits S.xxxxxxxxxxxxxxx, 5 coefficients. DET coefficients: Real, signed fraction, 16-bits S.xxxxxxxxxxxxxxx, 2 coefficients. INT coefficient: Real, signed fraction, 16-bits S.xxxxxxxxxxxxxxx, 1 coefficient. INT history: real, signed fraction, 16-bits S.xxxxxxxxxxxxxxx, 410 partial sums.
MU(19) INT partial sum histories	INT history: Real, signed fraction, 16-bits S.xxxxxxxxxxxxxxx, 410 partial sums.
MU(20) Spectral processing octave 4 output buffer	Output data; Real, signed fraction, 16-bits S.xxxxxxxxxxxxxxx, 410 2-byte outputs per STRING execution 810 2-byte output buffer.

STRING 7: Spectral Processing Octave 3
    Same as STRING 6 using MU(6) as the input buffer and MU(21) as output buffer.
    Execution rate = one half STRING 6 (Table 10.7).
STRING 8: Spectral Processing Octave 2
    Same as STRING 6 using MU(7) as the input buffer and MU(22) as output buffer.
    Execution rate = one half STRING 7 (Table 10.7).
STRING 9: Spectral Processing Octave 1
    Same as STRING 6 using MU(8) as the input buffer and MU(23) as output buffer.
    Execution rate = one half STRING 8 (Table 10.7).

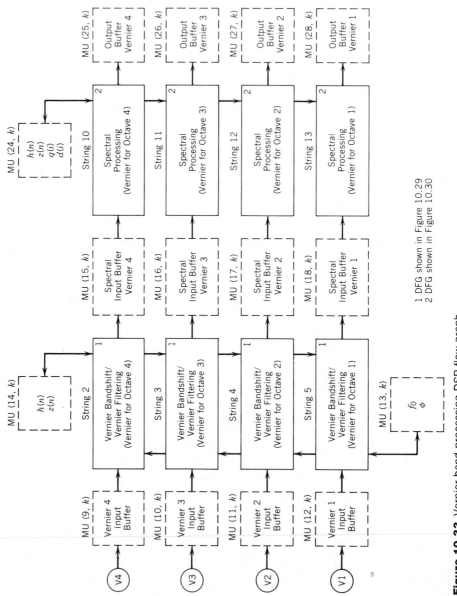

Figure 10.33 Vernier band processing DSP flow graph.

segment is used to perform the frequency translation and filtering (Fig. 10.29), and one segment is used for the spectral analysis (Fig. 10.30), which is identical to the octave spectral analysis segment except for the rate at which they are processed.

The total number of vernier segments required depends on the vernier assignments across the 40 channels. The maximum number is 320 when verniers are assigned for each of the octaves on each channel. Table 10.9 specifies the graph parameters to be used in the initial resource calculations. These parameters provide a starting point for the resource analysis and are varied to assess the dependencies of the resource requirements on the parameters of the DFGs.

The initial graphs developed defined three unique executable segments:

1. Complex bandshift/Octave band filtering (Fig. 10.32b).

2. Vernier bandshift/Vernier filtering (Fig. 10.29).

3. Spectral processing (Fig. 10.30).

The definition of groups of primitive operations as segments (strings or chains) was performed to reduce system level control, data transfers, and MU data queues that would be required if each primitive operation was independently scheduled and dispatched by the system controller. This reduces the overall storage requirements, system level control requirements, overhead, and data transfer bandwidths required by the design. The definition of long strings with significant data decimation along the string can cause inefficiencies due to smaller and smaller batch sizes being available for the back-end primitive operations. Also, the processing elements must have increased control capability to handle the intermediate control during string executions. Therefore, the designer should tradeoff processing string choices and input data batch sizes to arrive at the best implementation.

Next, an initial estimate of the resource requirements for each primitive/segment is made using the baseline segments and data batches selected previously. For these calculations a personal computer spreadsheet program provides a convenient tool for performing this estimate and can be subsequently used for tradeoff studies. The throughput estimate spreadsheet results are presented in Table 10.10. Each column in Table 10.10 is defined here.

10.1. MACRO–Contains the primitive name.

10.2. STR #–Defines the string number.

10.3. R/C–Defines the primitive input data as real or complex.

10.4. F–Defines the primitive input sampling rate.

10.5. INPUT BATCH–Defines the primitive input data queue size.

10.6. DATA DEC–Refers to the primitive output data decimation.

TABLE 10.9
VERNIER BAND PROCESSING GRAPH PARAMETERS (FIG. 10.33)
FOR ONE INPUT CHANNEL
(STRING 2: Vernier Bandshift/Vernier Filtering octave 4 channel k—execution rate = once
every 1024 input samples)

| | Memory Unit Buffers | | |
	Primitive	Input Buffers	Output Buffers
Primitive	VBS	MU(9), MU(13)	MU(13)
operations	HFIR11	MU(14)	MU(14)
(Fig. 10.29)	HFIR11	MU(14)	MU(14)
	HFIR11	MU(14)	MU(14)
	HFIR27	MU(14)	MU(15), MU(14)

Buffers	Memory Unit Specifications
MU(9) Octave 4 output data	Filter output data: Complex, signed fraction, 16-bit S.xxxxxxxxxxxxxxx per real and imaginary outputs, 256 4-byte outputs per execution of STRING 1. 2048 4-byte buffer
MU(13) VBS parameters	Frequency code: Real, positive integer, 16-bits Phase Code: Real, positive integer, 16-bits, 1 each per vernier

MU(14)
Vernier
filtering
filter
coefficients
and filter
histories
	Filter coefficients:
	    HFIR11: Real, signed fraction, 16-bits S.xx.....x
	        4 coefficients
	    HFIR27: Real, signed fraction, 16-bits S.xx.....x
	        8 coefficients
	Filter histories: Real, signed fraction, 16-bits
	    S.xxxxxxxxxxxxxxx
	    HFIR11: 4 Register values per filter
	    HFIR27: 8 Register values per filter

MU(15) Vernier 4 output data	Filter output data: Complex, signed fraction, 16-bit S.xxxxxxxxxxxxxxx per real and imaginary outputs, 256 4-byte outputs per execution of STRING 1. 1024 4-byte buffer

STRING 3: Vernier Bandshift/Vernier Filtering Octave 3 Channel k
Execution rate = Once every 1024 input samples
Primitive operations same as STRING 2
Memory units: MU(10) = Vernier 3 input
              MU(13) and MU(14) same as STRING 2
              MU(16) = Vernier 3 output data
STRING 4: Vernier Bandshift/Vernier Filtering Octave 2 Channel k
  Execution rate = Once every 1024 input samples
  Primitive operations same as STRING 2
  Memory units: MU(11) = Vernier 2 input
                MU(13) and MU(14) same as STRING 2
                MU(17) = Vernier 2 output data

STRING 5: Vernier Bandshift/Vernier Filtering Octave 1 Channel k
    Execution rate = Once every 1024 input samples
    Primitive operations same as STRING 2
    Memory units: MU(12) = Vernier 1 input
                  MU(13) and MU(14) same as STRING 2
                  MU(18) = Vernier 3 output data

STRING 10: Spectral Processing Vernier 4 (Fig. 10.30) (execution rate = once every 256 new input samples) (Table 10.7)

| | **Memory Unit Buffers** | | |
	Primitive	**Input Buffers**	**Output Buffers**
Primitive	FFT	MU(15), MU(24)	
operations	WEGT	MU(24)	
(Fig. 10.30)	DET	MU(24)	
	INT	MU(24)	MU(24), MU(25)

Buffers	**Memory Unit Specifications**

MU(15)
  FFT Vernier 4
  input buffer
	As defined above for vernier 4 filter output

MU(24)
  STRING 10
  parameters
	FFT length: Real integer, 16-bits
	WEGT coefficients: Real, signed fraction, 16-bits
	S.xxxxxxxxxxxxxxx, 5 coefficients
	DET coefficients: Real, signed fraction, 16-bits
	S.xxxxxxxxxxxxxxx, 2 coefficients
	INT coefficient: Real, signed fraction, 16-bits
	S.xxxxxxxxxxxxxxx, 1 coefficient
	INT history: Real, signed fraction, 16-bits
	S.xxxxxxxxxxxxxxx, 410 partial sums

MU(24)
  INT partial
  sum histories
	INT history: Real, signed fraction, 16-bits
	S.xxxxxxxxxxxxxxx, 410 partial sums

MU(25)
  Spectral
  processing
  vernier 4
  output buffer
	Output data: Real, signed fraction, 16-bits
	S.xxxxxxxxxxxxxx,
	410 2-byte outputs per STRING execution,
	810 2-byte output buffer

STRING 11: Spectral Processing Vernier 3
  Same as STRING 10 using MU(16) as the input buffer and MU(26) as output buffer.
  Execution rate = one half STRING 10 (Table 10.7).
STRING 8: Spectral Processing Vernier 2
  Same as STRING 10 using MU(17) as the input buffer and MU(27) as output buffer.
  Execution rate = one half STRING 11 (Table 10.7).
STRING 9: Spectral Processing Vernier 1
  Same as STRING 10 using MU(18) as the input buffer and MU(28) as output buffer.
  Execution rate = one half STRING 10 (Table 10.7).

TABLE 10.10
THROUGHPUT ESTIMATES PER STRING

(a) OCTAVE STRING

MACRO	STR #	R/C	F	INPUT BATCH	DATA DEC	#OF EXEC	SETUP TIME	MACRO EXEC TIME	BATCH TOTAL TIME	OUTPUT YES = 1	BATCH* MEM BW	MULT/ BATCH	ADDS/ BATCH
CBS	1	R	2500	1024	1	1	5	819.2	824.2		5500	2048	0
HFIR-11	1	C	2500	1024	2	2	7	409.6	833.2			4096	7168
HFIR-27	1	C	1250	512	2	2	12	307.2	638.4	1	2750	4096	7680
HFIR-27	1	R	2500	1024	2	1	12	614.4	626.4			4096	7680
CBS	1	R	1250	512	1	1	5	409.6	414.6			1024	0
HFIR-11	1	C	1250	512	2	2	7	204.8	423.6			2048	3584
HFIR-27	1	C	625	256	2	2	12	153.6	331.2	1	1375	2048	3840
HFIR-27	1	R	1250	512	2	1	12	307.2	319.2			2048	3840
CBS	1	R	625	256	1	1	5	204.8	209.8			512	0
HFIR-11	1	C	625	256	2	2	7	102.4	218.8			1024	1792
HFIR-27	1	C	312.5	128	2	2	12	76.8	177.6	1	687.5	1024	1920
HFIR-27	1	R	625	256	2	1	12	153.6	165.6			1024	1920
CBS	1	R	312.5	128	1	1	5	102.4	107.4			256	0
HFIR-11	1	C	312.5	128	2	2	7	51.2	116.4			512	896
HFIR-27	1	C	156.25	64	2	2	12	38.4	100.8	1	343.75	512	960
	1		2500	1024					5507.2		10656.	26368	41280

(b) VERNIER STRINGS

MACRO	STR #	R/C	F	INPUT BATCH	DATA DEC	#OF EXEC	SETUP TIME	MACRO EXEC TIME	BATCH TOTAL TIME	OUTPUT YES = 1	BATCH* MEM BW	MULT/ BATCH	ADDS/ BATCH
VBS	2	C	625	1024	1	1	10	1638.4	1648.4		2750	4096	2048
HFIR-11	2	C	625	1024	2	2	7	409.6	833.2			4096	7168

MACRO	STR #	R/C	F	INPUT BATCH	DATA DEC	# OF EXEC	SETUP TIME	MACRO EXEC TIME	BATCH TOTAL TIME	OUTPUT YES = 1	BATCH* MEM BW	MULT/ BATCH	ADDS/ BATCH
HFIR-11	2	C	312.5	512	2	2	7	204.8	423.6			2048	3584
HFIR-11	2	C	156.25	256	2	2	7	102.4	218.8			1024	1792
HFIR-27	2	C	78.125	128	2	2	12	76.8	177.6	1	171.88	1024	1920
"	2	C	625	1024					3301.6	1	2921.9	12288	16512
"	3	C	312.5	1024					3301.6	1	1460.9	12288	16512
"	4	C	156.25	1024					3301.6	1	730.47	12288	16512
"	5	C	78.125	1024					3301.6	1	365.23	12288	16512

(c) FFT STRINGS

MACRO	STR #	R/C	F	INPUT BATCH	DATA DEC	# OF EXEC	SETUP TIME	MACRO EXEC TIME	BATCH TOTAL TIME	OUTPUT YES = 1	BATCH* MEM BW	MULT/ BATCH	ADDS/ BATCH
FFT	6	C	625	512	1.2488	2	13.8	1945.6	3918.8		5500	18432	27648
WEGT	6	C	500.49	410	1	2	2	1435	2874			2460	1640
DET	6	C	500.49	410	1	2	4	738	1484			1640	820
INT	6	R	500.49	410	1	2	3	492	990	1	2202.1	820	820
"	6	R	625	512					9266.8		7702.1	23352	30928
"	7	R	312.5	512					9266.8	1	3851.1	23352	30928
"	8	R	156.25	512					9266.8	1	1925.5	23352	30928
"	9	R	78.125	512					9266.8	1	962.77	23352	30928
"	10	R	39.063	512					9266.8	1	481.38	23352	30928
"	11	R	19.531	512					9266.8	1	240.69	23352	30928
"	12	R	9.7656	512					9266.8	1	120.35	23352	30928
"	13	R	4.8228	512					9266.8	1	60.173	23352	30928

*BATCH MEM BW includes a 10% overhead.

10.7.   # OF EXEC–Defines the number of executions of the primitive listed (e.g., for FIR filters operating on complex inputs, the value is 2—one FIR for the real channel and one for the imaginary channel).

10.8.   SETUP TIME–Constant value per primitive execution given in Table 10.1. Value in microseconds.

10.9.   MACRO EXEC TIME–Product of multiplying factor from Table 10.1 with the input batch size from column 5 or the output batch size depending on the macro definition (e.g., for FIR macros the output batch size is computed by dividing the input batch, column 5, by the decimation factor, column 6).

10.10.   TOTAL TIME–Product of column 7 with the sum of columns 8 and 9.

$$\text{TOTAL TIME} = (\text{\# OF EXEC})$$
$$* (\text{SETUP TIME} + \text{MACRO EXEC TIME})$$

10.11.   OUTPUT–If this column is 1, then the data from this primitive output will be output from the string to mass memory.

10.12.   MEM BW–Calculates the bandwidth rate required to transfer the data to/from the mass memory for each of the strings.

$$\text{MEM BW} = (\text{Bytes} * \text{Sampling Rate}) \text{ bytes/second (BPS)}$$

10.13.   MULT–The number of multiplies required for each of the primitive operations/strings is calculated. This value uses the multiply requirements from Table 10.1 and performs the product with the batch size (column 10.5) for the operation. The number is divided by the data decimation factor for operations performed at the output rate, such as FIR filters.

$$\text{MULT} = (\text{Mult/operation}) * (\text{Batch size})$$

The designer should be careful in using multiplies in the analysis. As stated in Section 10.2, the multiply requirements provide only a rough-order estimate and could be misleading for complex operations. They are included here for the purpose of comparison and since they are easier to visualize.

10.14.   ADDS–The number of additions is given for each primitive operation/segment using the values from Table 10.1.

$$\text{ADDS} = (\text{Adds/operation}) * (\text{Batch size})$$

Again this number can be misleading and should be used only as a rough-order requirement.

The execution times to process one batch of data through each string is computed and listed in microseconds in Table 10.10. The octave segment is listed in Table 10.10a. Remember that the times listed are instantaneous times for processing the input batch through the segment. No time has been associated with the loading and unloading of data. Comparison of the raw batch processing time to the time it takes to collect the input batch will provide a good indication of the segment processing throughput requirements

$$\text{Segment}(i)\ \text{load} = \frac{\text{Segment}(i)\ \text{batch total time}}{\text{Input batch}/\text{Input sampling rate}}$$

The octave segment load is equal to 0.0134, which represents a 1.34% load.

The vernier frequency translation and filtering instantaneous processing times for segments 2 through 5 are presented in Table 10.10b. Since the segments are identical, the batch total time is equal to 3301.6 μsec for each of the segments. The segment load varies owing to the decrease in the input sampling rate. Therefore, the segment 2 load equals 0.002 or 0.2% load, and segments 3 through 5 each require one half the preceding segment load.

The spectral processing segment instantaneous processing times for the octave and vernier frequency bands are given in Table 10.10c. Again, each segment is identical and requires a batch total time of 9266.8 μsec. The segment load varies owing to the sampling rate variation. Segment 6 operates at the highest input rate of 625 samples per second, which results in a load of 0.0113 or 1.13%. The load for segments 7 through 13 is reduced by one half from the preceding segment load.

Given these results, the next step is to calculate the system load for throughput, control, memory storage, memory bandwidth, and program storage as a function of the total system requirements. Table 10.11 presents the spreadsheet for the system load computations. The column definitions and the required calculations for Table 10.11 are described here.

11.1. NUMBER OF CHs–Defines the system processing requirements for the total number of real-time channels to be processed concurrently.

11.2. STR #–Defines the string number per the Table 10.10 definition.

11.3. F–Defines the string input sampling rate, as given in Table 10.10.

11.4. INPUT BATCH–Defines the string input batch size as given in Table 10.10.

11.5. CNTROL INSTRs–Defines the control requirements in terms of the number of instructions that are executed by the controller in order to initiate the string operation. For this example, values have been assumed for each string. In an actual design the different control requirements would be broken down into instructions and a detailed estimate derived from the trial coded control programs.

TABLE 10.11
SYSTEM LOAD

NUMBER OF CHs	STR #	F	INPUT BATCH	CNTROL INSTRs	TOTAL TIME	OVERHD FACTOR	THRUPUT LOAD %
40	1	2500	1024	400	5507.2	1.3	69.916
40	2	625	1024	250	3301.6	1.3	10.479
40	3	312.5	1024	250	3301.6	1.3	5.2394
40	4	156.25	1024	250	3301.6	1.3	2.6197
40	5	78.125	1024	250	3301.6	1.3	1.3098
40	6	625	512	200	9266.8	1.5	67.872
40	7	312.5	512	200	9266.8	1.5	33.936
40	8	156.25	512	200	9266.8	1.5	16.968
40	9	78.125	512	200	9266.8	1.5	8.4840
40	10	39.063	512	200	9266.8	1.5	4.2420
40	11	19.531	512	200	9266.8	1.5	2.1210
40	12	9.7656	512	200	9266.8	1.5	1.0605
40	13	4.8828	512	200	9266.8	1.5	0.53025
Total				Percentage loads		=	224.78

11.6. TOTAL TIME–Defines the value in microseconds from Table 10.10 for the total processing time required to execute the segment the number of times required to process one batch of input data.

11.7. OVERHD FACTOR–Defines the overhead multiplier for each string, which approximates the additional time required to process multiple segments for multiple channels. Generally, these factors can be estimated with good accuracy based on past measurement with an existing processor. For a new processor, the architecture should be simulated to assess the inefficiencies. For this example, a value of 1.3 (30% overhead) has been used for the filtering portion and a value of 1.5 (50% overhead) has been assumed for the spectral processing.

11.8. THRUPUT LOAD–The throughput load is given by

$$\text{THRUPUT LOAD \%} = \text{Total time used per second} * 100$$

which is calculated from the following table entries.

$$\text{THRUPUT LOAD \%} = (\text{NUMBER OF CHs})$$
$$* (\text{F/INPUT BATCH}) * (\text{TOTAL TIME}) * 100$$

Therefore, the total percentage load is the sum of the individual segment values.

CNTROL LOAD %	MEMORY LOAD %	MEMORY BW %	THRUPUT MILLIONS / SEC		THRUPUT LOAD (%)	
			MULT	ADDS	MULT	ADDS
3.9063	9.0112	10.656	2.575	4.0313	51.5	40.313
0.61035	18.022	2.9219	0.3	0.40313	6	4.0313
0.30518	18.022	1.4609	0.15	0.20156	3	2.0156
0.15259	18.022	0.73047	0.075	0.10078	1.5	1.0078
0.07629	18.022	0.36523	0.0375	0.05039	0.75	0.50391
0.97656	23.443	7.7021	1.1402	1.5102	22.805	15.102
0.48828	23.443	3.8511	0.57012	0.75508	11.402	7.5508
0.24414	23.443	1.9255	0.28506	0.37754	5.7012	3.7754
0.12207	23.443	0.96277	0.14253	0.18877	2.8506	1.8877
0.06104	23.443	0.48138	0.07126	0.09438	1.4253	0.94385
0.03052	23.443	0.24069	0.03563	0.04719	0.71265	0.47192
0.01526	23.443	0.12035	0.01782	0.02360	0.35632	0.23596
0.00763	23.443	0.06017	0.00891	0.01180	0.17816	0.11798
6.9962	268.65	31.479	MMPS 5.4091	MAPS 7.7956	108.18	77.956

11.9. CNTROL LOAD–The control load is equal to the total instructions used divided by the processor capacity. The control load is given by

$$\text{CNTROL LOAD} = (\# \text{ of CHs}) * (\text{F/Input Batch})$$
$$* (\text{CNTROL INSTRS})/(\text{CNTROL CAPACITY}) * 100$$

and the total control load is the sum over the segments.

11.10. MEMORY LOAD–The memory load is given by

$$\text{MEMORY LOAD} = (\text{Total Storage Required})$$
$$/(\text{Total Available Storage}) * 100$$

This requires calculation of all input/output data queues plus all other mass memory storage such as history queues and coefficient queues and all miscellaneous memory. For this problem it is assumed that all data queues will be double buffered. Also, each queue will be assumed to be 16 bits or 2 bytes. Then the storage requirement is computed from Table 10.10 using the output column entries that are equal to 1. For each primitive/segment where an output is indicated, the number of bytes is estimated and multiplied by 2 for the double buffering. The sum of all the memory storage required divided by the 2-million byte capacity of the mass memory equals the load. For this problem the history queues were estimated by multiplying by a factor of 1.1. Generally, a 10% value represents a conservative estimate, but the detailed analysis should include calculations of the actual application requirements based upon the graph parameter tables developed.

11.11. MEMORY BW–The memory bandwidth is computed from the total data transfer rate required into and out of the memory. This is computed from the data provided in Table 10.10. Again the data must be adjusted to a common reference, such as bytes per second, which was chosen for this analysis.

11.12. MMPS–The number of million multiplications per second is calculated by multiplying the multiplications per batch (column 10.13), the number of channels (column 11.1), the sampling rate (column 11.3), and dividing the resultant product by the batch size (column 11.4). The value is normalized to millions of multiplications.

$$\text{MMPS} = \text{MULT} * (\# \text{ of CHs}) * F / (\text{Batch Size}/1000000)$$

11.13. MAPS–The number of million additions per second is calculated by multiplying the additions per batch (column 10.14) by the same factors as described for MMPS.

$$\text{MAPS} = \text{Adds} * (\# \text{ of CHs}) * F / (\text{Batch Size}/1000000)$$

11.14. MULT LOAD–Is the percentage load based on the processor's raw capability of 5 MMPS.

$$\text{MULT LOAD} = 100 * \text{MMPS}/5$$

11.15. ADDS LOAD–Is the percentage load based on the processor's raw addition capability of 10 MAPS.

$$\text{ADDS LOAD} = 100 * \text{MAPS}/10$$

From Table 10.11 it is readily seen that the throughput and memory storage loads are well above 100%. Otherwise, the control load and memory bandwidth requirements are relatively low. Other parameters, such as latency and program storage requirements, should be analyzed as defined in Section 10.2. The designer must be aware of all resource parameters and assure that all have been adequately analyzed in order to meet the application requirements. It has been assumed at this point that the finite arithmetic analysis per Chapter 9 has been performed as part of the analysis to ensure that the arithmetic requirements are satisfied.

Using Table 10.10 and 10.11 results, we can either repeat the analysis using different segment definitions and/or input batch sizes or establish a baseline processor configuration using Table 10.11 results. For this example, we will assume that the processing design has been optimized and go to step 5, configuration analysis, of the system design procedure.

10.3.5 Spectral Detection System Configuration Analysis — Step 5

The first configuration analysis operation is to define a configuration based on the resource analysis results and to compute the resources for the configuration. Then we can assess the impact of the nonsignal processing requirements on the configuration.

Based on the throughput load and memory load from Table 10.11 of greater than 200%, a configuration using three processor engines and three memory elements is selected. Several alternatives for allocating the processing to the configuration exist. Two have been chosen and will be analyzed.

A. Alterative A Partitioning

14 channels to processing element 1 and mass memory 1

13 channels to processing element 2 and mass memory 2

13 channels to processing element 3 and mass memory 3

B. Alternative B Partitioning

20 channels of octave processing to PE1 and mass memory 1

20 channels of octave processing to PE2 and mass memory 2

40 channels of vernier processing to PE3 and mass memory 3

Alternative A was easily derived from the viewpoint that all channels are being processed identically, and therefore this provides an even division of the requirements. Also, it avoids string partitions, which require data transfers between the multiple signal processing elements. Another feature to be considered in cases such as this is degraded modes that would permit the processing of several channels if one of the processing elements fails. The fault tolerance requirement in Section 10.3.1 specified that at least 20 channels of processing must be maintained. This requires any two of the three processing elements and mass memories to be functional. Table 10.11 can be easily recomputed for the 14 channels and the 13 channels giving the loads for the configuration selected. These are given in Table 10.12.

For alternative B, the totals for the 20 channels of octave processing and the 40 vernier channels must be computed from Table 10.10. These are given in Table 10.13. Note that with this implementation the vernier input data queues and the FFT input data queues must be directed to different mass memories and degraded modes are not as easily defined. Also, the 20-channel octave processing allocation requires 98.6% throughput load, which does not

TABLE 10.12
ALTERNATIVE A PROCESSING PARTITIONED TO CONFIGURATION

(*a*) 14 CHANNELS PARTITIONED TO PE1

NUMBER OF CHs	STR #	F	INPUT BATCH	CNTROL INSTRs	TOTAL TIME	OVERHD FACTOR	THRUPUT LOAD %
14	1	2500	1024	400	5507.2	1.3	24.470
14	2	625	1024	250	3301.6	1.3	3.6675
14	3	312.5	1024	250	3301.6	1.3	1.8338
14	4	156.25	1024	250	3301.6	1.3	0.91689
14	5	78.125	1024	250	3301.6	1.3	0.45844
14	6	625	512	200	9266.8	1.5	23.755
14	7	312.5	512	200	9266.8	1.5	11.878
14	8	156.25	512	200	9266.8	1.5	5.9388
14	9	78.125	512	200	9266.8	1.5	2.9694
14	10	39.063	512	200	9266.8	1.5	1.4847
14	11	19.531	512	200	9266.8	1.5	0.74235
14	12	9.7656	512	200	9266.8	1.5	0.37118
14	13	4.8828	512	200	9266.8	1.5	0.28559
Total				Percentage Loads		=	78.672

(*b*) 13 CHANNELS PARTITIONED TO PE2 AND PE3 EACH

NUMBER OF CHs	STR #	F	INPUT BATCH	CNTROL INSTRs	TOTAL TIME	OVERHD FACTOR	THRUPUT LOAD %
13	1	2500	1024	400	5507.2	1.3	22.723
13	2	625	1024	250	3301.6	1.3	3.4056
13	3	312.5	1024	250	3301.6	1.3	1.7028
13	4	156.25	1024	250	3301.6	1.3	0.85140
13	5	78.125	1024	250	3301.6	1.3	0.42570
13	6	625	512	200	9266.8	1.5	22.058
13	7	312.5	512	200	9266.8	1.5	11.029
13	8	156.25	512	200	9266.8	1.5	5.5146
13	9	78.125	512	200	9266.8	1.5	2.7573
13	10	39.063	512	200	9266.8	1.5	1.3787
13	11	19.531	512	200	9266.8	1.5	0.68933
13	12	9.7656	512	200	9266.8	1.5	0.34466
13	13	4.8828	512	200	9266.8	1.5	0.17233
Total				Percentage loads		=	73.053

CNTROL LOAD %	MEMORY LOAD %	MEMORY BW %	THRUPUT MILLIONS / SEC		THRUPUT LOAD (%)	
			MULT	ADDS	MULT	ADDS
1.3672	3.1539	3.7297	0.90125	1.4109	18.025	14.109
0.21362	6.3078	1.0227	0.105	0.14109	2.1	1.4109
0.10681	6.3078	0.51133	0.0525	0.07055	1.05	0.70547
0.05341	6.3078	0.25566	0.02625	0.03527	0.525	0.35273
0.02670	6.3078	0.12783	0.01313	0.01764	0.2625	0.17637
0.34180	8.2051	2.6958	0.39908	0.52855	7.9816	5.2855
0.17090	8.2051	1.3479	0.19954	0.26428	3.9908	2.6428
0.08545	8.2051	0.67394	0.09977	0.13214	1.9954	1.3214
0.04272	8.2051	0.33697	0.04989	0.06607	0.99771	0.66069
0.02136	8.2051	0.16848	0.02494	0.03303	0.49885	0.33035
0.01068	8.2051	0.08424	0.01247	0.01652	0.24943	0.16517
0.00534	8.2051	0.04212	0.00624	0.00826	0.12471	0.08259
0.00267	8.2051	0.02106	0.00312	0.00413	0.06236	0.04129
2.4487	94.026	11.018	MMPS 1.8932	MAPS 2.7285	37.863	27.285

CNTROL LOAD %	MEMORY LOAD %	MEMORY BW %	THRUPUT MILLIONS / SEC		THRUPUT LOAD (%)	
			MULT	ADDS	MULT	ADDS
1.2695	2.9286	3.4633	0.83688	1.3102	16.738	13.102
0.19836	5.8573	0.94961	0.0975	0.13102	1.95	1.3102
0.09918	5.8573	0.47480	0.04875	0.06551	0.975	0.65508
0.04959	5.8573	0.23740	0.02438	0.03275	0.4875	0.32754
0.02480	5.8573	0.11870	0.01219	0.01638	0.24375	0.16377
0.31738	7.6190	2.5032	0.37058	0.49080	7.4115	4.9080
0.15869	7.6190	1.2516	0.18529	0.24540	3.7058	2.4540
0.07935	7.6190	0.62580	0.09264	0.12270	1.8529	1.2270
0.03967	7.6190	0.31290	0.04632	0.06135	0.92644	0.61350
0.01984	7.6190	0.15645	0.02316	0.03068	0.46322	0.30675
0.00992	7.6190	0.07822	0.01158	0.01534	0.23161	0.15338
0.00496	7.6190	0.03911	0.00579	0.00767	0.11581	0.07669
0.00248	7.6190	0.01956	0.00290	0.00383	0.05790	0.03834
2.2738	87.310	10.231	MMPS 1.7579	MAPS 2.5336	35.159	25.336

TABLE 10.13
ALTERNATIVE B PROCESSING PARTITIONED TO CONFIGURATION

(*a*) **20 CHANNELS OF OCTAVE PROCESSING PARTITIONED TO PE1 & PE2**

NUMBER OF CHs	STR #	F	INPUT BATCH	CNTROL INSTRs	TOTAL TIME	OVERHD FACTOR	THRUPUT LOAD %
20	1	2500	1024	400	5507.2	1.3	34.958
0	2	625	1024	250	3301.6	1.3	0
0	3	312.5	1024	250	3301.6	1.3	0
0	4	156.25	1024	250	3301.6	1.3	0
0	5	78.125	1024	250	3301.6	1.3	0
20	6	625	512	200	9266.8	1.5	33.936
20	7	312.5	512	200	9266.8	1.5	16.968
20	8	156.25	512	200	9266.8	1.5	8.4840
20	9	78.125	512	200	9266.8	1.5	4.2420
0	10	39.063	512	200	9266.8	1.5	0
0	11	19.531	512	200	9266.8	1.5	0
0	12	9.7656	512	200	9266.8	1.5	0
0	13	4.8828	512	200	9266.8	1.5	0
Total				Percentage loads		=	98.588

(*b*) **40 CHANNELS OF VERNIER PROCESSING PARTITIONED TO PE3**

NUMBER OF CHs	STR #	F	INPUT BATCH	CNTROL INSTRs	TOTAL TIME	OVERHD FACTOR	THRUPUT LOAD %
0	1	2500	1024	400	5507.2	1.3	0
40	2	625	1024	250	3301.6	1.3	10.479
40	3	312.5	1024	250	3301.6	1.3	5.2394
40	4	156.25	1024	250	3301.6	1.3	2.6197
40	5	78.125	1024	250	3301.6	1.3	1.3098
0	6	625	512	200	9266.8	1.5	0
0	7	312.5	512	200	9266.8	1.5	0
0	8	156.25	512	200	9266.8	1.5	0
0	9	78.125	512	200	9266.8	1.5	0
40	10	39.063	512	200	9266.8	1.5	4.2420
40	11	19.531	512	200	9266.8	1.5	2.1210
40	12	9.7656	512	200	9266.8	1.5	1.0605
40	13	4.8828	512	200	9266.8	1.5	0.53025
Total				Percentage loads		=	27.601

CNTROL LOAD %	MEMORY LOAD %	MEMORY BW %	THRUPUT MILLIONS / SEC		THRUPUT LOAD (%)	
			MULT	ADDS	MULT	ADDS
1.9531	4.5056	5.3281	1.2875	2.0156	25.75	20.156
0	0	0	0	0	0	0
0	0	0	0	0	0	0
0	0	0	0	0	0	0
0	0	0	0	0	0	0
0.48828	11.722	3.8511	0.57012	0.75508	11.402	7.5508
0.24414	11.722	1.9255	0.28506	0.37754	5.7012	3.7754
0.12207	11.722	0.96277	0.14253	0.18877	2.8506	1.8877
0.06104	11.722	0.48138	0.07126	0.09438	1.4253	0.94385
0	0	0	0	0	0	0
0	0	0	0	0	0	0
0	0	0	0	0	0	0
0	0	0	0	0	0	0
2.8687	51.392	12.549	MMPS 2.3565	MAPS 3.4314	47.129	34.314

CNTROL LOAD %	MEMORY LOAD %	MEMORY BW %	THRUPUT MILLIONS / SEC		THRUPUT LOAD (%)	
			MULT	ADDS	MULT	ADDS
0	0	0	0	0	0	0
0.61035	18.022	2.9219	0.3	0.40313	6	4.0313
0.30518	18.022	1.4609	0.15	0.20156	3	2.0156
0.15259	18.022	0.73047	0.075	0.10078	1.5	1.0078
0.07629	18.022	0.36523	0.0375	0.05039	0.75	0.50391
0	0	0	0	0	0	0
0	0	0	0	0	0	0
0	0	0	0	0	0	0
0	0	0	0	0	0	0
0.06104	23.443	0.48138	0.07126	0.09438	1.4253	0.94385
0.03052	23.443	0.24069	0.03563	0.04719	0.71265	0.47192
0.01526	23.443	0.12035	0.01782	0.02360	0.35632	0.23596
0.00763	23.443	0.06017	0.00891	0.01180	0.17816	0.11798
1.2589	165.86	6.3811	MMPS 0.69612	MAPS 0.93283	13.922	9.3283

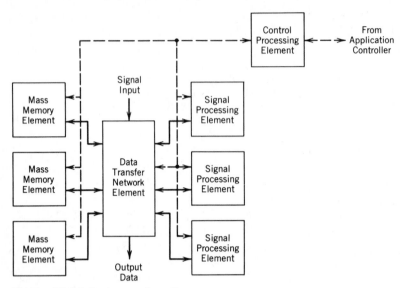

Figure 10.34 System configuration.

provide enough margin for growth during development, and the 40 vernier processing allocation requires only 27.6% throughput load but exceeds one mass memory by 65.8%. Therefore further partitioning of the processing would be required in order for alternative B to be acceptable.

For this problem, alternative A is selected as the design to be implemented. The system configuration is shown in Figure 10.34. At this point, all resource parameter requirements would be compared to the configuration capabilities to assure that the design provides acceptable performance.

As stated in Section 10.2, the configuration analysis includes all the nonsignal processing requirements as well as the final selection of the signal processing configuration. We will only address the signal processor specification herein. From Table 10.12 we see that the resource estimates are within the configuration capabilities. It is usually desirable to provide sufficient margin within each element to accommodate future changes and/or new requirements. The militaries generally specify reserve requirements that force the user to maintain at least 20% reserve and up to 50% reserve for key parameters. In some instances, additional reserves are imposed. Therefore, a case could be made for providing an additional memory element and possibly one additional processor engine based on reserve requirements.

Reliability and availability play a key role in the operation performance of a real-time signal processor application. If the processing is associated with navigation of an aircraft, then the safety of the crew is dependent on the processor operation. This could lead to adding additional elements for redundancy purposes improving the fault tolerance of the configuration.

All other nonsignal processing requirements must be assessed and the configuration modified accordingly to meet the overall system requirements. In cases where the system significantly exceeds the requirements cost performance, tradeoffs may be appropriate to assure that the proposed design is a good match to best solution to the problem.

10.4 SUMMARY

A procedure for performing the signal processor system design was presented in Section 10.2. The procedure described five key steps to be performed.

1. System requirements definition.
2. Signal analysis.
3. Signal processing design.
4. Resource analysis.
5. Configuration analysis.

Each step was discussed, and procedures were developed for the signal processing design and resource analysis steps 3 and 4, respectively. A brief discussion of a signal processor architecture and the associated resource parameters that must be computed were presented. Next, a narrowband detection spectral analysis application was described and a set of requirements were defined in Section 10.3. A system design was developed using the methodology from Section 10.2. Personal computer spreadsheets were developed to perform the analysis.

It is hoped that this chapter has provided you with an appreciation for the analysis tasks required to perform a signal processing system design. The design process requires expert knowledge in signal processing, processor architectures (both hardware and software), and the key nonsignal processing parameters discussed. Also, detailed signal processor primitive implementations must be performed, along with estimating the application resource requirements. Therefore, a multidiscipline team will be necessary to perform the overall system design as described in Chapter 1.

Following a systematic approach as outlined herein should assure that the design meets the intended requirements.

References

1. L. R. Rabiner and B. Gold, *Theory and Application of Digital Signal Processing*, Prentice–Hall, Englewood Cliffs, N.J., 1975.

2. G. L. Kratz, W. W. Sproul, and E. T. Walendziewicz, "A Microprogrammed Approach to Signal Processing," *IEEE C-23*, 1974, pp. 808–816.

3. P. R. Cappello and K. Steiglitz, "Completely-Pipelined Architectures for Digital Signal Processing," *IEEE Trans. ASSP*, ASSP-31 (4): 1016–1023, 1983.

4. S. Y. Kung, "VLSI Array Processors," *IEEE ASSP*, July 1985, pp. 4–22.

5. B. A. Bowan and W. R. Brown, *VLSI Systems Design for Digital Signal Processing, Vol. 1: Signal Processing and Signal Processors*, Prentice–Hall, Englewood Cliffs, N.J., 1982.

6. B. A. Bowan and W. R. Brown, *VLSI Systems Design for Digital Signal Processing, Vol. 2: Signal Processing and Signal Processors*, Prentice–Hall, Englewood Cliffs, N.J., 1985.

7. R. J. Urick, *Principles of Underwater Sound*, 2nd ed., McGraw–Hill, New York, 1975.

8. W. S. Burdic, *Underwater Acoustic System Analysis*, Prentice–Hall, Englewood Cliffs, N.J., 1984.

9. A. D. Whalen, *Detection of Signals in Noise*, Academic Press, New York, 1971.

10. W. B. Davenport, Jr. and W. L. Root, *Random Signals and Noise*, McGraw–Hill, New York, 1958.

11. G. M. Jenkins and D. G. Watts, *Spectral Analysis and Applications*, Holden–Day, San Francisco, 1968.

12. W. J. Albersheim, "A Closed-Form Approximation to Robertson's Detection Characteristics," *Proc. IEEE*, 69 (No. 7), July 1981.

Additional Readings

ALIPHAS, A. and FELDMAN, J. A, "The Versatility of Digital Signal Processing Chips," *IEEE Spectrum*, June 1987, pp. 40–45.

FALCONER, J. H. "An Architectural Review of a Self-contained/Standalone Acoustic Signal Processor." *IEEE* 0094-2898, 1982, pp. 206–209.

HARTIMO, I., KRONLOEF, K., SIMULA, O., and SKYTTAE, J. "DFSP: A Data Flow Signal Processor." Helsinki University of Technology, Espoo, Finland, Department of Technical Physics, TKK-F-A520, June 1983.

KNIGHT, W. C., PRIDHAM, R. G, and KAY, S. M. "Digital Signal Processing for Sonar." *Proc. IEEE*, Vol. 69, No. 11, November 1981.

KORAL, W. and SCHIRM, L. "Floating-Point Arithmetic for Digital Signal Processing." *IEEE* CH-1559, April 1980, pp. 380–382.

MEAD, C. and CONWAY, L. *Introduction to VLSI Systems*. Addison–Wesley, Reading, Mass., 1980.

MOORE, K. A. "A Distributive Signal Processor (The Realtime Signal Processor)." *IEEE* 0094-2898, 1982, pp. 210–214.

NISHITANI, T., KAWAKAMI, Y., MARUTA, R., and SAWAI, A. "LSI Signal Processor Development for Communications Equipment." *IEEE* CH-1559, April 1980, pp. 386–389.

PELED, A., and LIU, B. *Digital Signal Processing: Theory Design, and Implementation*. Wiley, New York, 1976.

SWARTZLANDER, E. E., JR. "Signal Processing Architectures with VLSI." *IEEE* CH-1559, April 1980, pp. 368–371.

THOMPSON, J. S. and BODDIE, J. R. "An LSI Digital Signal Processor." *IEEE* CH-1559, April 1980, pp. 383–385.

WALENDZIEWICZ, E. T. "Design and Application Considerations for a Programmable Digital Signal Processor." IBM No. 76TPA0002, IBM Federal Systems Division, Owego, N.Y., 1976.

WASER, S. "Survey of VLSI for Digital Signal Processing." *IEEE* CH-1559, April 1980, pp. 376–379.

WILEY, P. "A Parallel Architecture Comes of Age at Last." *IEEE Spectrum*, June 1987, pp. 46–50.

"Digital Signal Processing Applications with the TMS320 Family: Theory, Algorithms, and Implementations." *Digital Signal Processing Semiconductor Group*, Texas Instruments Inc., 1986.

"TMS320C25 User's Guide." *Digital Signal Processor Products*, Texas Instruments Inc., 1986.

PROBLEMS

10.3.1. Perform a trade-off between the Chebyshev filter selected in Section 10.3.3 (see Table 10.5), and the use of an elliptic filter. What elliptic filter order is required to maintain the same antialiasing performance?

10.3.2. Given the following processing requirements:

	Interfering Tone Level (dB)	Processing Band (Hz)	MDSNR (dB)	Input Sampling Frequency (Hz)	Decimation Factor (D)
a.	25	0–20	−8	400	5
b.	30	0–10	−6		
		10–20	−8	400	5
c.	20	0–100	−6		
		100–200	−4		
		200–500	−2	10,000	5

Determine the filter stopband frequencies and levels to limit the aliased frequency component to at least 8 dB below the MDSNR level.

10.3.3. Develop a spreadsheet template to calculate the resource estimates per string given in Table 10.10. Verify the results presented in Table 10.10 using the processor capabilities defined in Table 10.1.

10.3.4. Develop a spreadsheet template to calculate the processor resource requirements given in Table 10.11. Verify the results presented in Table 10.11 using the processor capabilities defined in Table 10.1 and the results from Table 10.10 verified in Problem 10.3.3.

10.3.5. Reduce the input batch sizes of the octave and vernier strings to 512 for the analysis performed in Tables 10.10 and 10.11 and calculate the processor resource requirements. Comment on the results compared to those obtained with the 1024 input batch sizes.

10.3.6. If each complex band shift and filtering operation associated with an octave band in Table 10.10*a* is defined as a string, what is the impact on the resources given in Table 10.11 for the octave string?

10.3.7. For the application example presented, design the digital signal conditioning portion assuming that the frequency translation is done at the input rate 2500 Hz for each octave and vernier band. Use a FIR design approach. Comment on the alternative design choices. How does this compare with the approach presented in the chapter?

10.3.8. Perform a tradeoff between the octave filtering FIR implementations of Section 10.3.3, and the use of IIR filters. Use elliptic digital filters in the trade-off. What are your conclusions relative to throughput requirements and system load for IIR implementations versus FIR implementations? Table 10.1 provides execution times for second-order and fourth-order stages. For the purpose of this exercise, assume that finite arithmetic effects are negligible.

10.3.9. Repeat Problem 8 using Chebyshev IIR filters. What are your conclusions relative to FIR and elliptic implementations?

TABLE P10.3-1
SPECTRAL ANALYSIS APPLICATION REQUIREMENTS

Parameter	Specification / Requirement
Input signal $x(t)$	$x(t) = n(t) + s(t)$
$n(t)$	Gaussian $(0, \delta)$
$s(t)$	Sinusoidal $A \sin(wt + p)$
Bandwidth	0–10,000 Hz uniform;
	10,000 Hz and above; 6 dB/octave rolloff
Max interfering tone level	30 dB/Hz
Dynamic range	50 dB
Performance P_d, P_{fa}, T_i	0.5, 0.0001, 5 min
Antialiasing P_d, P_{fa}, T_i	$\leq 0.01, 0.0001, 5$ min
Processing bandwidth	250–4000 Hz
Spectral analysis	
Analysis resolutions	Within $\pm 10\%$ of
Octaves, (Δf)	2.5 Hz from 2000 to 4000 Hz
	1.25 Hz from 1000 to 2000 Hz
	0.625 Hz from 500 to 1000 Hz
	0.3125 Hz from 250 to 500 Hz
	0.15625 Hz from 125 to 250 Hz
Verniers	1/4th of octaves.
Redundancy	$\geq 50\%$
Noise bandwidth	≤ 1.5 analysis resolution
Scalloping loss	
Maximum	≤ 2 dB
Average	≤ 0.7 dB
Highest sidelobe level	≤ -26 dB
Other losses	≤ 1 dB

*Requirements based on P_d, P_{fa}, T_i for signal levels at the system input.

10.3.10. Repeat Problem 8 for the vernier filtering implementations.

10.3.11. Compare alternative FIR implementations for the vernier FIR filtering implementation analysis of Section 10.3.3. Try one-stage and two-stage implementations. What are your conclusions?

10.3.12. Design the vernier filters for the application problem with no dependency on the octave band processing. Comment on the computational requirements and the resultant vernier flexibility versus the design chosen in the chapter.

10.3.13. Perform a signal processing design to meet the requirements of Table P10.3-1. Assume that the signal processor has been specified with the capabilities defined in Table 10.1. If other FIR filters are required, make an estimation based on the orders given in the table. What is the proposed configuration and the corresponding system load for 10 channels and 20 channels of processing? One vernier is required in each octave. If a computer is not available for performing the analysis, use the number of multiplies and adds for a rough approximation. Compute the system MDSNR using Section 8.3, and describe the effect of this on the filter requirements.

11

Adaptive Filtering

Filters that adapt to a changing environment are discussed. The stochastic Wiener filtering and deterministic least-squares problems are set up, and the similarity between them is pointed out. First, a block-processing approach is taken in the solution to these two problems. Second, a fading memory (recursive) approach is taken. Then the application of these algorithms to the adaptive beamforming (ABF) problem is considered. The chapter appendix provides background material on conventional beamforming (CBF).

11.0 INTRODUCTION

Numerous applications exist that require a linear filtering operation. Often, the nature of that filtering task is time-varying in some nondeterministic fashion owing to nonstationarity of the underlying time series. In such situations, a filter that can adapt to a changing environment is needed. Examples include adaptive noise canceling, line enhancing, frequency tracking, and channel equalization. Beyond the discussion here, additional material on adaptive filtering can be found in references 1 through 16.

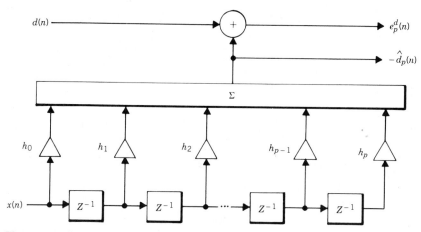

Figure 11.1 Direct-form realization of the joint process structure.

11.1 THE STOCHASTIC WIENER FILTERING AND DETERMINISTIC LEAST SQUARES PROBLEMS

The problem of interest is described in Figure 11.1. The goal is to filter the time series $x(n)$ with a causal, finite impulse response (FIR) digital filter to yield an estimate of the time series $d(n)$. The error in that estimate is denoted by $e^d(n)$

$$e^d(n) = d(n) + \sum_{k=0}^{p} h_k x(n-k) \tag{11.1}$$

Written in matrix form, Eq. 11.1 becomes

$$e^d(n) = d(n) + \begin{bmatrix} h_0 & h_1 & \cdots & h_p \end{bmatrix} \begin{bmatrix} x(n) \\ x(n-1) \\ \vdots \\ x(n-p) \end{bmatrix} \tag{11.2}$$

or, in vector notation,

$$e^d(n) = d(n) + \mathbf{h}^T \mathbf{x} \tag{11.3}$$

The solution to this problem for the unknown filter vector $\mathbf{h}$ can be approached in two ways, depending on the optimality criterion chosen

$$\underset{\mathbf{h}}{\text{minimize }} E\left[|e^d(n)|^2\right] \quad \text{Wiener filtering problem} \tag{11.4}$$

$$\underset{\mathbf{h}}{\text{minimize }} \sum_n |e^d(n)|^2 \quad \text{Least squares problem} \tag{11.5}$$

The Wiener filtering problem takes a stochastic viewpoint where the filter is optimized in an expected value sense (minimum mean-squared error). As an alternative, the least-squares problem views the time series involved as deterministic and the filter is optimized for the specific sequences of record (minimum sum-squared error). Both viewpoints give rise to similar mathematical expressions for $\mathbf{h}$.

The distinction between the Wiener filtering and least squares problems is made in the definitions of ϕ_{00}^d, $\mathbf{g}$, and $\mathbf{\Phi}$ where

$$\mathbf{g} = \begin{bmatrix} g_0 \\ g_1 \\ \vdots \\ g_p \end{bmatrix} \tag{11.6}$$

$$\mathbf{\Phi} = \begin{bmatrix} \phi_{00} & \cdots & \phi_{0p} \\ \vdots & & \vdots \\ \phi_{p0} & \cdots & \phi_{pp} \end{bmatrix} = \begin{bmatrix} \phi_{00} & \cdots & \phi_{0p} \\ \vdots & & \vdots \\ \phi_{0p}^* & \cdots & \phi_{pp} \end{bmatrix} \tag{11.7}$$

For the Wiener filtering problem

$$\phi_{00}^d = E[d(n)\, d(n)^*] \tag{11.8}$$

$$\mathbf{g} = \{ E[d(n)x(n-j)^*] \} = \{ g_j \} \tag{11.9}$$

$$\mathbf{\Phi} = \{ E[x(n-k)x(n-j)^*] \} = \{ \phi_{kj} \} \tag{11.10}$$

And, for the least squares problem

$$\phi_{00}^d = \sum_n d(n)\, d(n)^* \tag{11.11}$$

$$\mathbf{g} = \left\{ \sum_n d(n)x(n-j)^* \right\} = \{ g_j \} \tag{11.12}$$

$$\mathbf{\Phi} = \left\{ \sum_n x(n-k)x(n-j)^* \right\} = \{ \phi_{kj} \} \tag{11.13}$$

The next step in solving for the unknown filter vector $\mathbf{h}$ is to write an expression for squared error based on Eq. 11.3

$$|e^d(n)|^2 = d(n)^*\, d(n) + d(n)^*\, \mathbf{h}^T\mathbf{x} + \mathbf{h}^H\mathbf{x}^*\, d(n) + \mathbf{h}^H\mathbf{x}^*\mathbf{h}^T\mathbf{x} \tag{11.14}$$

By making use of the definitions of ϕ_{00}^d, $\mathbf{g}$, and $\mathbf{\Phi}$ the average squared error $E_p^d(\mathbf{h})$ can be derived from Eq. 11.14 for both the Wiener filtering and least squares problems as

$$E_p^d(\mathbf{h}) = \phi_{00}^d + \mathbf{g}^H\mathbf{h} + \mathbf{h}^H\mathbf{g} + \mathbf{h}^H\mathbf{\Phi}^T\mathbf{h} \tag{11.15}$$

Now, minimizing $E_p^d(\mathbf{h})$ with respect to $\mathbf{h}$ leads to the following equation.

$$0 = \mathbf{g} + \mathbf{\Phi}^T \mathbf{h} \tag{11.16}$$

or

$$\mathbf{\Phi}^T \mathbf{h} = -\mathbf{g} \tag{11.17}$$

Expressing Eq. 11.17 in expanded form will be of interest in the next section

$$
\begin{bmatrix}
\phi_{00} & \cdots & \phi_{0p}^* \\
\vdots & & \vdots \\
\phi_{0p} & \cdots & \phi_{pp}
\end{bmatrix}
\begin{bmatrix}
h_0 \\
\vdots \\
h_p
\end{bmatrix}
= -
\begin{bmatrix}
g_0 \\
\vdots \\
g_p
\end{bmatrix}
\tag{11.18}
$$

or

$$\sum_{k=0}^{p} \phi_{kj} h_k = -g_j \qquad j = 0, 1, \ldots, p \tag{11.19}$$

Now, solving Eq. 11.17 for $\mathbf{h}$ yields

$$\mathbf{h} = -(\mathbf{\Phi}^T)^{-1} \mathbf{g} \tag{11.20}$$

By substituting the solution for $\mathbf{h}$ back into Eq. 11.15, the minimum value of average squared error can be obtained as

$$E_p^d = \phi_{00}^d + \mathbf{g}^H \mathbf{h} \tag{11.21}$$

Expressing 11.21 in expanded form also will be of interest in the next section.

$$E_p^d = \phi_{00}^d + [g_0^* \cdots g_p^*]
\begin{bmatrix}
h_0 \\
\vdots \\
h_p
\end{bmatrix}
\tag{11.22}$$

or

$$E_p^d = \phi_{00}^d + \sum_{k=0}^{p} g_k^* h_k \tag{11.23}$$

An important property of the solution for $\mathbf{h}$ in Eq. 11.20 is known as the orthogonality principle. From Eq. 11.3, the product of the error $e^d(n)$ and the complex conjugate of the data available to the filter $\mathbf{x}^*$ can be written as

$$e^d(n)\mathbf{x}^* = d(n)\mathbf{x}^* + \mathbf{h}^T \mathbf{x}\mathbf{x}^*$$

$$= d(n)\mathbf{x}^* + \mathbf{x}^* \mathbf{x}^T \mathbf{h} \tag{11.24}$$

Making use of the definitions of **g** and **Φ** in Eq. 11.6 through Eq. 11.13 and the expression for **h** in Eq. 11.17, the average value of Eq. 11.24 is

$$\mathbf{g} + \mathbf{\Phi}^T\mathbf{h} = 0 \tag{11.25}$$

Thus, when the filter vector **h** is selected so that the average squared error is minimized, the error is orthogonal to the data.

The expressions of **Φ** in Eq. 11.10 and Eq. 11.13 have been left general. When the time series in question are stationary (Wiener filtering problem) or have been windowed (least squares problem), **Φ** will be Toeplitz

$$\phi_{kj} = \phi_{j-k} \tag{11.26}$$

Thus, all entries along a given diagonal of **Φ** are equal. Advantage can be taken of this special structure in the solution of Eq. 11.7, which avoids the necessity of directly inverting **Φ** as is done in Eq. 11.20.

11.2 BLOCK PROCESSING APPROACH

A common approach to the processing of time series is to split the observation interval into subintervals (perhaps overlapped by some factor). The collection of samples in any given subinterval then are processed as a block. Here we will assume that over the duration of a block, the time series in question are stationary (Wiener filtering problem) or have been windowed (least squares problem). The Toeplitz structure of **Φ** will be taken advantage of in solving for the filter. As will be seen, the general filter solution $\{h_0, h_1, \ldots, h_p\}$ has the solution to the linear prediction problem embedded in it.

11.2.1 Linear Prediction

One-step forward and backward linear predictors of order p are shown in Figure 11.2. The one-step forward linear predictor weights the most recent p samples $\{x(n-1), \ldots, x(n-p)\}$ of the time series with coefficients $\{a_1, \ldots, a_p\}$ to form an estimate of $-x(n)$. Similarly, the one-step backward linear predictor weights the current and most recent $p-1$ samples $\{x(n), \ldots, x(n-p+1)\}$ of the time series with coefficients $\{b_0, \ldots, b_{p-1}\}$ to form an estimate of $-x(n-p)$. Note that linear predictors are simply special cases of the general filtering problem shown in Figure 11.1 where $x(n)$ and $d(n)$ are drawn from the same underlying time series.

Written explicitly in matrix form, the equation that must be solved to determine the forward linear predictor coefficients is

$$\begin{bmatrix} \phi_0 & \cdots & \phi_{p-1}^* \\ \vdots & & \vdots \\ \phi_{p-1} & \cdots & \phi_0 \end{bmatrix} \begin{bmatrix} a_1^p \\ \vdots \\ a_p^p \end{bmatrix} = - \begin{bmatrix} \phi_1 \\ \vdots \\ \phi_p \end{bmatrix} \tag{11.27}$$

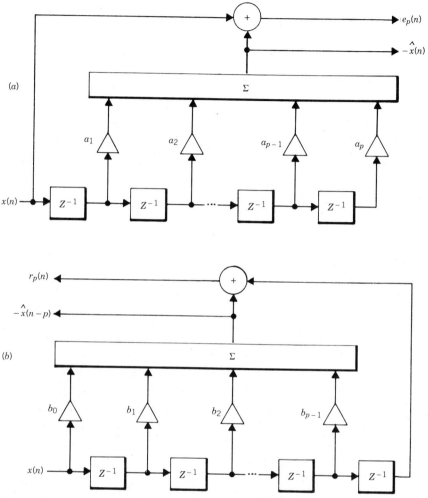

Figure 11.2 (*a*) One-step forward prediction error filter. (*b*) One-step backward prediction error filter.

Adding $[\phi_1 \ldots \phi_p]^T$ to both sides of Eq. 11.27 yields

$$\begin{bmatrix} \phi_1 & \phi_0 & \cdots & \phi_{p-1}^* \\ \vdots & \vdots & & \vdots \\ \phi_p & \phi_{p-1} & \cdots & \phi_0 \end{bmatrix} \begin{bmatrix} 1 \\ a_1^p \\ \vdots \\ a_p^p \end{bmatrix} = \begin{bmatrix} 0 \\ \vdots \\ 0 \end{bmatrix} \qquad (11.28)$$

Now, augmenting Eq. 11.28 with the expression for minimum forward prediction error

$$E_p^e = \phi_{00} + \sum_{k=1}^{p} \phi_k^* a_k^p \qquad (11.29)$$

yields

$$
\begin{bmatrix}
\phi_0 & & \cdots & \phi_p^* \\
\phi_1 & \phi_0 & \cdots & \phi_{p-1}^* \\
\vdots & \vdots & & \vdots \\
\phi_p & \phi_{p-1} & \cdots & \phi_0
\end{bmatrix}
\begin{bmatrix}
1 \\
a_1^p \\
\vdots \\
a_p^p
\end{bmatrix}
=
\begin{bmatrix}
E_p^e \\
0 \\
\vdots \\
0
\end{bmatrix}
\qquad (11.30)
$$

Thus, the solution to Eq. 11.30 provides both the p forward prediction error filter coefficients $\{a_1^p, \ldots, a_p^p\}$ as well as the minimum error E_p^e. Following these same steps for the backward prediction error filter yields

$$
\begin{bmatrix}
\phi_0 & \cdots & \phi_{p-1}^* & \phi_p^* \\
\vdots & & \vdots & \vdots \\
\phi_{p-1} & \cdots & \phi_0 & \phi_1^* \\
\phi_p & \cdots & & \phi_0
\end{bmatrix}
\begin{bmatrix}
b_0^p \\
\vdots \\
b_{p-1}^p \\
1
\end{bmatrix}
=
\begin{bmatrix}
0 \\
\vdots \\
0 \\
E_p^r
\end{bmatrix}
\qquad (11.31)
$$

The solution to Eq. 11.31 provides both the p backward prediction error filter coefficients $\{b_0^p, \ldots, b_{p-1}^p\}$ as well as the minimum error E_p^r. Although the backward prediction error filter will be carried along explicitly, since Φ is Toeplitz

$$
\begin{bmatrix}
b_0^p \\
\vdots \\
b_{p-1}^p \\
1
\end{bmatrix}
=
\begin{bmatrix}
a_p^{p*} \\
\vdots \\
a_1^{p*} \\
1
\end{bmatrix}
\qquad (11.32)
$$

and

$$E_p^r = E_p^e \qquad (11.33)$$

Given that the forward and backward prediction error filters of order p have been determined, it now is of interest to derive order update recursions that will permit generating the prediction error filters of order $p + 1$ from those of order p. Starting with the forward prediction error filter, simply guess that the correct filter of order $p + 1$ is $[1\ a_1^p \ldots a_p^p\ 0]^T$. Now Eq. 11.30 becomes

$$
\begin{bmatrix}
\phi_0 & \cdots & \phi_p^* & \phi_{p+1}^* \\
\vdots & & \vdots & \vdots \\
\phi_p & \cdots & \phi_0 & \phi_1^* \\
\phi_{p+1} & \cdots & \phi_1 & \phi_0
\end{bmatrix}
\begin{bmatrix}
1 \\
a_1^p \\
\vdots \\
a_p^p \\
0
\end{bmatrix}
=
\begin{bmatrix}
E_p^e \\
0 \\
\vdots \\
0 \\
\Delta_{p+1}^e
\end{bmatrix}
\qquad (11.34)
$$

where

$$\Delta_{p+1}^e = \phi_{p+1} + \phi_p a_1^p + \cdots + \phi_1 a_p^p \tag{11.35}$$

Making a similar guess for the backward prediction error filter of order $p + 1$ yields

$$\begin{bmatrix} \phi_0 & \phi_1^* & \cdots & \phi_{p+1}^* \\ \phi_1 & \phi_0 & \cdots & \phi_p^* \\ \vdots & \vdots & & \vdots \\ \phi_{p+1} & \phi_p & \cdots & \phi_0 \end{bmatrix} \begin{bmatrix} 0 \\ b_0^p \\ \vdots \\ b_{p-1}^p \\ 1 \end{bmatrix} = \begin{bmatrix} \Delta_{p+1}^r \\ 0 \\ \vdots \\ 0 \\ E_p^r \end{bmatrix} \tag{11.36}$$

where

$$\Delta_{p+1}^r = \phi_1^* b_0^p + \cdots + \phi_p^* b_{p-1}^p + \phi_{p+1}^* \tag{11.37}$$

Notice that if the discrepancy terms Δ_{p+1}^e and Δ_{p+1}^r were 0, the proposed forward and backward prediction error filters of order $p + 1$ would be correct. Since the discrepancy terms generally are not 0, the next step involves modifying the proposed filter coefficients so as to cancel the discrepancy. The modification involves adding a weighted version of the proposed backward prediction error filter vector to the proposed forward prediction error filter vector and vice versa

$$\begin{bmatrix} 1 \\ a_1^{p+1} \\ \vdots \\ a_p^{p+1} \\ a_{p+1}^{p+1} \end{bmatrix} = \begin{bmatrix} 1 \\ a_1^p \\ \vdots \\ a_p^p \\ 0 \end{bmatrix} + K_{p+1}^r \begin{bmatrix} 0 \\ b_0^p \\ \vdots \\ b_{p-1}^p \\ 1 \end{bmatrix} \tag{11.38}$$

yielding

$$\begin{bmatrix} E_{p+1}^e \\ 0 \\ \vdots \\ 0 \\ 0 \end{bmatrix} = \begin{bmatrix} E_p^e \\ 0 \\ \vdots \\ 0 \\ \Delta_{p+1}^e \end{bmatrix} + K_{p+1}^r \begin{bmatrix} \Delta_{p+1}^r \\ 0 \\ \vdots \\ 0 \\ E_p^r \end{bmatrix} \tag{11.39}$$

and

$$\begin{bmatrix} b_0^{p+1} \\ b_1^{p+1} \\ \vdots \\ b_p^{p+1} \\ 1 \end{bmatrix} = K_{p+1}^e \begin{bmatrix} 1 \\ a_1^p \\ \vdots \\ a_p^p \\ 0 \end{bmatrix} + \begin{bmatrix} 0 \\ b_0^p \\ \vdots \\ b_{p-1}^p \\ 1 \end{bmatrix} \tag{11.40}$$

yielding

$$\begin{bmatrix} 0 \\ 0 \\ \vdots \\ 0 \\ E^r_{p+1} \end{bmatrix} = K^e_{p+1} \begin{bmatrix} E^e_p \\ 0 \\ \vdots \\ 0 \\ \Delta^e_{p+1} \end{bmatrix} + \begin{bmatrix} \Delta^r_{p+1} \\ 0 \\ \vdots \\ 0 \\ E^r_p \end{bmatrix} \qquad (11.41)$$

In order to cancel the discrepancy terms

$$K^r_{p+1} = -\frac{\Delta^e_{p+1}}{E^r_p} \qquad (11.42)$$

and

$$K^e_{p+1} = -\frac{\Delta^r_{p+1}}{E^e_p} \qquad (11.43)$$

Thus, the weighting factors each consist of the discrepancy normalized by the prediction error from the filter of order p. Since Φ is Toeplitz,

$$\Delta^r_{p+1} = \Delta^{e*}_{p+1} \qquad (11.44)$$

and

$$K^e_{p+1} = K^{r*}_{p+1} \qquad (11.45)$$

Now, by substituting Eqs. 11.42 and 11.43 into the update equations for the forward and backward prediction errors,

$$E^e_{p+1} = E^e_p\left(1 - K^r_{p+1}K^e_{p+1}\right) \qquad (11.46)$$

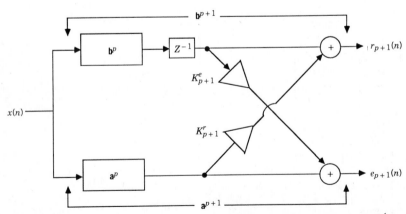

Figure 11.3 Order update procedure for the forward and backward prediction error filters.

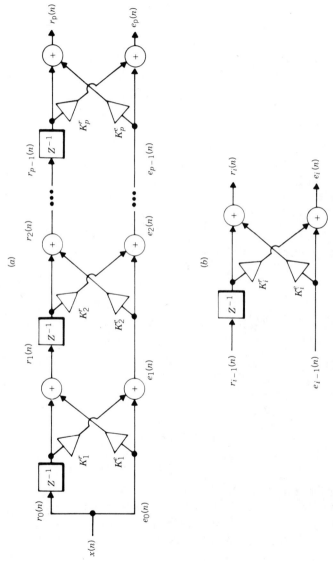

Figure 11.4 (*a*) Forward and backward prediction error filters. (*b*) The *i*th stage of the lattice.

and

$$E^r_{p+1} = E^r_p \left(1 - K^e_{p+1} K^r_{p+1}\right) \tag{11.47}$$

Since Φ is Toeplitz,

$$E^r_{p+1} = E^e_{p+1} \tag{11.48}$$

$$= E^e_p \left(1 - |K^r_{p+1}|^2\right) \tag{11.49}$$

The order update sequence is described graphically in Figure 11.3. The forward and backward prediction error filters of order p are updated using the weighting factors K^r_{p+1} and K^e_{p+1} to yield the corresponding prediction error filters of order $p + 1$. The lattice structure shown in Figure 11.4 is the result of building the forward and backward prediction error filters of order p completely by this order update procedure.

11.2.2 Relationship between the Direct-Form and Lattice Prediction Error Filter Coefficients

As shown in Section 11.2.1, an order-recursive procedure evolves out of the solution to the one-step linear prediction problem when Φ is Toeplitz. This procedure is known as the Levinson–Durbin algorithm.[17,18] Summarizing (see Figs. 11.2 through 11.4 with $K_i = K^r_i = K^{e*}_i$ and $E_i = E^r_i = E^e_i$):

Initialization

$$E_0 = \phi_0 \tag{11.50a}$$

Order update ($i = 1, 2, \ldots, p$)

$$K_i = -\frac{1}{E_{i-1}} \sum_{k=0}^{i-1} a_k^{(i-1)} \phi_{i-k}, \quad a_0 = 1 \tag{11.50b}$$

$$a_i^{(i)} = K_i \tag{11.50c}$$

$$a_k^{(i)} = a_k^{(i-1)} + K_i a_{i-k}^{(i-1)*}, \quad 1 \le k \le i - 1 \tag{11.50d}$$

$$E_i = \left(1 - |K_i|^2\right) E_{i-1} \tag{11.50e}$$

In the order-update recursion, Eqs. 11.50c and 11.50d are sometimes referred to as the step-up algorithm, where the equivalent direct form prediction error filter of order p can be generated from a set of p lattice coefficients.

The transfer functions of the forward and backward prediction error filters are defined as

$$A_p(z) = \sum_{k=0}^{p} a_k z^{-k}, \quad a_0 = 1 \tag{11.51a}$$

$$B_p(z) = \sum_{k=0}^{p} b_k z^{-k}, \quad b_p = 1 \tag{11.51b}$$

Since the forward and backward prediction error filters are related

$$\begin{bmatrix} b_0^p \\ \vdots \\ b_{p-1}^p \\ 1 \end{bmatrix} = \begin{bmatrix} a_p^{p*} \\ \vdots \\ a_1^{p*} \\ 1 \end{bmatrix} \tag{11.52}$$

their transfer functions also are related

$$B_p(z) = z^{-p} A_p^* \left(\frac{1}{z^*} \right) \tag{11.53}$$

Now, in terms of transfer functions, the step-up portion of the Levinson–Durbin algorithm can be expressed as

Initialization

$$A_0(z) = B_0(z) \tag{11.54a}$$

Order update ($i = 1, 2, \ldots, p$)

$$A_i(z) = A_{i-1}(z) + K_i z^{-1} B_{i-1}(z) \tag{11.54b}$$

$$B_i(z) = K_i^* A_{i-1}(z) + z^{-1} B_{i-1}(z) \tag{11.54c}$$

or, writing Eqs. 11.54b and 11.54c in matrix form

$$\begin{bmatrix} A_i(z) \\ B_i(z) \end{bmatrix} = \begin{bmatrix} 1 & z^{-1}K_i \\ K_i^* & z^{-1} \end{bmatrix} \begin{bmatrix} A_{i-1}(z) \\ B_{i-1}(z) \end{bmatrix} \tag{11.54d}$$

From Eq. 11.54d, the inverse procedure yields

$$A_{i-1}(z) = \frac{A_i(z) - K_i B_i(z)}{1 - |K_i|^2} \tag{11.55}$$

and the counterpart to Eqs. 11.50c and 11.50d is

Order update ($i = p, \ldots, 2, 1$)

$$K_i = a_i^{(i)} \tag{11.56a}$$

$$a_k^{(i-1)} = \frac{a_k^{(i)} - K_i a_{i-k}^{(i)*}}{1 - |K_i|^2}, \qquad 1 \le k \le i - 1 \tag{11.56b}$$

The recursion of Eq. 11.56 sometimes is referred to as the step-down algorithm where the equivalent lattice prediction error filter of order p can be generated from a set of p direct-form filter coefficients.

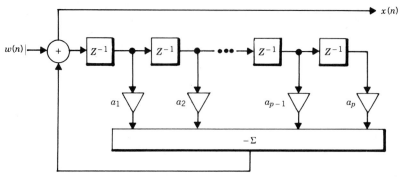

Figure 11.5 AR process generation model.

11.2.3 Spectral Estimation

In recent years, a great deal of interest has been shown in high-resolution spectral estimation techniques.[7,9,11,13] This interest typically has been motivated by the desire to resolve narrowband frequency components in data records too short for adequate frequency separation via standard FFT techniques. By incorporating into the estimation problem assumptions about how the observed data were generated, rather remarkable results have been obtained.

For a number of physical reasons, modeling the observed data as an AR (autoregressive) process has been accepted widely in several applications of time series analysis. As shown in Figure 11.5, an AR process of order p is obtained by passing the white noise sequence $x(n)$ having 0 mean and variance σ_x^2 through a p-pole filter. The corresponding input-output relationships are

$$x(n) = -a_1 x(n-1) - a_2 x(n-2) - \cdots - a_p x(n-p) + w(n)$$

$$= -\sum_{k=1}^{p} a_k x(n-k) + w(n) \tag{11.57}$$

where $w(n)$ is called the innovation of the process. By evaluating the all-pole filter's z-transform on the unit circle, the power spectrum of the AR process $x(n)$ is obtained as

$$S_x(\omega) = \left. \frac{\sigma_w^2}{|A_p(z)|^2} \right|_{z=e^{j\omega}}$$

$$= \frac{\sigma_w^2}{\left| 1 + \sum_{k=1}^{p} a_k \exp(-j\omega k) \right|^2} \tag{11.58}$$

Notice that the all-zero filter of order p that will recover $w(n)$ from $x(n)$ is $A_p(z)$. Appropriately, this filter has been called the whitening or inverse filter for the AR process and is simply obtained by the one-step forward linear predictor in Figure 11.2a. Removing the predictable components from $x(n)$ [or, correspondingly, the coloring from $S_x(\omega)$] yields the white forward prediction error sequence $e_p(n) = w(n)$. Now, given that the time series of interest is pth-order AR, the spectral estimation task becomes equivalently a problem of estimating the pth-order linear predictor, $A_p(z)$, along with the power, σ_w^2, of its prediction error output sequence.

11.2.4 General Filter Solution

The solution for the unknown filter vector **h** in Eq. 11.18 can be rewritten and augmented with the expression for minimum average squared error in Eq. 11.22 as follows.

$$
\begin{bmatrix}
\phi_0^d & g_0^* & \cdots & g_p^* \\
g_0 & \phi_0 & \cdots & \phi_p^* \\
\vdots & \vdots & & \vdots \\
g_0 & \phi_p & \cdots & \phi_0
\end{bmatrix}
\begin{bmatrix}
1 \\
h_0^p \\
\vdots \\
h_p^p
\end{bmatrix}
=
\begin{bmatrix}
E_p^d \\
0 \\
\vdots \\
0
\end{bmatrix}
\tag{11.59}
$$

Note the similarity of Eq. 11.59 and the corresponding expression in Eq. 11.30 for the forward prediction error filter.

Given the general filter of order p has been determined, it now is of interest to derive an order update recursion that will permit generating the general filter of order $p + 1$ from that of order p. Simply guess that the correct filter of order $p + 1$ is $[1 \ h_0^p \ldots h_p^p \ 0]^T$. Now, Eq. 11.59 becomes

$$
\begin{bmatrix}
\phi_0^d & g_0^* & \cdots & g_p^* & g_{p+1}^* \\
g_0 & \phi_0 & & \phi_p & \phi_{p+1} \\
\vdots & & & \vdots & \vdots \\
g_p & \phi_p & \cdots & \phi_0 & \phi_1^* \\
g_{p+1} & \phi_{p+1} & \cdots & \phi_1 & \phi_0
\end{bmatrix}
\cdot
\begin{bmatrix}
1 \\
h_0^p \\
\vdots \\
h_p^p \\
0
\end{bmatrix}
=
\begin{bmatrix}
E_p^d \\
0 \\
\vdots \\
0 \\
\Delta_{p+1}^d
\end{bmatrix}
\tag{11.60}
$$

where

$$
\Delta_{p+1}^d = g_{p+1} + \phi_{p+1} h_0^p + \cdots + \phi_1 h_p^p
\tag{11.61}
$$

If the discrepancy term Δ_{p+1}^d was 0, the proposed general filter of order $p + 1$ would be correct. Since the discrepancy term generally is not 0, the next step involves modifying the proposed general filter coefficients so as to cancel the discrepancy. The modification involves adding a weighted version of the backward prediction error filter vector of order $p + 1$ to the proposed general filter vector

$$
\begin{bmatrix} 1 \\ h_0^{p+1} \\ \vdots \\ h_p^{p+1} \\ h_{p+1}^{p+1} \end{bmatrix} = \begin{bmatrix} 1 \\ h_0^{p} \\ \vdots \\ h_p^{p} \\ 0 \end{bmatrix} + K_{p+1}^d \begin{bmatrix} 0 \\ b_0^{p+1} \\ \vdots \\ b_p^{p+1} \\ 1 \end{bmatrix}
\tag{11.62}
$$

yielding

$$
\begin{bmatrix} E_{p+1}^d \\ 0 \\ \vdots \\ 0 \\ 0 \end{bmatrix} = \begin{bmatrix} E_p^d \\ 0 \\ \vdots \\ 0 \\ \Delta_{p+1}^d \end{bmatrix} + K_{p+1}^d \begin{bmatrix} \nabla_{p+1}^d \\ 0 \\ \vdots \\ 0 \\ E_{p+1}^r \end{bmatrix}
\tag{11.63}
$$

where

$$
\nabla_{p+1}^d = g_0^* b_0^{p+1} + \cdots + g_p^* b_p^{p+1} + g_{p+1}^*
\tag{11.64}
$$

In order to cancel the discrepancy term

$$
K_{p+1}^d = -\frac{\Delta_{p+1}^d}{E_{p+1}^r}
\tag{11.65}
$$

Thus, the weighting factor consists of the discrepancy normalized by the backward prediction error from the prediction error filter of order $p + 1$. Now, the update equation for average squared error is

$$
E_{p+1}^d = E_p^d + K_{p+1}^d \nabla_{p+1}^d
\tag{11.66}
$$

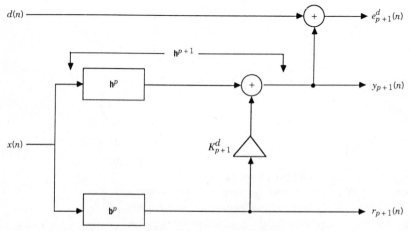

Figure 11.6 Order update procedure for the general filter.

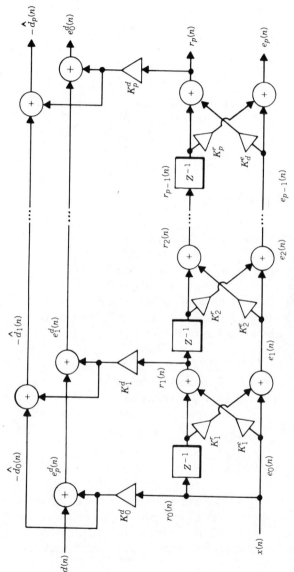

Figure 11.7 Lattice realization of the joint process structure.

Since Φ is Toeplitz

$$\nabla_{p+1}^d = \Delta_{p+1}^{d*} \tag{11.67}$$

and by substituting Eq. 11.65 into the update equation for average squared error,

$$E_{p+1}^d = E_p^d + K_{p+1}^d \Delta_{p+1}^{d*} \tag{11.68}$$

$$= E_p^d - \left| K_{p+1}^d \right|^2 E_{p+1}^r \tag{11.69}$$

The order update sequence is described graphically in Figure 11.6. The general filter of order p is updated using the weighting coefficient K_{p+1}^d to yield the corresponding general filter of order $p + 1$. The lattice structure shown in Figure 11.7 is the result of building the prediction error filters and the general filter of order p completely by this order update procedure.

11.3 FADING MEMORY (RECURSIVE) APPROACH

Instead of block processing time series data, the linear predictor and general filter structures discussed in Section 11.2 can be implemented as continuously updated, adaptive filters. By incorporating into their algorithms a fading memory of past data, these filters are capable of adapting to a changing environment. This is similar (but not equivalent) to splitting the observation interval into subintervals or blocks and processing each block independently.

Recall the direct-form implementation of the joint process structure as illustrated in Figure 11.1. The current and most recent p samples of the reference channel process $\{x(n), x(n-1), \ldots, x(n-p)\}$ are linearly combined to form an estimate of $-d(n)$. Of particular interest in adaptive filtering applications is the residual $e_p^d(n)$ obtained by adding the filtered reference channel to the primary channel. The corresponding transfer function of the reference channel filter is given by

$$H(z) = \sum_{k=0}^{p} h_k z^{-k} \tag{11.70}$$

Note that if $h_0 = 0$ and $d(n) = x(n)$, the filter in Figure 11.1 becomes the one-step forward linear predictor illustrated in Figure 11.2a. Removing the predictable components from $x(n)$ yields the forward prediction error sequence $e_p(n)$. The corresponding transfer function of the forward prediction error filter is given by

$$A(z) = \sum_{k=0}^{p} a_k z^{-k} \qquad a_0 = 1 \tag{11.71}$$

A companion to Figure 11.2a is the one-step backward linear predictor (coefficients $b_0, b_1, \ldots, b_{p-1}$) shown in Figure 11.2b along with the backward prediction error sequence $r_p(n)$.

The forward and backward prediction error filters can be realized equivalently in the form of a lattice structure. Shown in Figure 11.4, the lattice parameters K_i^e and K_i^r are known either as reflection coefficients or partial correlation coefficients. The $e_i(n)$ and $r_i(n)$ are the ith order forward and backward prediction error sequences, respectively. Thus, the pth-order prediction error filters are created on a stage-by-stage basis a single order at a time.

As shown in Figure 11.7, the lattice realization of the forward and backward prediction error filters can be embedded in the joint process structure. Additional parameters K_i^d weight the backward prediction error residuals $r_i(n)$. As with the lattice itself, the pth-order joint process structure is created on a stage-by-stage basis a single order at a time.

11.3.1 Complex Adaptive Joint Process Algorithms

Three continuously adapting approaches to the realization of the time-varying, complex, joint process structure are summarized in this section: (1) the joint complex gradient transversal filter (JCLMS), (2) the joint complex gradient lattice (JCGL), and (3) the joint complex least squares lattice (JCLSL). Each is based on a recursive-in-time solution for the filter whose transfer function is $H(z)$. Bounded by $[0, 1]$, a parameter α is found in all the algorithms that adjusts the time constants of their exponentially decaying memory of past data. Large α's imply short time constants.

JOINT COMPLEX GRADIENT TRANSVERSAL FILTER (JCLMS)

A key result from Wiener filtering theory is that the error sequence $e_p^d(n)$ must be orthogonal to the data $\{x(n), x(n-1), \ldots, x(n-p)\}$. The gradient transversal filter uses a coefficient update algorithm that attempts to implement this orthogonality condition directly

$$h_k(n) = h_k(n-1) - 2\mu e_p^d(n-1)x^*(n-k-1) \qquad (11.72)$$

where

$$2e_p^d(n-1)x^*(n-k-1) = \frac{\partial e_p^2(n-1)}{\partial h_k} \qquad (11.73)$$

is an instantaneous estimate of the gradient of the minimum mean-square error performance surface along the h_k coordinate. The algorithm follows a steepest descent path as it iteratively approaches satisfaction of the optimality criterion. The gradient step size parameter μ, common to all filter coefficients, governs the rate at which the $h_k(n)$ adapt. Given that the time series $x(n)$ is stationary, convergence in the mean of the $h_k(n)$ to their optimal values is guaranteed under relatively mild assumptions when

$$0 < \mu < \frac{1}{\lambda_{\max}} \qquad (11.74)$$

where λ_{max} is the maximum eigenvalue of the $(p + 1) \times (p + 1)$ data auto-correlation matrix Φ in Eq. 11.10. The variances of the $h_k(n)$ about their means converge to values that are proportional to μ.

It is convenient to introduce a new parameter α_{JCLMS} such that

$$\mu = \frac{\alpha_{JCLMS}}{(p + 1)E_0^e} \tag{11.75}$$

where E_0^e is the power of the sequence $x(n)$. Thus, Eq. 11.74 becomes

$$0 < \alpha_{JCLMS} < \frac{(p + 1)E_0^e}{\lambda_{max}} \tag{11.76}$$

Since

$$(p + 1)E_0^e = \sum_{i=0}^{p} \phi_0 = \sum_{i=0}^{p} \lambda_i \geq \lambda_{max} \tag{11.77}$$

Eq. 11.76 will be satisfied when

$$0 < \alpha_{JCLMS} < 1 \tag{11.78}$$

More extensive discussions related to the derivation of Eq. 11.72 and its convergence properties can be found in references 1 through 5.

From a practical standpoint, E_0^e is replaced by the estimate $E_0^e(n)$, which is generated recursively by

$$E_0^e(n) = (1 - \alpha_{JCLMS})E_0^e(n - 1) + \alpha_{JCLMS}|x(n)|^2 \tag{11.79}$$

Thus, Eq. 11.75 becomes

$$\mu(n) = \frac{\alpha_{JCLMS}}{(p + 1)E_0^e(n)} \tag{11.80}$$

and the gradient transversal filter is provided the capability of adapting to power-level fluctuations in the time series $x(n)$. The time constant of the recursive power estimator in Eq. 11.79 need not necessarily be tied to α_{JCLMS}.

The major drawback in the use of gradient transversal filters is that their convergence characteristics are determined by the eigenvalue spread of the time series $x(n)$. As shown in Eq. 11.76, λ_{max} determines the largest permissible value of the gradient step-size parameter. The system formed by the $h_k(n)$ can be broken into its natural modes by a decomposition based on the eigenvectors of the $(p + 1) \times (p + 1)$ data autocorrelation matrix. Each mode converges with a time constant inversely proportional to

$$\mu\lambda_i = \frac{\alpha_{JCLMS}\lambda_i}{(p + 1)E_0^e} \tag{11.81}$$

Thus, the small eigenvalue modes govern the total time to convergence. When

$x(n)$ consists of sinusoids and a rough correspondence can be made between the powers of individual sinusoids and eigenvalues of the $(p + 1) \times (p + 1)$ data autocorrelation matrix Φ,[3,19] it is seen that the gradient transversal filter will adapt to each sinusoidal component at a rate inversely proportional to its power.

In summary, the joint complex gradient transversal filter algorithm is expressed as follows (see Fig. 11.1).

Initialization $(k = 0, 1, \ldots, p)$

$$h_k(-1) = 0 \tag{11.82a}$$

$$E_0^e(-1) = \epsilon_{\text{JCLMS}}, \qquad \epsilon_{\text{JCLMS}} = E_0^e(0) \tag{11.82b}$$

$$x(-k) = x(-p - 1) = 0 \tag{11.82c}$$

$$e_p^d(-1) = 0 \tag{11.82d}$$

Time update $(n \geq 0$ and $k = 0, 1, \ldots, p)$

$$h_k(n) = h_k(n - 1) - \frac{2\alpha_{\text{JCLMS}}}{(p + 1)E_0^e(n - 1)}e_p^d(n - 1)x^*(n - k - 1) \tag{11.82e}$$

$$e_p^d(n) = \sum_{k=0}^{p} h_k(n)x(n - k) + d(n) \tag{11.82f}$$

$$E_0^e(n) = (1 - \alpha_{\text{JCLMS}})E_0^e(n - 1) + \alpha_{\text{JCLMS}}|x(n)|^2 \tag{11.82g}$$

Slightly simplifying Eqs. 11.82a–g, the one-step forward linear predictor complex gradient transversal filter algorithm can be written as (see Fig. 11.2a)

Initialization $(k = 1, 2, \ldots, p)$

$$a_k(-1) = 0 \tag{11.83a}$$

$$E_0^e(-1) = \epsilon_{\text{JCLMS}}, \qquad \epsilon_{\text{JCLMS}} = E_0^e(0) \tag{11.83b}$$

$$x(-k) = x(-p - 1) = 0 \tag{11.83c}$$

$$e_p(-1) = 0 \tag{11.83d}$$

Time update $(n \geq 0$ and $k = 1, 2, \ldots, p)$

$$a_k(n) = a_k(n - 1) - \frac{2\alpha_{\text{JCLMS}}}{pE_0^e(n - 1)}e_p(n - 1)x^*(n - k - 1) \tag{11.83e}$$

$$e_p(n) = \sum_{k=0}^{p} h_k(n)x(n - k), \qquad a_0 = 1 \tag{11.83f}$$

$$E_0^e(n) = (1 - \alpha_{\text{JCLMS}})E_0^e(n - 1) + \alpha_{\text{JCLMS}}|x(n)|^2 \tag{11.83g}$$

Correspondingly, the one-step backward linear predictor complex gradient transversal filter algorithm then takes on the form (see Fig. 11.2b)

Initialization ($k = 0, 1, \ldots, p - 1$)

$$b_k(-1) = 0 \tag{11.84a}$$

$$E_0^r(-1) = \epsilon_{\text{JCLMS}}, \qquad \epsilon_{\text{JCLMS}} = E_0^r(0) \tag{11.84b}$$

$$x(-k) = x(-p - 1) = 0 \tag{11.84c}$$

$$r_p(-1) = 0 \tag{11.84d}$$

Time update ($n \geq 0$ and $k = 0, 1, \ldots, p - 1$)

$$b_k(n) = b_k(n - 1) - \frac{2\alpha_{\text{JCLMS}}}{pE_0^r(n - 1)} r_p(n - 1)x^*(n - k - 1) \tag{11.84e}$$

$$r_p(n) = \sum_{k=0}^{p} b_k(n)x(n - k), \qquad b_p = 1 \tag{11.84f}$$

$$E_0^r(n) = (1 - \alpha_{\text{JCLMS}})E_0^r(n - 1) + \alpha_{\text{JCLMS}}|x(n)|^2 \tag{11.84g}$$

JOINT COMPLEX GRADIENT LATTICE (JCGL)

The all-zero forward prediction error filter $A(z)$ in Eq. 11.71 can be realized in direct form (Fig. 11.2a) or as a lattice structure (Fig. 11.4). When the time series $x(n)$ is stationary (Wiener filtering problem) or has been windowed (least-squares problem),

$$K_i = K_i^r = K_i^{e*} \tag{11.85}$$

The $e_i(n)$ and $r_i(n)$ are the ith-order forward and backward prediction error sequences, respectively. The sequence pair $e_i(n)$ and $r_{i-1}(n - 1)$ as well as the sequence pair $r_i(n)$ and $e_{i-1}(n)$ are orthogonal. Thus assuming stationarity or windowing, local satisfaction of the orthogonality principle on a stage-by-stage basis leads to global optimality of the entire structure as a pth-order prediction error filter. In the time-varying situation actually of interest, the lattice structure in Fig. 11.4 with $K_i = K_i^r = K_i^{e*}$ still will be assumed appropriate. The lattice coefficients are updated by a gradient steepest descent algorithm that attempts to satisfy locally the orthogonality condition between the two sequence pairs previously mentioned.

In the joint process case, the prediction error filters are realized in adaptive lattice form as already described. The backward prediction error residuals $r_i(n)$ are further weighted by coefficients K_i^d as shown in Fig. 11.7. These coefficients also are updated by a gradient steepest descent algorithm that attempts to satisfy locally the orthogonality between $r_i(n)$ and $e_i^d(n)$. Additional discussion related to the gradient lattice structure and its convergence properties can be found in references 4 and 5.

In summary, the joint complex gradient lattice algorithm is expressed as follows (see Fig. 11.7 with $K_i = K_i^r = K_i^{e*}$).

Initialization ($i = 0, 1, \ldots, p$)

$$K_i(-1) = 0, \qquad i \neq 0 \tag{11.86a}$$

$$E_i^r(-1) = E_i^e(-1) = \epsilon_{\text{JCGL}}, \qquad \epsilon_{\text{JCGL}} = E_0^e(0) \tag{11.86b}$$

$$E_i^e(-2) = \epsilon_{\text{JCGL}}, \qquad i \neq p \tag{11.86c}$$

$$e_i(-1) = r_i(-1) = 0 \tag{11.86d}$$

$$r_i(-2) = 0, \qquad i \neq p \tag{11.86e}$$

$$K_i^d(-1) = 0 \tag{11.86f}$$

$$e_i^d(-1) = 0 \tag{11.86g}$$

Time update ($n \geq 0$)

$$e_0(n) = r_0(n) = x(n) \tag{11.86h}$$

$$e_{-1}^d(n) = d(n) \tag{11.86i}$$

Order update ($i = 0, 1, \ldots, p$)

$$K_i(n) = K_i(n-1) - \frac{2\alpha_{\text{JCGL}}}{E_{i-1}^r(n-2) + E_{i-1}^e(n-1)}$$
$$\cdot |e_i(n-1)r_{i-1}^*(n-2) + r_i^*(n-1)e_{i-1}(n-1)|, \qquad i \neq 0 \tag{11.86j}$$

$$e_i(n) = e_{i-1}(n) + K_i(n)r_{i-1}(n-1), \qquad i \neq 0 \tag{11.86k}$$

$$r_i(n) = r_{i-1}(n-1) + K_i^*(n)e_{i-1}(n), \qquad i \neq 0 \tag{11.86l}$$

$$E_{i-1}^r(n-1) = (1 - \alpha_{\text{JCGL}})E_{i-1}^r(n-2) + \alpha_{\text{JCGL}}|r_{i-1}(n-1)|^2, \quad i \neq 0 \tag{11.86m}$$

$$E_{i-1}^e(n) = (1 - \alpha_{\text{JCGL}})E_{i-1}^e(n-1) + \alpha_{\text{JCGL}}|e_{i-1}(n)|^2, \quad i \neq 0 \tag{11.86n}$$

$$K_i^d(n) = K_i^d(n-1) - \frac{2\alpha_{\text{JCGL}}}{E_i^r(n-1)}e_i^d(n-1)r_i^*(n-1) \tag{11.86o}$$

$$e_i^d(n) = e_{i-1}^d(n) + K_i^d(n)r_i(n) \tag{11.86p}$$

Predictors ($b_k^{(i)}(n) = a_{p-k}^{(i)*}(n), k = 0, 1, \ldots, p$)

$$a_i^{(i)}(n) = K_i(n) \tag{11.86q}$$

$$a_k^{(i)}(n) = a_k^{(i-1)}(n) + K_i(n)a_{i-k}^{(i-1)*}(n), \qquad 1 \leq k \leq i - 1 \tag{11.86r}$$

Notice that $E_{i-1}^e(n)$ and $E_{i-1}^r(n-1)$ are simply estimates of the input power to the ith stage of the lattice. Assuming these estimates are equal to the actual power, the coefficients $K_i(n)$ and $K_i^d(n)$ will converge in the mean approximately to their optimal values provided

$$0 < \alpha_{\text{JCGL}} < 1 \tag{11.87}$$

The variances of $K_i(n)$ and $K_i^d(n)$ about their means converge to values that are proportional to α_{JCGL}.

The natural stage-by-stage decoupling that occurs in the lattice under stationary conditions suggests that the adaptive realization in Eq. 11.86 would have a high insensitivity to eigenvalue spread. This can be understood intuitively as follows. Notice that the adaptive computation of the lattice coefficients at any particular stage does not depend on the computation of $K_i(n)$ and $K_i^d(n)$ at succeeding stages. Unlike the gradient transversal filter where the pth-order error residual $e_p^d(n)$ is coupled back into the update expression (Eq. 11.82e) for each coefficient $h_k(n)$, in the JCGL the ith-stage error residuals required in Eq. 11.86j and Eq. 11.86o are not functions of the $K_i(n)$ and $K_i^d(n)$ in succeeding stages as can be seen from Fig. 11.7. Essentially, convergence of the lattice occurs progressively on a stage-by-stage basis. The time constants of each stage are approximately the same, which also is in contrast to the individual mode behavior in the JCLMS algorithm. Unlike Eq. 11.82e or Eq. 11.72 where the gradient step-size parameter is constant for each $h_k(n)$, Eq. 11.86j and Eq. 11.86o indicate that the gradient step-size parameter is not the same at each stage of the adaptive lattice, but is inversely proportional to the estimate of power entering that stage. And, as discussed previously, the convergence rate of the steepest descent algorithm is determined by the product of a power quantity and the gradient step size parameter.

JOINT COMPLEX LEAST SQUARES LATTICE (JCLSL)

The adaptive lattice given in Eq. 11.86 presupposes a certain structure for the prediction-error filter ($K_i^e = K_i^{r*}$) and uses a gradient-descent approach in an attempt to satisfy locally a minimum power optimality criterion by a direct implementation of the orthogonality principle. Although intuitively appealing, it is not clear that such an implementation will be globally optimal in a least squares sense when operating in a nonstationary environment. A global approach allows the structure to evolve out of the mathematics of the temporal and order recursive solution to the globally optimal pth-order general filter. The lattice forms in Figures 11.4 and 11.7 still appear in a natural fashion, but, in general, $K_i^e \neq K_i^{r*}$.

At every point in time n, the JCLSL solves the pth-order general filter by minimizing the following modification of Eq. 11.5.

$$\underset{\mathbf{h}}{\text{minimize}} \sum_{m=0}^{n} (1 - \alpha_{\text{JCLSL}})^{n-m} \left| e_p^d(m) \right|^2 \tag{11.88}$$

The fade factor, $(1 - \alpha_{\text{JCLSL}})$, enables the filter to adapt to a nonstationary environment by weighting recent errors more heavily than those that occurred in the distant past. The expression in Eq. 11.88 defines the prewindowed case where $x(n)$ is assumed identically 0 for $n < 0$ and no assumptions are made about the values of future data. More extensive discussion related to the derivation of the least squares lattice structure and its convergence properties can be found in references 4 through 8.

In summary, the least squares lattice algorithm is expressed as follows (see Fig. 11.7)

Initialization $(i = 0, 1, \ldots, p)$

$$r_i(-1) = 0, \qquad i \neq p \tag{11.89a}$$

$$E_i^r(-1) = \epsilon_{\text{JCLSL}}, \qquad \epsilon_{\text{JCLSL}} = 0.001 \quad \text{and} \quad i \neq p \tag{11.89b}$$

$$\Delta_i(-1) = 0, \qquad i \neq 0 \tag{11.89c}$$

$$b_k^{(i)}(-1) = 0, \qquad 0 \leq k \leq i - 1, \qquad i \neq 0, \qquad \text{and } i \neq p \tag{11.89d}$$

$$\gamma_{i-1}(-1) = 0, \qquad i \neq p \tag{11.89e}$$

$$K_i^d(i - 1) = 0 \tag{11.89f}$$

Time update $(n \geq 0)$

$$e_0(n) = r_0(n) = x(n) \tag{11.89g}$$

$$E_0^e(n) = E_0^r(n) = (1 - \alpha_{\text{JCLSL}}) E_0^r(n - 1) + |x(n)|^2 \tag{11.89h}$$

$$\gamma_{-1}(n) = 0 \tag{11.89i}$$

$$e_{-1}^d(n) = d(n) \tag{11.89j}$$

Order update $(i = 0, 1, \ldots, p)$

Lattice

$$\Delta_i(n) = (1 - \alpha_{\text{JCLSL}}) \Delta_i(n - 1) - \frac{e_{i-1}(n) r_{i-1}^*(n - 1)}{1 - \gamma_{i-2}(n - 1)}, \qquad i \neq 0 \tag{11.89k}$$

$$K_i^e(n) = \Delta_i^*(n) / E_{i-1}^e(n), \qquad i \neq 0 \tag{11.89l}$$

$$K_i^r(n) = \Delta_i(n) / E_{i-1}^r(n - 1), \qquad i \neq 0 \tag{11.89m}$$

$$e_i(n) = e_{i-1}(n) + K_i^r(n) r_{i-1}(n - 1), \qquad i \neq 0 \tag{11.89n}$$

$$r_i(n) = r_{i-1}(n - 1) + K_i^e(n) e_{i-1}(n), \qquad i \neq 0 \tag{11.89o}$$

$$E_i^e(n) = E_{i-1}^e(n) - |\Delta_i(n)|^2 / E_{i-1}^r(n - 1), \qquad i \neq 0 \tag{11.89p}$$

$$E_i^r(n) = E_{i-1}^r(n - 1) - |\Delta_i(n)|^2 / E_{i-1}^e(n), \qquad i \neq 0 \tag{11.89q}$$

$$\gamma_{i-1}(n) = \gamma_{i-2}(n) + |r_{i-1}(n)|^2 / E_{i-1}^r(n), \qquad i \neq 0 \qquad (11.89r)$$

$$\Delta_i^d(n) = (1 - \alpha_{\mathrm{JCLSL}}) \Delta_i^e(n-1) - \frac{e_{i-1}^d(n) r_i^*(n)}{1 - \gamma_{i-1}(n)} \qquad (11.89s)$$

$$K_i^d(n) = \frac{\Delta_i^d(n)}{E_i^r(n)} \qquad (11.89t)$$

$$e_i^d(n) = e_{i-1}^d(n) + K_i^d(n) r_i(n) \qquad (11.89u)$$

Predictors

$$a_i^{(i)}(n) = K_i^r(n) \qquad (11.89v)$$

$$b_0^{(i)}(n) = K_i^e(n) \qquad (11.89w)$$

$$a_k^{(i)}(n) = a_k^{(i-1)}(n) + K_i^r(n) b_{k-1}^{(i-1)}(n-1), \qquad 1 \leq k \leq i-1 \quad (11.89x)$$

$$b_k^{(i)}(n) = b_{k-1}^{(i-1)}(n-1) + K_i^e(n) a_k^{(i-1)}(n), \qquad 1 \leq k \leq i-1 \quad (11.89y)$$

Although the JCLSL algorithm is fairly complex, it is interesting to note that the major difference between it and the JCGL [after substituting Eqs. 11.86k and l into Eq. 11.86j) is the presence of the gain factor $[1 - \gamma_{i-2}(n-1)]^{-1}$ in Eq. 11.89k and a similar gain factor in Eq. 11.89s]. The data-dependent variable $\gamma_{i-2}(n-1)$ takes on the values $[0, 1]$, where the lower bound is reached for $\alpha_{\mathrm{JCLSL}} = 0$ and the upper bound is reached for $\alpha_{\mathrm{JCLSL}} = 1$. Under the assumption of 0 mean Gaussian statistics, it has been interpreted as a measure of the likelihood that the $(i - 1)$ samples, $\{x(n-1), \ldots, x(n-i+1)\}$, deviate from the $(i-1)$st-order multivariate Gaussian distribution parameterized by the sample covariance matrix whose elements are similar to those in Eq. 11.13 with the summation entrants weighted by $(1 - \alpha_{\mathrm{JCLSL}})^{n-m-1}$.[6] Thus, $\gamma_{i-2}(n-1)$ enables the JCLSL to quickly modify the $K_i^e(n)$ and $K_i^r(n)$ [and similarly $K_i^d(n)$] when data of a significantly different character than that of the recent past are encountered.

11.3.2 Simulation Examples

The convergence characteristics of the three complex adaptive joint process algorithms just summarized will now be compared. Explored is a problem where the reference channel time series consists of dual constant frequency sinusoids with power levels that are widely separated ($\mathrm{SNR}_1 = 20$ dB, $\mathrm{SNR}_2 = 10$ dB) and that undergo an instantaneous step downward in frequency. The primary channel time series consists of dual constant frequency sinusoids of equal power ($\mathrm{SNR}_1 = \mathrm{SNR}_2 = 20$ dB) with frequencies that coincide with those of the reference channel after the step. This problem is a complex joint process extension of the dual step simulations considered in reference 4 ($\omega_1 = 7\pi/8$, $\omega_2 = \pi/4$, $\Delta\omega_1 = \Delta\omega_2 = -\pi/8$).

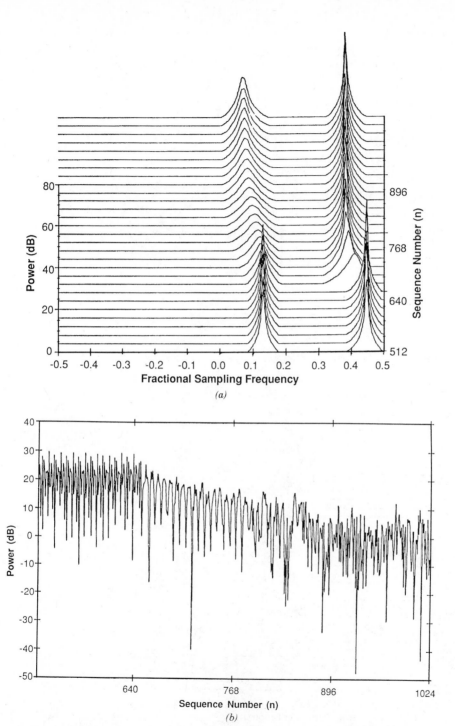

Figure 11.8 Joint complex gradient transversal filter (JCLMS): (*a*) Time-evolving high-resolution spectral estimate of the reference channel time series $x(n)$; (*b*) Residual time series $e_p^d(n)$; (*c*) Time-evolving conventional spectral estimate of the residual time series $e_p^d(n)$.

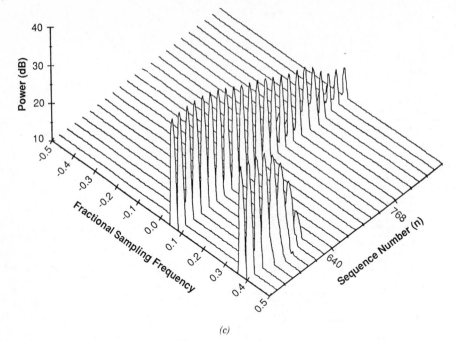

(c)

Figure 11.8 Continued

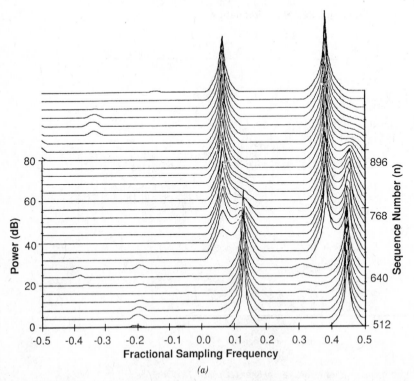

(a)

Figure 11.9 Joint complex gradient lattice (JCGL): (*a*) Time-evolving high-resolution spectral estimate of the reference channel time series $x(n)$; (*b*) Residual time series $e_p^d(n)$; (*c*) Time-evolving conventional spectral estimate of the residual time series $e_p^d(n)$.

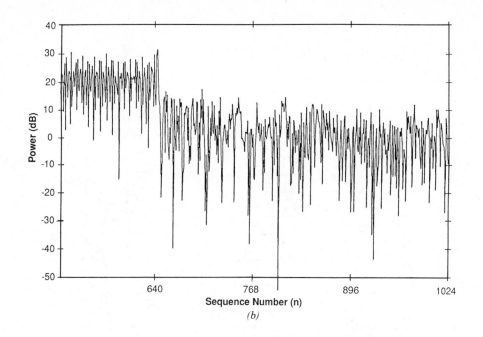

(b)

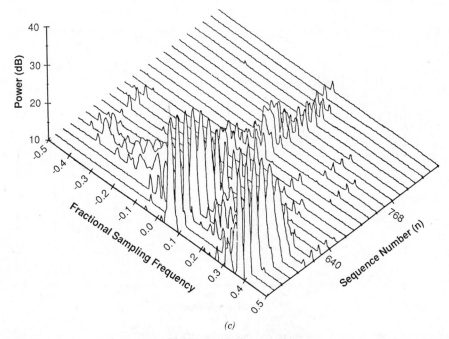

(c)

Figure 11.9 Continued

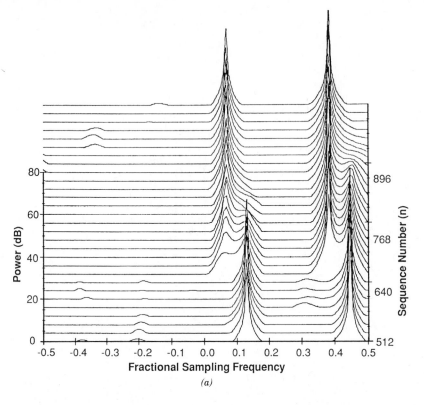

(a)

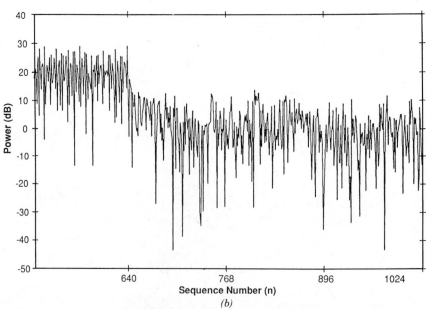

(b)

Figure 11.10 Joint complex least squares lattice (JCLSL): (*a*) Time-evolving high-resolution spectral estimate of the reference channel time series $x(n)$; (*b*) Residual time series $e_p^d(n)$; (*c*) Time-evolving conventional spectral estimate of the residual time series $e_p^d(n)$.

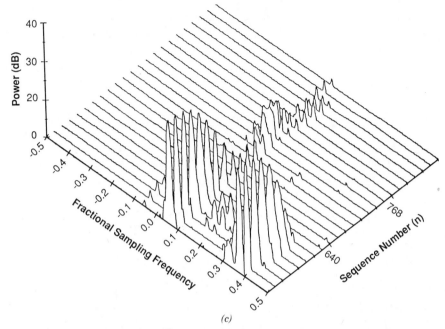

Figure 11.10 Continued

The adaptive filters under investigation were of order $p = 6$ with $\alpha_{JCLMS} = 0.04$, $\alpha_{JCGL} = 0.01$, and $\alpha_{JCLSL} = 0.02$. The adaptation rate parameters, α, were chosen such that the response to the higher power sinusoid by all three algorithms was similar. Their convergence characteristics are illustrated graphically in Figures 11.8 through 11.10 via three types of plots. First, a waterfall of high-resolution spectral estimates of the reference channel time series is obtained from the inverse of the forward prediction error filter transfer function $A(z)$ in Eq. 11.71. This particular spectral estimation technique is appropriate if the time series is correctly modeled as white noise $w(n)$ passed through an all-pole filter (see Section 11.2.3). Each slice of the waterfall is indexed by a sequence number corresponding to the instant when the snapshot of the forward prediction error filter's weights was taken. Second, the residual time series $e_p^d(n)$ is of interest as a gross characterization of the dynamics of the adaptation process. Third, a more detailed view of filter convergence is provided via a pseudo-three-dimensional map of the time-evolving spectrum of the residual time series. In this case, the map is derived from overlapped conventional FFTs [NFFT = 128, Kaiser–Bessel window ($\alpha = 2.5$), $\Delta = 16$)]. Each slice of the map is indexed by the sequence number corresponding to the first data point loaded into the FFT. The adaptive filters are initialized at $n = 0$. The reference channel step occurs at $n = 640$.

Of note in Figs. 11.8c, 11.9c, and 11.10c is the relative adaptation rates of the two sinusoids after the step. For the JCGL and JCLSL algorithms, the adaptation rates essentially are equivalent. The JCLMS algorithm, however, yielded a substantially larger adaptation time for the weaker reference channel sinusoid. From the discussion in Section 11.3.1, the significant difference in adaptation rate for the JCLMS algorithm is seen to be a manifestation of the approximate factor of 10 spread in eigenvalues corresponding to the two sinusoids.

11.4 ADAPTIVE BEAMFORMING (ABF)

The scalar adaptive algorithms discussed in Section 11.3 can be extended to operate on vector time series. The processing of multichannel data arises naturally when manipulating data from an array of spatially distributed sensors. The problem of coherently summing the outputs from such a collection of sensors is known as **beamforming**. A beamformer permits one to listen preferentially to wave fronts propagating from one direction over another. With the addition of a filter on the output of each sensor prior to the summation as shown in Figure 11.11, the conventional beamformer provides both spatial and spectral filtering on the incoming wave field. Additional background material on conventional beamforming (CBF) is included in the Appendix. Here, our focus will be on adaptive beamforming (ABF).

11.4.1 Constrained Adaptation

The early development of adaptive beamforming was motivated by a need to reduce radar system response to strong, spatially localized jamming signals arriving in the sidelobe region of the array. Since this noise environment was constantly changing in geometry, no one array shading always was adequate to prevent masking of weak signals in the desired look direction. An adaptive scheme was sought that would enable the system to continuously maintain specified beam characteristics in a particular region of space while simultaneously minimizing its response to the highly nonisotropic and dynamic noise background. The literature is rich with discussions of this and the related underwater sonar problem.[22-45]

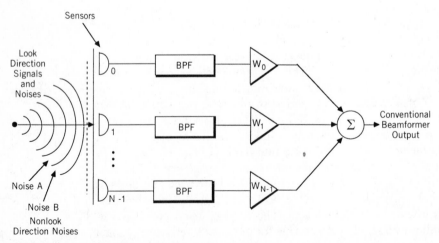

Figure 11.11 Conventional beamformer.

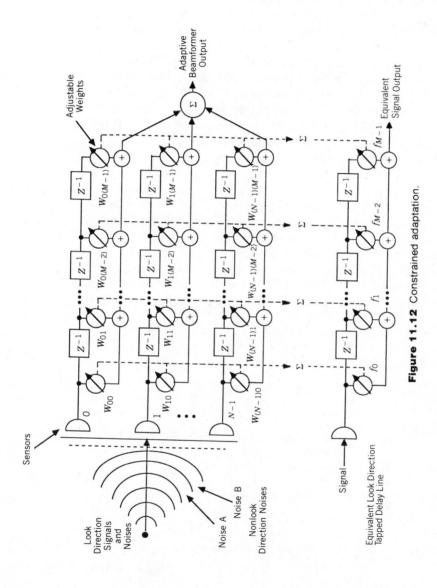

Figure 11.12 Constrained adaptation.

The approach taken in this section will be to view the adaptive beamforming problem as a problem of constrained minimization.[24] The solution to the constrained minimization problem is able to maintain a chosen frequency response in the desired look direction while simultaneously minimizing output noise power from all other directions. With a broadside look direction, the array processor illustrated in Figure 11.12 appears to the desired signal as a single-tapped delay line in which each weight $\{f_0, f_1, \ldots, f_{M-1}\}$ is equal to the corresponding vertical column sum in the figure. The values of these equivalent weights are constrained so as to give the desired frequency response to signals arriving broadside to the array. For look directions other than perpendicular to the line of sensors, appropriate time delays can be added just prior to the tapped delay lines. An adaptive array processor with N sensors and M taps per delay line requires M constraints (the column sums of Fig. 11.12) to constrain its look direction frequency response. The remaining $NM - M$ degrees of freedom are used in adjusting the weights to minimize the total power in the array output subject to the look direction constraint. Assuming the desired signal arrives from the constrained look direction and is matched to the specified frequency response, this minimization is equivalent to minimizing the system's response to nonlook direction noise power. Thus, the array shading adapts to the local noise environment.

The array processor output in Figure 11.12 can be expressed in the form of Eq. 11.3 as the error $e^d(m)$ between the desired response to nonlook direction noise $[d(m) = 0]$ and the actual output of this multichannel FIR filter

$$e^d(m) = d(m) + \mathbf{w}^T \mathbf{x} \qquad (11.90)$$

where

$$\mathbf{w} = \begin{bmatrix} \mathbf{w}_0 \\ \vdots \\ \mathbf{w}_{M-1} \end{bmatrix} \qquad (11.91)$$

$$\mathbf{w}_m = \begin{bmatrix} w_{0m} \\ \vdots \\ w_{(N-1)m} \end{bmatrix} \qquad (11.92)$$

and

$$\mathbf{x} = \begin{bmatrix} \mathbf{x}_0 \\ \vdots \\ \mathbf{x}_{M-1} \end{bmatrix} \qquad (11.93)$$

$$\mathbf{x}_m = \begin{bmatrix} x_{0m} \\ \vdots \\ x_{(N-1)m} \end{bmatrix} \qquad (11.94)$$

Thus, x_m represents a snapshot in time across the mth column of sensors weighted by w_m, and x is the stack of M such snapshots weighted by w. In this section, m will be used both as the general time index (when appearing in parentheses) and as relative position along the sensor tapped delay lines in Figure 11.12 (when appearing as a subscript).

Since $d(m) = 0$, the expression for average squared error from Eq. 11.15 simplifies to

$$E^d(\mathbf{w}) = \mathbf{w}^H \mathbf{\Phi}^T \mathbf{w} \tag{11.95}$$

Unconstrained, the solution for minimum average squared error from Eq. 11.17 is

$$\mathbf{\Phi}^T \mathbf{w} = 0 \tag{11.96}$$

or

$$\mathbf{w} = 0 \tag{11.97}$$

The constrained minimization problem produces a more useful result than Eq. 11.97 does.

$$\text{Minimize:} \quad \mathbf{w}^H \mathbf{\Phi}^T \mathbf{w} \tag{11.98a}$$
$$\underset{\mathbf{w}}{}$$

$$\text{Subject to:} \quad \mathbf{C}^H \mathbf{w} = \mathbf{f} \tag{11.98b}$$

The constraint equation Eq. 11.98b expresses the requirement that a specific response be maintained in the desired look direction. For the case depicted in Figure 11.12, the constraint consists of requiring the weights in the mth vertical column of taps to sum to f_m

$$\mathbf{c}_m^H \mathbf{w} = f_m \quad m = 0, \ldots, M - 1 \tag{11.99}$$

where the NM-dimensional vector $\mathbf{c}_m$ has the form

$$\mathbf{c}_m = \begin{bmatrix} 0 \\ \vdots \\ 0 \\ \vdots \\ 1 \\ \vdots \\ 1 \\ \vdots \\ 0 \\ \vdots \\ 0 \end{bmatrix} \tag{11.100}$$

and

$$C = [\mathbf{c}_0, \dots, \mathbf{c}_{M-1}] \tag{11.101}$$

The constraint vectors $\mathbf{c}_m$ are linearly independent. The M-dimensional vector $\mathbf{f}$ specifies the frequency response of the look direction equivalent FIR filter

$$\mathbf{f} = \begin{bmatrix} f_0 \\ \vdots \\ f_{M-1} \end{bmatrix} \tag{11.102}$$

Constraining the weight vector to satisfy the M equations of Eq. 11.99 restricts $\mathbf{w}$ to a $(NM - M)$-dimensional plane (i.e., $NM - M$ degrees of freedom).

The set of filter weights solving Eq. 11.98 is found by the method of Lagrange multipliers.[24] Adjoining the constraint equation (11.98b) to the quadratic form being minimized (11.98a) by the vector $\boldsymbol{\lambda}$ of undetermined Lagrange multipliers yields

$$\text{Minimize:} \quad \mathbf{w}^H \boldsymbol{\Phi}^T \mathbf{w} + [\mathbf{C}^H \mathbf{w} - \mathbf{f}]^H \boldsymbol{\lambda} \tag{11.103}$$
$$\mathbf{w}$$

Minimizing Eq. 11.103 with respect to $\mathbf{w}$ leads to the following equation

$$\boldsymbol{\Phi}^T \mathbf{w} + \mathbf{C} \boldsymbol{\lambda} = \mathbf{0} \tag{11.104}$$

or

$$\boldsymbol{\Phi}^T \mathbf{w} = -\mathbf{C} \boldsymbol{\lambda} \tag{11.105}$$

Now, solving Eq. 11.105 for $\mathbf{w}$ yields

$$\mathbf{w} = -(\boldsymbol{\Phi}^T)^{-1} \mathbf{C} \boldsymbol{\lambda} \tag{11.106}$$

The vector of Lagrange multipliers is evaluated from the constraint equation by substituting Eq. 11.106 into Eq. 11.98b

$$\mathbf{C}^H \left[-(\boldsymbol{\Phi}^T)^{-1} \mathbf{C} \boldsymbol{\lambda} \right] = \mathbf{f} \tag{11.107}$$

or

$$\boldsymbol{\lambda} = -\left[\mathbf{C}^H (\boldsymbol{\Phi}^T)^{-1} \mathbf{C} \right]^{-1} \mathbf{f} \tag{11.108}$$

Substituting Eq. 11.108 back into Eq. 11.106 completes the solution to the constrained minimization problem

$$\mathbf{w} = (\boldsymbol{\Phi}^T)^{-1} \mathbf{C} \left[\mathbf{C}^H (\boldsymbol{\Phi}^T)^{-1} \mathbf{C} \right]^{-1} \mathbf{f} \tag{11.109}$$

With this value of **w**, the average squared error in Eq. 11.95 becomes

$$E^d = \mathbf{f}^H\left[\mathbf{C}^H(\mathbf{\Phi}^T)^{-1}\mathbf{C}\right]^{-1}\mathbf{f} \tag{11.110}$$

A constrained gradient descent approach similar to the algorithm in Eq. 11.72 can be used to iteratively approach Eq. 11.109 and thus yield a continuously updated adaptive beamformer.[24] In place of Eq. 11.73, the following instantaneous estimate of the gradient of the performance surface in Eq. 11.103 is used

$$2e^d(m-1)\mathbf{x}^*(m-1) + 2\mathbf{C}\lambda(m-1) \tag{11.111}$$

The Lagrange multipliers $\lambda(m-1)$ are evaluated by requiring the constraint equation (11.98b) to be satisfied at every iteration of the algorithm

$$\mathbf{C}^H\mathbf{w}(m) = \mathbf{f} \tag{11.112}$$

After simplification, the constrained gradient descent algorithm can be written as

$$\mathbf{w}(0) = \mathbf{F} \tag{11.113a}$$

$$\mathbf{w}(m) = \mathbf{P}\left[\mathbf{w}(m-1) - 2\mu e^d(m-1)\mathbf{x}^*(m-1)\right] + \mathbf{F} \tag{11.113b}$$

where

$$\mathbf{F} = \mathbf{C}[\mathbf{C}^H\mathbf{C}]^{-1}\mathbf{f} \tag{11.114}$$

$$\mathbf{P} = \mathbf{I} - \mathbf{C}[\mathbf{C}^H\mathbf{C}]^{-1}\mathbf{C}^H \tag{11.115}$$

and μ is the gradient step size parameter.

A geometric interpretation of Eq. 11.113 is useful.[24] Figure 11.13 provides a representation of the performance surface in Eq. 11.98 where the contours represent constant average squared error $\mathbf{w}^H\mathbf{\Phi}^T\mathbf{w}$. Also indicated is

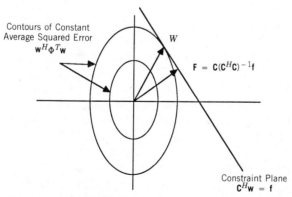

Figure 11.13 Geometric interpretation of the constrained minimization problem.

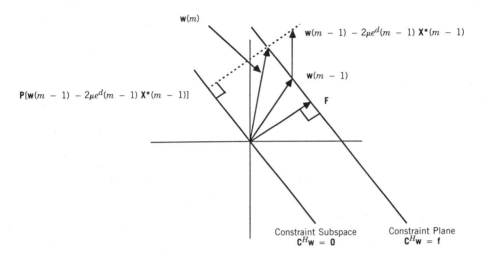

Figure 11.14 Operation of the constrained gradient descent algorithm.

the constraint plane $\mathbf{C}^H\mathbf{w} = \mathbf{f}$ on which the solution to the constrained minimization problem must lie. The operation of the constrained gradient descent algorithm in Eq. 11.113 is illustrated in Figure 11.14. The unconstrained negative instantaneous gradient estimate $-2e^d(m-1)\mathbf{x}^*(m-1)$ is scaled by μ and added to the weight vector $\mathbf{w}(m-1)$. This is a noisy attempt to change the weight vector in a direction that minimizes average squared error (11.98a) (see also Eqs. 11.72 and 11.73). In general, this change moves the resulting vector off the constraint plane (11.98b). The unconstrained solution first is projected onto the constraint subspace ($\mathbf{C}^H\mathbf{w} = \mathbf{0}$) by the projection operator $\mathbf{P}$. Thus, any components perpendicular to this ($NM - M$)-dimensional plane are removed. The resulting vector then is returned to the constraint plane by adding $\mathbf{F}$. The new weight vector $\mathbf{w}(m)$ satisfies the constraint to within the accuracy of the arithmetic used in implementing the algorithm.

The array processor depicted in Figure 11.12 implements a single spatial point constraint that is maintained by the projection operator $\mathbf{P}$ and vector $\mathbf{F}$. Frequently, constraining just the look direction does not exert sufficient influence on the spatial response over a wide enough angular region. Through the use of multiple constraints, it is possible to widen this region either by constraining the spatial transfer function at additional points near boresight or by controlling the first few derivatives of the spatial transfer function in the constrained look direction. An additional set of M constraints is added for each specified point or derivative of the spatial transfer function. The effects of multiple-point and derivative constraints are illustrated nicely in references 27, 30, and 34.

From a practical standpoint, multiple constraints often may be necessary to avoid degenerate system performance in the event of array element failure. For example, a single-column constraint as depicted in Figure 11.12 easily is satisfied by weighting the taps corresponding to a dead element to whatever value is necessary to satisfy the constraint. Since the element contributes nothing to the system output, any weight values are admissible. The protection offered by the constraint against removing signals from the desired look direction effectively has been removed.

11.4.2 Unconstrained Adaptation

The adaptive array processor can be reconfigured into two parts consisting of a conventional beamformer in one part and a matrix filter followed by an unconstrained adaptive system in the other.

As a preliminary, consider the structure illustrated in Figure 11.15. When the weight column sums are constrained to equal zero, the response characteristics of this array processor are identical to those of the array processor in Figure 11.12.[28] The zero-column sum requirement forces the adaptive part of the system to maintain a spatial null in the desired look direction. The frequency response constraint is implemented by the fixed FIR filter following the conventional beamformer. In this configuration, the adaptive array processor is cast as a constrained adaptive noise-canceling structure where the filtered output of the conventional beamformer takes on the role of $d(m)$ in Eq. 11.90. Since $d(m) \neq 0$, the constrained minimization problem takes on a different form than Eq. 11.98. A desirable feature of the structure in Figure 11.15 is that both the conventional and adaptive beamformer outputs are available simultaneously.

Now, Figure 11.16 illustrates how the column constraints can be removed from the adaptive portion of the system and incorporated into a matrix filter, $\mathbf{A}$.[27] The $N' \leq N - 1$ outputs of $\mathbf{A}$ can be thought of as beams, each of which satisfies the spatial constraints on the system. In Eq. 11.113, these spatial constraints are maintained algorithmically by the projection operator $\mathbf{P}$ and vector $\mathbf{F}$. If $N' = N - 1$, $\mathbf{A}$ has full rank, and each beam forms a spatial null in the desired look direction, the response characteristics of this array processor converge in the mean to the same steady-state values as those of the array processors in Figures 11.12 and 11.15. In addition, if the rows of $\mathbf{A}$ are mutually orthogonal, the sequence of response characteristics during adaptation also will be identical.[29, 32]

The distinctive aspect of the structure in Figure 11.16 is that the adaptive portion is completely unconstrained. Thus, the complexity of implementing the constrained adaptive algorithms required by the structures in Figures 11.12 and 11.15 has been alleviated. In this configuration, the adaptive array processor is cast as an unconstrained adaptive noise canceler. Similar to the case in Figure 11.15, the filtered output of the conventional beamformer takes on the role of $d(m)$ in Eq. 11.90. In this case, each output of the matrix filter is a signal-free reference to the contaminating noise. Thus, multichannel versions of the unconstrained adaptive algorithms discussed in Section 11.3 can be applied to the adaptive beamforming problem. Additional discussion

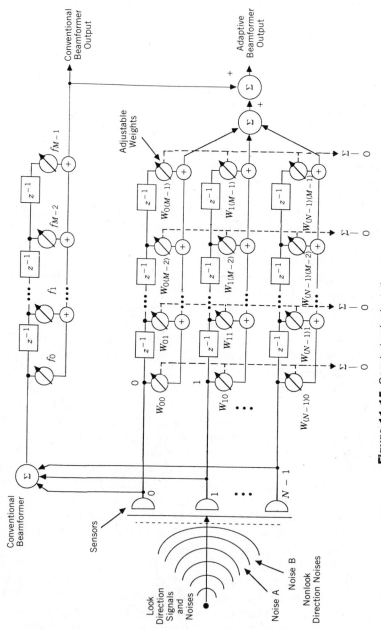

Figure 11.15 Constrained adaptation: two part structure.

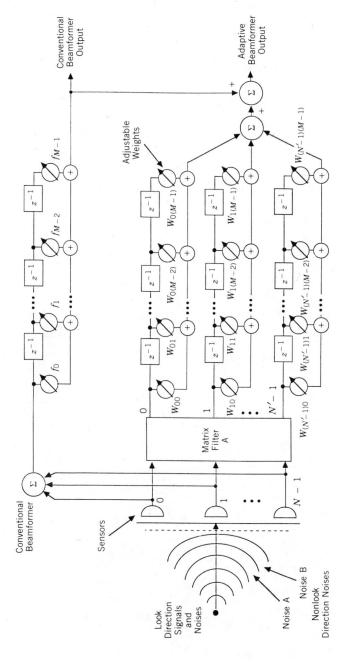

Figure 11.16 Unconstrained adaptation.

on the partitioned adaptive beamformer (also known as the generalized sidelobe canceler, including the implementation of point and derivative constraints through design of the matrix filter **A**, can be found in reference, 33 and 35 through 39.

11.4.3 Adaptive Beamforming Example

As an example of the adaptive beamforming structure shown in Figure 11.16, consider the following active sonar system.[46] An array of hydrophones is mounted on the bow of a surface ship and pointed 45° downward. The ship is traveling at 10 knots (5.1 m/sec), and the ocean depth is 300 m. The sonar operates at $f_c = 75$ kHz and transmits a 100 msec, rectangularly windowed pulse.

The 6×9 rectangular array has a sensor spacing of $\lambda/2$ at the operating frequency of the sonar. The conventional beam in Figure 11.16 is formed by simply summing the outputs of all elements. The corresponding beam pattern normal to the array plane (limited to between $\pm 30°$ in bearing and $\pm 40°$ in elevation) is shown in Figure 11.17. A reference beam (a single output of the matrix filter in Figure 11.16) is formed by first generating row sums, then weighting the rows with the sequence $\{0, +1, -1, +1, -1, 0\}$ and summing. The corresponding beam pattern similar to Figure 11.17 is shown in Figure 11.18. A transmit beam is assumed, which uniformly illuminates the medium over a sector $\pm 30°$ in bearing and $\pm 40°$ in elevation normal to the array plane.

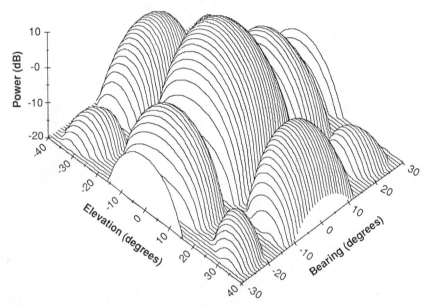

Figure 11.17 Sum beam pattern.

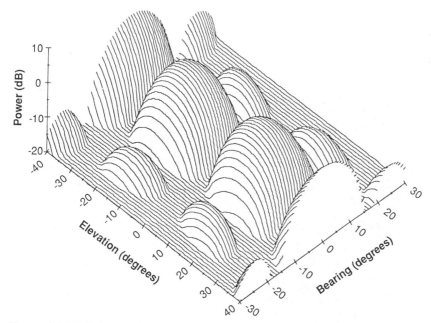

Figure 11.18 Reference beam pattern.

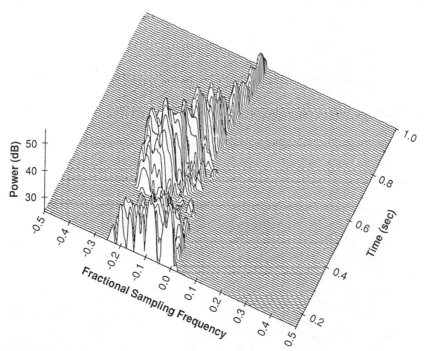

Figure 11.19 Conventional beamformer output.

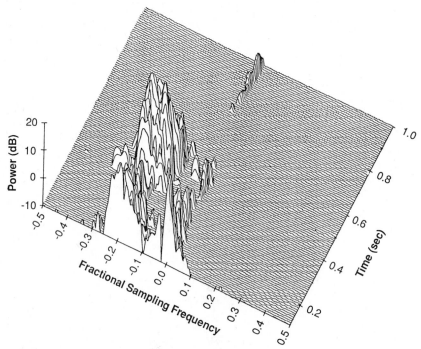

Figure 11.20 Reference beamformer output.

The backscattered return to this sonar was simulated using REVGEN (REVerberation GENerator).[47] REVGEN is a direct implementation of the point-scattering model of oceanic reverberation where the returns from a large number of discrete scatterers distributed randomly throughout the volume and on the boundaries are summed coherently to obtain the (complex basebanded) reverberation time series at the output of each hydrophone element. The backscattering coefficient (strength) for each reverberation type (surface, volume, and bottom) is specified along with random scatterer motion, platform trajectory, absorption, boundary reflection losses, and transmitting/receiving beam patterns.[48-50] In this case, an isovelocity sound speed profile was assumed ($c = 1500$ m/sec) along with the following parameters: (a) volume backscattering coefficient, $s_v = -70$ dB, (b) bottom backscattering coefficient, $s_b = -20$ dB, and (c) attenuation due to absorption, $\alpha = 23$ dB/km.

Range–Doppler maps of the returning reverberation for the conventional (sum) and reference beams are shown in Figures 11.19 and 11.20 ($f_s = 2$ kHz). These were generated by taking successive 128-point FFTs (Kaiser–Bessel window, $\alpha = 2.5$) overlapped by 87.5% (16 points). In addition, the range-Doppler maps have been left-shifted to compensate for ship velocity (a 5.1 m/sec velocity imparts a 515 Hz Doppler shift to stationary targets in line with the direction of ship motion at $f_c = 75$ kHz). In Figure 11.19 (sum

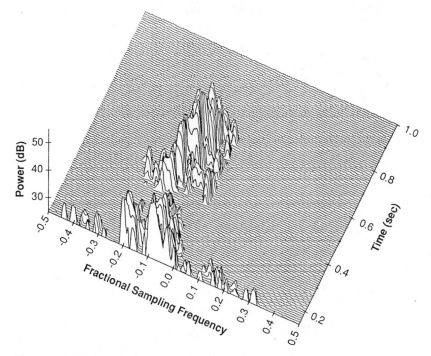

Figure 11.21 Adaptive beamformer output (JCLSL, $p = 3$, $\alpha = 0.02$).

beam), volume reverberation dominates early in the range-Doppler map. Later (at approximately 0.4 sec in range), the onset of bottom reverberation arriving through the lower edge of the mainlobe is seen. Bottom reverberation at approximately 0.6 sec in range corresponds to backscatter arriving through the center of the mainlobe. In Figure 11.20 (reference beam), bottom reverberation shows a primary peak at approximately 0.5 sec in range and a much smaller peak at approximately 0.9 sec in range. These peaks correspond to backscatter arriving through the two main lobes of the reference beam.

The results of adaptive beamforming using the JCLSL algorithm with $p = 3$ and $\alpha = 0.02$ are shown in Figure 11.21. Bottom reverberation is canceled except in the region where the look-direction constraint (implemented via the reference beam null normal to the array plane) allows arrivals to pass through the beamformer. Additional examples of boundary reverberation rejection are presented in references 51 and 52.

Appendix 11.A

Conventional Beamforming (CBF)

The problem of coherently summing the outputs from a number of spatially distributed sensors is discussed. After considering the general distribution of sensors in three-dimensions, the results are specialized to line arrays where the elements are equally spaced. The connection is made between such a finite-aperture, spatial sampling scheme and FIR digital filters. Lastly, the use of FFTs to implement the beamforming process is described.

11.A.0 INTRODUCTION

The problem of coherently summing the outputs from a number of spatially distributed sensors is known as beamforming. Beyond the directivity provided by an individual sensor, a beamformer permits one to listen preferentially to wavefronts propagating from one direction over another. Thus, a beamformer implements a spatial filter.

The intent of this appendix is to provide background material on conventional beamforming (CBF). Additional material can be found in references 46, 48, and 53 through 69. Applications of these concepts can be found in numerous areas—seismology, underwater acoustics, biomedical engineering, industrial testing, and the reception of the full spectrum of radio frequency transmissions.

11.A.1 BEAMFORMING FROM AN ARBITRARY DISTRIBUTION OF ARRAY ELEMENTS

Although we will later focus on line arrays, the elements of an antenna array often are distributed in two or three dimensions. Included are planar, conformal, towed, random, and distributed arrays. As an example, Figure 11.A.1 shows a random sonobuoy array deployed in a planar configuration.

11.A.1.1 Plane-Wave Beamforming

Consider a coordinate system defined as in Figure 11.A.2. The beamforming task consists of generating the waveform $b_m(t)$ (or its corresponding sampled sequence) for each desired steered beam direction B_m. Each $b_m(t)$ consists of the sum of suitably time-delayed replicas of the individual element signals $e_n(t)$. The time delays compensate for the assumed differential travel times between sensors for a signal from the desired beam direction.

Let the output of an element located at the origin of coordinates due to the lth source be $s_l(t)$. Under the assumption of plane-wave propagation, a source from direction S_l produces the following sensor outputs

$$e_n(t) = s_l\left(t + \frac{\mathbf{E}_n \cdot \mathbf{S}_l}{c}\right) \tag{11.A.1}$$

where c is the speed of propagation ($c \approx 1500$ m/sec for acoustic waves in the ocean). Appropriately delaying the individual element signals to point a beam

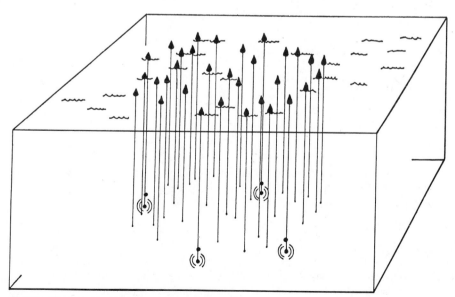

Figure 11.A.1 Random sonobuoy array.

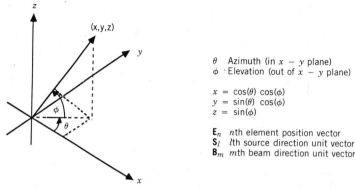

θ Azimuth (in $x - y$ plane)
ϕ Elevation (out of $x - y$ plane)

$x = \cos(\theta)\ \cos(\phi)$
$y = \sin(\theta)\ \cos(\phi)$
$z = \sin(\phi)$

$\mathbf{E}_n$ nth element position vector
$\mathbf{S}_l$ lth source direction unit vector
$\mathbf{B}_m$ mth beam direction unit vector

Figure 11.A.2 Coordinate system definition and vector notation.

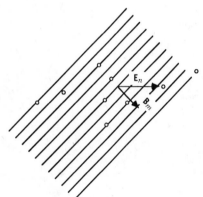

Figure 11.A.3 Element time-delay calculation for plane-wave beamforming ($\mathbf{B}_m = \mathbf{S}_l$).

in the direction $\mathbf{B}_m$ yields the beamformer output

$$b_m(t) = \sum_{n=0}^{N-1} w_n e_n\left(t - \frac{\mathbf{E}_n \cdot \mathbf{B}_m}{c}\right) \qquad (11.\text{A}.2)$$

$$= \sum_{n=0}^{N-1} w_n s_l\left(t + \frac{\mathbf{E}_n \cdot (\mathbf{S}_l - \mathbf{B}_m)}{c}\right) \qquad (11.\text{A}.3)$$

where the w_n are weights that have been applied to each element signal. The vectors involved in the scalar product of one component of Eq. 11.A.2 are illustrated in Figure 11.A.3. Portrayed is an array of sensors in a plane-wave acoustic field where $\mathbf{B}_m = \mathbf{S}_l$.

11.A.1.2 Curved Wavefront Beamforming

Consider again the coordinate system defined in Figure 11.A.2. The curved wavefront or focused beamforming task consists of generating the waveform $b_m(t)$ (or its corresponding sampled sequence) for each desired steered beam direction $\mathbf{B}_m$ and range of focus r_m. Each $b_m(t)$ consists of the sum of suitably delayed replicas of the individual element signals $e_n(t)$. The time delays compensate for the assumed differential travel times between sensors for a signal from the desired beam direction and range.

Let the output of an element located at the origin of coordinates due to the lth source be $s_l(t)$. A source from direction S_l and range r_l produces the following sensor outputs.

$$
\begin{aligned}
e_n(t) &= s_l\!\left(t + \frac{r_l - |r_l\mathbf{S}_l - \mathbf{E}_n|}{c} \right) \\
&= s_l\!\left(t + \frac{r_l - \left(r_l^2 + |\mathbf{E}_n|^2 - 2r_l\mathbf{S}_l \cdot \mathbf{E}_n \right)^{1/2}}{c} \right)
\end{aligned}
\tag{11.A.4}
$$

where c is the speed of propagation. Appropriately delaying the individual element signals to focus a beam in the direction $\mathbf{B}_m$ and range r_m yields the focused beamformer output.

$$
\begin{aligned}
b_m(t) &= \sum_{n=0}^{N-1} w_n e_n\!\left(t - \frac{r_m - |r_m\mathbf{B}_m - \mathbf{E}_m|}{c} \right) \\
&= \sum_{n=0}^{N-1} w_n e_n\!\left(t - \frac{r_m - \left(r_m^2 + |\mathbf{E}_n|^2 - 2r_m\mathbf{B}_m \cdot \mathbf{E}_n \right)^{1/2}}{c} \right) \\
&= \sum_{n=0}^{N-1} w_n s_l\!\left(t + \frac{r_l - r_m + |r_m\mathbf{B}_m - \mathbf{E}_n| - |r_l\mathbf{S}_l - \mathbf{E}_n|}{c} \right)
\end{aligned}
\tag{11.A.5}
$$

$$
= \sum_{n=0}^{N-1} w_n s_l\!\left(t + \frac{r_l - r_m + \left(r_m^2 + |\mathbf{E}_n|^2 - 2r_m\mathbf{B}_m \cdot \mathbf{E}_n \right)^{1/2} - \left(r_l^2 + |\mathbf{E}_n|^2 - 2r_l\mathbf{S}_l \cdot \mathbf{E}_n \right)^{1/2}}{c} \right)
\tag{11.A.6}
$$

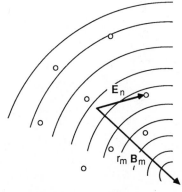

Figure 11.A.4 Element time-delay calculation for curved wavefront beamforming ($\mathbf{B}_m = \mathbf{S}_l$).

where the w_n are weights that have been applied to each element signal. The vectors involved in the calculation in Eq. 11.A.5 are illustrated in Figure 11.A.4.

11.A.2 LINE ARRAYS

The general discussion in the previous section on plane-wave beamforming will now be specialized to line arrays where the elements are equally spaced.

11.A.2.1 Plane-Wave Beamforming

Consider the geometry depicted in Figure 11.A.5. A line array of equally spaced elements is positioned along the y-axis. Let the output of an element located at the origin of coordinates due to the lth source be $s_l(t)$. Under the assumption of plane-wave propagation, a source from direction S_l produces the following sensor outputs:

$$e_n(t) = s\left(t + \frac{nd \sin(\theta_l)}{c}\right) \qquad (11.A.7)$$

where n is the array element index, d is the interlement spacing, θ_l is the angle of arrival of the lth source, and c is the speed of propagation. As in the general discussion, beamforming is accomplished by applying weights and time delays to the individual element signals, then summing them. In this case, since the array elements are equally spaced, the time delays appropriate for aligning signals arriving from angle θ_m are integer multiples of the propagation

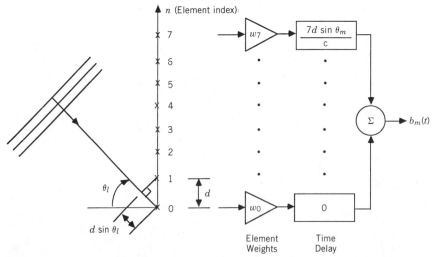

Figure 11.A.5 Line-array geometry.

time between adjacent elements:

$$b_m(t) = \sum_{n=0}^{N-1} w_n e_n \left(t - \frac{nd \sin(\theta_m)}{c} \right) \tag{11.A.8}$$

$$= \sum_{n=0}^{N-1} w_n s_l \left(t + \frac{nd(\sin(\theta_l) - \sin(\theta_m))}{c} \right) \tag{11.A.9}$$

Given a set of element time delays and weights, a beamformer forms a spatial filter. The directional response characteristics of the beamformer can be derived by considering a complex exponential, plane-wave signal propagating across the array

$$s_l(t) = e^{j\omega_l t} \tag{11.A.10}$$

Based on the angle of arrival θ_l, the source produces the following sensor outputs.

$$e_n(t) = e^{j\omega_l(t + nd \sin(\theta_l)/c)}$$
$$= e^{j\omega_l t} e^{j\omega_l nd \sin(\theta_l)/c}$$
$$= e^{j\omega_l t} e^{jnk_l d \sin(\theta_l)} \tag{11.A.11}$$

where $k_l = 2\pi/\lambda_l = \omega_l/c$ is wave number or spatial frequency (λ_l is wavelength). Appropriately delaying the individual element signals to point a beam in the direction $\mathbf{B}_m$ yields the beamformer output

$$b_m(t) = \sum_{n=0}^{N-1} w_n e^{j\omega_l(t - nd \sin(\theta_m)/c)} e^{j\omega_l nd \sin(\theta_l)/c}$$

$$= \sum_{n=0}^{N-1} w_n e^{j\omega_l t} e^{j\omega_l nd(\sin(\theta_l) - \sin(\theta_m))/c}$$

$$= e^{j\omega_l t} \sum_{n=0}^{N-1} w_n e^{-jnk_l d(\sin(\theta_m) - \sin(\theta_l))} \tag{11.A.12}$$

$$= e^{j\omega_l t} W(\theta_l) \tag{11.A.13}$$

where

$$W(\theta_l) = \sum_{n=0}^{N-1} w_n e^{-jnk_l d(\sin(\theta_m) - \sin(\theta_l))} \tag{11.A.14}$$

11.A.2.2 Spatial Transfer Function

As indicated in the previous section, the beamformer output due to a complex exponential plane wavefront signal can be written as the product of the original signal and a function $W(\theta_l)$

$$W(\theta_l) = \sum_{n=0}^{N-1} w_n e^{-jnk_l d(\sin(\theta_m) - \sin(\theta_l))}$$

$$= \sum_{n=0}^{N-1} w_n z^{-n} \tag{11.A.15}$$

where

$$z = e^{jk_l d(\sin(\theta_m) - \sin(\theta_l))} \tag{11.A.16}$$

$$k_l = \frac{2\pi}{\lambda_l} = \frac{\omega_l}{c} \tag{11.A.17}$$

$W(\theta_l)$ summarizes the spatial filtering characteristics of the array. As indicated, it can be written as the z-transform of the array element weights evaluated on the unit circle (i.e., the Fourier transform). Thus, $W(\theta_l)$ can be viewed as the spatial transfer function of the array. Often, the spatial transfer function is expressed as a function of an alternative argument

$$W(u_l) = \sum_{n=0}^{N-1} w_n z^{-n} \tag{11.A.18}$$

where

$$z = e^{jk_l d(u_m - u_l)} \tag{11.A.19}$$

$$k_l = \frac{2\pi}{\lambda_l} = \frac{\omega_l}{c} \tag{11.A.20}$$

$$u_m = \sin(\theta_m) \tag{11.A.21}$$

$$u_l = \sin(\theta_l) \tag{11.A.22}$$

Thus, as θ_l sweeps from $-90°$ through $90°$, u_l goes from -1 to 1. The nonlinear mapping between them indicates how a physical angle of arrival θ_l is perceived by the array in terms of a phase progression element-to-element (n increasing).

As an example, consider the beamformer shown in Figure 11.A.6a, where $N = 8$, $w_n = 1$ ($n = 0, \ldots, N - 1$), and $\theta_m = 0°$ (broadside beam). By selecting $d = \lambda_l/2$ (half-wavelength array element spacing)

$$W(\theta_l) = \sum_{n=0}^{N-1} w_n z^{-n}$$

$$= \sum_{n=0}^{N-1} z^{-n} \tag{11.A.23}$$

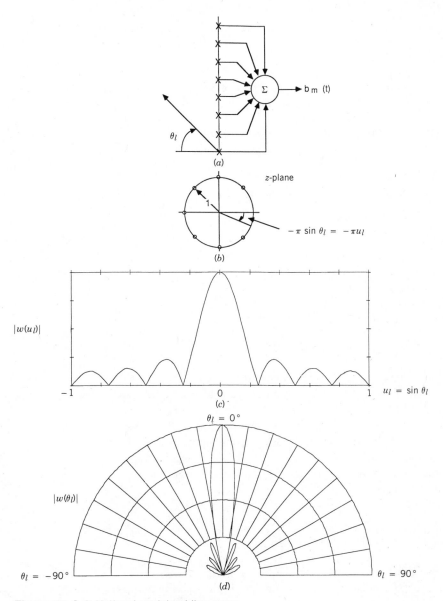

Figure 11.A.6 Uniformly weighted line array.

where

$$z = e^{jk_l d(\sin(\theta_m) - \sin(\theta_l))}$$
$$= e^{-j\pi \sin(\theta_l)} \tag{11.A.24}$$

Or, in terms of u_l

$$W(u_l) = \sum_{n=0}^{N-1} z^{-n} \tag{11.A.25}$$

where

$$z = e^{-j\pi u_l} \qquad (11.A.26)$$

The zeros of the spatial transfer function in Eq. (11.A.23) are displayed in Figure 11.A.6b. They are located at equally spaced intervals $(2\pi/N)$ around the unit circle. By solving the summation in Eq. 11.A.23, the spatial filtering characteristics of the beamformer can be written explicitly as

$$W(\theta_l) = \frac{\sin\left(\dfrac{N}{2}\left[\pi\sin(\theta_l)\right]\right)}{\sin\left(\dfrac{1}{2}\left[\pi\sin(\theta_l)\right]\right)} e^{j(N-1/2)\pi\sin(\theta_l)} \qquad (11.A.27)$$

Or, in terms of u_l

$$W(u_l) = \frac{\sin\left(\dfrac{N}{2}\pi u_l\right)}{\sin\left(\dfrac{1}{2}\pi u_l\right)} e^{j(N-1/2)\pi u_l} \qquad (11.A.28)$$

The magnitudes of these expressions are displayed in Figures 11.A.6d and 11.A.6c, respectively. Note that Eq. 11.A.28 and Figure 11.A.6c correspond to the Fourier transform of a N-point rectangular window.

The beamwidth of a line array is the angular distance between the mainlobe points of $W(\theta_l)$, which are 3 dB down from the maximum. Due to the nonlinear fashion in which θ_l appears in Eq. 11.A.15, beamwidth varies with θ_m from a minimum at $\theta_m = 0°$ (broadside) to a maximum at $\theta_m = \pm 90°$ (end-fire). At or near broadside

$$BW_{3dB} \approx 49.6\left(\frac{\lambda_l}{Nd}\right)\sec(\theta_m) \qquad (11.A.29)$$

for a uniformly weighted line array when $Nd \gg \lambda_l$. Figure 11.A.7 shows how beamwidth varies as a function of array length (Nd) and steering angle from broadside (θ_m).[48] In the same way that window functions are used to modify the characteristics of FFTs used for spectral analysis (i.e., mainlobe width and sidelobe level), weighting functions are used to alter the directional response or spatial filtering characteristics of a beamformer.[54-57]

11.A.2.3 Grating Lobes and Invisible Space

The nonlinear mapping between physical angle of arrival θ_l and phase angle in the z-plane $-k_l du_l$ ($u_l = \sin(\theta_l)$) is a function of array element separation d. The example in the previous section leading to Eqs. 11.A.23 through 11.A.28

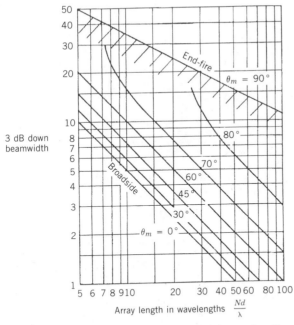

Figure 11.A.7 Beamwidth as a function of array length (Nd) and steering angle (θ_m).

considered array elements spaced at half-wavelength intervals ($d = \lambda_l/2$). Thus, $k_l d = \pi$. Writing Eq. 11.A.26 in general yields

$$z = e^{-jk_l d u_l} \qquad (11.A.30)$$

Now, when $u_l = \sin(\theta_l)$ sweeps from -1 to 1, the phase angle in the z-plane sweeps from $k_l d$ to $-k_l d$. Only when the array elements are half-wavelength spaced is the unit circle traversed exactly once.

The impact of varying array element spacing is illustrated in Figure 11.A.8 for the same example treated in the previous section. The top portion of the figure shows the spatial transfer function zeros for a $N = 8$, uniformly weighted ($w_n = 1$, $n = 1, \ldots, N - 1$), line array. The remaining panels present plots of $W(\theta_l)$ magnitude for three cases of array element spacing. In each, two horizontal axes are provided. The first is $u_l = \sin(\theta_l)$ and represents the range of physical wavefront arrival angles from $-90°$ to $90°$. The second, $k_l d u_l$, represents the negative of the phase angle in the z-plane. Only in the first case ($d = \lambda_l/2$) is the unit circle in the z-plane traversed once. In the second case ($d = \lambda_l$), the unit circle is traversed twice. A second peak occurs in the spatial response characteristics of the beamformer ($u_l = \pm 1$), which is equal to the response to a broadside source ($u_l = 0$). Such secondary peaks are known as grating lobes. In the third case ($d = \lambda_l/4$), only half the unit circle is traversed. Z-plane phase angles greater than $\pm \pi/2$ do not correspond to physically realizable arrival angles, and that region of the z-plane is known as invisible space.

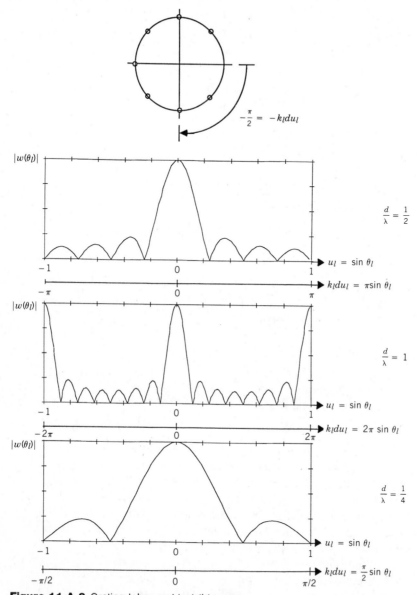

Figure 11.A.8 Grating lobes and invisible space.

11.A.2.4 Time-Delay Beamforming Example

As an application of the time-delay beamforming expression in Eq. 11.A.8, consider the following example. In April 1981, the Marine Physical Laboratory of the Scripps Institution of Oceanography, conducted an experiment investigating acoustic signal propagation in thick sediments. The experiment involved the deployment of a 20-element array from the Floating Laboratory

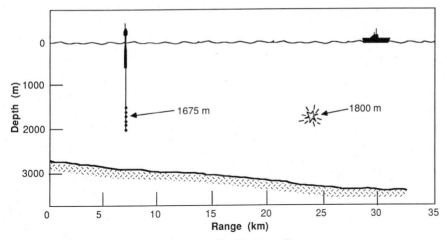

Figure 11.A.9 Refraction experiment on the Monterey Fan.

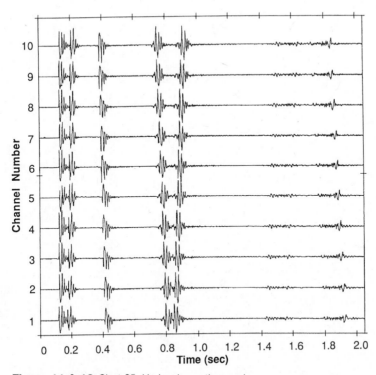

Figure 11.A.10 Shot 25: Hydrophone time series.

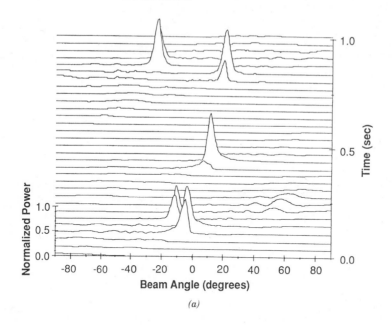

(a)

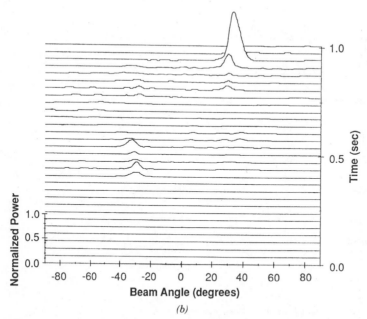

(b)

Figure 11.A.11 Shot 25: Beamformed hydrophone time series: (a) 0 sec $\leq t \leq$ 1.024 sec; (b) 1.024 sec $\leq t \leq$ 2.048 sec.

Instrument Platform (FLIP) in an area of thick sediments on the Monterey Fan. The hydrophone array, which had equal sensor spacing and a total aperture of approximately 500 m was placed at middepth in the 3-km deep water column. Using deep explosive sources (256 lb TNT detonating at approximately 1800 m depth), shooting runs were made at ranges out to 35 km from FLIP. Figure 11.A.9 summarizes the experimental scenario.

The time series collected from 10 hydrophones of the array during shot 25 are displayed in Figure 11.A.10 (hydrophones are numbered starting at the bottom of the array). The multipath nature of the shot propagation is clearly evident. The channel-to-channel offsets in arrival times of the various components are indicative of their wavefront angles of arrival.

The results from beamforming these hydrophone time series as indicated in Eq. 11.A.8 are shown in Figure 11.A.11 as a display of time-varying average power. Each slice in the waterfall display represents beam output power averaged over 32 points for each of 256 beams between $\pm 90°$. Beam angles are reported with respect to the horizontal (positive angles refer to beams pointing upward). The angles of arrival of the various multipath components can be determined directly from this type of display.

11.A.3 FFT BEAMFORMING

A two-dimensional FFT can be used to implement the beamforming operation for an equally spaced line array.[58,59] On a single frequency basis, the time-delay operations required in line-array beamforming are equivalent to phase shifts that are a linear function of the element index n. A FFT can be used to implement these phase shifts.

11.A.3.1 Plane-Wave Beamforming

The procedure followed in FFT beamforming is shown in Figure 11.A.12. Consider an equally spaced line array with $N = 8$ elements (element index $n = 0, \ldots, 7$). The signals from each array element are sampled simultaneously at rate f_s, yielding a multichannel time series $x_{n,m}$. M-point FFTs are taken along each channel (generally, the M-point segments would be overlapped 50 to 75% and weighted with a good window function, w_m). The results from one set of these along-channel FFTs will be denoted $X_n(i)$, where n is the array element index (channel number) and i is an integer frequency index, $0 \leq i \leq M - 1$

$$X_n(i) = \sum_{m=0}^{M-1} w_m x_{n,m} e^{-j(2\pi/M)mi} \tag{11.A.31}$$

The $X_n(i)$ are represented in the figure as N horizontal rows of M FFT bins.

Now, each along-channel FFT bin index will be considered separately. The 8-point sequence of complex values representing the same bin index for all channels is weighted and processed with a cross-channel FFT (often, the

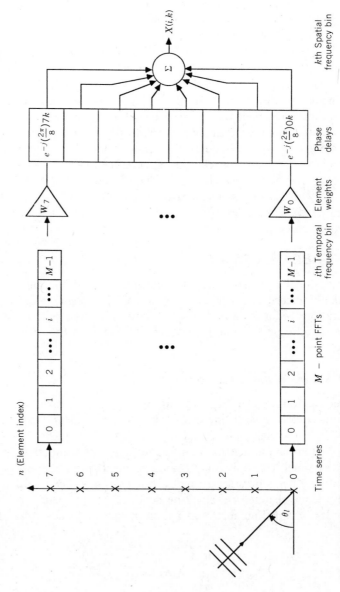

Figure 11.A.12 FFT beamforming computations.

Frequency versus wavenumber

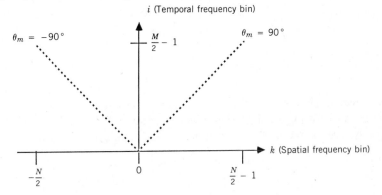

Frequency versus azimuth

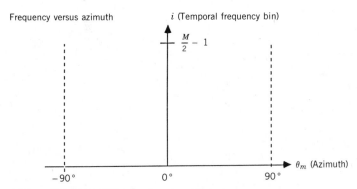

Figure 11.A.13 FFT beamforming displays.

weighted sequence is zero-padded prior the FFT)

$$X(i, k) = \sum_{n=0}^{N-1} w_n X_n(i) e^{-j(2\pi/N)nk} \tag{11.A.32}$$

The N bins resulting from a cross-channel FFT $0 \leq k \leq N - 1$ represent different incremental phase shifts being applied across the array. Thus, beams are formed at the frequency represented by the along-channel FFT bin index i in the directions represented by the cross-channel FFT bin indices k.

The results from FFT beamforming need to be properly interpreted in terms of physical angle (azimuth) θ_m. Two different methods of displaying $X(i, k)$ are shown in Figure 11.A.13. Both assume that the magnitude of $X(i, k)$ is represented on an axis that is normal to the page (e.g., as a contour map, grayscale image, or pseudo-3D plot).

The top panel shows the display obtained by directly plotting the output of the FFT beamformer. The vertical axis is temporal frequency as represented by along-channel FFT bin index i.

$$\omega = \left(\frac{2\pi}{M}\right) f_s i \tag{11.A.33}$$

The horizontal axis is apparent wavenumber or apparent spatial frequency (i.e., wavenumber or spatial frequency projected onto the line containing the array elements) as represented by cross-channel FFT bin index k

$$\frac{2\pi}{\lambda}\sin(\theta_m) = \left(\frac{2\pi}{N}\right)\left(\frac{1}{d}\right)k \qquad (11.A.34)$$

where $2\pi/\lambda$ is wavenumber or spatial frequency, $\lambda = c/(\omega/2\pi)$ is wavelength, and c is the wavefront speed of propagation. From Eq. 11.A.34, the relationship between cross-channel FFT bin index k and azimuth θ_m can be derived

$$\theta_m = \sin^{-1}\left(\frac{\lambda}{Nd}k\right) \qquad (11.A.35)$$

$$= \sin^{-1}\left(\frac{c}{Nd}\frac{2\pi}{\omega}k\right) \qquad (11.A.36)$$

$$= \sin^{-1}\left(\frac{c}{Nd}\frac{M}{f_s}\frac{k}{i}\right) \qquad (11.A.37)$$

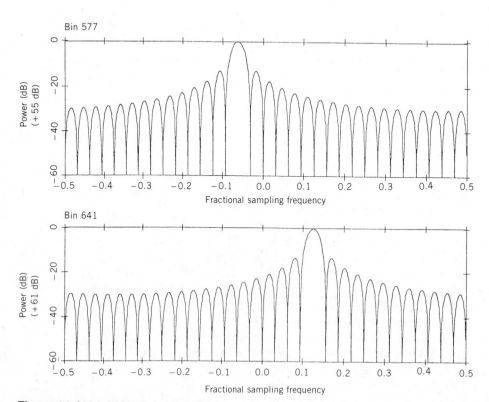

Figure 11.A.14 FFT beamforming results.

Thus, constant values of azimuth run along radial lines. The region between the $-90°$ and $90°$ radial lines corresponds to physically meaningful arrival angles for wavefronts propagating with speed c.

The bottom panel shows the display obtained by interpolating the FFT beamformer output along radial lines of constant azimuth. The vertical axis is temporal frequency as represented by along-channel FFT bin index i and the horizontal axis is azimuth θ_m. With this type of display, one can easily ascertain the spectral content of signals arriving from any given direction.

11.A.3.2 FFT Beamforming Example

As an illustration of FFT beamforming, consider the following simulation. Two plane wavefront, sinusoidal signals are arriving at a 32-element vertical array. Their fractional sampling frequencies are $f_1 = 0.0625$ and $f_2 = 0.1250$, and the ratio of their amplitudes is 2 (6 dB in terms of power). The along-channel FFT length is $M = 1024$. Thus, the positive frequency component of f_1 appears in bin 577 and the positive frequency component of f_2 appears in bin 641 of a FFT that has been rotated circularly by $M/2$ bins (here, bins are numbers starting with 1 rather than 0).

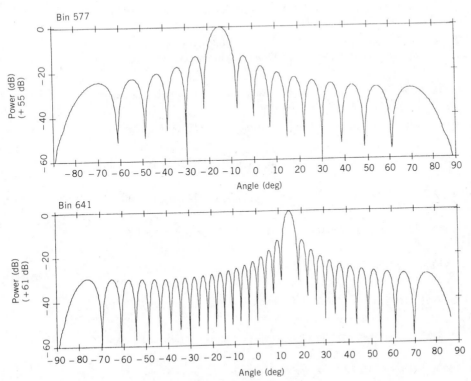

Figure 11.A.15 FFT beamforming results interpreted in terms of physical angle.

The cross-channel FFT length is $N = 1024$ (zero-padded). The arrival angles of the two wavefronts are such that the sinusoid at f_1 is time-delayed and the sinusoid at f_2 is time-advanced one sampling interval element-to-element (the reference array element is at the bottom of the array). Thus, as shown in Figure 11.A.14, a display of the cross-channel FFT taken in bin 577 and bin 641 will peak at fractional sampling frequency -0.0625 (sinusoid at f_1) and 0.01250 (sinusoid at f_2).

The array elements are half-wavelength spaced at f_2. Properly interpreting the FFT beamformer results in terms of physical angle via Eq. 11.A.37 results in the display shown in Figure 11.A.15. The two arrivals peak at $-14.5°$ (sinusoid at f_1) and $14.5°$ (sinusoid at f_2). Positive angles refer to upward looking beams.

References

1. B. Widrow, J. Glover, J. McCool, et al, "Adaptive Noise Cancelling: Principles and Applications," *Proc. IEEE* 63: 1692–1716, 1975.

2. B. Widrow, J. McCool, M. Larimore, R. Johnson, "Stationary and Nonstationary Learning Characteristics of the LMS Adaptive Filter," *Proc. IEEE* 64: 1151–1162, 1976.

3. J. Treichler, "Transient and Convergent Behavior of the Adaptive Line Enhancer," *IEEE Trans. ASSP*, ASSP-27: 53–62, 1979.

4. W. S. Hodgkiss, J. A. Presley, "Adaptive Tracking and Multiple Sinusoids Whose Power Levels are Widely Separated," *IEEE Trans. ASSP*, ASSP-29: 710–721, 1981.

5. W. S. Hodgkiss, J. A. Presley, "The Complex Adaptive Least-Squares Lattice," *IEEE Trans. ASSP*, ASSP-30: 330–333, 1982.

6. B. Friedlander, "Lattice Filters for Adaptive Processing," *Proc. IEEE* 70: 829–867, 1982.

7. B. Friedlander, "Lattice Methods for Spectral Estimation," *Proc. IEEE* 70: 990–1017, 1982.

8. B. Friedlander, "System Identification Techniques for Adaptive Noise Cancelling," *IEEE Trans. ASSP*, ASSP-30: 699–709, 1982.

9. S. Kay, S. L. Marple, "Spectrum Analysis—A Modern Perspective," *Proc. IEEE* 69: 1380–1419, 1981.

10. S. Parker, L. Griffiths (eds.), "Special Issue on Adaptive Signal Processing," *IEEE Trans. ASSP*, ASSP-29(3), 1981.

11. S. Haykin, J. A. Cadzow (eds.), "Special Issue on Spectral Estimation," *Proc. IEEE* 70(9), 1982.

12. C. Lawson, R. Hanson, *Solving Least Squares Problems*. Prentice–Hall, Englewood Cliffs, N.J., 1974.

13. S. Haykin (ed.), *Nonlinear Methods of Spectral Analysis*. Springer-Verlag, New York, 1979.

14. C. Cowan, P. Grant (eds.), *Adaptive Filters*. Prentice–Hall, Englewood Cliffs, N.J., 1985.

15. S. Haykin, *Adaptive Filter Theory*. Prentice–Hall, Englewood Cliffs, N.J., 1986.

16. B. Widrow, S. Stearns, *Adaptive Signal Processing*. Prentice–Hall, Englewood Cliffs, N.J., 1986.

17. N. Levinson, "The Wiener RMS (Root-Mean-Square) Error Criterion in Filter Design and Prediction," *J. Math. Phys.* 25: 261–278, 1947.

18. J. Durbin, "The Fitting of Time Series Models," *Rev. Intern. Statist. Inst.* 28: 233–244, 1960.

19. R. Gray, "On the Asymptotic Eigenvalue Distribution of Toeplitz Matrices," *IEEE Trans. Infor. Theory*, IT-18: 725–730, 1972.

20. A. Gray, J. Markel, "Digital Lattice and Ladder Filter Synthesis," *IEEE Trans. Audio Electroacoust.*, AU-21: 491–500, 1973.

21. J. Claerbout. *Fundamentals of Geophysical Data Processing*. McGraw–Hill, New York, 1976.

22. B. Widrow, P. Mantey, L. Griffiths, B. Goode, "Adaptive Antenna Systems," *Proc. IEEE* 55: 2143–2159, 1967.

23. L. Griffiths, "A Simple Adaptive Algorithm for Real-Time Processing in Antenna Arrays," *Proc. IEEE* 57: 1696–1704, 1969.

24. O. L. Frost, "An Algorithm for Linearly Constrained Adaptive Array Processing," *Proc. IEEE* 60: 926–935, 1972.

25. L. E. Brennan, I. S. Reed, "Theory of Adaptive Radar," *IEEE Trans. AES*, AES-9: 237–252, 1973.

26. W. F. Gabriel, "Adaptive Arrays: An Introduction," *Proc. IEEE* 64: 239–272, 1976.

27. S. Applebaum, D. Chapman, "Adaptive Arrays with Main Beam Constraints," *IEEE Trans. Ant. Prop.*, AP-24, 650–662, 1976.

28. L. J. Griffiths, "Adaptive Monopulse Beamforming," *Proc. IEEE* 64: 1260–1261, 1976.

29. C. W. Jim, "A Comparison of Two LMS Constrained Optimal Array Structures," *Proc. IEEE* 65: 1730–1731, 1977.

30. A. M. Vural, "Effects of Perturbations on the Performance of Optimum/Adaptive Arrays," *IEEE Trans. AES*, AES-15: 76–87, 1979.

31. V. C. Anderson, "Sidelobe Interference Suppression with an Adaptive Null Processor," *J. Acoust. Soc. Am.* 69: 185–190, 1981.

32. L. Griffiths, C. Jim, "An Alternative Approach to Linearly Constrained Adaptive Beamforming," *IEEE Trans. Ant. Prop.*, AP-30: 27–34, 1982.

33. A. Cantoni, L. Godara, "Fast Algorithms for Time Domain Broadband Adaptive Array Processing," *IEEE Trans. AES*, AES-18: 682–699, 1982.

34. M. H. Er, A. Cantoni, "Derivative Constraints for Broad-Band Element Space Antenna Array Processors," *IEEE Trans. ASSP*, ASSP-31: 1378–1393, 1983.

35. N. Jablon, "Steady State Analysis of the Generalized Sidelobe Canceller by Adaptive Noise Cancelling Techniques," *IEEE Trans. Ant. Prop.*, AP-32: 330–337, 1986.

36. K. M. Buckley, L. J. Griffiths, "An Adaptive Generalized Sidelobe Canceller with Derivative Constraints," *IEEE Trans. AP*, AP-34: 311–319, 1986.

37. M. H. Er, A. Cantoni, "A New Set of Linear Constraints for Broad-Band Time Domain Element Space Processors," *IEEE Trans. AP*, AP-34: 320–329, 1986.

38. M. Er, and A. Cantoni, "An Unconstrained Partitioned Realization for Derivative Constrained Broad-Band Antenna Array Processors," *IEEE Trans. ASSP*, ASSP-34: 1376–1379, 1986.

39. L. Griffiths, K. Buckley, "Quiescent Pattern Control in Linearly Constrained Adaptive Arrays," *IEEE Trans. ASSP*, ASSP-35: 917–926, 1987.

40. W. Gabriel (ed.), "Special Issue on Active and Adaptive Antennas," *IEEE Trans. AP*, AP-12, 1964.

41. W. Gabriel (ed.), "Special Issue on Adaptive Antennas," *IEEE Trans. AP*, AP-24(5), 1976.

42. W. Gabriel (ed.), "Special Issue on Adaptive Processing Antenna Systems," *IEEE Trans. AP*, AP-34(3), 1986.

43. M. Grossi, G. Tacconi (eds.), "Special Issue on Beam Forming," *IEEE J. OE*, OE-10(3), 1985.

44. R. Monzingo, T. Miller. *Introduction to Adaptive Arrays*, Wiley, New York, 1980.

45. S. Haykin (ed.), *Array Signal Processing*, Prentice–Hall, Englewood Cliffs, N.J., 1985.

46. W. Knight, R. Pridham, S. Kay, "Digital Signal Processing for Sonar," *Proc. IEEE* 69: 1451–1506, 1981.

47. D. W. Princehouse, "REVGEN, A Real Time Reverberation Generator," *Proc. 1977 IEEE Int. Conf. on Acoustics, Speech, and Signal Processing*: 827–835, 1977.

48. R. J. Urick, *Principles of Underwater Sound*, McGraw-Hill, New York, 1983.

49. S. Chamberlain, J. Galli, "A Model for Numerical Simulation of Nonstationary Sonar Reverberation Using Linear Spectral Prediction," *IEEE J. OE*, OE-8: 21–36, 1983.

50. W. Hodgkiss, "An Oceanic Reverberation Model," *IEEE J. OE.*, OE-9: 63–72, 1984.

51. W. S. Hodgkiss, D. Alexandrou, "Under-Ice Reverberation Rejection," *IEEE J. OE*, OE-10: 285–289, 1985.

52. W. S. Hodgkiss, D. Alexandrou, "An Adaptive Algorithm for Array Processing," *IEEE Trans. AP*, AP-34: 454–458, 1986.

53. V. Anderson, "Digital Array Phasing," *J. Acoust. Soc. Am*. 32: 867–870, 1960.

54. R. A. Mucci, D. W. Tufts, J. T. Lewis, "Constrained Least-Squares Synthesis of Coefficients for Arrays of Sensors and FIR Digital Filters," *IEEE Trans. AES*, AES-12(2): 195–202, 1976.

55. J. J. Anton, A. J. Rockmore, "A Unified Approach to Array-Factor Synthesis for Line Arrays with Nonuniformly Positioned Elements," *IEEE J. OE.*, OE-1(1): 14–21, 1976.

56. E. J. Sullivan, "Amplitude Shading of Irregular Acoustic Arrays," *J. Acoust. Soc. Am*. 63(6): 1873–1877, 1978.

57. F. J. Harris, "On the Use of Windows for Harmonic Analysis with the Discrete Fourier Transform," *Proc. IEEE* 66(1): 51–83, 1978.

58. J. R. Williams, "Fast Beam-Forming Algorithm," *J. Acoust. Soc. Am.* 44(5): 1454–1455, 1968.

59. G. DeMuth, "Frequency Domain Beamforming Techniques," *IEEE Int. Conf. Acoust., Speech, Signal Proc.*: 713–715, 1977.

60. R. Pridham, R. Mucci, "Digital Interpolation Beamforming for Low-Pass and Bandpass Signals," *Proc. IEEE* 67(6): 904–919, 1979.

61. R. Pridham, R. Mucci, "Shifted Sideband Beamformer," *IEEE Trans. ASSP*, ASSP-27(6): 713–722, 1979.

62. J. V. Thorn, N. O. Booth, J. C. Lockwood, "Random and Partially Random Acoustic Arrays," *J. Acoust. Soc. Am.* 67(4): 1277–1286, 1980.

63. W. S. Hodgkiss, V. C. Anderson, "Hardware Dynamic Beamforming," *J. Acoust. Soc. Am.* 69(4): 1075–1083, 1981.

64. D. J. Ramsdale, R. A. Howerton, "Effect of Element Failure and Random Errors in Amplitude and Phase on the Sidelobe Level Attainable with a Linear Array," *J. Acoust. Soc. Am.* 68(3): 901–906, 1980.

65. R. Mucci, R. Pridham, "Impact and beam steering errors on shifted sideband and phase shift beamforming techniques," *J. Acoust. Soc. Am.* 69: 1360–1368, 1981.

66. W. S. Hodgkiss, "The Effects of Array Shape Perturbation on Beamforming and Passive Ranging," *IEEE J. OE*, OE-8(3): 120–130, 1983.

67. R. Collin, F. Zucker, *Antenna Theory*. McGraw–Hill, New York, 1969.

68. B. Steinberg, *Principles of Aperture and Array System Design*. Wiley, New York, 1976.

69. W. Stutzman, G. Thiele. *Antenna Theory and Design*. Wiley, New York, 1981.

PROBLEMS

11.1 When Φ is not Toeplitz owing to nonstationarity of the time series (Wiener filtering problem) or lack of a window (least squares problem), Eq. 11.17 typically is solved using a more computationally expensive square-root method or Cholesky decomposition. Since Φ still is a Hermetian, positive semidefinite matrix, Eq. 11.17 can be expressed as

$$LDL^H \mathbf{h} = -\mathbf{g} \qquad (11.P.1)$$

where L is a lower triangular matrix whose main diagonal elements are all unity and D is a diagonal matrix. The solution for $\mathbf{h}$ proceeds in a two-step fashion. First, Eq. 11.P.1 is rewritten as

$$L\mathbf{y} = -\mathbf{g} \qquad (11.P.2)$$

where

$$\mathbf{y} = DL^H \mathbf{h} \qquad (11.P.3)$$

Show that since L is lower triangular, a recursive solution for **y** can be generated. Second, since D is diagonal, the following modification of Eq. 11.P.3 is readily computed

$$L^H \mathbf{h} = D^{-1} \mathbf{y} \tag{11.P.4}$$

Show that since L^H is upper triangular, a recursive solution for **h** can be generated. Lastly, it remains to determine the elements of L and D from

$$\Phi^T = LDL^H \tag{11.P.5}$$

Show that a recursive solution for D and L can be obtained by explicitly solving for the (i, j)th elements of both sides of (11.P.5) (separate and explicit solution for the off-diagonal elements from the explicit solution for the on-diagonal elements).

11.2 As a consequence of the orthogonality principle, both the forward and backward prediction errors [$e_p(n)$ and $r_p(n)$ in Figs. 11.2 and 11.4] are orthogonal to the data ($\{x(n-1), \ldots, x(n-p)\}$ in the case of $e_p(n)$ and $\{x(n), \ldots, x(n-p+1)\}$ in the case of $r_p(n)$). Show that the backward prediction errors also are orthogonal to each other.

11.3 In Eq. 11.42, the discrepancy Δ^e_{p+1} normalized by E^r_p defines the negative of the lattice parameter E^r_{p+1}. Show that Δ^e_{p+1} defined in Eq. 11.35 is equal to the average value (cross-correlation) of $e_p(n)r_p^*(n-1)$. For this reason, the lattice parameters sometimes are referred to as partial correlation coefficients.

11.4 By making use of the order update expression for $A_i(z)$ in Eq. 11.54b and substituting the equivalent form by $B_{i-1}(z)$ from Eq. 11.53, show that all the zeros of $A_p(z)$ are inside the unit circle [i.e., $A_p(z)$ is minimum phase] when $|K_i| < 1$, $i = 1, 2, \ldots, p$.[21]

11.5 In Section 11.2.2, step-up and step-down recursions were presented relating equivalent direct form and lattice FIR filter structures. Section 11.2.3 then discussed the generation of AR (autoregressive) time series by passing white noise through a direct-form, all-pole, IIR filter. By making use of the following modified version of Eqs. 11.54b and c,

$$A_{i-1}(z) = A_i(z) - K_i z^{-1} B_{i-1}(z) \tag{11.P.6a}$$

$$z^{-1} B_{i-1}(z) = B_i(z) - K_i^* A_{i-1}(z) \tag{11.P.6b}$$

show that the equivalent IIR lattice is given by Figure 11.P.1.[20]

11.6 In an expected value sense, the JCLMS algorithm in Eq. 11.72 exhibits adaptation characteristics that can be described as the sum of exponentials. A modal decomposition of the algorithm permits one to focus on the simple first-order characteristic of each individual mode. With regard to the average

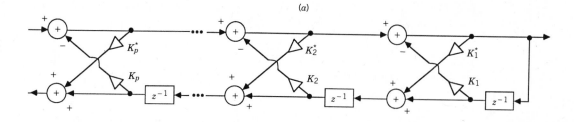

(a)

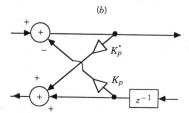

(b)

Figure 11.P.1 (a) All-pole IIR lattice structure. (b) The ith stage of the lattice.

squared error $E_p^d(h)$, the modes exhibit time constants

$$\tau_k = \frac{1}{4\mu\lambda_k} \tag{11.P.7}$$

or, using the parameters α_{JCLMS} introduced in Eq. 11.75,

$$\tau_k = \frac{(p+1)E_0^e}{4\alpha_{\text{JCLMS}}\lambda_k} \tag{11.P.8}$$

Noting that

$$\text{tr } \boldsymbol{\Phi} = \sum_{k=0}^{p} \lambda_k \tag{11.P.9}$$

show that for sinusoidal data, τ_k is inversely proportional to α_{JCLMS} and independent of filter length while for white noise data (i.e., data that are independent sample-to-sample), τ_k is proportional to filter length $(p+1)$ in addition to being inversely proportional to α_{JCLMS}.

11.7 Using the method of Lagrange multipliers, derive the solution for the more general constrained minimization problem (e.g., see Fig. 11.15 and the related discussion in Section 11.4.2)

$$\underset{\mathbf{w}}{\text{Minimize}} \quad \phi_{00}^d + \mathbf{g}^H\mathbf{w} + \mathbf{w}^H\mathbf{g} + \mathbf{w}^H\boldsymbol{\Phi}^T\mathbf{w} \tag{11.P.10a}$$

$$\text{Subject to} \quad \mathbf{C}^H\mathbf{w} = \mathbf{f} \tag{11.P.10b}$$

where $\mathbf{g}$ is defined in Eqs. 11.9 and 11.12.

11.8 In Figure 11.16, column constraints have been removed from the adaptive portion of the system and incorporated into the matrix filter $\mathbf{A}$. Viewing each row of $\mathbf{A}$ as the array element weights of a beamformer, show

how the look direction constraint can be interpreted as specifying the location of a zero in each corresponding spatial transfer function.

11.9 The spatial transfer function $W(\theta_l)$ of an equally spaced line array is given in Eqs. 11.A.15 and 11.A.18. Similar to FIR filter design, the problem of selecting array element weighting coefficients w_n can be viewed equivalently in the z-plane as the problem of locating the zeros of the spatial transfer function. Discuss the characteristics of the spatial transfer function in the vicinity of multiple zeros at the same location on the unit circle in the z-plane.

11.10 Array element weighting functions typically are designed to provide sidelobe control at the expense of broadening the mainlobe beamwidth. For the FFT beamforming example discussed in Section 11.A.3.2, investigate the effect of a dead element on the spatial transfer function. Does it make any difference if the dead element is near the array center or near the ends of the array? Consider uniform weighting of the array elements and at least one other weighting function (e.g., Hanning, Hamming, Kaiser–Bessel, Chebyshev, or Taylor).

11.11 The following parameters are relevant to the time-delay beamforming example described in Section 11.A.2.4.

$$d = 25 \text{ m}$$
$$c = 1500 \text{ m/s}$$
$$f_s = 1 \text{ kHz}$$

At what frequency are the array elements space $\lambda/2$? Without interpolation, what angle of arrival nearest to broadside can be beamformed (i.e., one sample delay/advance element-to-element)?

11.12 In conventional spectral analysis performed using overlapped FFTs, proper normalization of the computation to yield the two-sided power spectral density function is achieved as follows:

$$\hat{S}(f_i) = \frac{1}{f_s MU} |X(i)|^2_{\text{avg}} \tag{11.P.11a}$$

where

$$X(i) = \sum_{m=0}^{M-1} w_m x_m e^{-j(\frac{2\pi}{M})mi} \tag{11.P.11b}$$

$$f_i = \left(\frac{i}{M}\right) f_s \tag{11.P.11c}$$

$$U = \frac{1}{M} \sum_{m=0}^{M-1} w_m^2 \tag{11.P.11d}$$

and f_s is the sampling rate in Hz. Derive a similar expression for the proper normalization of the output from a FFT beamformer where the two-dimensional power spectral function is a function of both frequency (f) and arrival angle (θ).

Index